PENGUIN BOOKS

DARWIN'S DANGEROUS IDEA

'Dennett's dangerous idea: to use his gift for lucid explanation and his twinkling wit to cure the strange allergy to Darwin in modern intellectual life … essential – and pleasurable – reading for any thinking person' – Steven Pinker

'A discussion of Darwin and contemporary evolutionary biological theory that is exemplary in its presentation of ideas crucial to our understanding of ourselves' – Joyce Carol Oates in *The Times Literary Supplement*

'In this clear and rigorous testing of Darwinian theory across modern science, Dennett persuades us that evolution by natural selection is vital to the future of philosophy' – Edward O. Wilson

'A challenging, provocative read, especially the sections on ethics, language and meaning … Above all, the final chapters are a great rallying call in favour of diversity, of tolerance and accomodation, of honesty. The mood is one almost of spirituality, but there is an underlying tone of realism not all of which is easy medicine' – Mary Mulvihill in the *Irish Times*

'Excellent and stimulating. Most people who think they already know all about Darwinism will learn something new' – Anthony Gottlieb in the *Spectator*

ABOUT THE AUTHOR

Daniel C. Dennett is Distinguished Professor of Arts and Sciences and Director of the Center for Cognitive Studies at Tufts University, Massachusetts. He is also the author of *Content and Consciousness* (1969); *Brainstorms* (1978; Penguin, 1997); *Elbow Room* (1984); *The Intentional Stance* (1987); *Consciousness Explained* (1992; Penguin, 1993); and *Kinds of Minds* (1996).

DARWIN'S DANGEROUS IDEA

EVOLUTION AND THE MEANINGS OF LIFE

Daniel C. Dennett

PENGUIN BOOKS

Published by the Penguin Group
Penguin Books Ltd, 27 Wrights Lane, London W8 5TZ, England
Penguin Books USA Inc., 375 Hudson Street, New York, New York 10014, USA
Penguin Books Australia Ltd, Ringwood, Victoria, Australia
Penguin Books Canada Ltd, 10 Alcorn Avenue, Toronto, Ontario, Canada M4V 3B2
Penguin Books (NZ) Ltd, 182–190 Wairau Road, Auckland 10, New Zealand

Penguin Books Ltd, Registered Offices: Harmondsworth, Middlesex, England

First published in the USA by Simon & Schuster 1995
First published in Great Britain by Allen Lane The Penguin Press 1995
Published in Penguin Books 1996
3 5 7 9 10 8 6 4

The acknowledgements on p. 587 constitute an extension of this copyright page

The moral right of the author has been asserted

Printed in England by Clays Ltd, St Ives plc

To Van Quine
teacher and friend

Contents

PART II: DARWINIAN THINKING IN BIOLOGY

PART III: MIND, MEANING, MATHEMATICS, AND MORALITY

Preface

Darwin's theory of evolution by natural selection has always fascinated me, but over the years I have found a surprising variety of thinkers who cannot conceal their discomfort with his great idea, ranging from nagging skepticism to outright hostility. I have found not just lay people and religious thinkers, but secular philosophers, psychologists, physicists, and even biologists who would prefer, it seems, that Darwin were wrong. This book is about why Darwin's idea is so powerful, and why it promises—not threatens—to put our most cherished visions of life on a new foundation.

A few words about method. This book is largely about science but is not itself a work of science. Science is not done by quoting authorities, however eloquent and eminent, and then evaluating their arguments. Scientists do, however, quite properly persist in holding forth, in popular and not-so-popular books and essays, putting forward their interpretations of the work in the lab and the field, and trying to influence their fellow scientists. When I quote them, rhetoric and all, I am doing what they are doing: engaging in persuasion. There is no such thing as a sound Argument from Authority, but authorities can be persuasive, sometimes rightly and sometimes wrongly. I try to sort this all out, and I myself do not understand all the science that is relevant to the theories I discuss, but, then, neither do the scientists (with perhaps a few polymath exceptions). Interdisciplinary work has its risks. I have gone into the details of the various scientific issues far enough, I hope, to let the uninformed reader see just what the issues are, and why I put the interpretation on them that I do, and I have provided plenty of references.

Names with dates refer to full references given in the bibliography at the back of the book. Instead of providing a glossary of the technical terms used, I define them briefly when I first use them, and then often clarify their meaning in later discussion, so there is a very extensive index, which will let you survey all occurrences of any term or idea in the book. Footnotes are for digressions that some but not all readers will appreciate or require.

One thing I have tried to do in this book is to make it possible for you to read the scientific literature I cite, by providing a unified vision of the field, along with suggestions about the importance or non-importance of the controversies that rage. Some of the disputes I boldly adjudicate, and others I leave wide open but place in a framework so that you can see what the issues are, and whether it matters—to you—how they come out. I hope you will read this literature, for it is packed with wonderful ideas. Some of the books I cite are among the most difficult books I have ever read. I think of the books by Stuart Kauffman and Roger Penrose, for instance, but they are pedagogical *tours de force* of highly advanced materials, and they can and should be read by anyone who wants to have an informed opinion about the important issues they raise. Others are less demanding—clear, informative, well worth some serious effort—and still others are not just easy to read but a great delight—superb examples of Art in the service of Science. Since you are reading this book, you have probably already read several of them, so my grouping them together here will be recommendation enough: the books by Graham Cairns-Smith, Bill Calvin, Richard Dawkins, Jared Diamond, Manfred Eigen, Steve Gould, John Maynard Smith, Steve Pinker, Mark Ridley, and Matt Ridley. No area of science has been better served by its writers than evolutionary theory.

Highly technical philosophical arguments of the sort many philosophers favor are absent here. That is because I have a prior problem to deal with. I have learned that arguments, no matter how watertight, often fall on deaf ears. I am myself the author of arguments that I consider rigorous and unanswerable but that are often not so much rebutted or even dismissed as simply ignored. I am not complaining about injustice—we all must ignore arguments, and no doubt we all ignore arguments that history will tell us we should have taken seriously. Rather, I want to play a more direct role in changing what is ignorable by whom. I want to get thinkers in other disciplines to take evolutionary thinking seriously, to show them how they have been underestimating it, and to show them why they have been listening to the wrong sirens. For this, I have to use more artful methods. I have to tell a story. You don't want to be swayed by a story? Well, I *know* you won't be swayed by a formal argument; you won't even *listen* to a formal argument for my conclusion, so I start where I have to start.

The story I tell is mostly new, but it also pulls together bits and pieces from a wide assortment of analyses I've written over the last twenty-five years, directed at various controversies and quandaries. Some of these pieces are incorporated into the book almost whole, with improvements, and others are only alluded to. What I have made visible here is enough of the tip of the iceberg, I hope, to inform and even persuade the newcomer and at least challenge my opponents fairly and crisply. I have tried to navigate between the Scylla of glib dismissal and the Charybdis of grindingly detailed infighting, and whenever I glide swiftly by a controversy, I warn that I am doing so, and give the reader references to the opposition. The bibliography could easily have been doubled, but I have chosen on the principle that any serious reader needs only one or two entry points into the literature and can find the rest from there.

꘏꘏꘏

In the front of his marvelous new book, *Metaphysical Myths, Mathematical Practices: The Ontology and Epistemology of the Exact Sciences* (Cambridge: Cambridge University Press, 1994), my colleague Jody Azzouni thanks "the philosophy department at Tufts University for providing a near-perfect environment in which to *do* philosophy." I want to second both the thanks and the evaluation. At many universities, philosophy is studied but not done—"philosophy appreciation," one might call it—and at many other universities, philosophical research is an arcane activity conducted out of sight of the undergraduates and all but the most advanced postgraduates. At Tufts, we *do* philosophy, in the classroom and among our colleagues, and the results, I think, show that Azzouni's assessment is correct. Tufts has provided me with excellent students and colleagues, and an ideal setting in which to work with them. In recent years I have taught an undergraduate seminar on Darwin and philosophy, in which most of the ideas in this book were hammered out. The penultimate draft was probed, criticized, and polished by a particularly strong seminar of graduate and undergraduate students, for whose help I am grateful: Karen Bailey, Pascal Buckley, John Cabral, Brian Cavoto, Tim Chambers, Shiraz Cupala, Jennifer Fox, Angela Giles, Patrick Hawley, Dien Ho, Matthew Kessler, Chris Lerner, Kristin McGuire, Michael Ridge, John Roberts, Lee Rosenberg, Stacey Schmidt, Rhett Smith, Laura Spiliatakou, and Scott Tanona. The seminar was also enriched by frequent visitors: Marcel Kinsbourne, Bo Dahlbom, David Haig, Cynthia Schossberger, Jeff McConnell, David Stipp. I also want to thank my colleagues, especially Hugo Bedau, George Smith, and Stephen White, for a variety of valuable suggestions. And I must especially thank Alicia Smith, the secretary at the Center for Cognitive Studies, whose virtuoso performance as a reference-finder, fact-checker, permission-seeker, draft-updater/printer/mailer, and general coordinator of the whole project put wings on my heels.

I have also benefited from detailed comments from those who read most or all the penultimate-draft chapters: Bo Dahlbom, Richard Dawkins, David Haig, Doug Hofstadter, Nick Humphrey, Ray Jackendoff, Philip Kitcher, Justin Leiber, Ernst Mayr, Jeff McConnell, Steve Pinker, Sue Stafford, and Kim Sterelny. As usual, they are not responsible for any errors they failed to dissuade me from. (And if you can't write a good book about evolution with the help of this sterling group of editors, you should give up!)

Many others answered crucial questions, and clarified my thinking in dozens of conversations: Ron Amundsen, Robert Axelrod, Jonathan Bennett, Robert Brandon, Madeline Caviness, Tim Clutton-Brock, Leda Cosmides, Helena Cronin, Arthur Danto, Mark De Voto, Marc Feldman, Murray Gell-Mann, Peter Godfrey-Smith, Steve Gould, Danny Hillis, John Holland, Alastair Houston, David Hoy, Bredo Johnsen, Stu Kauffman, Chris Langton, Dick Lewontin, John Maynard Smith, Jim Moore, Roger Penrose, Joanne Phillips, Robert Richards, Mark and Matt (the Ridley conspecifics), Dick Schacht, Jeff Schank, Elliot Sober, John Tooby, Robert Trivers, Peter Van Inwagen, George Williams, David Sloan Wilson, Edward O. Wilson, and Bill Wimsatt.

I want to thank my agent, John Brockman, for steering this big project past many shoals, and helping me see ways of making it a better book. Thanks also go to Terry Zaroff, whose expert copyediting caught many slips and inconsistencies, and clarified and unified the expression of many points. And Ilavenil Subbiah, who drew the figures, except for Figures 10.3 and 10.4, which were created by Mark McConnell on a Hewlett-Packard Apollo workstation, using I-dea.

Last and most important: thanks and love to my wife, Susan, for her advice, love, and support.

DANIEL DENNETT
September 1994

PART I

STARTING IN THE MIDDLE

Neurath has likened science to a boat which, if we are to rebuild it, we must rebuild plank by plank while staying afloat in it. The philosopher and the scientist are in the same boat. . . .

Analyze theory-building how we will, we all must start in the middle. Our conceptual firsts are middle-sized, middle-distanced objects, and our introduction to them and to everything comes midway in the cultural evolution of the race. In assimilating this cultural fare we are little more aware of a distinction between report and invention, substance and style, cues and conceptualization, than we are of a distinction between the proteins and the carbohydrates of our material intake. Retrospectively we may distinguish the components of theory-building, as we distinguish the proteins and carbohydrates while subsisting on them.

—WILLARD VAN ORMAN QUINE 1960, pp. 4–6

CHAPTER ONE

Tell Me Why

⚎

1. IS NOTHING SACRED?

We used to sing a lot when I was a child, around the campfire at summer camp, at school and Sunday school, or gathered around the piano at home. One of my favorite songs was "Tell Me Why." (For those whose personal memories don't already embrace this little treasure, the music is provided in the appendix. The simple melody and easy harmony line are surprisingly beautiful.)

> Tell me why the stars do shine,
> Tell me why the ivy twines,
> Tell me why the sky's so blue.
> Then I will tell you just why I love you.

> Because God made the stars to shine,
> Because God made the ivy twine,
> Because God made the sky so blue.
> Because God made you, that's why I love you.

This straightforward, sentimental declaration still brings a lump to my throat—so sweet, so innocent, so reassuring a vision of life!

And then along comes Darwin and spoils the picnic. Or does he? That is the topic of this book. From the moment of the publication of *Origin of Species* in 1859, Charles Darwin's fundamental idea has inspired intense reactions ranging from ferocious condemnation to ecstatic allegiance, sometimes tantamount to religious zeal. Darwin's theory has been abused and misrepresented by friend and foe alike. It has been misappropriated to lend scientific respectability to appalling political and social doctrines. It has been pilloried in caricature by opponents, some of whom would have it compete in our children's schools with "creation science," a pathetic hodge-podge of pious pseudo-science.[1]

Almost no one is indifferent to Darwin, and no one should be. The Darwinian theory is a scientific theory, and a great one, but that is not all it is. The creationists who oppose it so bitterly are right about one thing: Darwin's dangerous idea cuts much deeper into the fabric of our most fundamental beliefs than many of its sophisticated apologists have yet admitted, even to themselves.

The sweet, simple vision of the song, taken literally, is one that most of us have outgrown, however fondly we may recall it. The kindly God who lovingly fashioned each and every one of us (all creatures great and small) and sprinkled the sky with shining stars for our delight—*that* God is, like Santa Claus, a myth of childhood, not anything a sane, undeluded adult could literally believe in. *That* God must either be turned into a symbol for something less concrete or abandoned altogether.

Not all scientists and philosophers are atheists, and many who are believers declare that their idea of God can live in peaceful coexistence with, or even find support from, the Darwinian framework of ideas. Theirs is not an anthropomorphic Handicrafter God, but still a God worthy of worship in their eyes, capable of giving consolation and meaning to their lives. Others ground their highest concerns in entirely secular philosophies, views of the meaning of life that stave off despair without the aid of any concept of a Supreme Being—other than the Universe itself. Something *is* sacred to these thinkers, but they do not call it God; they call it, perhaps, Life, or Love, or Goodness, or Intelligence, or Beauty, or Humanity. What both groups share, in spite of the differences in their deepest creeds, is a conviction that life does have meaning, that goodness matters.

But can *any* version of this attitude of wonder and purpose be sustained in the face of Darwinism? From the outset, there have been those who thought they saw Darwin letting the worst possible cat out of the bag: nihilism. They thought that if Darwin was right, the implication would be that nothing could be sacred. To put it bluntly, nothing could have any point. Is this just an overreaction? What exactly are the implications of Darwin's idea—and, in any case, has it been scientifically proven or is it still "just a theory"?

Perhaps, you may think, we could make a useful division: there are the parts of Darwin's idea that really are established beyond any reasonable doubt, and then there are the speculative extensions of the scientifically

---

1. I will not devote any space in this book to cataloguing the deep flaws in creationism, or supporting my peremptory condemnation of it. I take that job to have been admirably done by Kitcher 1982, Futuyma 1983, Gilkey 1985, and others.

irresistible parts. Then—if we were lucky—perhaps the rock-solid scientific facts would have no stunning implications about religion, or human nature, or the meaning of life, while the parts of Darwin's idea that get people all upset could be put into quarantine as highly controversial extensions of, or mere interpretations of, the scientifically irresistible parts. That would be reassuring.

But alas, that is just about backwards. There are vigorous controversies swirling around in evolutionary theory, but those who feel threatened by Darwinism should not take heart from this fact. Most—if not quite all—of the controversies concern issues that are "just science"; no matter which side wins, the outcome will not undo the basic Darwinian idea. That idea, which is about as secure as any in science, really does have far-reaching implications for our vision of what the meaning of life is or could be.

In 1543, Copernicus proposed that the Earth was not the center of the universe but in fact revolved around the Sun. It took over a century for the idea to sink in, a gradual and actually rather painless transformation. (The religious reformer Philipp Melanchthon, a collaborator of Martin Luther, opined that "some Christian prince" should suppress this madman, but aside from a few such salvos, the world was not particularly shaken by Copernicus himself.) The Copernican Revolution did eventually have its own "shot heard round the world": Galileo's *Dialogue Concerning the Two Chief World Systems*, but it was not published until 1632, when the issue was no longer controversial among scientists. Galileo's projectile provoked an infamous response by the Roman Catholic Church, setting up a shock wave whose reverberations are only now dying out. But in spite of the drama of that epic confrontation, the idea that our planet is not the center of creation has sat rather lightly in people's minds. Every schoolchild today accepts this as the matter of fact it is, without tears or terror.

In due course, the Darwinian Revolution will come to occupy a similarly secure and untroubled place in the minds—and hearts—of every educated person on the globe, but today, more than a century after Darwin's death, we still have not come to terms with its mind-boggling implications. Unlike the Copernican Revolution, which did not engage widespread public attention until the scientific details had been largely sorted out, the Darwinian Revolution has had anxious lay spectators and cheerleaders taking sides from the outset, tugging at the sleeves of the participants and encouraging grandstanding. The scientists themselves have been moved by the same hopes and fears, so it is not surprising that the relatively narrow conflicts among theorists have often been not just blown up out of proportion by their adherents, but seriously distorted in the process. Everybody has seen, dimly, that a lot is at stake.

Moreover, although Darwin's own articulation of his theory was monumental, and its powers were immediately recognized by many of the scientists and other thinkers of his day, there really were large gaps in his theory that have only recently begun to be properly filled in. The biggest gap looks almost comical in retrospect. In all his brilliant musings, Darwin never hit upon the central concept, without which the theory of evolution is hopeless: the concept of a *gene*. Darwin had no proper *unit* of heredity, and so his account of the process of natural selection was plagued with entirely reasonable doubts about whether it would work. Darwin supposed that offspring would always exhibit a sort of blend or average of their parents' features. Wouldn't such "blending inheritance" always simply average out all differences, turning everything into uniform gray? How could diversity survive such relentless averaging? Darwin recognized the seriousness of this challenge, and neither he nor his many ardent supporters succeeded in responding with a description of a convincing and well-documented mechanism of heredity that could combine traits of parents while maintaining an underlying and unchanged identity. The idea they needed was right at hand, uncovered ("formulated" would be too strong) by the monk Gregor Mendel and published in a relatively obscure Austrian journal in 1865, but, in the best-savored irony in the history of science, it lay there unnoticed until its importance was appreciated (at first dimly) around 1900. Its triumphant establishment at the heart of the "Modern Synthesis" (in effect, the synthesis of Mendel and Darwin) was eventually made secure in the 1940s, thanks to the work of Theodosius Dobzhansky, Julian Huxley, Ernst Mayr, and others. It has taken another half-century to iron out most of the wrinkles of that new fabric.

The fundamental core of contemporary Darwinism, the theory of DNA-based reproduction and evolution, is now beyond dispute among scientists. It demonstrates its power every day, contributing crucially to the explanation of planet-sized facts of geology and meteorology, through middle-sized facts of ecology and agronomy, down to the latest microscopic facts of genetic engineering. It unifies all of biology and the history of our planet into a single grand story. Like Gulliver tied down in Lilliput, it is unbudgeable, not because of some one or two huge chains of argument that might— hope against hope—have weak links in them, but because it is securely tied by hundreds of thousands of threads of evidence anchoring it to virtually every other area of human knowledge. New discoveries may conceivably lead to dramatic, even "revolutionary" *shifts* in the Darwinian theory, but the hope that it will be "refuted" by some shattering breakthrough is about as reasonable as the hope that we will return to a geocentric vision and discard Copernicus.

Still, the theory is embroiled in remarkably hot-tempered controversy, and one of the reasons for this incandescence is that these debates about scientific matters are usually distorted by fears that the "wrong" answer would have intolerable moral implications. So great are these fears that they are carefully left unarticulated, displaced from attention by several layers of distracting rebuttal and counter-rebuttal. The disputants are forever changing the subject slightly, conveniently keeping the bogeys in the shadows. It is this misdirection that is mainly responsible for postponing the day when we can all live as comfortably with our new biological perspective as we do with the astronomical perspective Copernicus gave us.

Whenever Darwinism is the topic, the temperature rises, because more is at stake than just the empirical facts about how life on Earth evolved, or the correct logic of the theory that accounts for those facts. One of the precious things that is at stake is a vision of what it means to ask, and answer, the question "Why?" Darwin's new perspective turns several traditional assumptions upside down, undermining our standard ideas about what ought to count as satisfying answers to this ancient and inescapable question. Here science and philosophy get completely intertwined. Scientists sometimes deceive themselves into thinking that philosophical ideas are only, at best, decorations or parasitic commentaries on the hard, objective triumphs of science, and that they themselves are immune to the confusions that philosophers devote their lives to dissolving. But there is no such thing as philosophy-free science; there is only science whose philosophical baggage is taken on board without examination.

The Darwinian Revolution is both a scientific and a philosophical revolution, and neither revolution could have occurred without the other. As we shall see, it was the philosophical prejudices of the scientists, more than their lack of scientific evidence, that prevented them from seeing how the theory could actually work, but those philosophical prejudices that had to be overthrown were too deeply entrenched to be dislodged by mere philosophical brilliance. It took an irresistible parade of hard-won scientific facts to force thinkers to take seriously the weird new outlook that Darwin proposed. Those who are still ill-acquainted with that beautiful procession can be forgiven their continued allegiance to the pre-Darwinian ideas. And the battle is not yet over; even among the scientists, there are pockets of resistance.

Let me lay my cards on the table. If I were to give an award for the single best idea anyone has ever had, I'd give it to Darwin, ahead of Newton and Einstein and everyone else. In a single stroke, the idea of evolution by natural selection unifies the realm of life, meaning, and purpose with the realm of space and time, cause and effect, mechanism and physical law. But it is not just a wonderful scientific idea. It is a dangerous idea. My admiration for Darwin's magnificent idea is unbounded, but I, too, cherish many of the ideas and ideals that it *seems* to challenge, and want to protect them. For instance, I want to protect the campfire song, and what is beautiful and true in it, for my little grandson and his friends, and for their children when they grow up. There are many more magnificent ideas that are also jeopardized, it seems, by Darwin's idea, and they, too, may need protection. The only good way to do this—the only way that has a chance in the long run—is to cut through the smokescreens and look at the idea as unflinchingly, as dispassionately, as possible.

On this occasion, we are not going to settle for "There, there, it will all come out all right." Our examination will take a certain amount of nerve. Feelings may get hurt. Writers on evolution usually steer clear of this apparent clash between science and religion. Fools rush in, Alexander Pope said, where angels fear to tread. Do you want to follow me? Don't you really want to know what survives this confrontation? What if it turns out that the sweet vision—or a better one—survives intact, strengthened and deepened by the encounter? Wouldn't it be a shame to forgo the opportunity for a strengthened, renewed creed, settling instead for a fragile, sickbed faith that you mistakenly supposed must not be disturbed?

There is no future in a sacred myth. Why not? Because of our curiosity. Because, as the song reminds us, *we want to know why*. We may have outgrown the song's answer, but we will never outgrow the question. Whatever we hold precious, we cannot protect it from our curiosity, because being who we are, one of the things we deem precious is the truth. Our love of truth is surely a central element in the meaning we find in our lives. In any case, the idea that we might preserve meaning by kidding ourselves is a more pessimistic, more nihilistic idea than I for one can stomach. If that were the best that could be done, I would conclude that nothing mattered after all.

This book, then, is for those who agree that the only meaning of life worth caring about is one that can withstand our best efforts to examine it. Others are advised to close the book now and tiptoe away.

For those who stay, here is the plan. Part I of the book locates the Darwinian Revolution in the larger scheme of things, showing how it can transform the world-view of those who know its details. This first chapter sets out the background of philosophical ideas that dominated our thought before Darwin. Chapter 2 introduces Darwin's central idea in a somewhat new guise, as the idea of evolution as an *algorithmic process*, and clears up some common misunderstandings of it. Chapter 3 shows how this idea overturns the tradition encountered in chapter 1. Chapters 4 and 5 explore some of the striking—and unsettling—perspectives that the Darwinian way of thinking opens up.

Part II examines the challenges to Darwin's idea—to neo-Darwinism or the Modern Synthesis—that have arisen within biology itself, showing that contrary to what some of its opponents have declared, Darwin's idea survives these controversies not just intact but strengthened. Part III then shows what happens when the same thinking is extended to the species we care about most: *Homo sapiens*. Darwin himself fully recognized that this was going to be the sticking point for many people, and he did what he could to break the news gently. More than a century later, there are still those who want to dig a moat separating us from most if not all of the dreadful implications they think they see in Darwinism. Part III shows that this is an error of both fact and strategy; not only does Darwin's dangerous idea apply to us directly and at many levels, but the proper application of Darwinian thinking to human issues—of mind, language, knowledge, and ethics, for instance—illuminates them in ways that have always eluded the traditional approaches, recasting ancient problems and pointing to their solution. Finally, we can assess the bargain we get when we trade in pre-Darwinian for Darwinian thinking, identifying both its uses and abuses, and showing how what really matters to us—and ought to matter to us—shines through, transformed but enhanced by its passage through the Darwinian Revolution.

2. WHAT, WHERE, WHEN, WHY—AND HOW?

Our curiosity about things takes different forms, as Aristotle noted at the dawn of human science. His pioneering effort to classify them still makes a lot of sense. He identified four basic questions we might want answered about anything, and called their answers the four *aitia*, a truly untranslatable Greek term traditionally but awkwardly translated the four "causes."

(1) We may be curious about what something is made of, its matter or *material cause*.

(2) We may be curious about the form (or structure or shape) that that matter takes, its *formal cause*.

(3) We may be curious about its beginning, how it got started, or its *efficient cause*.

(4) We may be curious about its *purpose* or *goal* or *end* (as in "Do the ends justify the means?"), which Aristotle called its *telos*, sometimes translated in English, awkwardly, as "final cause."

It takes some pinching and shoving to make these four Aristotelian *aitia* line up as the answers to the standard English questions "what, where, when, and why." The fit is only fitfully good. Questions beginning with "why," however, do standardly ask for Aristotle's fourth "cause," the *telos* of a thing. Why this? we ask. What is it *for*? As the French say, what is its *raison d'être*, or reason for being? For hundreds of years, these "why" questions have been recognized as problematic by philosophers and scientists, so distinct that the topic they raise deserves a name: teleology.

A *teleological* explanation is one that explains the existence or occurrence of something by citing a goal or purpose that is served by the thing. Artifacts are the most obvious cases; the goal or purpose of an artifact is the function it was designed to serve by its creator. There is no controversy about the *telos* of a hammer: it is for hammering in and pulling out nails. The *telos* of more complicated artifacts, such as camcorders or tow trucks or CT scanners, is if anything more obvious. But even in simple cases, a problem can be seen to loom in the background:

"Why are you sawing that board?"
"To make a door."
"And what is the door for?"
"To secure my house."
"And why do you want a secure house?"
"So I can sleep nights."
"And why do you want to sleep nights?"
"Go run along and stop asking such silly questions."

This exchange reveals one of the troubles with teleology: where does it all stop? What *final* final cause can be cited to bring this hierarchy of reasons to a close? Aristotle had an answer: God, the Prime Mover, the *for-which* to end all *for-whiches*. The idea, which is taken up by the Christian, Jewish, and Islamic traditions, is that all *our* purposes are ultimately God's purposes. The idea is certainly natural and attractive. If we look at a pocket watch and wonder *why* it has a clear glass crystal on its face, the answer obviously harks back to the needs and desires of the users of watches, who want to tell time, by looking at the hands through the transparent, protective glass, and so forth. If it weren't for these facts about *us*, for whom the watch was created, there would be no explanation of the "why" of its crystal. If the universe was created by God, for God's purposes, then all the purposes we can find in it must ultimately be due to God's purposes. But what are God's purposes? That is something of a mystery.

One way of deflecting discomfort about that mystery is to switch the topic slightly. Instead of responding to the "why" question with a "because"-type answer (the sort of answer it seems to demand), people often substitute a "how" question for the "why" question, and attempt to answer it by telling a story about *how it came to be* that God created us and the rest of the universe, without dwelling overmuch on just why God might want to have done that. The "how" question does not get separate billing on Aristotle's list, but it was a popular question and answer long before Aristotle undertook his analysis. The answers to the biggest "how" questions are *cosmogonies*, stories about how the *cosmos*, the whole universe and all its denizens, came into existence. The book of Genesis is a cosmogony, but there are many others. Cosmologists exploring the hypothesis of the Big Bang, and speculating about black holes and super-strings, are present-day creators of cosmogonies. Not all ancient cosmogonies follow the pattern of an artifact-maker. Some involve a "world egg" laid in "the Deep" by one mythic bird or another, and some involve seeds' being sown and tended. Human imagination has only a few resources to draw upon when faced with such a mind-boggling question. One early creation myth speaks of a "self-existent Lord" who, "with a thought, cre-ated the waters, and deposited in them a seed which became a golden egg, in which egg he himself is born as Brahma, the progenitor of the worlds" (Muir 1972, vol. IV, p. 26).

And what's the point of all this egg-laying or seed-sowing or world-building? Or, for that matter, what's the point of the Big Bang? Today's cosmologists, like many of their predecessors throughout history, tell a diverting story, but prefer to sidestep the "why" question of teleology. Does the universe exist for any reason? Do reasons play any intelligible role in explanations of the cosmos? Could something exist for a reason without its being *somebody's* reason? Or are reasons—Aristotle's type (4) causes—only appropriate in explanations of the works and deeds of people or other rational agents? If God is not a person, a rational agent, an Intelligent Arti-ficer, what possible sense could the biggest "why" question make? And if the biggest "why" question doesn't make any sense, how could any smaller, more parochial, "why" questions make sense?

One of Darwin's most fundamental contributions is showing us a new way to make sense of "why" questions. Like it or not, Darwin's idea offers one way—a clear, cogent, astonishingly versatile way—of dissolving these old conundrums. It takes some getting used to, and is often misapplied, even by its staunchest friends. Gradually exposing and clarifying this way of thinking is a central project of the present book. Darwinian thinking must be carefully distinguished from some oversimplified and all-too-popular im-postors, and this will take us into some technicalities, but it is worth it. The prize is, for the first time, a stable system of explanation that does not go round and round in circles or spiral off in an infinite regress of mysteries. Some people would much prefer the infinite regress of mysteries, appar-ently, but in this day and age the cost is prohibitive: you have to get yourself deceived. You can either deceive yourself or let others do the dirty work, but there is no intellectually defensible way of rebuilding the mighty bar-riers to comprehension that Darwin smashed.

The first step to appreciating this aspect of Darwin's contribution is to see how the world looked before he inverted it. By looking through the eyes of two of his countrymen, John Locke and David Hume, we can get a clear vision of an alternative world-view—still very much with us in many quar-ters—that Darwin rendered obsolete.

3. LOCKE'S "PROOF" OF THE PRIMACY OF MIND

John Locke invented common sense, and only Englishmen have had it ever since!

—BERTRAND RUSSELL[2]

John Locke, a contemporary of "the incomparable Mr. Newton," was one of the founding fathers of British Empiricism, and, as befits an Empiricist, he was not much given to deductive arguments of the rationalist sort, but one of his uncharacteristic forays into "proof" deserves to be quoted in full, since it perfectly illustrates the blockade to imagination that was in place before the Darwinian Revolution. The argument may seem strange and stilted to modern minds, but bear with it—consider it a sign of how far we have come since then. Locke himself thought that he was just reminding people of something obvious! In this passage from his *Essay Concerning Human Understanding* (1690, IV, x, 10), Locke wanted to *prove* something that he thought all people knew in their hearts in any case: that "in the beginning" there was Mind. He began by asking himself what, if anything, was eternal:

> If, then, there must be something eternal, let us see what sort of Being it must be. And to that it is very obvious to Reason, that it must necessarily be a cogitative Being. For it is as impossible to conceive that ever bare incogitative Matter should produce a thinking intelligent Being, as that nothing should of itself produce Matter. . . .

Locke begins his proof by alluding to one of philosophy's most ancient and oft-used maxims, *Ex nihilo nihil fit*: nothing can come from nothing. Since this is to be a deductive argument, he must set his sights high: it is not just unlikely or implausible or hard to fathom but *impossible to conceive* that "bare incogitative Matter should produce a thinking intelligent Being." The argument proceeds by a series of mounting steps:

---

2. Gilbert Ryle recounted this typical bit of Russellian hyperbole to me. In spite of Ryle's own distinguished career as Waynflete Professor of Philosophy at Oxford, he and Russell had seldom met, he told me, in large measure because Russell steered clear of academic philosophy after the Second World War. Once, however, Ryle found himself sharing a compartment with Russell on a tedious train journey, and, trying desperately to make conversation with his world-famous fellow traveler, Ryle asked him why he thought Locke, who was neither as original nor as good a writer as Berkeley, Hume, or Reid, had been so much more influential than they in the English-speaking philosophical world. This had been his reply, and the beginning of the only good conversation, Ryle said, that he ever had with Russell.

> Let us suppose any parcel of Matter eternal, great or small, we shall find it, in itself, able to produce nothing.... Matter then, by its own strength, cannot produce in itself so much as Motion: the Motion it has, must also be from Eternity, or else be produced, and added to Matter by some other Being more powerful than Matter.... But let us suppose Motion eternal too: yet Matter, incogitative Matter and Motion, whatever changes it might produce of Figure and Bulk, could never produce Thought: Knowledge will still be as far beyond the power of Motion and Matter to produce, as Matter is beyond the power of nothing or nonentity to produce. And I appeal to everyone's own thoughts, whether he cannot as easily conceive Matter produced by nothing, as Thought produced by pure Matter, when before there was no such thing as Thought, or an intelligent Being existing....

It is interesting to note that Locke decides he may safely "appeal to everyone's own thoughts" to secure this "conclusion." He was sure that *his* "common sense" was truly *common* sense. Don't we see how obvious it is that whereas matter and motion could produce changes of "Figure and Bulk," they could *never* produce "Thought"? Wouldn't this rule out the prospect of robots—or at least robots that would claim to have genuine Thoughts among the motions in their material heads? Certainly in Locke's day—which was also Descartes's day—the very idea of Artificial Intelligence was so close to unthinkable that Locke could confidently expect unanimous endorsement of this appeal to his audience, an appeal that would risk hoots of derision today.[3] And as we shall see, the field of Artificial Intelligence is a quite direct descendant of Darwin's idea. Its birth, which was all but prophesied by Darwin himself, was attended by one of the first truly impressive demonstrations of the formal power of natural selection (Art Samuel's legendary checkers-playing program, which will be described in some detail later). And both evolution and AI inspire the same loathing in many people who should know better, as we shall see in later chapters. But back to Locke's conclusion:

> So if we will suppose nothing first, or eternal: Matter can never begin to be: If we suppose bare Matter, without Motion, eternal: Motion can never begin to be: If we suppose only Matter and Motion first, or eternal: Thought can never begin to be. For it is impossible to conceive that Matter either with or without Motion could have originally in and from itself Sense,

---

3. Descartes's inability to think of Thought as Matter in Motion is discussed at length in my book *Consciousness Explained* (1991a). John Haugeland's aptly titled book, *Artificial Intelligence: The Very Idea* (1985), is a fine introduction to the philosophical paths that make this idea thinkable after all.

Perception, and Knowledge, as is evident from hence, that then Sense, Perception, and Knowledge must be a property eternally inseparable from Matter and every particle of it.

So, if Locke is right, Mind must come first—or at least tied for first. It could not come into existence at some later date, as an effect of some confluence of more modest, mindless phenomena. This purports to be an entirely secular, logical—one might almost say mathematical—vindication of a central aspect of Judeo-Christian (and also Islamic) cosmogony: in the beginning was something with Mind—"a cogitative Being," as Locke says. The traditional idea that God is a rational, thinking agent, a Designer and Builder of the world, is here given the highest stamp of scientific approval: like a mathematical theorem, its denial is supposedly impossible to conceive.

And so it seemed to many brilliant and skeptical thinkers before Darwin. Almost a hundred years after Locke, another great British Empiricist, David Hume, confronted the issue again, in one of the masterpieces of Western philosophy, his *Dialogues Concerning Natural Religion* (1779).

4. HUME'S CLOSE ENCOUNTER

Natural religion, in Hume's day, meant a religion that was supported by the natural sciences, as opposed to a "revealed" religion, which would depend on revelation—on mystical experience or some other uncheckable source of conviction. If your only grounds for your religious belief is "God told me so in a dream," your religion is not natural religion. The distinction would not have made much sense before the dawn of modern science in the seventeenth century, when science created a new, and competitive, standard of evidence for all belief. It opened up the question:

Can you give us any *scientific* grounds for your religious beliefs?

Many religious thinkers, appreciating that the prestige of scientific thought was—other things being equal—a worthy aspiration, took up the challenge. It is hard to see why anybody would want to shun scientific confirmation of one's creed, if it were there to be had. The overwhelming favorite among purportedly scientific arguments for religious conclusions, then and now, was one version or another of the Argument from Design: among the effects we can objectively observe in the world, there are many that are not (cannot be, for various reasons) mere accidents; they must have been designed to be as they are, and there cannot be design without a Designer; therefore, a Designer, God, must exist (or have existed), as the source of all these wonderful effects.

Such an argument can be seen as an attempt at an alternate route to Locke's conclusion, a route that will take us through somewhat more empirical detail instead of relying so bluntly and directly on what is deemed inconceivable. The actual features of the observed designs may be analyzed, for instance, to secure the grounds for our appreciation of the wisdom of the Designer, and our conviction that mere chance could not be responsible for these marvels.

In Hume's *Dialogues*, three fictional characters pursue the debate with consummate wit and vigor. Cleanthes defends the Argument from Design, and gives it one of its most eloquent expressions.[4] Here is his opening statement of it:

> Look round the world: Contemplate the whole and every part of it: You will find it to be nothing but one great machine, subdivided into an infinite number of lesser machines, which again admit of subdivisions to a degree beyond what human senses and faculties can trace and explain. All these various machines, and even their most minute parts, are adjusted to each other with an accuracy which ravishes into admiration all men who have ever contemplated them. The curious adapting of means to ends, throughout all nature, resembles, exactly, though it much exceeds, the productions of human contrivance—of human design, thought, wisdom, and intelligence. Since therefore the effects resemble each other, we are led to infer, by all the rules of analogy, that the causes also resemble, and that the Author of Nature is somewhat similar to the mind of man, though possessed of much larger faculties, proportioned to the grandeur of the work which he has executed. By this argument *a posteriori*, and by this argument alone, do we prove at once the existence of a Deity and his similarity to human mind and intelligence. [Pt. II.]

Philo, a skeptical challenger to Cleanthes, elaborates the argument, setting it up for demolition. Anticipating Paley's famous example, Philo notes: "Throw several pieces of steel together, without shape or form; they will never arrange themselves so as to compose a watch."[5] He goes on: "Stone, and mortar, and wood, without an architect, never erect a house. But the

---

4. William Paley carried the Argument from Design into much greater biological detail in his 1803 book, *Natural Theology*, adding many ingenious flourishes. Paley's influential version was the actual inspiration and target of Darwin's rebuttal, but Hume's Cleanthes catches all of the argument's logical and rhetorical force.

5. Gjertsen points out that two millennia earlier, Cicero used the same example for the same purpose: "When you see a sundial or a water-clock, you see that it tells the time by design and not by chance. How then can you imagine that the universe as a whole is devoid of purpose and intelligence, when it embraces everything, including these artifacts themselves and their artificers?" (Gjertsen 1989, p. 199).

ideas in a human mind, we see, by an unknown, inexplicable economy, arrange themselves so as to form the plan of a watch or house. Experience, therefore, proves, that there is an original principle of order in mind, not in matter" (Pt. II).

Note that the Argument from Design depends on an inductive inference: where there's smoke, there's fire; and where there's design, there's mind. But this is a dubious inference, Philo observes: human intelligence is

> no more than one of the springs and principles of the universe, as well as heat or cold, attraction or repulsion, and a hundred others, which fall under daily observation.... But can a conclusion, with any propriety, be transferred from parts to the whole?... From observing the growth of a hair, can we learn any thing concerning the generation of a man?... What peculiar privilege has this little agitation of the brain which we call thought, that we must thus make it the model of the whole universe?... Admirable conclusion! Stone, wood, brick, iron, brass have not, at this time, in this minute globe of earth, an order or arrangement without human art and contrivance: Therefore the universe could not originally attain its order and arrangement, without something similar to human art. [Pt. II.]

Besides, Philo observes, if we put mind as the first cause, with its "unknown, inexplicable economy," this only postpones the problem:

> We are still obliged to mount higher, in order to find the cause of this cause, which you had assigned as satisfactory and conclusive.... How therefore shall we satisfy ourselves concerning the cause of that Being, whom you suppose the Author of nature, or, according to your system of anthropomorphism, the ideal world, into which you trace the material? Have we not the same reason to trace that ideal world into another ideal world, or new intelligent principle? But if we stop, and go no farther; why go so far? Why not stop at the material world? How can we satisfy ourselves without going on *in infinitum*? And after all, what satisfaction is there in that infinite progression? [Pt. IV.]

Cleanthes has no satisfactory responses to these rhetorical questions, and there is worse to come. Cleanthes insists that God's mind is *like the human*—and agrees when Philo adds "the liker the better." But, then, Philo presses on, is God's mind perfect, "free from every error, mistake, or incoherence in his undertakings" (Pt. V)? There is a rival hypothesis to rule out:

> And what surprise must we entertain, when we find him a stupid mechanic, who imitated others, and copied an art, which, through a long succession of ages, after multiplied trials, mistakes, corrections, deliberations, and controversies, had been gradually improving? Many worlds might have

been botched and bungled, throughout an eternity, ere this system was struck out: Much labour lost: Many fruitless trials made: And a slow, but continued improvement carried on during infinite ages of world-making. [Pt. V.]

When Philo presents this fanciful alternative, with its breathtaking anticipations of Darwin's insight, he doesn't take it seriously except as a debating foil to Cleanthes' vision of an all-wise Artificer. Hume uses it only to make a point about what he saw as the limitations on our knowledge: "In such subjects, who can determine, where the truth; nay, who can conjecture where the probability, lies; amidst a great number of hypotheses which may be proposed, and a still greater number which may be imagined" (Pt. V).

Imagination runs riot, and, exploiting that fecundity, Philo ties Cleanthes up in knots, devising weird and comical variations on Cleanthes' own hypotheses, defying Cleanthes to show why his own version should be preferred. "Why may not several Deities combine in contriving and framing a world? . . . And why not become a perfect anthropomorphite? Why not assert the Deity or Deities to be corporeal, and to have eyes, a nose, mouth, ears, etc.?" (Pt. V). At one point, Philo anticipates the Gaia hypothesis: the universe

bears a great resemblance to an animal or organized body, and seems actuated with a like principle of life and motion. A continual circulation of matter in it produces no disorder. . . . The world, therefore, I infer, is an animal, and the Deity is the SOUL of the world, actuating it and actuated by it. [Pt. VI.]

Or perhaps isn't the world really more like a vegetable than an animal?

In like manner as a tree sheds its seed into the neighboring fields, and produces other trees; so the great vegetable, the world, or this planetary system, produces within itself certain seeds, which, being scattered into the surrounding chaos, vegetate into new worlds. A comet, for instance, is the seed of a world. . . . [Pt. VII.]

One more wild possibility for good measure:

The Brahmins assert, that the world arose from an infinite spider, who spun this whole complicated mass from his bowels, and annihilates afterwards the whole or any part of it, by absorbing it again, and resolving it into his own essence. Here is a species of cosmogony, which appears to us ridiculous; because a spider is a little contemptible animal, whose operation we are never likely to take for a model of the whole universe. But still here is a new species of analogy, even in our globe. And were there a planet wholly inhabited by spiders (which is very possible), this inference would there appear as natural and irrefragable as that which in our planet ascribes the origin of all things to design and intelligence, as explained by Cleanthes. Why an orderly system may not be spun from the belly as well as from the brain, it will be difficult for him to give a satisfactory reason. [Pt. VII.]

Cleanthes resists these onslaughts gamely, but Philo shows fatal flaws in every version of the argument that Cleanthes can devise. At the very end of the *Dialogues*, however, Philo surprises us by agreeing with Cleanthes:

... the legitimate conclusion is that ... if we are not contented with calling the first and supreme cause a *God* or *Deity*, but desire to vary the expression, what can we call him but *Mind* or *Thought* to which he is justly supposed to bear a considerable resemblance? [Pt. XII.]

Philo is surely Hume's mouthpiece in the *Dialogues*. Why did Hume cave in? Out of fear of reprisal from the establishment? No. Hume knew he had shown that the Argument from Design was an irreparably flawed bridge between science and religion, and he arranged to have his *Dialogues* published after his death in 1776 precisely in order to save himself from persecution. He caved in because he *just couldn't imagine* any other explanation of the origin of the manifest design in nature. Hume could not see how the "curious adapting of means to ends, throughout all nature" could be due to chance—and if not chance, what?

What could possibly account for this high-quality design if not an intelligent God? Philo is one of the most ingenious and resourceful competitors in any philosophical debate, real or imaginary, and he makes some wonderful stabs in the dark, hunting for an alternative. In Part VIII, he dreams up some speculations that come tantalizingly close to scooping Darwin (and some more recent Darwinian elaborations) by nearly a century.

Instead of supposing matter infinite, as Epicurus did, let us suppose it finite. A finite number of particles is only susceptible of finite transpositions: And it must happen, in an eternal duration, that every possible order or position must be tried an infinite number of times. . . . Is there a system, an order, an economy of things, by which matter can preserve that perpetual agitation, which seems essential to it, and yet maintain a constancy in the forms, which it produces? There certainly is such an economy: For this is actually the case with the present world. The continual motion of matter, therefore, in less than infinite transpositions, must produce this economy or order; and by its very nature, that order, when once established, supports itself, for many ages, if not to eternity. But wherever matter is so poised, arranged, and adjusted as to continue in perpetual motion, and yet preserve a constancy in the forms, its situation must, of necessity, have all the same appearance of art and contrivance which we observe at present.... A defect in any of these particulars destroys the form; and the matter, of which it is composed, is again set loose, and is thrown into irregular motions and fermentations, till it unite itself to some other regular form....

Suppose ... that matter were thrown into any position, by a blind, unguided force; it is evident that this first position must in all probability be the most confused and most disorderly imaginable, without any resemblance to those works of human contrivance, which, along with a symmetry of parts, discover an adjustment of means to ends and a tendency to self-preservation.... Suppose, that the actuating force, whatever it be, still continues in matter.... Thus the universe goes on for many ages in a continued succession of chaos and disorder. But is it not possible that it may settle at last ...? May we not hope for such a position, or rather be assured of it, from the eternal revolutions of unguided matter, and may not this account for all the appearing wisdom and contrivance which is in the universe?

Hmm, it seems that something like this might work ... but Hume couldn't quite take Philo's daring foray seriously. His final verdict: "A total suspense of judgment is here our only reasonable resource" (Pt. VIII). A few years before him, Denis Diderot had also written some speculations that tantalizingly foreshadowed Darwin: "I can maintain to you ... that monsters annihilated one another in succession; that all the defective combinations of matter have disappeared, and that there have only survived those in which the organization did not involve any important contradiction, and which could subsist by themselves and perpetuate themselves" (Diderot 1749). Cute ideas about evolution had been floating around for millennia, but, like most philosophical ideas, although they did seem to offer a solution of sorts to the problem at hand, they didn't promise to go any farther, to open up new investigations or generate surprising predictions that could be tested, or explain any facts they weren't expressly designed to explain. The evolution revolution had to wait until Charles Darwin saw how to weave an evolutionary hypothesis into an explanatory fabric composed of literally thousands of hard-won and often surprising facts about nature. Darwin neither invented the wonderful idea out of whole cloth all by himself, nor understood it in its entirety even when he had formulated it. But he did such a monumental job of clarifying the idea, and tying it down so it would never again float away, that he deserves the credit if anyone does. The next chapter reviews his basic accomplishment.

CHAPTER 1: *Before Darwin, a "Mind-first" view of the universe reigned unchallenged; an intelligent God was seen as the ultimate source of all Design, the ultimate answer to any chain of "Why?" questions. Even David*

Hume, who deftly exposed the insoluble problems with this vision, and had glimpses of the Darwinian alternative, could not see how to take it seriously.

CHAPTER 2: *Darwin, setting out to answer a relatively modest question about the origin of species, described a process he called natural selection, a mindless, purposeless, mechanical process. This turns out to be the seed of an answer to a much grander question: how does Design come into existence?*

CHAPTER TWO

An Idea Is Born

~~~

1. WHAT IS SO SPECIAL ABOUT SPECIES?

Charles Darwin did not set out to concoct an antidote to John Locke's conceptual paralysis, or to pin down the grand cosmological alternative that had barely eluded Hume. Once his great idea occurred to him, he saw that it would indeed have these truly revolutionary consequences, but at the outset he was not trying to explain the meaning of life, or even its origin. His aim was slightly more modest: he wanted to explain the origin of *species*.

In his day, naturalists had amassed mountains of tantalizing facts about living things and had succeeded in systematizing these facts along several dimensions. Two great sources of wonder emerged from this work (Mayr 1982). First, there were all the discoveries about the *adaptations* of organisms that had enthralled Hume's Cleanthes: "All these various machines, and even their most minute parts, are adjusted to each other with an accuracy which ravishes into admiration all men who have ever contemplated them" (Pt. II). Second, there was the prolific *diversity* of living things—literally millions of different kinds of plants and animals. Why were there so many?

This diversity of design of organisms was as striking, in some regards, as their excellence of design, and even more striking were the patterns discernible within that diversity. Thousands of gradations and variations between organisms could be observed, but there were also huge gaps between them. There were birds and mammals that swam like fish, but none with gills; there were dogs of many sizes and shapes, but no dogcats or dogcows or feathered dogs. The patterns called out for classification, and by Darwin's time the work of the great taxonomists (who began by adopting and correcting Aristotle's ancient classifications) had created a detailed hierarchy of two kingdoms (plants and animals), divided into phyla, which divided into classes, which divided into orders, which divided into families, which divided into genera (the plural of "genus"), which divided into species.

Species could also be subdivided, of course, into subspecies or varieties—cocker spaniels and basset hounds are different varieties of a single species: dogs, or *Canis familiaris*.

How many different kinds of organisms were there? Since no two organisms are exactly alike—not even identical twins—there were as many different kinds of organisms as there were organisms, but it seemed obvious that the differences could be graded, sorted into minor and major, or *accidental* and *essential*. Thus Aristotle had taught, and this was one bit of philosophy that had permeated the thinking of just about everybody, from cardinals to chemists to costermongers. All things—not just living things—had two kinds of properties: essential properties, without which they wouldn't be the particular *kind* of thing they were, and accidental properties, which were free to vary within the kind. A lump of gold could change shape *ad lib* and still be gold; what made it gold were its essential properties, not its accidents. With each kind went an essence. Essences were definitive, and as such they were timeless, unchanging, and all-or-nothing. A thing couldn't be *rather* silver or *quasi*-gold or a *semi*-mammal.

Aristotle had developed his theory of essences as an improvement on Plato's theory of Ideas, according to which every earthly thing is a sort of imperfect copy or reflection of an ideal exemplar or Form that existed timelessly in the Platonic realm of Ideas, reigned over by God. This Platonic heaven of abstractions was not visible, of course, but was accessible to Mind through deductive thought. What geometers thought about, and proved theorems about, for instance, were the Forms of the circle and the triangle. Since there were also Forms for the eagle and the elephant, a deductive science of nature was also worth a try. But just as no earthly circle, no matter how carefully drawn with a compass, or thrown on a potter's wheel, could actually be one of the perfect circles of Euclidean geometry, so no actual eagle could perfectly manifest the essence of eaglehood, though every eagle strove to do so. Everything that existed had a divine specification, which captured its essence. The taxonomy of living things Darwin inherited was thus itself a direct descendant, via Aristotle, of Plato's essentialism. In fact, the word "species" was at one point a standard translation of Plato's Greek word for Form or Idea, *eidos*.

We post-Darwinians are so used to thinking in historical terms about the development of life forms that it takes a special effort to remind ourselves that in Darwin's day species of organisms were deemed to be as timeless as the perfect triangles and circles of Euclidean geometry. Their individual members came and went, but the species itself remained unchanged and unchangeable. This was part of a philosophical heritage, but it was not an idle or ill-motivated dogma. The triumphs of modern science, from Copernicus and Kepler, Descartes and Newton, had all involved the application of precise mathematics to the material world, and this apparently requires abstracting away from the grubby accidental properties of things to find their secret mathematical essences. It makes no difference what color or shape a thing is when it comes to the thing's obeying Newton's inverse-square law of gravitational attraction. All that matters is its mass. Similarly, alchemy had been succeeded by chemistry once chemists settled on their fundamental creed: There were a finite number of basic, *immutable* elements, such as carbon, oxygen, hydrogen, and iron. These might be mixed and united in endless combinations over time, but the fundamental building blocks were identifiable by their changeless essential properties.

The doctrine of essences looked like a powerful organizer of the world's phenomena in many areas, but was it true of every classification scheme one could devise? Were there *essential* differences between hills and mountains, snow and sleet, mansions and palaces, violins and violas? John Locke and others had developed elaborate doctrines distinguishing *real* essences from merely *nominal* essences; the latter were simply parasitic on the *names* or words we chose to use. You could set up any classification scheme you wanted; for instance, a kennel club could vote on a defining list of necessary conditions for a dog to be a genuine Ourkind Spaniel, but this would be a mere nominal essence, not a real essence. Real essences were discoverable by scientific investigation into the internal nature of things, where essence and accident could be distinguished according to principles. It was hard to say just what the *principled* principles were, but with chemistry and physics so handsomely falling into line, it seemed to stand to reason that there had to be defining marks of the real essences of living things as well.

From the perspective of this deliciously crisp and systematic vision of the hierarchy of living things, there were a considerable number of awkward and puzzling facts. These apparent exceptions were almost as troubling to naturalists as the discovery of a triangle whose angles didn't quite add up to 180 degrees would have been to a geometer. Although many of the taxonomic boundaries were sharp and apparently exceptionless, there were all manner of hard-to-classify intermediate creatures, who seemed to have portions of more than one essence. There were also the curious higher-order patterns of shared and unshared features: why should it be backbones rather than feathers that birds and fish shared, and why shouldn't *creature with eyes* or *carnivore* be as important a classifier as *warm-blooded creature*? Although the broad outlines and most of the specific rulings of taxonomy were undisputed (and remain so today, of course), there were heated controversies about the problem cases. Were all these lizards members of the same species, or of several different species? Which principle of classification should "count"? In Plato's famous image, which system "carved nature at the joints"?

Before Darwin, these controversies were fundamentally ill-formed, and could not yield a stable, well-motivated answer because there was no background theory of *why* one classification scheme would count as getting the joints right—the way things *really* were. Today bookstores face the same sort of ill-formed problem: how should the following categories be cross-organized: best-sellers, science fiction, horror, garden, biography, novels, collections, sports, illustrated books? If horror is a genus of fiction, then true tales of horror present a problem. Must all novels be fiction? Then the bookseller cannot honor Truman Capote's own description of *In Cold Blood* (1965) as a nonfiction novel, but the book doesn't sit comfortably amid either the biographies or the history books. In what section of the bookstore should the book you are reading be shelved? Obviously there is no one Right Way to categorize books—nominal essences are all we will ever find in this domain. But many naturalists were convinced on general principles that there were real essences to be found among the categories of their Natural System of living things. As Darwin put it, "They believe that it reveals the plan of the Creator; but unless it be specified whether order in time or space, or what else is meant by the plan of the Creator, it seems to me that nothing is thus added to our knowledge" (*Origin*, p. 413).

Problems in science are sometimes made easier by adding complications. The development of the science of geology and the discovery of fossils of manifestly extinct species gave the taxonomists further curiosities to confound them, but these curiosities were also the very pieces of the puzzle that enabled Darwin, working alongside hundreds of other scientists, to discover the key to its solution: species were *not* eternal and immutable; they had evolved over time. Unlike carbon atoms, which, for all one knew, had been around forever in exactly the form they now exhibited, species had births in time, could change over time, and could give birth to new species in turn. This idea itself was not new; many versions of it had been seriously discussed, going back to the ancient Greeks. But there was a powerful Platonic bias against it: essences were unchanging, and a thing couldn't change its essence, and new essences couldn't be born—except of course by God's command in episodes of Special Creation. Reptiles could no more *turn into* birds than copper could turn into gold.

It isn't easy today to sympathize with this conviction, but the effort can be helped along by a fantasy: consider what your attitude would be towards a theory that purported to show how the number 7 had once been an even number, long, long ago, and had gradually acquired its oddness through an arrangement whereby it exchanged some properties with the ancestors of the number 10 (which had once been a prime number). Utter nonsense, of course. Inconceivable. Darwin knew that a parallel attitude was deeply ingrained among his contemporaries, and that he would have to labor mightily to overcome it. Indeed, he more or less conceded that the elder authorities of his day would tend to be as immutable as the species they believed in, so in the conclusion of his book he went so far as to beseech the support of his younger readers: "Whoever is led to believe that species are mutable will do good service by conscientiously expressing his conviction; for only thus can the load of prejudice by which this subject is overwhelmed be removed" (*Origin*, p. 482).

Even today Darwin's overthrow of essentialism has not been completely assimilated. For instance, there is much discussion in philosophy these days about "natural kinds," an ancient term the philosopher W. V. O. Quine (1969) quite cautiously resurrected for limited use in distinguishing good scientific categories from bad ones. But in the writings of other philosophers, "natural kind" is often sheep's clothing for the wolf of real essence. The essentialist urge is still with us, and not always for bad reasons. Science does aspire to carve nature at its joints, and it often seems that we need essences, or something like essences, to do the job. On this one point, the two great kingdoms of philosophical thought, the Platonic and the Aristotelian, agree. But the Darwinian mutation, which at first seemed to be just a new way of thinking about kinds in biology, can spread to other phenomena and other disciplines, as we shall see. There are persistent problems both inside and outside biology that readily dissolve once we adopt the Darwinian perspective on what makes a thing the sort of thing it is, but the tradition-bound resistance to this idea persists.

2. NATURAL SELECTION—AN AWFUL STRETCHER

It is an awful stretcher to believe that a peacock's tail was thus formed; but, believing it, I believe in the same principle somewhat modified applied to man.

> —CHARLES DARWIN, letter quoted in Desmond and Moore 1991, p. 553

Darwin's project in *Origin* can be divided in two: to prove *that* modern species were revised descendants of earlier species—species had evolved—and to show *how* this process of "descent with modification" had occurred. If Darwin hadn't had a vision of a mechanism, natural selection, by which this well-nigh-inconceivable historical transformation could have been accomplished, he would probably not have had the motivation to assemble all the circumstantial evidence that it had actually occurred. Today we can readily enough imagine proving Darwin's first case—the brute historic fact of descent with modification—quite independently of any consideration of natural selection or indeed any other mechanism for bringing these brute events about, but for Darwin the idea of the mechanism was both the hunting license he needed, and an unwavering guide to the right questions to ask.[1]

The idea of natural selection was not itself a miraculously novel creation of Darwin's but, rather, the offspring of earlier ideas that had been vigorously discussed for years and even generations (for an excellent account of this intellectual history, see R. Richards 1987). Chief among these parent ideas was an insight Darwin gained from reflection on the 1798 *Essay on the Principle of Population* by Thomas Malthus, which argued that population explosion and famine were inevitable, given the excess fertility of human beings, unless drastic measures were taken. The grim Malthusian vision of the social and political forces that could act to check human overpopulation may have strongly flavored Darwin's thinking (and undoubtedly has flavored the shallow political attacks of many an anti-Darwinian), but the idea Darwin needed from Malthus is purely logical. It has nothing at all to do with political ideology, and can be expressed in very abstract and general terms.

Suppose a world in which organisms have many offspring. Since the offspring themselves will have many offspring, the population will grow and grow ("geometrically") until inevitably, sooner or later—surprisingly soon, in fact—it must grow too large for the available resources (of food, of space, of whatever the organisms need to survive long enough to reproduce). At that point, whenever it happens, not all organisms will have offspring. Many will die childless. It was Malthus who pointed out the mathematical inevitability of such a crunch in *any* population of long-term reproducers—people, animals, plants (or, for that matter, Martian clone-machines, not that such fanciful possibilities were discussed by Malthus). Those populations that reproduce at less than the replacement rate are headed for extinction unless they reverse the trend. Populations that maintain a stable population over long periods of time will do so by settling on a rate of overproduction of offspring that is balanced by the vicissitudes encountered. This is obvious, perhaps, for houseflies and other prodigious breeders, but Darwin drove the point home with a calculation of his own: "The elephant is reckoned to be the slowest breeder of all known animals, and I have taken some pains to estimate its probable minimum rate of natural increase: . . . at the end of the fifth century there would be alive fifteen million elephants, descended from the first pair" (*Origin*, p. 64).[2] Since elephants have been around for millions

---

1. This has often happened in science. For instance, for many years there was lots of evidence lying around in favor of the hypothesis that the continents have drifted—that Africa and South America were once adjacent and broke apart—but until the mechanisms of plate tectonics were conceived, it was hard to take the hypothesis seriously.

2. This sum as it appeared in the first edition is wrong, and when this was pointed out, Darwin revised his calculations for later editions, but the general principle is still unchallenged.

of years, we can be sure that only a fraction of the elephants born in any period have progeny of their own.

So the normal state of affairs for any sort of reproducers is one in which more offspring are produced in any one generation than will in turn reproduce in the next. In other words, it is almost always crunch time.[3] At such a crunch, which prospective parents will "win"? Will it be a fair lottery, in which every organism has an equal chance of being among the few that reproduce? In a political context, this is where invidious themes enter, about power, privilege, injustice, treachery, class warfare, and the like, but we can elevate the observation from its political birthplace and consider in the abstract, as Darwin did, what would—must—happen in nature. Darwin added two further logical points to the insight he had found in Malthus: the first was that at crunch time, if there was significant variation among the contestants, then any advantages enjoyed by any of the contestants would inevitably bias the sample that reproduced. However tiny the advantage in question, if it was actually an advantage (and thus not absolutely invisible to nature), it would tip the scales in favor of those who held it. The second was that *if* there was a "strong principle of inheritance"—if offspring tended to be more like their parents than like their parents' contemporaries—the biases created by advantages, however small, would become amplified over time, creating trends that could grow indefinitely. "More individuals are born than can possibly survive. A grain in the balance will determine which individual shall live and which shall die,—which variety or species shall increase in number, and which shall decrease, or finally become extinct" (*Origin*, p. 467).

What Darwin saw was that if one merely supposed these few general conditions to apply at crunch time—conditions for which he could supply ample evidence—the resulting process would *necessarily* lead in the direction of individuals in future generations who tended to be better equipped to deal with the problems of resource limitation that had been faced by the individuals of their parents' generation. This fundamental idea—Darwin's dangerous idea, the idea that generates so much insight, turmoil, confusion, anxiety—is thus actually quite simple. Darwin summarizes it in two long sentences at the end of chapter 4 of *Origin*:

> If during the long course of ages and under varying conditions of life, organic beings vary at all in the several parts of their organization, and I

---

3. A familiar example of Malthus' rule in action is the rapid expansion of yeast populations introduced into fresh bread dough or grape juice. Thanks to the feast of sugar and other nutrients, population explosions ensue that last for a few hours in the dough, or a few weeks in the juice, but soon the yeast populations hit the Malthusian ceiling, done in by their own voraciousness and the accumulation of their waste products—carbon dioxide (which forms the bubbles that make the bread rise, and the fizz in champagne) and alcohol being the two that we yeast-exploiters tend to value.

think this cannot be disputed; if there be, owing to the high geometric powers of increase of each species, at some age, season, or year, a severe struggle for life, and this certainly cannot be disputed; then, considering the infinite complexity of the relations of all organic beings to each other and to their conditions of existence, causing an infinite diversity in structure, constitution, and habits, to be advantageous to them, I think it would be a most extraordinary fact if no variation ever had occurred useful to each being's own welfare, in the same way as so many variations have occurred useful to man. But if variations useful to any organic being do occur, assuredly individuals thus characterized will have the best chance of being preserved in the struggle for life; and from the strong principle of inheritance they will tend to produce offspring similarly characterized. This principle of preservation, I have called, for the sake of brevity, Natural Selection. [*Origin*, p. 127.]

This was Darwin's great idea, not the idea of evolution, but the idea of evolution *by natural selection*, an idea he himself could never formulate with sufficient rigor and detail to prove, though he presented a brilliant case for it. The next two sections will concentrate on curious and crucial features of this summary statement of Darwin's.

3. DID DARWIN EXPLAIN THE ORIGIN OF SPECIES?

Darwin did wrestle brilliantly and triumphantly with the problem of adaptation, but he had limited success with the issue of diversity— even though he titled his book with reference to his relative failure: the origin of species.

—STEPHEN JAY GOULD 1992a, p. 54

Thus the grand fact in natural history of the subordination of group under group, which, from its familiarity, does not always sufficiently strike us, is in my judgment fully explained.

—CHARLES DARWIN, *Origin*, p. 413

Notice that Darwin's summary does not mention speciation at all. It is entirely about the adaptation of organisms, the *excellence* of their design, not the diversity. Moreover, on the face of it, this summary takes the diversity of species *as an assumption*: "the infinite [sic] complexity of the relations of all organic beings to each other and to their conditions of existence." What makes for this stupendous (if not actually infinite) complexity is the presence at one and the same time (and competing for the same living space) of so many different life forms, with so many different needs and strategies. Darwin doesn't even purport to offer an explanation of the origin of the *first* species, or of life itself; he begins in the middle, supposing many different species with many different talents already present, and claims that starting from such a mid-stage point, the process he has described will inevitably hone and diversify the talents of the species already existing. And will that process create still further species? The summary is silent on that score, but the book is not. In fact, Darwin saw his idea explaining both great sources of wonder in a single stroke. The generation of adaptations and the generation of diversity were different aspects of a single complex phenomenon, and the unifying insight, he claimed, was the principle of natural selection.

Natural selection would inevitably produce *adaptation*, as the summary makes clear, and under the right circumstances, he argued, accumulated adaptation would create speciation. Darwin knew full well that explaining variation is not explaining speciation. The animal-breeders he pumped so vigorously for their lore knew about how to breed *variety* within a single species, but had apparently never created a new *species*, and scoffed at the idea that their particular different breeds might have a common ancestor. "Ask, as I have asked, a celebrated raiser of Hereford cattle, whether his cattle might not have descended from longhorns, and he will laugh you to scorn." Why? Because "though they well know that each race varies slightly, for they win their prizes by selecting such slight differences, yet they ignore all general arguments and refuse to sum up in their minds slight differences accumulated during many successive generations" (*Origin*, p. 29).

The further diversification into species would occur, Darwin argued, because if there was a variety of heritable skills or equipment in a population (of a single species), these different skills or equipment would tend to have different payoffs for different subgroups of the population, and hence these subpopulations would tend to diverge, each one pursuing its favored sort of excellence, until eventually there would be a complete parting of the ways. Why, Darwin asked himself, would this divergence lead to separation or clumping of the variations instead of remaining a more or less continuous fan-out of slight differences? Simple geographical isolation was part of his answer; when a population got split by a major geological or climatic event, or by haphazard emigration to an isolated range such as an island, this discontinuity in the environment ought to become mirrored eventually in a discontinuity in the useful variations observable in the two populations. And once discontinuity got a foothold, it would be self-reinforcing, all the way to separation into distinct species. Another, rather different, idea of his was that in intraspecific infighting, a "winner take all" principle would tend to operate:

> For it should be remembered that the competition will generally be most severe between those forms which are most nearly related to each other in habits, constitution and structure. Hence all the intermediate forms

between the earlier and later states, that is between the less and more improved state of a species, as well as the original parent-species itself, will generally tend to become extinct. [*Origin*, p. 121.]

He formulated a variety of other ingenious and plausible speculations on how and why the relentless culling of natural selection would actually create species boundaries, but they remain speculations to this day. It has taken a century of further work to replace Darwin's brilliant but inconclusive musings on the mechanisms of speciation with accounts that are to some degree demonstrable. Controversy about the mechanisms and principles of speciation still persists, so in one sense neither Darwin nor any subsequent Darwinian has explained the origin of species. As the geneticist Steve Jones (1993) has remarked, had Darwin published his masterpiece under its existing title today, "he would have been in trouble with the Trades Description Act because if there is one thing which *Origin of Species* is not about, it is the origin of species. Darwin knew nothing about genetics. Now we know a great deal, and although the way in which species begin is still a mystery, it is one with the details filled in."

But the fact of speciation itself is incontestable, as Darwin showed, building an irresistible case out of literally hundreds of carefully studied and closely argued instances. That is how species originate: by "descent with modification" from earlier species—not by Special Creation. So in another sense Darwin undeniably did explain the origin of species. Whatever the mechanisms are that operate, they manifestly begin with the emergence of variety within a species, and end, after modifications have accumulated, with the birth of a new, descendant species. What start as "well-marked varieties" turn gradually into "the doubtful category of subspecies; but we have only to suppose the steps in the process of modification to be more numerous or greater in amount, to convert these . . . forms into well-defined species" (*Origin*, p. 120).

Notice that Darwin is careful to describe the eventual outcome as the creation of "well-defined" species. Eventually, he is saying, the divergence becomes so great that there is just no reason to deny that what we have are two different species, not merely two different varieties. But he declines to play the traditional game of declaring what the "essential" difference is:

. . . it will be seen that I look at the term species, as one arbitrarily given for the sake of convenience to a set of individuals closely resembling each other, and that it does not essentially differ from the term variety, which is given to less distinct and more fluctuating forms. [*Origin*, p. 52.]

One of the standard marks of species difference, as Darwin fully recognized, is reproductive isolation—there is no interbreeding. It is interbreeding that reunites the splitting groups, mixing their genes and "frustrating" the process of speciation. It is not that anything *wants* speciation to happen, of course (Dawkins 1986a, p. 237), but if the irreversible divorce that marks speciation is to happen, it must be preceded by a sort of trial separation period in which interbreeding ceases for one reason or another, so that the parting groups can move further apart. The criterion of reproductive isolation is vague at the edges. Do organisms belong to different species when they *can't* interbreed, or when they just *don't* interbreed? Wolves and coyotes and dogs are considered to be different species, and yet interbreeding does occur, and—unlike mules, the offspring of horse and donkey—their offspring are not in general sterile. Dachshunds and Irish wolfhounds are deemed to be of the same species, but unless their owners provide some distinctly unnatural arrangements, they are about as reproductively isolated as bats are from dolphins. The white-tailed deer in Maine don't in fact interbreed with the white-tailed deer in Massachusetts, since they don't travel that far, but they surely could if transported, and naturally they count as of the same species.

And finally—a true-life example seemingly made to order for philosophers—consider the herring gulls that live in the Northern Hemisphere, their range forming a broad ring around the North Pole.

> As we look at the herring gull, moving westwards from Great Britain to North America, we see gulls that are recognizably herring gulls, although they are a little different from the British form. We can follow them, as their appearance gradually changes, as far as Siberia. At about this point in the continuum, the gull looks more like the form that in Great Britain is called the lesser black-backed gull. From Siberia, across Russia, to northern Europe, the gull gradually changes to look more and more like the British lesser black-backed gull. Finally, in Europe, the ring is complete; the two geographically extreme forms meet, to form two perfectly good species: the herring and lesser black-backed gull can be both distinguished by their appearance and do not naturally interbreed. [Mark Ridley 1985, p. 5.]

"Well-defined" species certainly do exist—it is the purpose of Darwin's book to explain their origin—but he discourages us from trying to find a "principled" definition of the concept of a species. Varieties, Darwin keeps insisting, are just "incipient species," and what normally turns two varieties into two species is not the *presence* of something (a new essence for each group, for instance) but the *absence* of something: the intermediate cases, which used to be there—which were necessary stepping-stones, you might say—but have eventually gone extinct, leaving two groups that are *in fact* reproductively isolated as well as different in their characteristics.

Origin of Species presents an overwhelmingly persuasive case for Darwin's first thesis—the historical fact of evolution as the cause of the origin of species—and a tantalizing case in favor of his second thesis—that the fundamental mechanism responsible for "descent with modification" was natural selection.[4] Levelheaded readers of the book simply could no longer doubt that species had evolved over the eons, as Darwin said they had, but scrupulous skepticism about the power of his proposed mechanism of natural selection was harder to overcome. Intervening years have raised the confidence level for both theses, but not erased the difference (Ellegård [1958] provides a valuable account of this history). The evidence for evolution pours in, not only from geology, paleontology, biogeography, and anatomy (Darwin's chief sources), but of course from molecular biology and every other branch of the life sciences. To put it bluntly but fairly, anyone today who doubts that the variety of life on this planet was produced by a process of evolution is simply ignorant—inexcusably ignorant, in a world where three out of four people have learned to read and write. Doubts about the power of Darwin's idea of natural selection to explain this evolutionary process are still intellectually respectable, however, although the burden of proof for such skepticism has become immense, as we shall see.

So, although Darwin depended on his idea of the mechanism of natural selection to inspire and guide his research on evolution, the end result reversed the order of dependence: he showed so convincingly that species *had* to have evolved that he could then turn around and use this fact to support his more radical idea, natural selection. He had described a mechanism or process that, according to his arguments, *could* have produced all these effects. Skeptics were presented with a challenge: Could they show that his arguments were mistaken? Could they show how natural selection would be incapable of producing the effects?[5] Or could they even describe

---

4. As is often pointed out, Darwin didn't insist that natural selection explained everything: it was the "main but not exclusive means of modification" (*Origin*, p. 6).

5. It is sometimes suggested that Darwin's theory is systematically irrefutable (and hence scientifically vacuous), but Darwin was forthright about what sort of finding it would take to refute his theory. "Though nature grants vast periods of time for the work of natural selection, she does not grant an indefinite period" (*Origin*, p. 102), so, if the geological evidence mounted to show that not enough time had elapsed, his whole theory would be refuted. This still left a temporary loophole, for the theory wasn't formulatable in sufficiently rigorous detail to say just how many millions of years was the minimal amount required, but it was a temporary loophole that made sense, since at least some proposals about its size could be evaluated independently. (Kitcher [1985a, pp. 162–65], has a good discussion of the further subtleties of argument that kept Darwinian theory from being directly confirmed or disconfirmed.) Another famous instance: "If it could be demonstrated that any complex organ existed, which could not possibly have been formed by numerous, successive, slight modifications, my theory would absolutely break down" (*Origin*, p. 189). Many have risen to this challenge, but, as we shall see in chapter 11, there are good reasons why they have not succeeded in their attempted demonstrations.

another process that might achieve these effects? What *else* could account for evolution, if not the mechanism he had described?

This challenge effectively turned Hume's predicament inside out. Hume caved in because he could not imagine how anything other than an Intelligent Artificer could be the cause of the adaptations that anyone could observe. Or, more accurately, Hume's Philo imagined several different alternatives, *but Hume had no way of taking these imaginings seriously*. Darwin described how a Nonintelligent Artificer could produce those adaptations over vast amounts of time, and proved that many of the intermediate stages that would be needed by that proposed process had indeed occurred. Now the challenge to imagination was reversed: given all the telltale signs of the historical process that Darwin uncovered—all the brushmarks of the artist, you might say—could anyone imagine how any process *other* than natural selection could have produced all these effects? So complete has this reversal of the burden of proof been that scientists often find themselves in something like the mirror image of Hume's predicament. When they are confronted with a *prima facie* powerful and undismissable objection to natural selection (we will consider the strongest cases in due course), they are driven to reason as follows: I cannot (yet) see how to refute this objection, or overcome this difficulty, but since I cannot imagine how anything other than natural selection could be the cause of the effects, I will have to assume that the objection is spurious; *somehow* natural selection must be sufficient to explain the effects.

Before anyone jumps on this and pronounces that I have just conceded that Darwinism is just as much an unprovable faith as natural religion, it should be borne in mind that there is a fundamental difference: having declared their allegiance to natural selection, these scientists have then proceeded to take on the burden of showing how the difficulties with their view could be overcome, and, time and time again, they have succeeded in meeting the challenge. In the process, Darwin's fundamental idea of natural selection has been articulated, expanded, clarified, quantified, and deepened in many ways, becoming stronger every time it overcame a challenge. With every success, the scientists' conviction grows that they must be on the right track. It is reasonable to believe that an idea that was ultimately false would surely have succumbed by now to such an unremitting campaign of attacks. That is not a conclusive proof, of course, just a mighty persuasive consideration. One of the goals of this book is to explain why the idea of natural selection appears to be a clear winner, even while there are unresolved controversies about how it can handle some phenomena.

4. NATURAL SELECTION AS AN ALGORITHMIC PROCESS

What limit can be put to this power, acting during long ages and rigidly scrutinising the whole constitution, structure, and habits of each creature,—favouring the good and rejecting the bad? I can see no limit to this power, in slowly and beautifully adapting each form to the most complex relations of life.

—CHARLES DARWIN, *Origin*, p. 469

The second point to notice in Darwin's summary is that he presents his principle as deducible by a formal argument—*if* the conditions are met, a certain outcome is *assured*.[6] Here is the summary again, with some key terms in boldface.

If, during the long course of ages and under varying conditions of life, organic beings vary at all in the several parts of their organization, and I think this cannot be disputed; **if** there be, owing to the high geometric powers of increase of each species, at some age, season, or year, a severe struggle for life, and this certainly cannot be disputed; **then,** considering the infinite complexity of the relations of all organic beings to each other and to their conditions of existence, causing an infinite diversity in structure, constitution, and habits, to be advantageous to them, **I think it would be a most extraordinary fact if no variation ever had occurred useful to each being's own welfare,** in the same way as so many variations have occurred useful to man. But **if** variations useful to any organic being do occur, **assuredly** individuals thus characterized will have the best chance of being preserved in the struggle for life; and from the strong principle of inheritance they will tend to produce offspring similarly characterized. This principle of preservation, I have called, for the sake of brevity, Natural Selection. [*Origin,* p. 127 (facs. ed. of 1st ed.).]

The basic deductive argument is short and sweet, but Darwin himself described *Origin of Species* as "one long argument." That is because it

---

6. The ideal of a deductive (or "nomologico-deductive") science, modeled on Newtonian or Galilean physics, was quite standard until fairly recently in the philosophy of science, so it is not surprising that much effort has been devoted to devising and criticizing various axiomatizations of Darwin's theory—since it was presumed that in such a formalization lay scientific vindication. The idea, introduced in this section, that Darwin should be seen, rather, as postulating that evolution is an algorithmic process, permits us to do justice to the undeniable *a priori* flavor of Darwin's thinking without forcing it into the Procrustean (and obsolete) bed of the nomologico-deductive model. See Sober 1984a and Kitcher 1985a.

consists of two sorts of demonstrations: the logical demonstration that a certain *sort* of process would necessarily have a certain sort of outcome, and the empirical demonstration that the requisite conditions for that sort of process had in fact been met in nature. He bolsters up his logical demonstration with thought experiments—"imaginary instances" (*Origin*, p. 95)—that show *how* the meeting of these conditions *might* actually account for the effects he claimed to be explaining, but his whole argument extends to book length because he presents a wealth of hard-won empirical detail to convince the reader that these conditions have been met over and over again.

Stephen Jay Gould (1985) gives us a fine glimpse of the importance of this feature of Darwin's argument in an anecdote about Patrick Matthew, a Scottish naturalist who as a matter of curious historical fact had scooped Darwin's account of natural selection by many years—in an appendix to his 1831 book, *Naval Timber and Arboriculture*. In the wake of Darwin's ascent to fame, Matthew published a letter (in *Gardeners' Chronicle!*[7]) proclaiming his priority, which Darwin graciously conceded, excusing his ignorance by noting the obscurity of Matthew's choice of venue. Responding to Darwin's published apology, Matthew wrote:

> To me the conception of this law of Nature came intuitively as a self-evident fact, almost without an effort of concentrated thought. Mr. Darwin here seems to have more merit in the discovery than I have had—to me it did not appear a discovery. He seems to have worked it out by inductive reason, slowly and with due caution to have made his way synthetically from fact to fact onwards; while with me it was by a general glance at the scheme of Nature that I estimated this select production of species as an a priori recognizable fact—an axiom, requiring only to be pointed out to be admitted by unprejudiced minds of sufficient grasp. [Quoted in Gould 1985, pp. 345–46.]

Unprejudiced minds may well resist a new idea out of sound conservatism, however. Deductive arguments are notoriously treacherous; what seems to "stand to reason" can be betrayed by an overlooked detail. Darwin appreciated that only a relentlessly detailed survey of the evidence for the historical processes he was postulating would—or should—persuade scientists to abandon their traditional convictions and take on his revolutionary vision, even if it was in fact "deducible from first principles."

---

7. *Gardeners' Chronicle*, April 7, 1860. See Hardin 1964 for more details.

From the outset, there were those who viewed Darwin's novel mixture of detailed naturalism and abstract reasoning about processes as a dubious and inviable hybrid. It had a tremendous air of plausibility, but so do many get-rich-quick schemes that turn out to be empty tricks. Compare it to the following stock-market principle: Buy Low, Sell High. This is guaranteed to make you wealthy. You cannot fail to get rich *if* you follow this advice. Why doesn't it work? It does work—for everybody who is fortunate enough to act according to it, but, alas, there is no way of determining that the conditions are met until it is too late to act on them. Darwin was offering a skeptical world what we might call a get-rich-*slow* scheme, a scheme for creating Design out of Chaos without the aid of Mind.

The theoretical power of Darwin's abstract scheme was due to several features that Darwin quite firmly identified, and appreciated better than many of his supporters, but lacked the terminology to describe explicitly. Today we could capture these features under a single term. Darwin had discovered the power of an *algorithm*. An algorithm is a certain sort of formal process that can be counted on—logically—to yield a certain sort of result whenever it is "run" or instantiated. Algorithms are not new, and were not new in Darwin's day. Many familiar arithmetic procedures, such as long division or balancing your checkbook, are algorithms, and so are the decision procedures for playing perfect tic-tac-toe, and for putting a list of words into alphabetical order. What is relatively new—permitting us valuable hindsight on Darwin's discovery—is the theoretical reflection by mathematicians and logicians on the nature and power of algorithms in general, a twentieth-century development which led to the birth of the computer, which has led in turn, of course, to a much deeper and more lively understanding of the powers of algorithms in general.

The term *algorithm* descends, via Latin (*algorismus*) to early English (*algorisme* and, mistakenly therefrom, *algorithm*), from the name of a Persian mathematician, Mûusâ al-Khowârizm, whose book on arithmetical procedures, written about 835 A.D., was translated into Latin in the twelfth century by Adelard of Bath or Robert of Chester. The idea that an algorithm is a foolproof and somehow "mechanical" procedure has been present for centuries, but it was the pioneering work of Alan Turing, Kurt Gödel, and Alonzo Church in the 1930s that more or less fixed our current understanding of the term. Three key features of algorithms will be important to us, and each is somewhat difficult to define. Each, moreover, has given rise to confusions (and anxieties) that continue to beset our thinking about Darwin's revolutionary discovery, so we will have to revisit and reconsider these introductory characterizations several times before we are through:

(1) *substrate neutrality:* The procedure for long division works equally well with pencil or pen, paper or parchment, neon lights or skywrit-

ing, using any symbol system you like. The power of the procedure is due to its *logical* structure, not the causal powers of the materials used in the instantiation, just so long as those causal powers permit the prescribed steps to be followed exactly.

(2) *underlying mindlessness:* Although the overall design of the procedure may be brilliant, or yield brilliant results, each constituent step, as well as the transition between steps, is utterly simple. How simple? Simple enough for a dutiful idiot to perform—or for a straightforward mechanical device to perform. The standard textbook analogy notes that algorithms are *recipes* of sorts, designed to be followed by *novice* cooks. A recipe book written for great chefs might include the phrase "Poach the fish in a suitable wine until almost done," but an algorithm for the same process might begin, "Choose a white wine that says 'dry' on the label; take a corkscrew and open the bottle; pour an inch of wine in the bottom of a pan; turn the burner under the pan on high; ... "—a tedious breakdown of the process into dead-simple steps, requiring no wise decisions or delicate judgments or intuitions on the part of the recipe-reader.

(3) *guaranteed results:* Whatever it is that an algorithm does, it always does it, if it is executed without misstep. An algorithm is a foolproof recipe.

It is easy to see how these features made the computer possible. *Every computer program is an algorithm*, ultimately composed of simple steps that can be executed with stupendous reliability by one simple mechanism or another. Electronic circuits are the usual choice, but the power of computers owes nothing (save speed) to the causal peculiarities of electrons darting about on silicon chips. The very same algorithms can be performed (even faster) by devices shunting photons in glass fibers, or (much, much slower) by teams of people using paper and pencil. And as we shall see, the capacity of computers to run algorithms with tremendous speed and reliability is now permitting theoreticians to explore Darwin's dangerous idea in ways heretofore impossible, with fascinating results.

What Darwin discovered was not really *one* algorithm but, rather, a large class of related algorithms that he had no clear way to distinguish. We can now reformulate his fundamental idea as follows:

Life on Earth has been generated over billions of years in a single branching tree—the Tree of Life—by one algorithmic process or another.

What this claim means will become clear gradually, as we sort through the various ways people have tried to express it. In some versions it is utterly vacuous and uninformative; in others it is manifestly false. In between lie the versions that really do explain the origin of species and promise to explain much else besides. These versions are becoming clearer all the time, thanks as much to the determined criticisms of those who frankly hate the idea of evolution as an algorithm, as to the rebuttals of those who love it.

5. PROCESSES AS ALGORITHMS

When theorists think of algorithms, they often have in mind kinds of algorithms with properties that are *not* shared by the algorithms that will concern us. When mathematicians think about algorithms, for instance, they usually have in mind algorithms that can be proven to compute particular mathematical functions of interest to them. (Long division is a homely example. A procedure for breaking down a huge number into its prime factors is one that attracts attention in the exotic world of cryptography.) But the algorithms that will concern us have nothing particular to do with the number system or other mathematical objects; they are algorithms for sorting, winnowing, and building things.[8]

Because most mathematical discussions of algorithms focus on their guaranteed or mathematically provable powers, people sometimes make the elementary mistake of thinking that a process that makes use of chance or randomness is not an algorithm. But even long division makes good use of randomness!

$$47 \overline{)326574} 7?$$

Does the divisor go into the dividend six or seven or eight times? Who knows? Who cares? You don't have to know; you don't have to have any wit or discernment to do long division. The algorithm directs you just to choose a digit—at random, if you like—and check out the result. If the chosen number turns out to be too small, increase it by one and start over; if too large, decrease it. The good thing about long division is that it always works

---

8. Computer scientists sometimes restrict the term *algorithm* to programs that can be proven to *terminate*—that have no infinite loops in them, for instance. But this special sense, valuable as it is for some mathematical purposes, is not of much use to us. Indeed, few of the computer programs in daily use around the world would qualify as algorithms in this restricted sense; most are designed to cycle indefinitely, patiently waiting for instructions (including the instruction to terminate, without which they keep on going). Their subroutines, however, are algorithms in this strict sense—except where undetected "bugs" lurk that can cause the program to "hang."

eventually, even if you are maximally stupid in making your first choice, in which case it just takes a little longer. Achieving success on hard tasks in spite of utter stupidity is what makes computers seem magical—how could something as mindless as a machine do something as smart as that? Not surprisingly, then, the tactic of finessing ignorance by randomly generating a candidate and then testing it out mechanically is a ubiquitous feature of interesting algorithms. Not only does it not interfere with their provable powers as algorithms; it is often the key to their power. (See Dennett 1984, pp. 149–52, on the particularly interesting powers of Michael Rabin's random algorithms.)

We can begin zeroing in on the phylum of evolutionary algorithms by considering everyday algorithms that share important properties with them. Darwin draws our attention to repeated waves of competition and selection, so consider the standard algorithm for organizing an elimination tournament, such as a tennis tournament, which eventually culminates with quarter-finals, semi-finals, and then a final, determining the solitary winner.

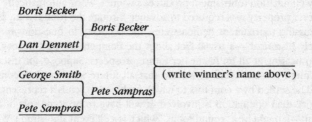

Notice that this procedure meets the three conditions. It is the same procedure whether drawn in chalk on a blackboard, or updated in a computer file, or—a weird possibility—not written down anywhere, but simply enforced by building a huge fan of fenced-off tennis courts each with two entrance gates and a single exit gate leading the winner to the court where the next match is to be played. (The losers are shot and buried where they fall.) It doesn't take a genius to march the contestants through the drill, filling in the blanks at the end of each match (or identifying and shooting the losers). And it always works.

But what, exactly, does this algorithm *do*? It takes as input a set of competitors and guarantees to terminate by identifying a single winner. But what is a winner? It all depends on the competition. Suppose the tournament in question is not tennis but coin-tossing. One player tosses and the other calls; the winner advances. The winner of this tournament will be that single player who has won n consecutive coin-tosses without a loss, depending on how many rounds it takes to complete the tournament.

There is something strange and trivial about this tournament, but what is it? The winner does have a rather remarkable property. How often have you ever met anyone who just won, say, ten consecutive coin-tosses without a loss? Probably never. The odds against there being such a person might seem enormous, and in the normal course of events, they surely are. If some gambler offered you ten-to-one odds that he could produce someone who before your very eyes would proceed to win ten consecutive coin-tosses using a fair coin, you might be inclined to think this a good bet. If so, you had better hope the gambler doesn't have 1,024 accomplices (they don't have to cheat—they play fair and square). For that is all it takes (2^{10} competitors) to form a ten-round tournament. The gambler wouldn't have a clue, as the tournament started, which person would end up being the exhibit A that would guarantee his winning the wager, but the tournament algorithm is sure to produce such a person in short order—it is a sucker bet with a surefire win for the gambler. (I am not responsible for any injuries you may sustain if you attempt to get rich by putting this tidbit of practical philosophy into use.)

Any elimination tournament produces a winner, who "automatically" has whatever property was required to advance through the rounds, but, as the coin-tossing tournament demonstrates, the property in question *may* be "merely historical"—a trivial fact about the competitor's past history that has no bearing at all on his or her future prospects. Suppose, for instance, the United Nations were to decide that all future international conflicts would be settled by a coin-toss to which each nation sends a representative (if more than one nation is involved, it will have to be some sort of tournament—it might be a "round robin," which is a different algorithm). Whom should we designate as our national representative? The best coin-toss caller in the land, obviously. Suppose we organized every man, woman, and child in the U.S.A. into a giant elimination tournament. Somebody would have to win, and that person would have just won twenty-eight consecutive coin-tosses without a loss! This would be an irrefutable historical fact about that person, but since calling a coin-toss is just a matter of luck, there is absolutely no reason to believe that the winner of such a tournament would do any better in international competition than somebody else who lost in an earlier round of the tournament. Chance has no memory. A person who holds the winning lottery ticket has certainly *been* lucky, and, thanks to the millions she has just won, she may never need to be lucky again—which is just as well, since there is no reason to think she is more likely than anyone else to win the lottery a second time, or to win the next coin-toss she calls. (Failing to appreciate the fact that chance has no memory is known as the Gambler's Fallacy; it is surprisingly popular—so popular that I should probably stress that it *is* a fallacy, beyond any doubt or controversy.)

In contrast to tournaments of pure luck, like the coin-toss tournament, there are tournaments of skill, like tennis tournaments. Here there *is* reason to believe that the players in the later rounds would do better *again* if they played the players who lost in the early rounds. There is reason to believe— but no guarantee—that the winner of such a tournament is the best player of them all, not just today but tomorrow. Yet, though any well-run tournament is guaranteed to produce a winner, there is no guarantee that a tournament of skill will identify the best player as the winner in any nontrivial sense. That's why we sometimes say, in the opening ceremonies, "May the best man win!"—because it is not guaranteed by the procedure. The best player—the one who is best by "engineering" standards (has the most reliable backhand, fastest serve, most stamina, etc.)—may have an off day, or sprain his ankle, or get hit by lightning. Then, trivially, he may be bested in competition by a player who is not really as good as he is. But nobody would bother organizing or entering tournaments of skill if it weren't the case that *in the long run*, tournaments of skill are won by the best players. *That* is guaranteed by the very definition of a fair tournament of skill; if there were no probability greater than half that the better players would win each round, it would be a tournament of luck, not of skill.

Skill and luck intermingle naturally and inevitably in any real competition, but their ratios may vary widely. A tennis tournament played on very bumpy courts would raise the luck ratio, as would an innovation in which the players were required to play Russian roulette with a loaded revolver before continuing after the first set. But even in such a luck-ridden contest, more of the better players would *tend*, statistically, to get to the late rounds. The power of a tournament to "discriminate" skill differences in the long run may be diminished by haphazard catastrophe, but it is not in general reduced to zero. This fact, which is as true of evolutionary algorithms in nature as of elimination tournaments in sports, is sometimes overlooked by commentators on evolution.

Skill, in contrast to luck, is *projectable*; in the same or similar circumstances, it can be counted on to give repeat performances. This relativity to circumstances shows us another way in which a tournament might be weird. What if the conditions of competition kept changing (like the croquet game in *Alice in Wonderland*)? If you play tennis the first round, chess in the second round, golf in the third round, and billiards in the fourth round, there is no reason to suppose the eventual winner will be particularly good, compared with the whole field, in *any* of these endeavors—all the good golfers may lose in the chess round and never get a chance to demonstrate their prowess, and even if luck plays no role in the fourth-round billiards final, the winner might turn out to be the second-*worst* billiards player in the whole field. Thus there has to be some measure of uniformity of the conditions of competition for there to be any *interesting* outcome to a tournament.

But does a tournament—or any algorithm—have to do something interesting? No. The algorithms we tend to talk about almost always do something interesting—that's why they attract our attention. But a procedure doesn't fail to be an algorithm just because it is of no conceivable use or value to anyone. Consider a variation on the elimination-tournament algorithm in which the *losers* of the semi-finals play in the finals. This is a stupid rule, destroying the *point* of the whole tournament, but the tournament would still be an algorithm. Algorithms don't have to have points or purposes. In addition to all the useful algorithms for alphabetizing lists of words, there are kazillions of algorithms for reliably *mis*alphabetizing words, and they work perfectly every time (as if anyone would care). Just as there is an algorithm (many, actually) for finding the square root of any number, so there are algorithms for finding the square root of any number except 18 or 703. Some algorithms do things so boringly irregular and pointless that there is no succinct way of saying what they are *for*. They just do what they do, and they do it every time.

We can now expose perhaps the most common misunderstanding of Darwinism: the idea that Darwin showed that evolution by natural selection is a procedure *for* producing Us. Ever since Darwin proposed his theory, people have often misguidedly tried to interpret it as showing that we are the destination, the goal, the point of all that winnowing and competition, and our arrival on the scene was guaranteed by the mere holding of the tournament. This confusion has been fostered by evolution's friends and foes alike, and it is parallel to the confusion of the coin-toss tournament winner who basks in the misconsidered glory of the idea that since the tournament had to have a winner, and since he is the winner, the tournament had to produce him as the winner. Evolution can be an algorithm, and evolution can have produced us by an algorithmic process, without its being true that evolution is an algorithm for producing us. The main conclusion of Stephen Jay Gould's *Wonderful Life: The Burgess Shale and the Nature of History* (1989a) is that if we were to "wind the tape of life back" and play it again and again, the likelihood is infinitesimal of *Us* being the product on any other run through the evolutionary mill. This is undoubtedly true (if by "Us" we mean the particular variety of *Homo sapiens* we are: hairless and upright, with five fingers on each of two hands, speaking English and French and playing tennis and chess). Evolution is not a process that was designed to produce us, but it does not follow from this that evolution is not an algorithmic process that has in fact produced us. (Chapter 10 will explore this issue in more detail.)

Evolutionary algorithms are manifestly interesting algorithms—interesting to us, at least—not because what they are guaranteed to do is interesting to us, but because what they are guaranteed to *tend* to do is interesting to us. They are like tournaments of skill in this regard. The power of an algorithm to yield something of interest or value is not at all limited to what the algorithm can be mathematically proven to yield in a foolproof way, and this is especially true of evolutionary algorithms. Most of the controversies about Darwinism, as we shall see, boil down to disagreements about just how powerful certain postulated evolutionary processes are—could they actually do all this or all that in the time available? These are typically investigations into what an evolutionary algorithm *might* produce, or *could* produce, or is *likely* to produce, and only indirectly into what such an algorithm would *inevitably* produce. Darwin himself sets the stage in the wording of his summary: his idea is a claim about what "assuredly" the process of natural selection will "tend" to yield.

All algorithms are guaranteed to do whatever they do, but it need not be anything interesting; some algorithms are further guaranteed to tend (with probability p) to do something—which may or may not be interesting. But if what an algorithm is guaranteed to do doesn't have to be "interesting" in any way, how are we going to distinguish algorithms from other processes? Won't *any* process be an algorithm? Is the surf pounding on the beach an algorithmic process? Is the sun baking the clay of a dried-up riverbed an algorithmic process? The answer is that there may be features of these processes that *are* best appreciated if we consider them as algorithms! Consider, for instance, the question of why the grains of sand on a beach are so uniform in size. This is due to a natural sorting process that occurs thanks to the repetitive launching of the grains by the surf—alphabetical order on a grand scale, you might say. The pattern of cracks that appear in the sun-baked clay may be best explained by looking at chains of events that are not unlike the successive rounds in a tournament.

Or consider the process of annealing a piece of metal to temper it. What could be a more physical, less "computational" process than that? The blacksmith repeatedly heats the metal and then lets it cool, and somehow in the process it becomes much stronger. How? What kind of an explanation can we give for this magical transformation? Does the heat create special toughness atoms that coat the surface? Or does it suck subatomic glue out of the atmosphere that binds all the iron atoms together? No, nothing like that happens. The right level of explanation is the algorithmic level: As the metal cools from its molten state, the solidification starts in many different spots at the same time, creating crystals that grow together until the whole is solid. But the first time this happens, the arrangement of the individual crystal structures is suboptimal—weakly held together, and with lots of internal stresses and strains. Heating it up again—but not all the way to melting—partially breaks down these structures, so that, when they are permitted to cool the next time, the broken-up bits will adhere to the still-solid bits in a different arrangement. It can be proven mathematically that these rearrangements will tend to get better and better, approaching the optimum or strongest total structure, provided the regime of heating and cooling has the right parameters. So powerful is this optimization procedure that it has been used as the inspiration for an entirely general problem-solving technique in computer science—"simulated annealing," which has nothing to do with metals or heat, but is just a way of getting a computer program to build, disassemble, and rebuild a data structure (such as another program), over and over, blindly groping towards a better— indeed, an optimal—version (Kirkpatrick, Gelatt and Vecchi 1983). This was one of the major insights leading to the development of "Boltzmann machines" and "Hopfield nets" and the other constraint-satisfaction schemes that are the basis for the Connectionist or "neural-net" architectures in Artificial Intelligence. (For overviews, see Smolensky 1983, Rumelhart 1989, Churchland and Sejnowski 1992, and, on a philosophical level, Dennett 1987a, Paul Churchland 1989.)

If you want a deep understanding of how annealing works in metallurgy, you have to learn the physics of all the forces operating at the atomic level, of course, but notice that the basic idea of how annealing works (and particularly *why* it *works*) can be lifted clear of those details—after all, I just explained it in simple lay terms (and I don't know the physics!). The explanation of annealing can be put in *substrate-neutral* terminology: we should expect optimization of a certain sort to occur in any "material" that has components that get put together by a certain sort of building process and that can be disassembled in a sequenced way by changing a single global parameter, etc. That is what is common to the processes going on in the glowing steel bar and the humming supercomputer.

Darwin's ideas about the powers of natural selection can also be lifted out of their home base in biology. Indeed, as we have already noted, Darwin himself had few inklings (and what inklings he had turned out to be wrong) about how the microscopic processes of genetic inheritance were accomplished. Not knowing any of the details about the physical substrate, he could nevertheless discern that if certain conditions were somehow met, certain effects would be wrought. This substrate neutrality has been crucial in permitting the basic Darwinian insights to float like a cork on the waves of subsequent research and controversy, for what has happened since Darwin has a curious flip-flop in it. Darwin, as we noted in the preceding chapter, never hit upon the utterly necessary idea of a gene, but along came Mendel's concept to provide just the right structure for making mathematical sense out of heredity (and solving Darwin's nasty problem of blending inheritance). And then, when DNA was identified as the actual physical vehicle of the genes, it looked at first (and still looks to many participants) as if Mendel's genes could be simply *identified* as particular hunks of DNA. But then complexities began to emerge; the more scientists have learned about the actual molecular biology of DNA and its role in reproduction, the clearer it becomes that the Mendelian story is at best a vast oversimplification. Some would go so far as to say that we have recently learned that there really *aren't* any Mendelian genes! Having climbed Mendel's ladder, we must now throw it away. But of course no one wants to throw away such a valuable tool, still proving itself daily in hundreds of scientific and medical contexts. The solution is to bump Mendel up a level, and declare that he, like Darwin, captured an *abstract* truth about inheritance. We may, if we like, talk of *virtual genes*, considering them to have their reality distributed around in the concrete materials of the DNA. (There is much to be said in favor of this option, which I will discuss further in chapters 5 and 12.)

But then, to return to the question raised above, are there any limits at all on what may be considered an algorithmic process? I guess the answer is No; if you wanted to, you could treat any process at the abstract level as an algorithmic process. So what? Only some processes yield interesting results when you do treat them as algorithms, but we don't have to try to define "algorithm" in such a way as to include only the *interesting* ones (a tall philosophical order!). The problem will take care of itself, since nobody will waste time examining the algorithms that aren't interesting for one reason or another. It all depends on what needs explaining. If what strikes you as puzzling is the uniformity of the sand grains or the strength of the blade, an algorithmic explanation is what will satisfy your curiosity—and it will be the truth. Other interesting features of the same phenomena, or the processes that created them, might not yield to an algorithmic treatment.

Here, then, is Darwin's dangerous idea: the algorithmic level *is* the level that best accounts for the speed of the antelope, the wing of the eagle, the shape of the orchid, the diversity of species, and all the other occasions for wonder in the world of nature. It is hard to believe that something as mindless and mechanical as an algorithm could produce such wonderful things. No matter how impressive the products of an algorithm, the underlying process always consists of nothing but a set of individually mindless steps succeeding each other without the help of any intelligent supervision; they are "automatic" by definition: the workings of an automaton. They feed on each other, or on blind chance—coin-flips, if you like—and on nothing else. Most algorithms we are familiar with have rather modest products: they do long division or alphabetize lists or figure out the income of the Average Taxpayer. Fancier algorithms produce the dazzling computer-animated graphics we see every day on television, transforming faces, creating herds of imaginary ice-skating polar bears, simulating whole virtual worlds of entities never seen or imagined before. But the actual biosphere is much fancier still, by many orders of magnitude. Can it really be the outcome of nothing but a cascade of algorithmic processes feeding on chance? And if so, who designed that cascade? Nobody. It is itself the product of a blind, algorithmic process. As Darwin himself put it, in a letter to the geologist Charles Lyell shortly after publication of *Origin*, "I would give absolutely nothing for the theory of Natural Selection, if it requires miraculous additions at any one stage of descent. . . . If I were convinced that I required such additions to the theory of natural selection, I would reject it as rubbish . . ." (F. Darwin 1911, vol. 2, pp. 6–7).

According to Darwin, then, evolution is an algorithmic process. Putting it this way is still controversial. One of the tugs-of-war going on within evolutionary biology is between those who are relentlessly pushing, pushing, pushing towards an algorithmic treatment, and those who, for various submerged reasons, are resisting this trend. It is rather as if there were metallurgists around who were disappointed by the algorithmic explanation of annealing. "You mean that's all there is to it? No submicroscopic Superglue specially created by the heating and cooling process?" Darwin has convinced all the scientists that evolution, like annealing, *works*. His radical vision of *how* and *why* it works is still somewhat embattled, largely because those who resist can dimly see that their skirmish is part of a larger campaign. If the game is lost in evolutionary biology, where will it all end?

CHAPTER 2: *Darwin conclusively demonstrated that, contrary to ancient tradition, species are not eternal and immutable; they evolve. The origin of new species was shown to be the result of "descent with modification." Less conclusively, Darwin introduced an idea of how this evolutionary process took place: via a mindless, mechanical—algorithmic—process he called "natural selection." This idea, that all the fruits of evolution can be explained as the products of an algorithmic process, is Darwin's dangerous idea.*

CHAPTER 3: *Many people, Darwin included, could dimly see that his idea of natural selection had revolutionary potential, but just what did it promise to overthrow? Darwin's idea can be used to dismantle and then rebuild a traditional structure of Western thought, which I call the Cosmic Pyramid. This provides a new explanation of the origin, by gradual accumulation, of all the Design in the universe. Ever since Darwin, skepticism has been aimed at his implicit claim that the various processes of natural selection, in spite of their underlying mindlessness, are powerful enough to have done all the design work that is manifest in the world.*

CHAPTER THREE

Universal Acid

〰〰

1. EARLY REACTIONS

Origin of man now proved. —Metaphysics must flourish. —He who understands baboon would do more towards metaphysics than Locke.

—CHARLES DARWIN, in a notebook not intended for publication, in P. H. Barrett et al. 1987, D26, M84

His subject is the 'Origin of Species,' & not the origin of Organization; & it seems a needless mischief to have opened the latter speculation at all.

—HARRIET MARTINEAU, a friend of Darwin's, in a letter to Fannie Wedgwood, March, 13, 1860, quoted in Desmond and Moore 1991, p. 486

Darwin began his explanation in the middle, or even, you might say, at the end: starting with the life forms we presently see, and showing how the patterns in today's biosphere could be explained as having arisen by the process of natural selection from the patterns in yesterday's biosphere, and so on, back into the very distant past. He started with facts that everyone knows: all of today's living things are the offspring of parents, who are the offspring of grandparents, and so forth, so everything that is alive today is a branch of a genealogical family, which is itself a branch of a larger clan. He went on to argue that, if you go back far enough, you find that all the branches of all the families eventually spring from common ancestral limbs, so that there is a single Tree of Life, all the limbs, branches, and twigs united by descent with modification. The fact that it has the branching organization of a tree is crucial to the explanation of the sort of process involved, for such a tree *could* be created by an automatic, recursive process: first build an x, then modify x's descendants, then modify those modifications, then modify the modifications of the modifications.... If Life is a Tree, it could all have arisen from an inexorable, automatic rebuilding process in which designs would accumulate over time.

Working backwards, starting at or near "the end" of a process, and solving the next-to-last step before asking how *it* could have been produced, is a tried and true method of computer programmers, particularly when creating programs that use recursion. Usually this is a matter of practical modesty: if you don't want to bite off more than you can chew, the right bite to start with is often the finishing bite, if you can find it. Darwin found it, and then very cautiously worked his way back, skirting around the many grand issues that his investigations stirred up, musing about them in his private notebooks, but postponing their publication indefinitely. (For instance, he deliberately avoided discussing human evolution in *Origin*; see the discussion in R. J. Richards 1987, pp. 160ff.) But he could see where all this was leading, and, in spite of his near-perfect silence on these troubling extrapolations, so could many of his readers. Some loved what they thought they saw, and others hated it.

Karl Marx was exultant: "Not only is a death blow dealt here for the first time to 'Teleology' in the natural sciences but their rational meaning is empirically explained" (quoted in Rachels 1991, p. 110). Friedrich Nietzsche saw—through the mists of his contempt for all things English—an even more cosmic message in Darwin: God is dead. If Nietzsche is the father of existentialism, then perhaps Darwin deserves the title of grandfather. Others were less enthralled with the thought that Darwin's views were utterly subversive to sacred tradition. Samuel Wilberforce, Bishop of Oxford, whose debate with Thomas Huxley in June 1860 was one of the most celebrated confrontations between Darwinism and the religious establishment (see chapter 12), said in an anonymous review:

> Man's derived supremacy over the earth; man's power of articulate speech; man's gift of reason; man's free-will and responsibility . . .—all are equally and utterly irreconcilable with the degrading notion of the brute origin of him who was created in the image of God.... [Wilberforce 1860.]

When speculation on these extensions of his view arose, Darwin wisely chose to retreat to the security of his base camp, the magnificently provisioned and defended thesis that began in the middle, with life already on the scene, and "merely" showed how, once this process of design accumulation was under way, it could proceed without any (further?) intervention from any Mind. But, as many of his readers appreciated, however comforting this modest disclaimer might be, it was not really a stable resting place.

Did you ever hear of universal acid? This fantasy used to amuse me and some of my schoolboy friends—I have no idea whether we invented or inherited it, along with Spanish fly and saltpeter, as a part of underground youth culture. Universal acid is a liquid so corrosive that it will eat through *anything*! The problem is: what do you keep it in? It dissolves glass bottles and stainless-steel canisters as readily as paper bags. What would happen if you somehow came upon or created a dollop of universal acid? Would the whole planet eventually be destroyed? What would it leave in its wake? After everything had been transformed by its encounter with universal acid, what would the world look like? Little did I realize that in a few years I would encounter an idea—Darwin's idea—bearing an unmistakable likeness to universal acid: it eats through just about every traditional concept, and leaves in its wake a revolutionized world-view, with most of the old landmarks still recognizable, but transformed in fundamental ways.

Darwin's idea had been born as an answer to questions in biology, but it threatened to leak out, offering answers—welcome or not—to questions in cosmology (going in one direction) and psychology (going in the other direction). If *re*design could be a mindless, algorithmic process of evolution, why couldn't that whole process itself be the product of evolution, and so forth, *all the way down*? And if mindless evolution could account for the breathtakingly clever artifacts of the biosphere, how could the products of our own "real" minds be exempt from an evolutionary explanation? Darwin's idea thus also threatened to spread *all the way up*, dissolving the illusion of our own authorship, our own divine spark of creativity and understanding.

Much of the controversy and anxiety that has enveloped Darwin's idea ever since can be understood as a series of failed campaigns in the struggle to contain Darwin's idea within some acceptably "safe" and merely partial revolution. Cede some or all of modern biology to Darwin, perhaps, but hold the line there! Keep Darwinian thinking out of cosmology, out of psychology, out of human culture, out of ethics, politics, and religion! In these campaigns, many battles have been won by the forces of containment: flawed applications of Darwin's idea have been exposed and discredited, beaten back by the champions of the pre-Darwinian tradition. But new waves of Darwinian thinking keep coming. They seem to be improved versions, not vulnerable to the refutations that defeated their predecessors, but are they sound extensions of the unquestionably sound Darwinian core idea, or might they, too, be perversions of it, and even more virulent, more dangerous, than the abuses of Darwin already refuted?

Opponents of the spread differ sharply over tactics. Just where should the protective dikes be built? Should we try to contain the idea within biology itself, with one post-Darwinian counterrevolution or another? Among those who have favored this tactic is Stephen Jay Gould, who has offered several different revolutions of containment. Or should we place the barriers farther out? To get our bearings in this series of campaigns, we should start with a crude map of the pre-Darwinian territory. As we shall see, it will have to be revised again and again to make accommodations as various skirmishes are lost.

2. DARWIN'S ASSAULT ON THE COSMIC PYRAMID

A prominent feature of Pre-Darwinian world-views is an overall top-to-bottom map of things. This is often described as a Ladder; God is at the top, with human beings a rung or two below (depending on whether angels are part of the scheme). At the bottom of the Ladder is Nothingness, or maybe Chaos, or maybe Locke's inert, motionless Matter. Alternatively, the scale is a Tower, or, in the intellectual historian Arthur Lovejoy's memorable phrase (1936), a Great Chain of Being composed of many links. John Locke's argument has already drawn our attention to a particularly abstract version of the hierarchy, which I will call the Cosmic Pyramid:

God

Mind

Design

Order

Chaos

Nothing

(Warning: each term in the pyramid must be understood in an old-fashioned, pre-Darwinian sense!)

Everything finds its place on one level or another of the Cosmic Pyramid, even blank nothingness, the ultimate foundation. Not all matter is Ordered, some is in Chaos; only some Ordered matter is also Designed; only some Designed things have Minds, and of course only one Mind is God. God, the first Mind, is the source and explanation of everything underneath. (Since everything thus *depends on* God, perhaps we should say it is a chandelier, hanging from God, rather than a pyramid, supporting Him.)

What is the difference between Order and Design? As a first stab, we might say that Order is mere regularity, mere pattern; Design is Aristotle's *telos,* an exploitation of Order for a purpose, such as we see in a cleverly designed artifact. The solar system exhibits stupendous Order, but does not (apparently) have a purpose—it isn't *for* anything. An eye, in contrast, is *for* seeing. Before Darwin, this distinction was not always clearly marked. Indeed, it was positively blurred:

> In the thirteenth century, Aquinas offered the view that natural bodies
> [such as planets, raindrops, volcanos] act as if guided toward a definite goal

or end "so as to obtain the best result." This fitting of means to ends implies, argued Aquinas, an intention. But, seeing as natural bodies lack consciousness, they cannot supply that intention themselves. "Therefore some intelligent being exists by whom all natural things are directed to their end; and this being we call God." [Davies 1992, p. 200.]

Hume's Cleanthes, following in this tradition, lumps the adapted marvels of the living world with the regularities of the heavens—it's *all* like a wonderful clockwork to him. But Darwin suggests a division: Give me Order, he says, and time, and I will give you Design. Let me start with regularity—the mere purposeless, mindless, pointless regularity of physics—and I will show you a process that eventually will yield products that exhibit not just regularity but purposive design. (This was just what Karl Marx thought he saw when he declared that Darwin had dealt a death blow to Teleology: Darwin had *reduced* teleology to nonteleology, Design to Order.)

Before Darwin, the difference between Order and Design didn't loom large, because in any case it all came down from God. The whole universe was His artifact, a product of His Intelligence, His Mind. Once Darwin jumped into the middle with his proposed answer to the question of how Design could arise from mere Order, the rest of the Cosmic Pyramid was put in jeopardy. Suppose we accept that Darwin has explained the Design of the bodies of plants and animals (including our own bodies—we have to admit that Darwin has placed us firmly in the animal kingdom). Looking up, if we concede to Darwin our bodies, can we keep him from taking our minds as well? (We will address this question, in many forms, in part III.) Looking down, Darwin asks us to give him Order as a premise, but is there anything to keep him from stepping down a level and giving himself an algorithmic account of the origin of Order out of mere Chaos? (We will address this question in chapter 6.)

The vertigo and revulsion this prospect provokes in many was perfectly expressed in an early attack on Darwin, published anonymously in 1868:

> In the theory with which we have to deal, Absolute Ignorance is the artificer; so that we may enunciate as the fundamental principle of the whole system, that, IN ORDER TO MAKE A PERFECT AND BEAUTIFUL MACHINE, IT IS NOT REQUISITE TO KNOW HOW TO MAKE IT. This proposition will be found, on careful examination, to express, in condensed form, the essential purport of the Theory, and to express in a few words all Mr. Darwin's meaning; who, by a strange inversion of reasoning, seems to think Absolute Ignorance fully qualified to take the place of Absolute Wisdom in all the achievements of creative skill. [MacKenzie 1868.]

Exactly! Darwin's "strange inversion of reasoning" was in fact a new and wonderful way of thinking, completely overturning the Mind-first way that

John Locke "proved" and David Hume could see no way around. John Dewey nicely described the inversion some years later, in his insightful book *The Influence of Darwin on Philosophy*: "Interest shifts ... from an intelligence that shaped things once for all to the particular intelligences which things are even now shaping" (Dewey 1910, p. 15). But the idea of treating Mind as an effect rather than as a First Cause is too revolutionary for some—an "awful stretcher" that their own minds cannot accommodate comfortably. This is as true today as it was in 1860, and it has always been as true of some of evolution's best friends as of its foes. For instance, the physicist Paul Davies, in his recent book *The Mind of God*, proclaims that the reflective power of human minds can be "no trivial detail, no minor by-product of mindless purposeless forces" (Davies 1992, p. 232). This is a most revealing way of expressing a familiar denial, for it betrays an ill-examined prejudice. Why, we might ask Davies, would its being a by-product of mindless, purposeless forces make it trivial? Why couldn't the most important thing of all be something that arose from unimportant things? Why should the importance or excellence of *anything* have to rain down on it from on high, from something more important, a gift from God? Darwin's inversion suggests that we abandon that presumption and look for sorts of excellence, of worth and purpose, that can emerge, bubbling up out of "mindless, purposeless forces."

Alfred Russel Wallace, whose own version of evolution by natural selection arrived on Darwin's desk while he was still delaying publication of *Origin*, and whom Darwin managed to treat as codiscoverer of the principle, never quite got the point.[1] Although at the outset Wallace was much more forthcoming on the subject of the evolution of the human mind than Darwin was willing to be, and stoutly maintained at first that human minds were no exception to the rule that all features of living things were products of evolution, he could not see the "strange inversion of reasoning" as the key to the greatness of the great idea. Echoing John Locke, Wallace proclaimed that "the marvelous complexity of forces which appear to control matter, if not actually to constitute it, are and must be mind-products" (Gould 1985, p. 397). When, later in his life, Wallace converted to spiritualism and exempted human consciousness altogether from the iron rule of

---

1. This fascinating and even excruciating story has been well told many times, but still the controversies rage. Why did Darwin delay publication in the first place? Was his treatment of Wallace generous or monstrously unfair? The unsettled relations between Darwin and Wallace are not just a matter of Darwin's uneasy conscience about how he handled Wallace's innocent claim-jumping correspondence; as we see here, the two were also separated by vast differences in insight and attitude about the idea they both discovered. For particularly good accounts, see Desmond and Moore 1991; Richards 1987, pp. 159–61.

evolution, Darwin saw the crack widen and wrote to him: "I hope you have not murdered too completely your own and my child" (Desmond and Moore 1991, p. 569).

But was it really so inevitable that Darwin's idea should lead to such revolution and subversion? "It is obvious that the critics did not wish to understand, and to some extent Darwin himself encouraged their wishful thinking" (Ellegård 1956). Wallace wanted to ask what the *purpose* of natural selection might be, and though this might seem in retrospect to be squandering the fortune he and Darwin had uncovered, it was an idea for which Darwin himself often expressed sympathy. Instead of reducing teleology all the way to purposeless Order, why couldn't we reduce all mundane teleology to a single purpose: God's purpose? Wasn't this an obvious and inviting way to plug the dike? Darwin was clear in his own mind that the variation on which the process of natural selection depended *had* to be unplanned and undesigned, but the process itself might have a purpose, mightn't it? In a letter in 1860 to the American naturalist Asa Gray, an early supporter, Darwin wrote, "I am inclined to look at everything as resulting from *designed* [emphasis added] laws, with the details whether good or bad, left to the working out of what we may call chance" (F. Darwin 1911, vol. 2, p. 105).

Automatic processes are themselves often creations of great brilliance. From today's vantage point, we can see that the inventors of the automatic transmission and the automatic door-opener were no idiots, and their genius lay in seeing how to create something that could do something "clever" without having to think about it. Indulging in some anachronism, we could say that, to some observers in Darwin's day, it seemed that he had left open the possibility that God did His handiwork by designing an automatic design-maker. And to some of these, the idea was not just a desperate stopgap but a positive improvement on tradition. The first chapter of Genesis describes the successive waves of Creation and ends each with the refrain "and God saw that it was good." Darwin had discovered a way to eliminate this retail application of Intelligent Quality Control; natural selection would take care of that without further intervention from God. (The seventeenth-century philosopher Gottfried Wilhelm Leibniz had defended a similar hands-off vision of God the Creator.) As Henry Ward Beecher put it, "Design by wholesale is grander than design by retail" (Rachels 1991, p. 99). Asa Gray, captivated by Darwin's new idea but trying to reconcile it with as much of his traditional religious creed as possible, came up with this marriage of convenience: God *intended* the "stream of variations" and *foresaw* just how the laws of nature He had laid down would prune this stream over the eons. As John Dewey later aptly remarked, invoking yet another mercantile metaphor, "Gray held to what may be called design on the installment plan" (Dewey 1910, p. 12).

It is not unusual to find such metaphors, redolent of capitalism, in evolutionary explanations. Examples are often gleefully recounted by those critics and interpreters of Darwin who see this language as revealing—or should we say betraying—the social and political environment in which Darwin developed his ideas, thereby (somehow) discrediting their claim to scientific objectivity. It is certainly true that Darwin, being an ordinary mortal, was the inheritor of a huge manifold of concepts, modes of expression, attitudes, biases, and visions that went with his station in life (as a Victorian Englishman might put it), but it is also true that the economic metaphors that come so naturally to mind when one is thinking about evolution get their power from one of the deepest features of Darwin's discovery.

3. THE PRINCIPLE OF THE ACCUMULATION OF DESIGN

The key to understanding Darwin's contribution is *granting* the premise of the Argument from Design. What conclusion ought one to draw if one found a watch lying on the heath in the wilderness? As Paley (and Hume's Cleanthes before him) insisted, a watch exhibits a tremendous amount of *work done*. Watches and other designed objects don't just happen; they have to be the product of what modern industry calls "R and D"—research and development—and R and D is costly, in both time and energy. Before Darwin, the only model we had of a process by which this sort of R-and-D work could be done was an Intelligent Artificer. What Darwin saw was that in principle the same work could be done by a different sort of process that *distributed* that work over huge amounts of time, by thriftily conserving the design work that had been accomplished at each stage, so that it didn't have to be done over again. In other words, Darwin had hit upon what we might call the Principle of Accumulation of Design. Things in the world (such as watches and organisms and who knows what else) may be seen as products embodying a certain amount of Design, and one way or another, that Design had to have been created by a process of R and D. Utter undesignedness—pure chaos in the old-fashioned sense—was the null or starting point.

A more recent idea about the difference—and tight relation—between Design and Order will help clarify the picture. This is the proposal, first popularized by the physicist Erwin Schrödinger (1967), that Life can be defined in terms of the Second Law of Thermodynamics. In physics, order or organization can be measured in terms of *heat differences* between regions of space time; *entropy* is simply disorder, the opposite of order, and according to the Second Law, the entropy of any isolated system increases with time. In other words, things run down, inevitably. According to the

Second Law, the universe is unwinding out of a more ordered state into the ultimately disordered state known as the heat death of the universe.[2]

What, then, are living things? They are things that defy this crumbling into dust, at least for a while, by not being isolated—by taking in from their environment the wherewithal to keep life and limb together. The psychologist Richard Gregory summarizes the idea crisply:

> Time's arrow given by Entropy—the loss of organization, or loss of temperature differences—is statistical and it is subject to local small-scale reversals. Most striking: life is a systematic reversal of Entropy, and intelligence creates structures and energy differences against the supposed gradual 'death' through Entropy of the physical Universe. [Gregory 1981, p. 136.]

Gregory goes on to credit Darwin with the fundamental enabling idea: "It is the measure of the concept of Natural Selection that increases in the complexity and order of organisms in biological time can now be understood." Not just individual organisms, but the whole process of evolution that creates them, thus can be seen as fundamental physical phenomena running contrary to the larger trend of cosmic time, a feature captured by William Calvin in one of the meanings of the title of his classic exploration of the relationship between evolution and cosmology, *The River That Flows Uphill: A Journey from the Big Bang to the Big Brain* (1986).

A *designed* thing, then, is either a living thing or a part of a living thing, or the artifact of a living thing, organized in any case in aid of this battle against disorder. It is not impossible to oppose the trend of the Second Law, but it is costly. Consider iron. Iron is a very useful element, essential for our bodily health, and also valuable as the major component of steel, that wonderful building material. Our planet used to have vast reserves of iron ore, but they are gradually being depleted. Does this mean that the Earth is running out of iron? Hardly. With the trivial exception of a few tons that have recently been launched out of Earth's effective gravitational field in the form of space-probe components, there is just as much iron on the planet today as there ever was. The trouble is that more and more of it is scattered about in the form of rust (molecules of iron oxide), and other low-grade, low-concentration materials. In principle, it could all be recovered, but that would take enormous amounts of energy, craftily focused on the particular project of extracting and reconcentrating the iron.

It is the organization of just such sophisticated processes that constitutes

---

2. And where did the initial order come from? The best discussion I have encountered of this good question is "Cosmology and the Arrow of Time," ch. 7 of Penrose 1989.

the hallmark of life. Gregory dramatizes this with an unforgettable example. A standard textbook expression of the directionality imposed by the Second Law of Thermodynamics is the claim that you can't *un*scramble an egg. Well, not that you absolutely can't, but that it would be an extremely costly, sophisticated task, uphill all the way against the Second Law. Now consider: how expensive would it be to make a device that would take scrambled eggs as input and deliver unscrambled eggs as output? There is one ready solution: put a live hen in the box! Feed it scrambled eggs, and it will be able to make eggs for you—for a while. Hens don't normally strike us as near-miraculously sophisticated entities, but here is one thing a hen can do, thanks to the Design that has organized it, that is still way beyond the reach of the devices created by human engineers.

The more Design a thing exhibits, the more R-and-D work had to have occurred to produce it. Like any good revolutionary, Darwin exploits as much as possible of the old system: the vertical dimension of the Cosmic Pyramid is retained, and becomes the measure of how much Design has gone into the items at that level. In Darwin's scheme, as in the traditional Pyramid, Minds do end up near the top, among the most designed of entities (in part because they are the self-redesigning things, as we shall see in chapter 13). But this means that they are among the most advanced *effects* (to date) of the creative process, not—as in the old version—its cause or source. Their products in turn—the human artifacts that were our initial model—must count as more designed still. This may seem counterintuitive at first. A Keats ode may seem to have some claim to having a grander R and D pedigree than a nightingale—at least it might seem so to a poet ignorant of biology—but what about a paper clip? Surely a paper clip is a trivial product of design compared with any living thing, however rudimentary. In one obvious sense, yes, but reflect for a moment. Put yourself in Paley's shoes, but walking along the apparently deserted beach on an alien planet. Which discovery would excite you the most: a clam or a clam-rake? Before the planet could make a clam-rake, it would have to make a clam-rake-maker, and that is a more designed thing by far than a clam.

Only a theory with the logical shape of Darwin's could *explain* how designed things came to exist, because any other sort of explanation would be either a vicious circle or an infinite regress (Dennett 1975). The old way, Locke's Mind-first way, endorsed the principle that it takes an Intelligence to make an intelligence. This idea must have always seemed self-evident to our ancestors, the artifact-makers, going back to *Homo habilis*, the "handy" man, from whom *Homo sapiens*, the "knowing" man, descended. Nobody ever saw a spear fashion a hunter out of raw materials. Children chant, "It takes one to know one," but an even more persuasive slogan would seem to be "It takes a greater one to make a lesser one." Any view inspired by this slogan immediately faces an embarrassing question, however, as Hume had noted: If God created and designed all these wonderful things, who created God? Supergod? And who created Supergod? Superdupergod? Or did God create Himself? Was it hard work? Did it take time? Don't ask! Well, then, we may ask instead whether this bland embrace of mystery is any improvement over just denying the principle that intelligence (or design) must spring from Intelligence. Darwin offered an explanatory path that actually honored Paley's insight: real work went into designing this watch, and work isn't free.

How much design does a thing exhibit? No one has yet offered a system of design quantification that meets all our needs. Theoretical work that bears on this interesting question is under way in several disciplines,[3] and in chapter 6 we will consider a natural metric that provides a neat solution to special cases—but in the meantime we have a powerful intuitive sense of different amounts of design. Automobiles contain more design than bicycles, sharks contain more design than amoebas, and even a short poem contains more design than a "Keep Off the Grass" sign. (I can hear the skeptical reader saying, "Whoa! Slow down! Is this supposed to be uncontroversial?" Not by a long shot. In due course I will attempt to justify these claims, but for the time being I want to draw attention to, and build on, some familiar—but admittedly unreliable—intuitions.)

Patent law, including the law of copyright, is a repository of our practical grasp of the question. How much novelty of design counts as enough to justify a patent? How much can one borrow from the intellectual products of others without recompense or acknowledgment? These are slippery slopes on which we have had to construct some rather arbitrary terraces, codifying what otherwise would be a matter of interminable dispute. The burden of proof in these disputes is fixed by our intuitive sense of how much design is too much design to be mere coincidence. Our intuitions here are very strong and, I promise to show, sound. Suppose an author is accused of plagiarism, and the evidence is, say, a single paragraph that is almost identical to a paragraph in the putative source. Might this be just a coincidence? It depends crucially on how mundane and formulaic the paragraph is, but most paragraph-length passages of text are "special" enough (in ways we will soon explore) to make independent creation highly unlikely. No reasonable jury would require the prosecutor in a plagiarism case to demonstrate exactly the causal pathway by which the alleged copying took place. The defendant would clearly have the burden of establishing that his work was, remarkably, an independent work rather than a copying of work already done.

A similar burden of proof falls on the defendant in an industrial-espionage

---

3. For accessible overviews of some of the ideas, see Pagels 1988, Stewart and Golubitsky 1992, and Langton et al. 1992.

case: the interior of the defendant's new line of widgets looks suspiciously similar in design to that of the plaintiff's line of widgets—is this an innocent case of convergent evolution of design? Really the only way to prove your innocence in such a case is to show clear evidence of actually having done the necessary R-and-D work (old blueprints, rough drafts, early models and mockups, memos about the problems encountered, etc.). In the absence of such evidence, but also in the absence of any physical evidence of your espionage activities, you would be convicted—and you'd deserve to be! Cosmic coincidences on such a scale just don't happen.

The same burden of proof now reigns in biology, thanks to Darwin. What I am calling the Principle of Accumulation of Design doesn't logically *require* that all design (on this planet) descend via one branch or another from a single trunk (or root or seed), but it says that since each new designed thing that appears must have a large design investment in its etiology somewhere, the cheapest hypothesis will always be that the design is largely copied from earlier designs, which are copied from earlier designs, and so forth, so that actual R-and-D innovation is minimized. We know for a fact, of course, that many designs have been independently re-invented many times—eyes, for instance, dozens of times—but every case of such convergent evolution must be proven against a background in which most of the design is copied. It is logically possible that all the life forms in South America were created independently of all the life forms in the rest of the world, but this is a wildly extravagant hypothesis that would need to be demonstrated, piece by piece. Suppose we discover, on some remote island, a novel species of bird. Even if we don't *yet* have direct confirmatory evidence that this bird is related to all the other birds in the world, that is our overpoweringly secure default assumption, after Darwin, because birds are very special designs.[4]

So the fact that organisms—and computers and books and other artifacts—are effects of very special chains of causation is not, after Darwin, a merely reliable generalization, but a deep fact out of which to build a theory. Hume recognized the point—"Throw several pieces of steel together, without shape or form; they will never arrange themselves to compose a watch"—but he and other, earlier, thinkers thought they had to ground this deep fact in Mind. Darwin came to see how to distribute it in vast spaces of Nonmind, thanks to his ideas about how design innovations could be conserved and reproduced, and hence accumulated.

The idea that Design is something that has taken work to create, and

---

4. Note, by the way, that it would not follow *logically* that the bird was related to other birds if we found that its DNA was almost identical in sequence to that of other birds! "Just a coincidence, not plagiarism," would be a logical possibility—but one that nobody would take seriously.

hence has value at least in the sense that it is something that might be conserved (and then stolen or sold), finds robust expression in economic terms. Had Darwin not had the benefit of being born into a mercantile world that had already created its Adam Smith and its Thomas Malthus, he would not have been in position to find ready-made pieces he could put together into a new, value-added product. (You see, the idea applies to itself very nicely.) The various sources of the Design that went into Darwin's grand idea give us important insights into the idea itself, but do no more to diminish its value or threaten its objectivity than the humble origins of methane diminish its BTUs when it is put to use as a fuel.

4. THE TOOLS FOR R AND D: SKYHOOKS OR CRANES?

The work of R and D is not like shoveling coal; it is somehow a sort of "intellectual" work, and this fact grounds the other family of metaphors that has both enticed and upset, enlightened and confused, the thinkers who have confronted Darwin's "strange inversion of reasoning": the apparent attribution of intelligence to the very process of natural selection that Darwin insisted was *not* intelligent.

Was it not unfortunate, in fact, that Darwin had chosen to call his principle "natural *selection*," with its anthropomorphic connotations? Wouldn't it have been better, as Asa Gray suggested to him, to replace the imagery about "nature's Guiding Hand" with a discussion of the different ways of winning life's race (Desmond and Moore 1991, p. 458)? Many people just didn't get it, and Darwin was inclined to blame himself: "I must be a very bad explainer," he said, conceding: "I suppose 'natural selection' was a bad term" (Desmond and Moore 1991, p. 492). Certainly this Janus-faced term has encouraged more than a century of heated argument. A recent opponent of Darwin sums it up:

> Life on Earth, initially thought to constitute a sort of prima facie case for a creator, was, as a result of Darwin's idea, envisioned merely as being the outcome of a process and a process that was, according to Dobzhansky, "blind, mechanical, automatic, impersonal," and, according to de Beer, was "wasteful, blind, and blundering." But as soon as these criticisms [sic] were leveled at natural selection, the "blind process" itself was compared to a poet, a composer, a sculptor, Shakespeare—to the very notion of creativity that the idea of natural selection had originally replaced. It is clear, I think, that there was something very, very wrong with such an idea. [Bethell 1976.]

Or something very, very right. It seems to skeptics like Bethell that there is something willfully paradoxical in calling the process of evolution the "blind watchmaker" (Dawkins 1986a), for this takes away with the left hand ("blind") the very discernment, purpose, and foresight it gives with the right hand. But others see that this manner of speaking—and we shall find that it is not just ubiquitous but irreplaceable in contemporary biology—is just the right way to express the myriads of detailed discoveries that Darwinian theory helps to expose. There is simply no denying the breathtaking brilliance of the designs to be found in nature. Time and again, biologists baffled by some apparently futile or maladroit bit of bad design in nature have eventually come to see that they have underestimated the ingenuity, the sheer brilliance, the depth of insight to be discovered in one of Mother Nature's creations. Francis Crick has mischievously baptized this trend in the name of his colleague Leslie Orgel, speaking of what he calls "Orgel's Second Rule: Evolution is cleverer than you are." (An alternative formulation: Evolution is cleverer than Leslie Orgel!)

Darwin shows us how to climb from "Absolute Ignorance" (as his outraged critic said) to creative genius without begging any questions, but we must tread very carefully, as we shall see. Among the controversies that swirl around us, most if not all consist of different challenges to Darwin's claim that he can take us all the way to *here* (the wonderful world we inhabit) from *there* (the world of chaos or utter undesignedness) in the time available without invoking anything beyond the mindless mechanicity of the algorithmic processes he had proposed. Since we have reserved the vertical dimension of the traditional Cosmic Pyramid as a measure of (intuitive) designedness, we can dramatize the challenge with the aid of another fantasy item drawn from folklore.

> *skyhook*, orig. Aeronaut. An imaginary contrivance for attachment to the sky; an imaginary means of suspension in the sky. [*Oxford English Dictionary*.]

The first use noted by the *OED* is from 1915: "an aeroplane pilot commanded to remain in place (aloft) for another hour, replies 'the machine is not fitted with skyhooks.' " The skyhook concept is perhaps a descendant of the *deus ex machina* of ancient Greek dramaturgy: when second-rate playwrights found their plots leading their heroes into inescapable difficulties, they were often tempted to crank down a god onto the scene, like Superman, to save the situation supernaturally. Or skyhooks may be an entirely independent creation of convergent folkloric evolution. Skyhooks would be wonderful things to have, great for lifting unwieldy objects out of difficult circumstances, and speeding up all sorts of construction projects. Sad to say, they are impossible.[5]

---

5. Well, not quite impossible. Geostationary satellites, orbiting in unison with the Earth's rotation, are a kind of real, nonmiraculous skyhook. What makes them so valuable—what

There are cranes, however. Cranes can do the lifting work our imaginary skyhooks might do, and they do it in an honest, non-question-begging fashion. They are expensive, however. They have to be designed and built, from everyday parts already on hand, and they have to be located on a firm base of existing ground. Skyhooks are miraculous lifters, unsupported and insupportable. Cranes are no less excellent as lifters, and they have the decided advantage of being real. Anyone who is, like me, a lifelong onlooker at construction sites will have noticed with some satisfaction that it sometimes takes a small crane to set up a big crane. And it must have occurred to many other onlookers that in principle this big crane could be used to enable or speed up the building of a still more spectacular crane. Cascading cranes is a tactic that seldom if ever gets used more than once in real-world construction projects, but in principle there is no limit to the number of cranes that could be organized in series to accomplish some mighty end.

Now imagine all the "lifting" that has to get done in Design Space to create the magnificent organisms and (other) artifacts we encounter in our world. Vast distances must have been traversed since the dawn of life with the earliest, simplest self-replicating entities, spreading outward (diversity) and upward (excellence). Darwin has offered us an account of the crudest, most rudimentary, stupidest imaginable lifting process—the wedge of natural selection. By taking tiny—the tiniest possible—steps, this process can gradually, over eons, traverse these huge distances. Or so he claims. At no point would anything miraculous—from on high—be needed. Each step has been accomplished by brute, mechanical, algorithmic climbing, from the base already built by the efforts of earlier climbing.

It does seem incredible. Could it really have happened? Or did the process need a "leg up" now and then (perhaps only at the very beginning) from one sort of skyhook or another? For over a century, skeptics have been trying to find a proof that Darwin's idea just can't work, at least not *all the way*. They have been hoping for, hunting for, praying for skyhooks, as exceptions to what they see as the bleak vision of Darwin's algorithm churning away. And time and again, they have come up with truly interesting challenges—leaps and gaps and other marvels that do seem, at first, to need

---

makes them financially sound investments—is that we often do want very much to attach something (such as an antenna or a camera or telescope) to a place high in the sky. Satellites are impractical for *lifting*, alas, because they have to be placed so high in the sky. The idea has been carefully explored. It turns out that a rope of the strongest artificial fiber yet made would have to be over a hundred meters in diameter at the top—it could taper to a nearly invisible fishing line on its way down—just to suspend its own weight, let alone any payload. Even if you could spin such a cable, you wouldn't want it falling out of orbit onto the city below!

skyhooks. But then along have come the cranes, discovered in many cases by the very skeptics who were hoping to find a skyhook.

It is time for some more careful definitions. Let us understand that a *skyhook* is a "mind-first" force or power or process, an exception to the principle that all design, and apparent design, is ultimately the result of mindless, motiveless mechanicity. A *crane*, in contrast, is a subprocess or special feature of a design process that can be demonstrated to permit the local speeding up of the basic, slow process of natural selection, *and* that can be demonstrated to be itself the predictable (or retrospectively explicable) product of the basic process. Some cranes are obvious and uncontroversial; others are still being argued about, very fruitfully. Just to give a general sense of the breadth and application of the concept, let me point to three very different examples.

It is now generally agreed among evolutionary theorists that *sex* is a crane. That is, species that reproduce sexually can move through Design Space at a much greater speed than that achieved by organisms that reproduce asexually. Moreover, they can "discern" design improvements along the way that are all but "invisible" to asexually reproducing organisms (Holland 1975). This cannot be the *raison d'être* of sex, however. Evolution cannot see way down the road, so anything it builds must have an immediate payoff to counterbalance the cost. As recent theorists have insisted, the "choice" of reproducing sexually carries a huge *immediate* cost: organisms send along only 50 percent of their genes in any one transaction (to say nothing of the effort and risk involved in securing a transaction in the first place). So the *long-term* payoff of heightened efficiency, acuity, and speed of the redesign process—the features that make sex a magnificent crane—is as nothing to the myopic, local competitions that must determine which organisms get favored in the very next generation. Some other, short-term, benefit must have maintained the positive selection pressure required to make sexual reproduction an offer few species could refuse. There are a variety of compelling—and competing—hypotheses that might solve this puzzle, which was first forcefully posed for biologists by John Maynard Smith (1978). For a lucid introduction to the current state of play, see Matt Ridley 1993. (More on this later.)

What we learn from the example of sex is that a crane of great power may exist that was not created *in order to exploit* that power, but for other reasons, although its power as a crane may help explain why it has been maintained ever since. A crane that was obviously created to be a crane is *genetic engineering*. Genetic engineers—human beings who engage in recombinant-DNA tinkering—can now unquestionably take huge leaps through Design Space, creating organisms that would never have evolved by "ordinary" means. This is no miracle—*provided that genetic engineers (and the artifacts they use in their trade) are themselves wholly the products of earlier, slower evolutionary processes.* If the creationists were right that mankind is a species unto itself, divine and inaccessible via brute Darwinian paths, then genetic engineering would not be a crane after all, having been created with the help of a major skyhook. I don't imagine that any genetic engineers think of themselves this way, but it is a logically available perch, however precarious. Less obviously silly is this idea: if the bodies of genetic engineers are products of evolution, but their *minds* can do creative things that are irreducibly nonalgorithmic or inaccessible by all algorithmic paths, then the leaps of genetic engineering might involve a skyhook. Exploring this prospect will be the central topic of chapter 15.

A crane with a particularly interesting history is the Baldwin Effect, named for one of its discoverers, James Mark Baldwin (1896), but more or less simultaneously discovered by two other early Darwinians, Conwy Lloyd Morgan (famed for Lloyd Morgan's Canon of Parsimony [for discussion, see Dennett 1983]) and H. F. Osborn. Baldwin was an enthusiastic Darwinian, but he was oppressed by the prospect that Darwin's theory would leave Mind with an insufficiently important and originating role in the (re)design of organisms. So he set out to demonstrate that animals, *by dint of their own clever activities in the world,* might hasten or guide the further evolution of their species. Here is what he asked himself: how could it be that individual animals, by solving problems in their own lifetimes, could change the conditions of competition for their own offspring, making those problems easier to solve in the future? And he came to realize that this was in fact possible, under certain conditions, which we can illustrate with a simple example (drawn, with revisions, from Dennett 1991a).

Consider a population of a species in which there is considerable variation at birth in the way their brains are wired up. Just one of the ways, we may suppose, endows its possessor with a Good Trick—a behavioral talent that protects it or enhances its chances dramatically. The standard way of representing such differences in fitness between individual members of a population is known as an "adaptive landscape" or a "fitness landscape" (S. Wright 1931). The altitude in such a diagram stands for fitness (higher is better), and the longitude and latitude stand for some factors of individual design—in this case, features of brain-wiring. Each different way a brain might be wired is represented by one of the rods that compose the landscape—each rod is a different *genotype.* The fact that just one of the combinations of features is any good—that is, any better than run-of-the-mill—is illustrated by the way it stands out like a telephone pole in the desert.

As figure 3.1 makes clear, only one wiring is favored; the others, no matter how "close" to being the good wiring, are about equal in fitness. So such an isolated peak is indeed a needle in the haystack: it will be practically invisible to natural selection. Those few individuals in the population that are lucky enough to have the Good Trick genotype will typically have difficulty

FIGURE 3.1

passing it on to their offspring, since under most circumstances their chances of finding a mate who also has the Good Trick genotype are remote, and a miss is as good as a mile.

But now we introduce just one "minor" change: suppose that although the individual organisms *start out* with different wirings (whichever wiring was ordered by their particular genotype or genetic recipe)—as shown by their scatter on the fitness landscape—they have some capacity to adjust or revise their wiring, depending on what they encounter during their lifetimes. (In the language of evolutionary theory, there is some "plasticity" in their *phenotypes*. The phenotype is the eventual body design created by the genotype in interaction with environment. Identical twins raised in different environments would share a genotype but might be dramatically different in phenotype.) Suppose, then, that these organisms can end up, after exploration, with a design different from the one they were born with. We may suppose their explorations are random, but they have an innate capacity to recognize (and stay with) a Good Trick when they stumble upon it. Then those individuals who begin life with a genotype that is closer to the Good Trick genotype—fewer redesign steps away from it—are more likely to come across it, and stick with it, than those that are born with a faraway design.

This head start in the race to redesign themselves can give them the edge in the Malthusian crunch—if the Good Trick is so good that those who never learn it, or who learn it "too late," are at a severe disadvantage. In populations with this sort of phenotypic plasticity, a near-miss is *better* than a mile. For such a population, the telephone pole in the desert becomes the summit of a gradual hill, as in figure 3.2; those perched near the summit, although they start out with a design that serves them no better than others, will tend to discover the summit design in short order.

In the long run, natural selection—redesign at the genotype level—will tend to *follow the lead of* and *confirm* the directions taken by the individual organisms' successful explorations—redesign at the individual or phenotype level.

The way I have just described the Baldwin Effect certainly keeps Mind to

FIGURE 3.2

a minimum, if not altogether out of the picture; all it requires is some brute, mechanical capacity to stop a random walk when a Good Thing comes along, a minimal capacity to "recognize" a tiny bit of progress, to "learn" something by blind trial and error. In fact, I have put it in *behavioristic* terms. What Baldwin discovered was that creatures capable of "reinforcement learning" not only do better individually than creatures that are entirely "hard-wired"; their species will *evolve faster* because of its greater capacity to discover design improvements in the neighborhood.[6] This is not how Baldwin described the effect he proposed. His temperament was the farthest thing from behaviorism. As Richards notes:

> The mechanism conformed to ultra-Darwinian assumptions, but nonetheless allowed consciousness and intelligence a role in directing evolution. By philosophic disposition and conviction, Baldwin was a spiritualistic metaphysician. He felt the beat of consciousness in the universe; it pulsed through all the levels of organic life. Yet he understood the power of mechanistic explanations of evolution. [R. J. Richards 1987, p. 480.][7]

The Baldwin Effect, under several different names, has been variously described, defended, and disallowed over the years, and recently independently rediscovered several more times (e.g., Hinton and Nowland 1987).

---

6. Schull (1990), is responsible for the perspective that allows us to see species as variably capable of "seeing" design improvements, thanks to their variable capacities for phenotypic exploration (for commentary, see Dennett 1990a).

7. Robert Richards' account of the history of the Baldwin Effect (1987, especially pp. 480–503 and discussion later in that book) has been one of the major provocations and guides to my thinking in this book. What I found particularly valuable (see my review, Dennett 1989a) was that Richards not only shares with Baldwin and many other Darwinians a submerged yearning for skyhooks—or at least a visceral dissatisfaction with theories that insist on cranes—but also has the intellectual honesty and courage to expose and examine his own discomfort with what he is obliged to call "ultra-Darwinism." Richards' heart is clearly with Baldwin, but his mind won't let him bluster, or try to paper over the cracks he sees in the dikes that others have tried to erect against universal acid.

Although it has been regularly described and acknowledged in biology textbooks, it has typically been shunned by overcautious thinkers, because they thought it smacked of the Lamarckian heresy (the presumed possibility of inheritance of acquired characteristics—see chapter 11 for a detailed discussion). This rejection is particularly ironic, since, as Richards notes, it was intended by Baldwin to be—and truly is—an acceptable *substitute* for Lamarckian mechanisms.

> The principle certainly seemed to dispatch Lamarckism, while supplying that positive factor in evolution for which even staunch Darwinists like Lloyd Morgan longed. And to those of metaphysical appetite, it revealed that under the clanking, mechanical vesture of Darwinian nature, mind could be found. [R. J. Richards 1987, p. 487.]

Well, not Mind—if by that we mean a full-fledged, intrinsic, original, skyhook-type Mind—but only a nifty mechanistic, behavioristic, crane-style mind. That is not nothing, however; Baldwin discovered an effect that genuinely increases the power—locally—of the underlying process of natural selection wherever it operates. It shows how the "blind" process of the basic phenomenon of natural selection can be abetted by a limited amount of "look-ahead" in the activities of individual organisms, which create fitness differences that natural selection can then act upon. This is a welcome complication, a wrinkle in evolutionary theory that removes one reasonable and compelling source of doubt, and enhances our vision of the power of Darwin's idea, especially when it is cascaded in multiple, nested applications. And it is typical of the outcome of other searches and controversies we will explore: the motivation, the passion that drove the research, was the hope of finding a skyhook; the triumph was finding how the same work could be done with a crane.

5. WHO'S AFRAID OF REDUCTIONISM?

Reductionism is a dirty word, and a kind of 'holistier than thou' self-righteousness has become fashionable.

—RICHARD DAWKINS 1982, p. 113

The term that is most often bandied about in these conflicts, typically as a term of abuse, is "reductionism." Those who yearn for skyhooks call those who eagerly settle for cranes "reductionists," and they can often make reductionism seem philistine and heartless, if not downright evil. But like most terms of abuse, "reductionism" has no fixed meaning. The central image is of somebody claiming that one science "reduces" to another: that chemistry reduces to physics, that biology reduces to chemistry, that the social sciences reduce to biology, for instance. The problem is that there are both bland readings and preposterous readings of any such claim. According to the bland readings, it is possible (and desirable) to *unify* chemistry and physics, biology and chemistry, and, yes, even the social sciences and biology. After all, societies are composed of human beings, who, as mammals, must fall under the principles of biology that cover all mammals. Mammals, in turn, are composed of molecules, which must obey the laws of chemistry, which in turn must answer to the regularities of the underlying physics. No sane scientist disputes this bland reading; the assembled Justices of the Supreme Court are as bound by the law of gravity as is any avalanche, because they are, in the end, also a collection of physical objects. According to the preposterous readings, reductionists want to abandon the principles, theories, vocabulary, laws of the higher-level sciences, in favor of the lower-level terms. A reductionist dream, on such a preposterous reading, might be to write "A Comparison of Keats and Shelley from the Molecular Point of View" or "The Role of Oxygen Atoms in Supply-Side Economics," or "Explaining the Decisions of the Rehnquist Court in Terms of Entropy Fluctuations." Probably nobody is a reductionist in the preposterous sense, and everybody should be a reductionist in the bland sense, so the "charge" of reductionism is too vague to merit a response. If somebody says to you, "But that's so reductionistic!" you would do well to respond, "That's such a quaint, old-fashioned complaint! What on Earth did you have in mind?"

I am happy to say that in recent years, some of the thinkers I most admire have come out in defense of one or another version of reductionism, carefully circumscribed. The cognitive scientist Douglas Hofstadter, in *Gödel Escher Bach,* composed a "Prelude . . . Ant Fugue" (Hofstadter 1979, pp. 275–336) that is an analytical hymn to the virtues of reductionism in its proper place. George C. Williams, one of the pre-eminent evolutionists of the day, published "A Defense of Reductionism in Evolutionary Biology" (1985). The zoologist Richard Dawkins has distinguished what he calls hierarchical or gradual reductionism from precipice reductionism; he rejects only the precipice version (Dawkins 1986b, p. 74).[8] More recently the physicist Steven Weinberg, in *Dreams of a Final Theory* (1992), has written a chapter entitled "Two Cheers for Reductionism," in which he distinguishes between uncompromising reductionism (a bad thing) and compromising reductionism (which he ringingly endorses). Here is my own version. We must distinguish reductionism, which is in general a good

8. See also his discussion of Lewontin, Rose, and Kamin's (1984) idiosyncratic version of reductionism—Dawkins aptly calls it their "private bogey"—in the second edition of *The Selfish Gene* (1989a), p. 331.

thing, from *greedy reductionism*, which is not. The difference, in the context of Darwin's theory, is simple: greedy reductionists think that everything can be explained without cranes; good reductionists think that everything can be explained without skyhooks.

There is no reason to be compromising about what I call good reductionism. It is simply the commitment to non-question-begging science without any cheating by embracing mysteries or miracles at the outset. (For another perspective on this, see Dennett 1991a, pp. 33–39.) *Three* cheers for that brand of reductionism—and I'm sure Weinberg would agree. But in their eagerness for a bargain, in their zeal to explain too much too fast, scientists and philosophers often underestimate the complexities, trying to skip whole layers or levels of theory in their rush to fasten everything securely and neatly to the foundation. That is the sin of greedy reductionism, but notice that it is only when overzealousness leads to falsification of the phenomena that we should condemn it. In itself, the desire to reduce, to unite, to explain it all in one big overarching theory, is no more to be condemned as immoral than the contrary urge that drove Baldwin to his discovery. It is not wrong to yearn for simple theories, or to yearn for phenomena that no simple (or complex!) theory could ever explain; what is wrong is zealous misrepresentation, in either direction.

Darwin's dangerous idea is reductionism incarnate,[9] promising to unite and explain just about everything in one magnificent vision. Its being the idea of an *algorithmic* process makes it all the more powerful, since the substrate neutrality it thereby possesses permits us to consider its application to just about anything. It is no respecter of material boundaries. It applies, as we have already begun to see, even to itself. The most common fear about Darwin's idea is that it will not just explain but *explain away* the Minds and Purposes and Meanings that we all hold dear. People fear that once this universal acid has passed through the monuments we cherish, they will cease to exist, dissolved in an unrecognizable and unlovable puddle of scientistic destruction. This cannot be a sound fear; a *proper* reductionistic explanation of these phenomena would leave them still standing but just demystified, unified, placed on more secure foundations. We might learn some surprising or even shocking things about these treasures, but unless our valuing these things was based all along on confusion or mistaken identity, how could increased understanding of them diminish their value in our eyes?[10]

---

9. Yes, incarnate. Think about it: would we want to say it was reductionism *in spirit*?

10. Everybody knows how to answer this rhetorical question with another: "Are you so in love with Truth at all costs that you would want to know if your lover were unfaithful to you?" We are back where we started. I for one answer that I love the world so much that I am sure I want to know the truth about it.

A more reasonable and realistic fear is that the greedy abuse of Darwinian reasoning might lead us to deny the existence of real levels, real complexities, real phenomena. By our own misguided efforts, we might indeed come to discard or destroy something valuable. We must work hard to keep these two fears separate, and we can begin by acknowledging the pressures that tend to distort the very description of the issues. For instance, there is a strong tendency among many who are uncomfortable with evolutionary theory to exaggerate the amount of disagreement among scientists ("It's just a theory, and there are many reputable scientists who don't accept this"), and I must try hard not to overstate the compensating case for what "science has shown." Along the way, we will encounter plenty of examples of genuine ongoing scientific disagreement, and unsettled questions of fact. There is no reason for me to conceal or downplay these quandaries, for no matter how they come out, a certain amount of corrosive work has already been done by Darwin's dangerous idea, and can never be undone.

We should be able to agree about one result already. Even if Darwin's relatively modest idea about the origin of species came to be *rejected* by science—yes, utterly discredited and replaced by some vastly more powerful (and currently unimaginable) vision—it would still have irremediably sapped conviction in any reflective defender of the tradition expressed by Locke. It has done this by opening up new possibilities of imagination, and thus utterly destroying any illusions anyone might have had about the soundness of an argument such as Locke's *a priori* proof of the *inconceivability* of Design without Mind. Before Darwin, this was inconceivable in the pejorative sense that no one knew how to take the hypothesis seriously. Proving it is another matter, but the evidence does in fact mount, and we certainly can and must take it seriously. So whatever else you may think of Locke's argument, it is now as obsolete as the quill pen with which it was written, a fascinating museum piece, a curiosity that can do no real work in the intellectual world today.

CHAPTER 3: *Darwin's dangerous idea is that Design can emerge from mere Order via an algorithmic process that makes no use of pre-existing Mind. Skeptics have hoped to show that at least somewhere in this process, a helping hand (more accurately, a helping Mind) must have been provided—a skyhook to do some of the lifting. In their attempts to prove a role for skyhooks, they have often discovered cranes: products of earlier algorithmic processes that can amplify the power of the basic Darwinian algorithm, making the process locally swifter and more efficient in a nonmiraculous way. Good reductionists suppose that all Design can be explained without skyhooks; greedy reductionists suppose it can all be explained without cranes.*

CHAPTER 4: *How did the historical process of evolution actually make the Tree of Life? In order to understand the controversies about the power of natural selection to explain the origins of all the Design, we must first learn how to visualize the Tree of Life, getting clear about some easily misunderstood features of its shape, and a few of the key moments in its history.*

CHAPTER FOUR

The Tree of Life

〜〜〜

1. HOW SHOULD WE VISUALIZE THE TREE OF LIFE?

*Extinction has only separated groups: it has by no means made them;
for if every form which has ever lived on this earth were suddenly to
reappear, though it would be quite impossible to give definitions by
which each group could be distinguished from other groups, as all
would blend together by steps as fine as those between the finest
existing varieties, nevertheless a natural classification, or at least a
natural arrangement, would be possible.*

—CHARLES DARWIN, *Origin*, p. 432

In the previous chapter, the idea of R-and-D work as analogous to moving
around in something I called Design Space was introduced on the fly, without proper attention to detail or a definition of terms. In order to sketch the
big picture, I helped myself to several controversial claims, promising to
defend them later. Since the idea of Design Space is going to be put to heavy
use, I must now secure it, and, following Darwin's lead, I will once more
begin in the middle, by looking first at some *actual* patterns in some relatively well-explored spaces. These will serve as guides, in the next chapter,
to a more general perspective on *possible* patterns, and the way in which
certain sorts of processes bring possibilities into reality.

Consider the Tree of Life, the graph that plots the time-line trajectories of
all the things that have ever lived on this planet—or, in other words, the
total fan-out of *offspring*. The rules for drawing the graph are simple. An
organism's time line begins when it is born and stops when it dies, and
either there are offspring lines emanating from it or there aren't. The
close-up view of an organism's offspring lines—if there are any—would vary
in appearance depending on several facts: whether the organism reproduces
by fission or budding, or giving birth to eggs or live young, and whether the

parent organism survives to coexist for a while with its offspring. But such microdetails of the fan-out will not in general concern us at this time. There is no serious controversy about the fact that all the diversity of life that has ever existed on this planet is derived from this single fan-out; the controversies arise about how to discover and describe *in general terms* the various forces, principles, constraints, etc., that permit us to give a scientific explanation of the patterns in all this diversity.

The Earth is about 4.5 billion years old, and the first life forms appeared quite "soon"; the simplest single-celled organisms—the *prokaryotes*—appeared at least 3.5 billion years ago, and for probably another 2 billion years, that was all the life there was: bacteria, blue-green algae, and their equally simple kin. Then, about 1.4 billion years ago, a major revolution happened: some of these simplest life forms literally joined forces, when some bacteria-like prokaryotes invaded the membranes of other prokaryotes, creating the *eukaryotes*—cells with nuclei and other specialized internal bodies (Margulis 1981). These internal bodies, called *organelles* or *plastids,* are the key design innovation opening up the regions of Design Space inhabited today. The *chloroplasts* in plants are responsible for photosynthesis, and *mitochondria*, which are to be found in every cell of every plant, animal, fungus—every organism with nucleated cells—are the fundamental oxygen-processing energy-factories that permit us all to fend off the Second Law of Thermodynamics by exploiting the materials and energy around us. The prefix "eu-" in Greek means "good," and from our point of view the eukaryotes were certainly an improvement, since, thanks to their internal complexity, they could specialize, and this eventually made possible the creation of multicelled organisms, such as ourselves.

That *second* revolution—the emergence of the first multicelled organisms—had to wait 700 million years or so. Once multicelled organisms were on the scene, the pace picked up. The subsequent fan-out of plants and animals—from ferns and flowers to insects, reptiles, birds, and mammals—has populated the world today with millions of different species. In the process, millions of other species have come and gone. Surely many more species have gone extinct than now exist—perhaps a hundred extinct species for every existent species.

What is the overall shape of this huge Tree of Life spreading its branches through 3.5 billion years? What would it look like if we could see it all at once from a God's-eye view, with all of time spread out before us in a spatial dimension? The usual practice in scientific graphing is to plot time on the horizontal axis, with *earlier* to the left and *later* to the right, but evolutionary diagrams have always been the exception, usually plotting time on the vertical dimension. Even more curiously, we have accustomed ourselves to two opposite conventions for labeling the vertical dimension, and with these conventions have come their associated metaphors. We can put *ear-*

lier on top and *later* on the bottom, in which case our diagram shows ancestors and their *descendants*. Darwin used this convention when he spoke of speciation as modification with *descent*, and of course in the title of his work on human evolution, *The Descent of Man, and Selection in Relation to Sex* (1871). Alternatively, we can draw a tree in normal orientation, so it looks like a tree, on which the later "descendants" compose the limbs and branches that *rise*, over time, from the trunk and the initial roots. Darwin also exploited this convention—for instance, in the only diagram in *Origin*—but also, along with everyone else, in uses of expressions that align *higher* with *later*. Both metaphor groups coexist with little turbulence in the language and diagrams of biology today. (This tolerance for topsy-turvy imagery is not restricted to biology. "Family trees" are more often than not drawn with the ancestors at the top, and generative linguists, among others, draw their derivational trees upside down, with the "root" at the top of the page.)

Since I have already proposed labeling the vertical dimension in Design Space as a measure of amount of Design, so that *higher* = *more designed*, we must be careful to note that in the Tree of Life (drawn right-side-up, as I propose to do) *higher* = *later* (and nothing else). It does not *necessarily* mean more designed. What is the relation between time and Design, or what could it be? Could things that are more designed come first and

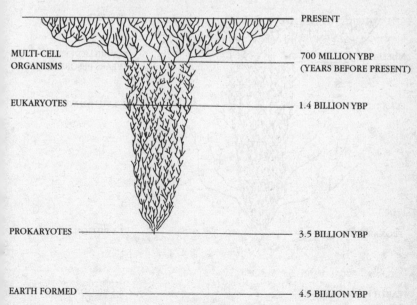

	PRESENT
MULTI-CELL ORGANISMS	700 MILLION YBP (YEARS BEFORE PRESENT)
EUKARYOTES	1.4 BILLION YBP
PROKARYOTES	3.5 BILLION YBP
EARTH FORMED	4.5 BILLION YBP

FIGURE 4.1

gradually lose Design? Is there a possible world in which bacteria are the descendants of mammals and not vice versa? These questions about possibility will be easier to answer if we first look a bit more closely at what has actually happened on our planet. So let us be clear that for the time being, the vertical dimension in the diagrams below stands for time, and time alone, with *early* at the bottom and *late* at the top. Following standard practice, the left-right dimension is taken as a sort of single-plane summary of diversity. Each individual organism has to have its time line, distinct from all others, so, even if two organisms are exact atom-for-atom duplicates of each other, they will have to appear side by side at best. How we line them all up, however, can be according to some measure or family of measures of difference in individual body shape—*morphology*, to use the technical term.

So, to return to our question, what would the overall shape of the entire Tree of Life look like, if we could take it all in at a glance? Wouldn't it look rather like a palm tree, as in figure 4.1?

This is the first of many trees, or *dendrograms,* we will consider, and of course the limited resolution of the ink on the page blurs quadrillions of separate lines together. I have left the "root" of the tree deliberately fuzzy and indistinct for the time being. We are still exploring the middle, saving the ultimate beginnings for a later chapter. If we were to zoom in on the trunk of this tree and look at any cross-section of it—an "instant" in

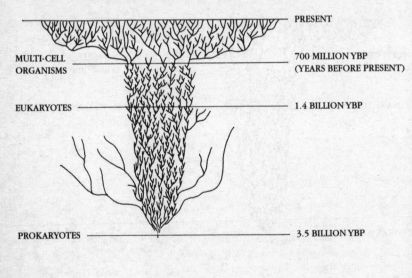

FIGURE 4.2

time—we would see billions upon billions of individual unicellular organisms, a fraction of which would have trails leading to progeny slightly higher up the trunk. (In those early days, reproduction was by budding or fission; somewhat later, a kind of unicellular sex evolved, but pollen-wafting and egg-laying and the other phenomena of our kind of sexual reproduction have to wait for the multicellular revolution in the fronds.) There would be some diversity, and some revision of design over time, so perhaps the whole trunk should be shown leaning left or right, or spreading more than I have shown. Is it just our ignorance that prevents us from differentiating this "trunk" of unicellular varieties into salient streams? Perhaps it should be shown with various dead-end branches large enough to be visible, as in figure 4.2, marking various hundred-million-year experiments in alternative unicellular design that eventually all ended in extinction.

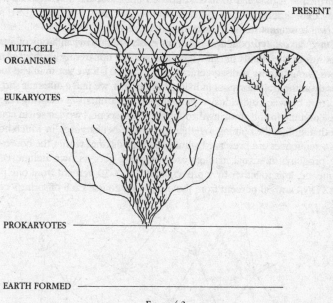

PRESENT

MULTI-CELL
ORGANISMS

EUKARYOTES

PROKARYOTES

EARTH FORMED

FIGURE 4.3

There must have been billions of failed design experiments, but perhaps none ever became very distant departures from a single unicellular norm. In any event, if we were to zoom way in on the trunk, we would see a luxuriant growth of short-lived alternatives, as in figure 4.3, all but invisible against the norm of conservative replication. How can we be sure of this? Because, as we shall see, the odds are heavily against any mutation's being more viable than the theme on which it is a variation.

Until sexual reproduction is invented, almost all the branches we observe, at any zoom level, diverge. The exceptions are remarkable, however. At the time of the eukaryotic revolution, if we look in just the right place, we will see a bacterium entering the rudimentary body of some other prokaryote to create the first eukaryote. Its progeny all have a dual inheritance—they contain two entirely independent DNA sequences, one for the host cell and another for the "parasite," sharing its fate with its host's, and linking the fate of all its descendants (now on their way to becoming benign resident mitochondria) to the fate of the cells they will inhabit, the descendants of the cell first invaded. It's an amazing feature of the microscopic geometry of the Tree of Life: whole lineages of mitochondria, tiny living things in their own right, with their own DNA, living their entire lives within the walls of the cells of larger organisms that compose other lineages. In principle it only has to have happened once, but we may suppose that many experiments in such radical symbiosis occurred (Margulis 1981; for accessible summaries, see Margulis and Sagan 1986, 1987).

Once sexual reproduction becomes established many millions of years later, up in the fronds of our Tree (and sex has apparently evolved many times, though there is disagreement on this score), if we zoom in and look closely at the trajectories of individual organisms, we find a different sort of juncture between individuals—matings—with starbursts of offspring resulting. Zooming in and "looking through the microscope," we can see in figure 4.4 that, unlike the coming together that created eukaryotes, in which both DNA sequences are preserved whole and kept distinct within the bodies of the progeny, in sexual matings each offspring gets its own unique DNA sequence, knit together by a process that draws 50 percent from one parent's DNA and 50 percent from the other's. Of course each offspring's cells

FIGURE 4.4

also contain mitochondria, and these always come from one parent only, the female. (If you are a male, all the mitochondria in your cells are in an evolutionary cul-de-sac; they will not get passed on to any offspring of yours, who will get all their mitochondria from their mother.) Now step back a pace from our close-up of matings-with-offspring and notice (in figure 4.4) that *most* of those offspring's trajectories terminate without mating, or at least without offspring of their own. This is the Malthusian crunch. Everywhere we look, the branches and twigs are covered with the short, terminal fuzz of birth-death without further issue.

It would be impossible to see at one time all the branch points and junctions in the whole Tree of Life, extending over 3.5 billion years, but if we backed way off from the details and looked for some large-scale shapes, we could recognize a few familiar landmarks. Early in the multicellular fan-out that began about 700 million years ago, we could see the forks that created two large branches—the kingdoms of plants and animals—and another for the fungi, departing from the trunk of the single-celled organisms. And if we looked closely, we would see that, once they become separated by some distance, no matings reunite any of the trajectories of their individual members. By this time, the groups had become reproductively isolated, and the gap grew wider and wider.[1] Further forks created the multicellular phyla, orders, classes, families, genera, and species.

2. Color-coding a Species on the Tree

What does a *species* look like in this Tree? Since the questions of what a species is, and how a species starts, continue to generate controversy, we can take advantage of the God's-eye perspective we have temporarily adopted to look closely at the whole Tree of Life and see what would happen if we tried to color-code a single species in it. One thing can be sure: whatever region we color in will be a single, connected region. No separated blobs of organisms, no matter how similar in appearance or morphology, could count as composed of members of a *single* species, which must be united by descent. The next point to make is that until sexual reproduction arrives on the scene, the hallmark of reproductive isolation can have no bearing at all. This handy boundary-making condition has no definition in the asexual world. In those ancient and contemporary strands in the Tree

---

1. There have been some remarkable symbiotic reunions, however, of organisms that belong to different kingdoms. The flatworm *Convoluta roscoffensis* has no mouth and never needs to eat, since it is filled with algae that photosynthesize its nourishment (Margulis and Sagan 1986)!

that reproduce asexually, groupings of one sort or another may interest us for various good reasons—groupings of shared morphology or behavior or of genetic similarity, for instance—and we might choose to call the resulting group a species, but there may very well be no theoretically important sharp edges that would delimit such a species. So let us concentrate on sexually reproducing species, all of which are to be found up in the multicellular fronds of the Tree. How might we go about coloring all the life-lines of a single such species red? We could start by looking at individuals at random until we found one with lots of descendants. Call her Lulu, and color her red. (Red is represented by the thick lines in figure 4.5.) Now move stepwise up the Tree, coloring all Lulu's descendants red; these will all be members of one species *unless* we find our red ink spreading into two distinct higher branches, none of whose members form junctions across the void. If that happens, we know there has been speciation, and we will have to back up and make several decisions. We must first choose whether to keep one of the branches red (the "parent" species continues red and the other branch is considered the new daughter species) or to stop the red ink altogether as soon as the branching happens (the "parent" species has gone extinct, fissioning into two daughter species).

If the organisms in the branch on the left are all pretty much the same in appearance, equipment, and habits as Lulu's contemporaries, while the organisms in the right branch almost all sport novel horns, or webbed feet, or stripes, then it is pretty obvious that we should label the left branch as the continuing, parent species, and the right branch the new offshoot. If both branches soon show major changes, our color-coding decision is not so obvious. There are no secret facts that could tell us which choice is right, which choice carves nature at the joints, for we are looking right at the places where the joints would have to be, and there aren't any. There is nothing more to being a species than being one of these branches of interbreeding organisms, and nothing more to being the *conspecific* of some other organism (contemporary or not) than being part of the same branch. The choice we make will then have to depend on pragmatic or aesthetic considerations: Is it *ungainly* to keep the same label for this branch as for its parent trunk? Would it be *misleading* for one reason or another to say the branch on the right rather than the branch on the left was the new species?[2]

---

2. The cladists (whose views will be briefly discussed later) are a school of taxonomists that reject, for various reasons, the concept of a "parent" species' persisting. Every speciation event, in their terms, results in a pair of daughter species and the extinction of their common parent, no matter how closely one surviving branch resembles the parent, compared with the other branch.

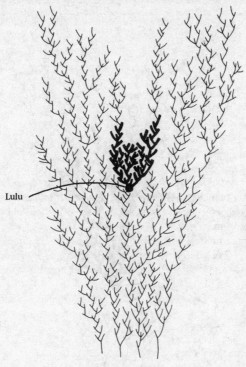

Lulu

FIGURE 4.5

The same sort of quandary faces us when we try to complete the task of color-coding the whole species by carrying our red ink down the Tree to include all Lulu's ancestors. We will encounter no gaps or joints on this downward path, which will take us all the way to the prokaryotes at the base of the Tree if we persist. But if we also color *sideways* as we go down, filling in the cousins, aunts, and uncles of Lulu and her ancestors, and then color up from these sideways spreaders, we will eventually fill in a whole branch on which Lulu resides down to the point where coloring any lower (earlier) nodes (for instance, at A in figure 4.6) causes "leakage" of red into neighboring branches that clearly belong to other species.

If we stop there, we can be sure that *only* members of Lulu's species have been colored red. It will be arguable that we have left out some that deserve to be colored, but only arguable, for there are, again, no hidden facts, no essences that could settle the issue. As Darwin pointed out, if it weren't for the separations that time and the extinction of the intermediate stepping-stones has created, although we could put the life forms into a "natural

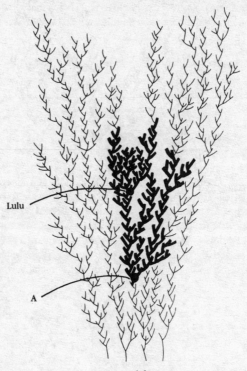

Lulu

A

FIGURE 4.6

arrangement" (of descent), we could not put them into a "natural classification"—we need the biggish gaps between *extant* forms to form the
"boundaries" of any such classes.

The theoretical concept of species that predates Darwin's theory had two
fundamental ideas: that species members have different essences, and that
"therefore" they don't/can't interbreed. What we have subsequently figured
out is that in principle there could be two subpopulations that were different *only* in that their pairings were sterile due to a tiny genetic incompat-
ibility. Would these be different species? They could look alike, feed alike,
live together in the same niche, and be genetically very, very similar, yet
reproductively isolated. They would not be different enough to count as
salient *varieties*, but they would satisfy the primary condition for being two
different *species*. In fact, there are cases of "cryptic sibling species" that
approximate this extreme. As we already noted, at the other extreme we
have the dogs, readily distinguished into morphological types by the naked
eye, adapted to vastly different environments, but not reproductively iso-

lated. Where should we draw the line? Darwin shows that we don't need to draw the line in an essentialist way in order to get on with our science. We have the best of reasons to realize that these extremes are improbable: in general, where there is genetic speciation there is marked morphological difference, or marked difference in geographical distribution, or (most likely) both. If this generalization weren't largely true, the concept of species would not be important, but we need not ask exactly how much difference (in addition to reproductive isolation) is *essential* for a case of *real* species-difference.[3]

Darwin shows us that questions like "What is the difference between a variety and a species?" are like the question "What is the difference between a peninsula and an island?"[4] Suppose you see an island half a mile offshore at high tide. If you can walk to it at low tide without getting your feet wet, is it still an island? If you build a bridge to it, does it cease to be an island? What if you build a solid causeway? If you cut a canal across a peninsula (like the Cape Cod Canal), do you turn it into an island? What if a hurricane does the excavation work? This sort of inquiry is familiar to philosophers. It is the Socratic activity of definition-mongering or essence-hunting: looking for the "necessary and sufficient conditions" for being-an-*X*. Sometimes almost everyone can see the pointlessness of the quest—islands obviously don't have real essences, but only nominal essences at best. But at other times there can still seem to be a serious scientific question that needs answering.

More than a century after Darwin, there are still serious debates among biologists (and even more so among philosophers of biology) about how to define *species*. Shouldn't scientists define their terms? Yes, of course, but only up to a point. It turns out that there are different species concepts with different uses in biology—what works for paleontologists is not much use to ecologists, for instance—and no clean way of uniting them or putting them in an order of importance that would crown one of them (the most important one) as *the* concept of species. So I am inclined to interpret the persisting debates as more a matter of vestigial Aristotelian tidiness than a useful disciplinary trait. (This is all controversial, but see Kitcher 1984 and G. C. Williams 1992 for further support and concurring arguments, and the recent anthology on the topic, Ereshefsky 1992, and Sterelny 1994, an insightful review essay on that anthology.)

---

3. The issues are further complicated by the existence of hybridization—in which members of two different species *do* have fertile offspring—a phenomenon that raises interesting issues that are off the track we are exploring.

4. The evolutionary epistemologist and psychologist Donald Campbell has been the most vigorous developer of the implications of this side of Darwin's legacy.

3. RETROSPECTIVE CORONATIONS: MITOCHONDRIAL EVE AND INVISIBLE BEGINNINGS

When we tried to see whether Lulu's descendants split into more than one species, we had to look ahead to see if any large branches appeared, and then *back up* if we deemed that somewhere along the line a speciation event must have happened. We never addressed the presumably important question of *exactly* when speciation should be said to occur. Speciation can now be seen to be a phenomenon in nature that has a curious property: you can't tell that it is occurring at the time it occurs! You can only tell much later that it has occurred, retrospectively crowning an event when you discover that its sequels have a certain property. This is not a point about our epistemic limitations—as if we *would* be able to tell when speciation occurs if only we had better microscopes, or even if we could get in a time machine and go back in time to observe the appropriate moments. This is a point about the objective property of being a speciation event. It is not a property that an event has simply by virtue of its spatio-temporally local properties.

Other concepts exhibit similar curiosities. I once read about a comically bad historical novel in which a French doctor came home to supper one evening in 1802 and said to his wife: "Guess what *I* did today! I assisted at the birth of Victor Hugo!" What is wrong with that story? Or consider the property of being a widow. A woman in New York City may suddenly acquire that property by virtue of the effects that a bullet has just had on some man's brain in Dodge City, over a thousand miles away. (In the days of the Wild West, there was a revolver nicknamed the Widowmaker. Whether a particular revolver lived up to its nickname on a particular occasion might be a fact that could not be settled by any spatio-temporally local examination of its effects.) This case gets its curious capacity to leap through space and time from the conventional nature of the relation of marriage, in which a past historical event, a wedding, is deemed to create a permanent relation—a *formal* relation—of interest in spite of subsequent wanderings and concrete misfortunes (the accidental loss of a ring, or the destruction of the marriage certificate, for instance.)

The systematicity of genetic reproduction is not conventional but natural, but that very systematicity permits us to think *formally* about causal chains extending over millions of years, causal chains that would otherwise be virtually impossible to designate or refer to or track. This permits us to become interested in, and reason rigorously about, even more distant and locally invisible relationships than the formal relationship of marriage. Speciation is, like marriage, a concept anchored within a tight, formally definable system of thought, but, unlike marriage, it has no conventional saliencies—weddings, rings, certificates—by which it can be observed. We can see this feature of speciation in a better light by looking first at another instance of retrospective crowning, the conferring of the title of Mitochondrial Eve.

Mitochondrial Eve is the woman who is the most recent direct ancestor, in the female line, of every human being alive today. People have a hard time thinking about this individual woman, so let's just review the reasoning. Consider the set A, of all human beings alive today. Each was born of one and only one mother, so consider next the set, B, of all the mothers of those alive today. B is of necessity smaller than A, since no one has more than one mother, and some mothers have more than one child. Continue with the set C, of mothers of all those mothers in set B. It is smaller still. Continue on with sets D and E and so forth. The sets must contract as we go back each generation. Notice that as we move back through the years, we exclude many women who were contemporaries of those in our set. Among these excluded women are those who either lived and died childless or whose female progeny did. Eventually, this set must funnel down to one—the woman who is the closest direct female ancestor of everybody alive on earth today. She is Mitochondrial Eve, so named (by Cann et al. 1987) because since the mitochondria in our cells are passed through the maternal line alone, all the mitochondria in all the cells in all the people alive today are direct descendants of the mitochondria in her cells!

The same logical argument establishes that there is—must be—an Adam as well: the closest direct male ancestor of everybody alive today. We could call him Y-Chromosome Adam, since all our Y-chromosomes pass down through the paternal line just the way our mitochondria pass through the maternal line.[5] Was Y-Chromosome Adam the husband or lover of Mitochondrial Eve? Almost certainly not. There is only a tiny probability that these two individuals were alive at the same time. (Paternity being a much less time-and-energy-consuming business than maternity, what is *logically* possible is that Y-Chromosome Adam lived very recently, and was very, very busy in the bedroom—leaving Errol Flynn in his, um, dust. He could, in principle, be the great-grandfather of us all. This is about as unlikely as the case in which Y-Chromosome Adam and Mitochondrial Eve were a couple.)

Mitochondrial Eve has been in the news recently because the scientists who christened her think they can analyze the patterns in the mitochondrial

---

5. Note one important difference between the legacies of Mitochondrial Eve and Y-Chromosome Adam: we all, male and female, have mitochondria in our cells, but they all come from our mothers; if you are male, you have a Y-chromosome and got it from your father, but most—virtually all, but not quite all—females have no Y-chromosome at all.

DNA of the different people alive today and deduce from that how recently Mitochondrial Eve lived, and even where she lived. According to their original calculations, Mitochondrial Eve lived in Africa, very, very recently— less than three hundred thousand years ago, and maybe less than half that. These methods of analysis are controversial, however, and the African Eve hypothesis may be fatally flawed. Deducing *where* and *when* is a far trickier task than deducing *that* there was a Mitochondrial Eve, something that nobody denies. Consider a few of the things we already know about Mitochondrial Eve, setting aside the recent controversies. We know that she had at least two daughters who had surviving children. (If she had just one daughter, her daughter would wear the crown of Mitochondrial Eve.) To distinguish her title from her proper name, let's call her Amy. Amy bears the title of Mitochondrial Eve; that is, she just happens to have been the maternal founder of today's line of people.[6] It is important to remind ourselves that *in all other regards*, there was probably nothing remarkable or special about Mitochondrial Eve; she was certainly not the First Woman, or the founder of the species *Homo sapiens*. Many earlier women were unquestionably of our species, but happen not to have any direct female lines of descendants leading to people living today. It is also true that Mitochondrial Eve was probably no stronger, faster, more beautiful, or more fecund than the other women of her day.

To bring out just how unspecial Mitochondrial Eve—that is, Amy—probably was, suppose that tomorrow, thousands of generations later, a virulent new virus were to spread around the Earth, wiping out 99 percent of the human race in a few years. The survivors, fortunate to have some innate resistance to the virus, would probably all be quite closely related. *Their* closest common direct female ancestor—call her Betty—would be some woman who lived hundreds or thousands of generations later than Amy, and the crown of Mitochondrial Eve would pass to her, retroactively. She may have been the source of the mutation that centuries later came into its own as a species-saver, but it didn't do *her* any good, since the virus against which it is to triumph didn't exist then. The point is that Mitochondrial Eve can only be *retrospectively* crowned. This historically pivotal role is determined not just by the accidents of Amy's own time, but by the accidents of later times as well. Talk about massive contingency! If Amy's uncle hadn't saved her from drowning when she was three, none of *us* (with our particular mitochondrial DNA, thanks ultimately to Amy) would ever have

---

6. Philosophers have often discussed strange examples of individuals known to us only via definite descriptions, but they have usually confined their attention to such boring—if real—individuals as the shortest spy. (There has to be one, doesn't there?) I suggest that Mitochondrial Eve is a much more delicious example, all the more so for being of some genuine theoretical interest in evolutionary biology.

existed! If Amy's granddaughters had all starved to death in infancy—as so many infants did in those days—the same oblivion would be ours.

The curious invisibility of the crown of Mitochondrial Eve in her own lifetime is easier to understand and accept than the near-invisibility of what every species must have: a beginning. If species aren't eternal, then all of time can be divided, somehow, into the times before the existence of species x, and all subsequent times. But what must have happened at the interface? It may help if we think of a similar puzzle that has baffled many people. Have you ever wondered, when hearing a new joke, where it came from? If you are like almost everybody else I have ever known or heard of, you never make up jokes; you pass on, perhaps with "improvements," something you heard from someone who heard it from someone, who . . . Now, we know the process cannot go on forever. A joke about President Clinton, for instance, cannot be more than a year or so old. So who makes up the jokes? Joke-authors (as contrasted with joke-purveyors) are invisible.[7] Nobody ever seems to catch them in the act of authorship. There is even folklore—an "urban legend"—to the effect that these jokes are all created in prison, by prisoners, those dangerous and unnatural folks, so unlike the rest of us, and with nothing better to do with their time than to fashion jokes in their secret underground joke-workshops. Nonsense. It is hard to believe—but it must be true—that the jokes we hear and pass on have evolved from earlier stories, picking up revisions and updates as they are passed along. A joke typically has no one author; its authorship is distributed over dozens or hundreds or thousands of tellers, solidifying for a while in some particularly topical and currently amusing version, before going dormant, like the ancestors from which it grew. Speciation is equally hard to witness, and for the same reason.

When has speciation occurred? In many cases (perhaps most, perhaps almost all—biologists disagree about how important the exceptions are), the speciation depends on a geographical split in which a small group—maybe a single mating pair—wander off and start a lineage that becomes reproductively isolated. This is *allopatric* speciation, in contrast to *sympatric* speciation, which does not involve any geographic barriers. Suppose we watch the departure and resettlement of the founding group. Time passes, and several generations come and go. Has speciation occurred? Not yet, certainly. We won't know until many generations later whether or not these individuals should be crowned as species-initiators.

There is not *and could not be* anything internal or intrinsic to the individuals—or even to the individuals-as-they-fit-into-their-environment—from

---

7. There are, of course, the writers who make their living writing funny lines for television comedians, and the comedians themselves, who create much of their own material, but, with negligible exceptions, these people are not the creators of the joke stories ("Did you hear the one about the guy who . . . ?") that get passed around.

which it followed that they were—as they later turn out to be—the founders of a new species. We can imagine, if we want, an extreme (and improbable) case in which a single mutation guarantees reproductive isolation in a single generation, but, of course, whether or not the individual who has that mutation counts as a species-founder or simply as a freak of nature depends on nothing in its individual makeup or biography, but on what happens to subsequent generations—if any—of its offspring.

Darwin was not able to present a single instance of speciation by natural selection in *Origin of Species*. His strategy in that book was to develop in detail the evidence that artificial selection by dog- and pigeon-breeders could build up large differences by a series of gradual changes. He then pointed out that *deliberate* choice by the animals' keepers was inessential; the runts of the litter tended not to be valued, and hence tended not to reproduce as much as their more valued siblings, so, without any conscious policy of breeding, human animal-keepers presided unwittingly over a steady process of design revision. He offered the nice example of the King Charles spaniel, "which has been unconsciously modified to a large extent since the time of that monarch" (*Origin*, p. 35)—as can be confirmed by a careful examination of the dogs in various portraits of King Charles. He called such cases "unconscious selection" by human domesticators, and he used it as a persuasive bridge to get his readers to the hypothesis of even more unconscious selection by the impersonal environment. But he had to admit, when challenged, that he could provide no cases of animal-breeders' producing a new species. Such breeding had definitely produced different *varieties*, but not a single new species. Dachshund and St. Bernard were not different species, however different in appearance. Darwin admitted as much, but he might quite correctly have gone on to point out that it was simply too early to tell whether he had given any examples of speciation accomplished by artificial selection. Any lady's lapdog could at some future date be discovered *to have been* the founding member of a species that split off from *Canis familiaris*.

The same moral applies to the creation of new genera, families, and even kingdoms, of course. The major branching that we would retrospectively crown as the parting of the plants from the animals began as a segregation of two gene pools every bit as inscrutable and unremarkable at the time as any other temporary drifting apart of members of a single population.

4. PATTERNS, OVERSIMPLIFICATION, AND EXPLANATION

Much more interesting than the question of how to draw the species boundary are all the questions about the shapes of the branches—and even more interesting, the shapes of the empty spaces between the branches. What trends, forces, principles—or historical events—have influenced these shapes or made them possible? Eyes have evolved independently in dozens of lineages, but feathers probably only once. As John Maynard Smith observes, mammals go in for horns but birds do not. "Why should the pattern of variation be limited in this way? The short answer is that we do not know" (Maynard Smith 1986, p. 41).

We *can't* rewind the tape of life and replay it to see what happens next time, alas, so the only way to answer questions about such huge and experimentally inaccessible patterns is to leap boldly into the void with the risky tactic of deliberate oversimplification. This tactic has a long and distinguished history in science, but it tends to provoke controversy, since scientists have different thresholds at which they get nervous about playing fast and loose with the recalcitrant details. Newtonian physics was overthrown by Einstein, but it is still a good approximation for almost all purposes. No physicist objects when NASA uses Newtonian physics to calculate the forces at liftoff and the orbital trajectory of the space shuttle, but, strictly speaking, this is a deliberate use of a false theory in order to make calculation feasible. In the same spirit, physiologists studying, say, mechanisms for changing the rate of metabolism try in general to avoid the bizarre complexities of subatomic quantum physics, hoping that any quantum effects will cancel out or in other ways be beneath the threshold of their models. In general, the tactic pays off handsomely, but one can never be sure when one scientist's grubby complication will be elevated into another scientist's Key to the Mystery. And it can just as well work the other way around: the Key is often discovered by climbing out of the trenches and going for the panoramic view.

I once got in a debate with Francis Crick about the virtues and vices of Connectionism—the movement in cognitive science that models psychological phenomena by building up patterns in the connection-strengths between the nodes in *very* unrealistic and oversimplified "neural nets" simulated on computers. "These people may be good engineers," Crick averred (as best I recall), "but what they are doing is terrible science! These people willfully turn their backs on what we *already* know about how neurons interact, so their models are utterly useless as models of brain function." This criticism somewhat surprised me, for Crick is famous for his own brilliant opportunism in uncovering the structure of DNA; while others struggled up the straight and narrow path of strict construction from the evidence, he and Watson took a few daring and optimistic sidesteps, with gratifying results. But in any case, I was curious to know how widely he would cast his denunciation. Would he say the same thing about population geneticists? The derogatory term for some of their models is "bean-bag genetics," for they pretend that genes for this and that are like so many color-coded beads on a string. What they call a gene (or an *allele* at a *locus*) bears only a passing resemblance to the intricate machinery of the codon sequences on DNA molecules. But thanks to these deliberate simplifications, their models are computationally tractable, enabling them to discover and confirm many large-scale patterns in gene flow that would otherwise be utterly invisible. Adding complications would tend to bring their research to a grinding halt. But is their research good science? Crick replied that he had himself thought about the comparison, and had to say that population genetics wasn't science either!

My tastes in science are more indulgent, as perhaps you would expect from a philosopher, but I do have my reasons: I think the case is strong that not only do "over"-simplified models often actually *explain* just what needs explaining, but no more complicated model could do the job. When what provokes our curiosity are the *large patterns* in phenomena, we need an explanation at the right level. In many instances this is obvious. If you want to know why traffic jams tend to happen at a certain hour every day, you will still be baffled after you have painstakingly reconstructed the steering, braking, and accelerating processes of the thousands of drivers whose various trajectories have summed to create those traffic jams.

Or imagine tracing all the electrons through a hand calculator as it multiplies two numbers together and gets the correct answer. You could be 100 percent sure you understood each of the millions of causal microsteps in the process and yet still be utterly baffled about *why* or even *how* it always got the *right* answer to the questions you posed it. If this is not obvious, imagine that somebody made—as a sort of expensive prank—a hand calculator that usually gave the wrong answers! It would obey exactly the same physical laws as the good calculator, and would cycle through the same sorts of microprocesses. You could have *perfect* explanations of how both calculators worked at the electronic level, and still be utterly unable to explain the intensely interesting fact that one of them got the answers right and the other got them wrong. This is the sort of case that shows what would be silly about the preposterous forms of reductionism; *of course* you can't explain all the patterns that interest us at the level of physics (or chemistry, or any one low level). This is undeniably true of such mundane and unperplexing phenomena as traffic jams and pocket calculators; we should expect it to be true of biological phenomena as well. (For more on this topic, see Dennett 1991b.)

Now consider a parallel question in biology, a textbook standard: why do giraffes have long necks? There is one answer that could in principle be "read off" the total Tree of Life, if we had it to look at: Each giraffe has a neck of the length it has because its parents had necks of the lengths they had, and so forth back through the generations. If you check them off one by one, you will see that the long neck of each living giraffe has been traced back through long-necked ancestors all the way back ... to ancestors who didn't even have necks. So that's how come giraffes have long necks. End of explanation. (And if that doesn't satisfy you, note that you will be even less satisfied if the answer throws in all the details about the individual developmental and nutritional history of each giraffe in the lineage.)

Any acceptable explanation of the patterns we observe in the Tree of Life must be contrastive: why do we see this actual pattern rather than that one—or no pattern at all? What are the nonactualized alternatives that need to be considered, and how are they organized? To answer such questions, we need to be able to talk about what is possible in addition to what is actual.

CHAPTER 4: *There are patterns in the unimaginably detailed Tree of Life, highlighting crucial events that made the later flourishing of the Tree possible. The eukaryotic revolution and the multicellular revolution are the most important, followed by the speciation events, invisible at the time, but later seen to mark even such major divisions as those between plants and animals. If science is to explain the patterns discernible in all this complexity, it must rise above the microscopic view to other levels, taking on idealizations when necessary so we can see the woods for the trees.*

CHAPTER 5: *The contrast between the actual and the possible is fundamental to all explanation in biology. It seems we need to distinguish different grades of possibility, and Darwin provides a framework for a unified treatment of biological possibility in terms of accessibility in "the Library of Mendel," the space of all genomes. In order to construct this useful idealization, we must acknowledge and then set aside certain complications in the relations between a genome and a viable organism.*

CHAPTER FIVE

The Possible and the Actual

~~~

1. GRADES OF POSSIBILITY?

However many ways there may be of being alive, it is certain that there are vastly more ways of being dead, or rather not alive.

—RICHARD DAWKINS 1986A, P. 9

Any particular non-existent form of life may owe its absence to one of two reasons. One is negative selection. The other is that the necessary mutations have never appeared.

—MARK RIDLEY 1985, P. 56

Take, for instance, the possible fat man in that doorway; and, again, the possible bald man in that doorway. Are they the same possible man, or two possible men? How do we decide? How many possible men are there in that doorway? Are there more possible thin ones than fat ones? How many of them are alike? Or would their being alike make them one? Are no two possible things alike? Is this the same as saying that it is impossible for two things to be alike? Or, finally, is the concept of identity simply inapplicable to unactualized possibles?

—WILLARD VAN ORMAN QUINE 1953, P. 4

There seem to be at least four different kinds or grades of possibility: logical, physical, biological, and historical, nested in that order. The most lenient is mere logical possibility, which according to philosophical tradition is simply a matter of being describable without contradiction. Superman, who flies faster than the speed of light, is *logically* possible, but Duperman, who flies faster than the speed of light *without moving anywhere*, is not even logically possible. Superman, however, is not *physically* possible, since a law of physics proclaims that nothing can move faster than the speed of light. There is no dearth of difficulties with this superficially straightforward distinction. How do we distinguish fundamental physical laws from logical laws? Is it physically or logically impossible to travel backwards in time, for instance? How could we tell for sure whether a description that is *apparently* coherent—such as the story in the film *Back to the Future*—is subtly self-contradictory or merely denies a very fundamental (but not logically necessary) assumption of physics? There is also no dearth of philosophy dealing with these difficulties, so we will just acknowledge them and pass on to the next grade.

Superman flies by simply leaping into the air and striking a gallant midair pose, a talent which is certainly physically impossible. Is a flying horse physically possible? The standard model from mythology would never get off the ground—a fact from physics (aerodynamics), not biology—but a horse with suitable wingspan could presumably stay aloft. It might have to be a tiny horse, something aeronautical engineers might calculate from considerations of weight-strength ratios, the density of air, and so forth. But now we are descending into the third grade of possibility, *biological possibility*, for once we begin considering the strength of bones, and the payload requirements for keeping the flapping machinery going, we concern ourselves with development and growth, metabolism, and other clearly biological phenomena. Still, the verdict may appear to be that of course flying horses are biologically possible, since bats are actual. Maybe even full-sized flying horses are possible, since there once were pteranodons and other flying creatures approaching that size. There is nothing to beat actuality, present or past, for clinching possibility. Whatever is or has been actual is obviously possible. Or is it?

The lessons of actuality are hard to read. Could such flying horses really be viable? Would they perhaps need to be carnivorous to store enough energy and carry it aloft? Perhaps—in spite of fruit-eating bats—only a carnivorous horse could get off the ground. Is a carnivorous horse possible? Perhaps a carnivorous horse would be biologically possible *if it could evolve*, but would such a diet shift be accessible from where horses would have to start? And, short of radical constructive surgery, could a horse-descendant have both front legs and wings? Bats, after all, make wings of their arms. Is there any possible evolutionary history of skeletal revision that would yield a six-limbed mammal?

This brings us to our fourth grade of possibility, *historical possibility*. There might have been a time, in the very distant past, when the possibility of six-limbed mammals on Earth had not yet been foreclosed, but it might also be true that once our four-finned fishy ancestors got selected for moving onto the land, the basic four-limbed architecture was so deeply anchored in our developmental routines that alteration at this time is *no longer possible*. But even that distinction may not be sharp-edged. Is such an alteration in fundamental building-plan flat impossible, or just highly unlikely, so resistant to change that only an astronomically improbable sequence of selective blows could drive it into existence? It seems there might be two kinds or grades of biological impossibility: violation of a biological *law of nature* (if there are any), and "mere" biohistorical consignment to oblivion.

Historical impossibility is simply a matter of opportunities passed up. There was a time when many of us worried about the possibility of President Barry Goldwater, but it didn't happen, and after 1964, the odds against such a thing's ever happening lengthened reassuringly. When lottery tickets are put on sale, this creates an opportunity for you: you may choose to buy one, provided you act by a certain date. If you buy one, this creates a further opportunity for you—the opportunity to win—but soon it slides into the past, and it is no longer possible for you to win *those* millions of dollars. Is this everyday vision we have of opportunities—*real* opportunities—an illusion? In what sense *could* you have won? Does it make a difference if the winning lottery number is chosen *after* you buy your ticket, or do you still have an opportunity to win, a real opportunity, if the winning number is sealed in a vault before the tickets are put on sale (Dennett 1984)? Is there *ever* really any opportunity at all? Could anything happen other than what actually happens? This dread hypothesis, the idea that *only* the actual is possible, has been called *actualism* (Ayers 1968). It is generally ignored, for good reasons, but these reasons are seldom discussed. (Dennett 1984, and Lewis 1986, pp. 36–38, offer good reasons for dismissing actualism.)

These familiar and *prima facie* reliable ideas about possibility can be summed up in a diagram, but every boundary in it is embattled. As Quine's questions suggest, there is something fishy about casual catalogues of merely possible objects, but since science cannot even express—let alone confirm—the sorts of explanations we crave without drawing such a distinction, there is little chance that we can simply renounce all such talk. When biologists wonder whether a horned bird—or even a giraffe with stripes instead of blotches—is possible, the questions they are addressing epitomize what we want biology to discover for us. Alerted by Quine, we can be struck by the dubious metaphysical implications of Richard Dawkins' vivid claim that there are many more ways of being dead than of being alive, but manifestly he is getting at something important. We should try to find a way of recasting such claims in a metaphysically more modest and less contentious framework—and Darwin's starting in the middle gives us just the foothold we need. *First* we can deal with the relation between historical and

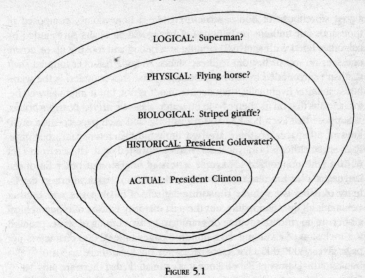

LOGICAL: Superman?

PHYSICAL: Flying horse?

BIOLOGICAL: Striped giraffe?

HISTORICAL: President Goldwater?

ACTUAL: President Clinton

FIGURE 5.1

biological possibility, and then perhaps it will suggest some payoffs for how to make sense of the grander varieties.[1]

2. The Library of Mendel

The Argentine poet Jorge Luis Borges is not typically classified as a philosopher, but in his short stories he has given philosophy some of its most valuable thought experiments, most of them gathered in the stunning collection *Labyrinths* (1962). Among the best is the fantasy—actually, it is more a philosophical reflection than a narrative—that describes the Library of Babel. For us, the Library of Babel will be an anchoring vision for helping to answer very difficult questions about the scope of biological possibility, so we will pause to explore it at some length. Borges tells of the forlorn explorations and speculations of some people who find themselves living in

---

1. Back in 1982, François Jacob, the Nobel laureate biologist, published a book entitled *The Possible and the Actual*, and I rushed to read it, expecting it to be an eye-opening essay on how biologists should think about some of these conundrums about possibility. To my disappointment, the book had very little to say on this topic. It is a fine book, and has a great title, but the two don't go together, in my humble opinion. The book I was eager to read hasn't yet been written, apparently, so I'll have to try to write part of it myself, in this chapter.

a vast storehouse of books, structured like a honeycomb, composed of thousands (or millions or billions) of hexagonal air shafts surrounded by balconies lined with shelves. Standing at a railing and looking up or down, one sees no top or bottom to these shafts. Nobody has ever found a shaft that isn't surrounded by six neighboring shafts. They wonder: is the warehouse infinite? Eventually, they decide that it is not, but it might as well be, for it seems that on its shelves—in no order, alas—lie *all the possible books*.

Suppose that each book is 500 pages long, and each page consists of 40 lines of 50 spaces, so there are two thousand character-spaces per page. Each space either is blank, or has a character printed on it, chosen from a set of 100 (the upper- and lowercase letters of English and other European languages, plus the blank and punctuation marks).[2] Somewhere in the Library of Babel is a volume consisting entirely of blank pages, and another volume is all question marks, but the vast majority consist of typographical gibberish; no rules of spelling or grammar, to say nothing of sense, prohibit the inclusion of a volume. Five hundred pages times 2,000 characters per page gives 1,000,000 character-spaces per book, so there are $100^{1,000,000}$ books in the Library of Babel. Since it is estimated[3] that there are only 100^{40} (give or take a few) *particles* (protons, neutrons, and electrons) in the region of the universe we can observe, the Library of Babel is not remotely a physically possible object, but, thanks to the strict rules with which Borges constructed it in his imagination, we can think about it clearly.

Is this truly the set of *all* possible books? Obviously not—since they are restricted to being printed from "only" 100 different characters, excluding, we may suppose, the characters of Greek, Russian, Chinese, Japanese, and Arabic, thereby overlooking many of the most important *actual* books. Of course, the Library does contain superb translations of all these actual books into English, French, German, Italian, . . . , as well as uncountable trillions of shoddy translations of each book. Books of more than 500 pages are there,

2. Borges chose slightly different figures: books 410 pages long, with 40 lines of 80 characters each. The total number of characters per book is close enough to mine (1,312,000 versus 1,000,000) to make no difference. I chose my rounder numbers for ease of handling. Borges chose a character set with only 25 members, which is enough for uppercase Spanish (with a blank, a comma, and a period as the only punctuation), but not for English. I chose the more commodious 100 to make room without any doubt for the upper- and lowercase letters and punctuation of all the Roman-alphabet languages.

3. Stephen Hawking (1988, p. 129) insists on putting it this way: "There are something like ten million million million million million million million million million million million million million (1 with eighty zeroes after it) particles in the region of the universe that we can observe." Denton (1985) provides the estimate of 10^{70} atoms in the observable universe. Eigen (1992, p. 10) calculates the volume of the universe as 10^{84} cubic centimeters.

beginning in one volume and continuing without a break in some other volume or volumes.

It is amusing to think about some of the volumes that must be in the Library of Babel somewhere. One of them is the best, most accurate 500-page biography of you, from the moment of your birth until the moment of your death. Locating it, however, would be all but impossible (that slippery word), since the Library also contains kazillions of volumes that are magnificently accurate biographies of you up till your tenth, twentieth, thirtieth, fortieth ... birthday, and completely false about subsequent events of your life—in a kazillion different and diverting ways. But even finding one readable volume in this huge storehouse is unlikely in the extreme.

We need some terms for the quantities involved. The Library of Babel is not infinite, so the chance of finding anything interesting in it is not literally infinitesimal.[4] These words exaggerate in a familiar way—we caught Darwin doing it in his summary, where he helped himself to an illicit "infinitely"—but we should avoid them. Unfortunately, all the standard metaphors—"astronomically large," "a needle in a haystack," "a drop in the ocean"—fall comically short. No *actual* astronomical quantity (such as the number of elementary particles in the universe, or the time since the Big Bang measured in nanoseconds) is even visible against the backdrop of these huge but finite numbers. If a readable volume in the Library were as easy to find as a particular drop in the ocean, we'd be in business! If you were dropped at random into the Library, your chance of ever encountering a volume with so much as a grammatical sentence in it would be so vanishingly small that we might do well to capitalize the term—"Vanishingly" small—and give it a mate, "Vastly," short for "Very-much-more-than-astronomically."[5]

Moby Dick is in the Library of Babel, of course, but so are 100,000,000 mutant impostors that differ from the canonical *Moby Dick* by a *single*

---

4. The Library of Babel is finite, but, curiously enough, it contains all the grammatical sentences of English within its walls. But that's an infinite set, and the library is finite! Still, any sentence of English, of whatever length, can be broken down into 500-page chunks, each of which is somewhere in the library! How is this possible? Some books may get used more than once. The most profligate case is the easiest to understand: since there are volumes that each contain a single character and are otherwise blank, repeated use of these 100 volumes will create any text of any length. As Quine points out in his informative and amusing essay "Universal Library" (in Quine 1987), if you avail yourself of this strategy of re-using volumes, and translate everything into the ASCII code your word-processor uses, you can store the whole Library of Babel in two extremely slender volumes, in one of which is printed a 0 and in the other of which appears a 1! (Quine also points out that the psychologist Theodor Fechner propounded the fantasy of the universal library long before Borges.)

5. Quine (1987) coins the term "hyperastronomic" for the same purpose.

typographical error. That's not yet a Vast number, but the total rises swiftly when we add the variants that differ by 2 or 10 or 1,000 typos. Even a volume with 1,000 typos—2 per page on average—would be unmistakably recognizable as *Moby Dick*, and there are Vastly many of those volumes. It wouldn't matter which of these volumes you found, if you could only find one of them. They would almost all be just about equally wonderful reading, and all tell the same story, except for truly negligible—almost indiscriminable—differences. Not quite all of them, however. Sometimes a single typo, in a crucial position, can be fatal. Peter De Vries, another philosophically delicious writer of fiction, once published a novel[6] that began:

"Call me, Ishmael."

Oh, what a single comma can do! Or consider the many mutants that begin: "Ball me Ishmael. . . ."

In Borges' story, the books are not shelved in any order, but even if we found them scrupulously alphabetized, we would have insoluble problems finding *the* book we were looking for (for instance, the "essential" version of *Moby Dick*). Imagine traveling by spaceship through the *Moby Dick* galaxy of the Library of Babel. This galaxy is in itself Vastly larger than the whole physical universe, so, no matter what direction you go in, for centuries on end, even if you travel at the speed of light, all you see are virtually indistinguishable copies of *Moby Dick*—you will never ever reach anything that looks like anything else. *David Copperfield* is unimaginably distant in this space, even though we know that there is a path—a shortest path, ignoring the kazillions of others—leading from one great book to the other by single typographical changes. (If you found yourself on this path, you would find it almost impossible to tell, by local inspection, which direction to go to move towards *David Copperfield*, even if you had texts of both target books in hand.)

In other words, this *logical* space is so Vast that many of our usual ideas about location, about searching and finding and other such mundane and practical activities, have no straightforward application. Borges put the books on the shelves in random order, a nice touch from which he drew several delectable reflections, but look at the problems he would have

---

6. *The Vale of Laughter* (1953). (It goes on: "Feel absolutely free to. Call me any hour of the day or night. . . .") De Vries also may have invented the game of seeing how large an effect (deleterious or not) you can achieve with a single typographical change. One of the best: "Whose woods are these, I think I know; his house is in the *V*illage though. . . ." Others have taken up the game: in the state of nature, mutant-Hobbes tells us, one finds "the *w*ife of man, solitary, poore, nasty, brutish, and short." Or consider the question: "Am I my brothe*l*'s keeper?"

created for himself if he'd tried to arrange them in alphabetical order in his honeycomb. Since there are only a hundred different alphabetic characters (in our version), we can treat some specific sequence of them as Alphabetical Order—e.g., a, A, b, B, c, C . . . z, Z, ?, ;, „ ., !,), (, %, . . . à, â, è, ê, é, . . . Then we can put all the books beginning with the same character on the same *floor*. Now our library is only 100 stories high, shorter than the World Trade Center. We can divide each floor into 100 *corridors*, each of which we line with the books whose second character is the same, one corridor for each character, in alphabetical order. On each corridor, we can place 100 *shelves*, one for each third-slot. Thus all the books that begin with "aardvarks love Mozart"—and how many there are!—are shelved on the same shelf (the "r" shelf) in the first corridor on the first floor. But that's a mighty long shelf, so perhaps we had better stack the books in file drawers at right angles to the shelf, one drawer for each fourth-letter position. That way, each shelf can be only, say, 100 feet long. But now the file drawers are awfully deep, and will run into the backs of the file drawers in the neighboring corridor, so . . . but we've run out of dimensions in which to line up the books. We need a million-dimensional space to store all the books neatly, and all we have is three dimensions: up-down, left-right, and front-back. So we will just have to pretend we can imagine a multidimensional space, each dimension running "at right angles" to all the others. We can conceive of such hyperspaces, as they are called, even if we can't visualize them. Scientists use them all the time to organize the expression of their theories. The geometry of such spaces (whether or not they count as only imaginary) is well behaved and well explored by mathematicians. We can confidently speak about locations, paths, trajectories, volumes (hypervolumes), distances, and directions in these logical spaces.

We are now prepared to consider a variation on Borges' theme, which I will call the *Library of Mendel*. This Library contains "all possible genomes"—DNA sequences. Richard Dawkins describes a similar space, which he calls "Biomorph Land," in *The Blind Watchmaker* (1986a). His discussion is the inspiration for mine, and our two accounts are entirely compatible, but I want to stress some points he chose to pass over lightly.

If we consider the Library of Mendel to be composed of *descriptions* of genomes, then it is already just a proper part of the Library of Babel. The standard code for describing DNA consists of only four characters, A, C, G, and T (standing for Adenine, Cytosine, Guanine, and Thymine, the four kinds of nucleotides that compose the letters of the DNA alphabet). All the 500-page permutations of these four letters are already in the Library of Babel. Typical genomes are much longer than ordinary books, however. Taking the current estimate of 3×10^9 nucleotides in the human genome, the exhaustive description of a single human genome—such as your own— would take approximately 3,000 of the 500-page volumes in the Library of

Babel (keeping print size the same).[7] The description of the genome for a horse (flying or not) or a cabbage or an octopus would be composed of the same letters, A, C, G, and T, and certainly not much longer, so we can suppose, arbitrarily, that the Library of Mendel consists of all the DNA strings described in all the 3,000-volume boxed sets consisting entirely of those four characters. This will capture enough of the "possible" genomes to serve any serious theoretical purpose.

I overstated the case in describing the Library of Mendel as containing "all possible" genomes, of course. Just as the Library of Babel ignored the Russian and Chinese languages, so the Library of Mendel ignores the (apparent) possibility of alternative genetic alphabets—based on different chemical constituents, for instance. We are *still* beginning in the middle, making sure we understand today's local, earthly circumstances before casting our nets wider. So any conclusions we come to regarding what is possible relative to *this* Library of Mendel may have to be reconsidered when we try to apply them to some broader notion of possibility. This is actually a strength rather than a weakness of our tactic, since we can keep close tabs on exactly what sort of modest, circumscribed possibility we are talking about.

One of the important features of DNA is that all the permutations of sequences of Adenine, Cytosine, Guanine, and Thymine are about equally stable, chemically. All could be constructed, in principle, in the gene-

---

7. The comparison of a human genome with the volumes in the galaxy of *Moby Dick* readily explains something that occasionally baffles people about the Human Genome Project. How can scientists speak of sequencing (copying down) *the* human genome if every human genome is different from every other in not just one but hundreds or thousands of places (*loci*, in the language of genetics)? Like the proverbial snowflakes, or fingerprints, no two actual human genomes are exactly alike, even those of identical twins (the chance of typos creeping in is always present, even in the cells of a single individual). Human DNA is readily distinguishable from the DNA of any other species, even that of the chimpanzee, which is over 90 percent the same at every locus. Every actual human genome that has ever existed is contained within a galaxy of possible human genomes that is Vastly distant from the galaxies of other species' genomes, yet within the galaxy there is plenty of room for no two human genomes to be alike. You have two versions of each of your genes, one from your mother and one from your father. They passed on to you exactly half of their own genes, randomly selected from those they received from their parents, your grandparents, but since your grandparents were all members of *Homo sapiens*, their genomes agree at almost all loci, so it makes no difference most of the time which grandparent provides either of your genes. But their genomes nevertheless differ at many thousands of loci, and in those slots, which genes you get is a matter of chance—a coin-toss built into the machinery for forming your parents' contributions to your DNA. Moreover, mutations accumulate at the rate of about 100 per genome per generation in mammals. "That is, your children will have one hundred differences from you and your spouse in their genes as a result of random copying errors by your enzymes or as a result of mutations in your ovaries or testicles caused by cosmic rays" (Matt Ridley 1993, p. 45).

splicing laboratory, and, once constructed, would have an indefinite shelf life, like a book in a library. But not every such sequence in the Library of Mendel corresponds to a viable organism. Most DNA sequences—the Vast majority—are surely gibberish, recipes for nothing living at all. That is what Dawkins means, of course, when he says there are many more ways of being dead (or not alive) than ways of being alive. But what kind of a fact is this, and why should it be so?

3. THE COMPLEX RELATION BETWEEN GENOME AND ORGANISM

If we are going to try to make progress by boldly oversimplifying, we should at least alert ourselves to some of the complications we are temporarily setting aside. I see three main sorts of complexity we should acknowledge and keep an eye on as we proceed, even if we are once again postponing their full discussion.

The first concerns the "reading" of the "recipe." The Library of Babel presupposed readers: the people who inhabited the Library. Without them, the very idea of the collection of volumes would make no sense at all; their pages might as well be smeared with jam or worse. If we are to make any sense of the Library of Mendel, we must also presuppose something analogous to readers, for without readers DNA sequences don't *specify* anything at all—not blue eyes or wings or anything else. Deconstructionists will tell you that no two readers of a text will come up with the same reading, and something similar is undoubtedly true when we consider the relationship between a genome and the embryonic environment—the chemical microenvironment as well as the surrounding support conditions—in which it has its informational effects. The immediate effect of the "reading" of DNA during the creation of a new organism is the fabrication of many different proteins out of amino acids (which have to be on hand in the vicinity, of course, ready to be linked together). There are Vastly many possible proteins, but which become actual depends on the DNA text. These proteins get created in strict sequence, and in amounts determined by the "words"—triplets of nucleotides—as they are "read." So, for a DNA sequence to specify what it is supposed to specify, there must be an elaborate reader-constructor, well stocked with amino-acid building blocks.[8] But that is just a small part of the process. Once the proteins get created, they have to be

---

8. This is an oversimplification, leaving out the role of messenger RNA and other complications.

brought into the right relations with each other. The process begins with a single fertilized cell, which then divides into two daughter cells, which divide again, and so forth (each with its own duplicate copy of all the DNA that is being read, of course). These newly formed cells, of many different varieties (depending on which proteins are jiggled into which places in which order), must in turn migrate to the right locations in the embryo, which grows by dividing and dividing, building, rebuilding, revising, extending, repeating, and so forth.

This is a process that is only partly controlled by the DNA, which in effect *presupposes* (and hence does not itself *specify*) the reader and the reading process. Compare genomes to musical scores. Does a written score of Beethoven's Fifth Symphony *specify* that piece of music? Not to Martians, it wouldn't, because it presupposes the existence of violins, violas, clarinets, trumpets. Suppose we take the score and attach a sheaf of directions and blueprints for making (and playing) all the instruments, and send the whole package to Mars. Now we are getting closer to a package that could in principle be used to re-create Beethoven's music on Mars. But the Martians would still have to be able to decipher the recipe, make the instruments, and then play them as the score directed.

This is what makes the story of Michael Crichton's novel *Jurassic Park* (1990)—and the Steven Spielberg movie made of it—a fantasy: even completely intact dinosaur DNA would be powerless to re-create a dinosaur without the aid of a dinosaur-DNA-reader, and those are just as extinct as dinosaurs (they are, after all, the ovaries of dinosaurs). If you *have* a (living) dinosaur ovary, then it, together with dinosaur DNA, can specify *another* dinosaur, another dinosaur ovary, and so forth indefinitely, but dinosaur DNA by itself, even complete dinosaur DNA, is only half (or, depending on how you count, maybe less than half) the equation. We might say that every species that has ever existed on this planet has had its own dialect of DNA-reading. Still, these dialects have had a lot in common with each other. The principles of DNA-reading are apparently uniform across all species, after all. That is what makes genetic engineering possible; the organismic effect of a particular permutation in DNA can often be predicted in practice. So the idea of bootstrapping our way back to a dinosaur-DNA-reader is a coherent idea, however improbable. With a helping of poetic license, the film-makers might pretend that acceptable substitute readers could be found (introduce the dinosaur-DNA text to the DNA-reader in a frog, and hope for the best).[9]

---

9. The film-makers never really address the problem of the DNA-*reader* at all, and use frog DNA just to patch the missing parts of the dinosaur DNA. David Haig has pointed out to me that this choice of a frog by the film-makers manifests an interesting error—an

We will cautiously help ourselves to some poetic license, too. Suppose we proceed *as if* the Library of Mendel were equipped with a single or standard DNA-reader that can equally well turn out a turnip or a tiger, depending on the recipe it finds in one of the genome volumes. This is a brutal oversimplification, but later we can reopen the question of the developmental or embryological complications.[10] Whatever standard DNA-reader we choose, relative to it the Vast majority of DNA sequences in the Library of Mendel will be utter gibberish. Any attempt to "execute" such a recipe for creating a viable organism would quickly terminate in absurdity. We wouldn't change this picture appreciably if instead we imagined there to be millions of different dialects of DNA-readers, analogous to the different actual languages represented in the Library of Babel. In that Library, the English books may be gibberish to the Polish readers and vice versa, but Vastly most of the volumes are gibberish to all readers. Take any one volume at random, and no doubt we can imagine that it is composed in a language, Babelish, in which it tells a wonderful tale. (Imagination is cheap if we don't have to bother with the details.) But if we remind ourselves that real languages have to be compact and *practical* things, with short, easily read sentences that depend on systematic regularity to get their messages across, we can assure ourselves that, compared with the Vast variety of texts in the Library, the possible languages are Vanishingly few. So we might as well pretend, for the time being, that there was just one language, just one sort of reader.

The second complexity we may acknowledge and postpone concerns viability. A tiger is viable *now*, in certain existing environments on our planet, but would not have been viable in most earlier days, and may become inviable in the future (as may all life on Earth, in fact). Viability is relative to the environment in which the organism must make its living. Without breathable atmosphere and edible prey—to take the most obvious conditions—the organic features that make tigers viable today would be to no avail. And since environments are to a great extent composed of, and by,

---

instance, he suggests, of the Great Chain of Being fallacy. "Humans, of course, are more closely related to dinosaurs than either is to frogs. Human DNA would have been better than frog DNA. Bird DNA would be better still."

10. A recent theme often heard among evolutionary theorists is that the "gene centrism" that is more or less standard these days has gone too far. According to this complaint, orthodoxy vastly overestimates the extent to which the DNA can be considered to be a recipe, composed of genes, specifying a phenotype or an organism. Those who make this claim are the deconstructionists of biology, elevating the reader to power by demoting the text. It is a useful theme as an antidote to oversimplified gene centrism, but in overdose it is about as silly as deconstructionism in literary studies. This will be given more attention in chapter 11.

the *other* organisms extant, viability is a constantly changing property, a moving target, not a fixed condition. This problem is minimized if we join Darwin in starting in the middle, with currently existing environments, and extrapolate cautiously to earlier and later possibilities. We can leave till later a consideration of the initial bootstrapping that may (or must) have happened to set this coevolution of organisms and their environments in motion.

The third complexity concerns the relationship between the texts of the genomes that do determine viable organisms, and the features those organisms exhibit. As we have already noted several times in passing, there is no *simple* mapping of nucleotide "words" onto Mendelian genes—putative carriers of the "specs" (as an engineer would say) *for* one feature or another. It is simply not the case that there is a sequence of nucleotides that spells "blue eyes" or "webbed feet" or "homosexual" in any descriptive language. And you can't spell "firm" or "flavorful" in the language of tomato DNA—even though you *can* revise the nucleotide sequence in that language so that the effect is firmer, more flavorful tomatoes.

When this complication is acknowledged, it is usually pointed out that genomes are not like descriptions or blueprints of finished products, but more like recipes for building them. This does not mean, as some critics have contended, that it is always—or even ever—a mistake to speak of a gene *for* this or that. The presence or absence of an instruction in a recipe can make a typical and important difference, and whatever difference it makes may be correctly described as what the instruction—the gene—is "for." This point has been so frequently and influentially missed by the critics that it is worth pausing to expose its error vividly. Richard Dawkins has come up with an example that does this so well that it is worth quoting in full (it also highlights the importance of the second of our complications, the relativity of viability to environment):

> Reading is a learned skill of prodigious complexity, but this provides no reason in itself for scepticism about the possible existence of a gene for reading. All we would need in order to establish the existence of a gene for reading is to discover a gene for not reading, say a gene which induced a brain lesion causing specific dyslexia. Such a dyslexic person might be normal and intelligent in all respects except that he could not read. No geneticist would be particularly surprised if this type of dyslexia turned out to breed true in some Mendelian fashion. Obviously, in this event, the gene would only exhibit its effect in an environment which included normal education. In a prehistoric environment it might have had no detectable effect, or it might have had some different effect and have been known to cave-dwelling geneticists as, say, a gene for inability to read animal footprints. In our educated environment it would properly be called a gene 'for' dyslexia, since dyslexia would be its most salient consequence.

Similarly, a gene which caused total blindness would also prevent reading, but it would not usefully be regarded as a gene for not reading. This is simply because preventing reading would not be its most obvious or debilitating phenotypic effect. [Dawkins 1982, p. 23. See also Dawkins 1989a, pp. 281–82, and Sterelny and Kitcher 1988.]

The indirect way in which groups of codons—triplets of DNA nucleotides—instruct the building process does not prohibit us, then, from speaking of a gene for *x* or for *y*, using the familiar geneticists' shorthand, and bearing in mind that that is what we are doing. But it does mean that there may be fundamental differences between the space of genomes and the space of "possible" organisms. The fact that *we* can consistently describe a finished product—say, a giraffe with green stripes instead of brown blotches—does not guarantee that there is a DNA recipe for making it. It may just be that, because of the peculiar requirements of development, there simply is no starting point in DNA that has such a giraffe as its destination.

This may seem very implausible. What could be impossible about a giraffe with green stripes? Zebras have stripes, drakes have green feathers on their heads—there is nothing biologically impossible about the properties in isolation, and surely they can be put together in one giraffe! So you'd think. But you must not count on it. You'd probably also think a striped animal with a spotted tail was possible, but it may well not be. James Murray (1989) has developed mathematical models that show how the developmental process of distributing color on animals could readily make a spotted animal with a striped tail, but not vice versa. This is suggestive, but not yet—as some have rashly said—a strict proof of impossibility. Anyone who had learned how to build a tiny ship in a bottle—a hard enough trick—might think it was flat impossible to put a whole fresh pear in a narrow-necked bottle, but it isn't; witness the bottles of Poire William liqueur. How is it done? Could the molten glass somehow be blown around a pear without scorching it? No, the bottles are hung on the trees in the spring so that the pears can grow inside them. Proving that there is no *straightforward* way for biology to accomplish some trick is never a proof of impossibility. Remember Orgel's Second Rule!

In his account of Biomorph Land, Dawkins stresses that a tiny—indeed minimal—change in the genotype (the recipe) can produce a strikingly large change in the phenotype (the resulting individual organism), but he tends to slight one of the major implications of this: if a single step in the genotype can produce a giant step in the phenotype, intermediate steps for the phenotype may be simply unavailable, given the mapping rules. To take a deliberately extreme and fanciful example, you might think that if a beast could have twenty-centimeter tusks and forty-centimeter tusks, it would stand to reason that it could also have thirty-centimeter tusks, but the rules for tusk-making in the recipe system may not allow for such a case. The species in question might have to "choose" between tusks ten centimeters "too short" or ten centimeters "too long." This means that arguments that proceed from engineering assumptions about which design would be the optimal or best design must be extremely cautious in assuming that what seems intuitively to be available or possible is actually accessible in the organism's design space, given the way it reads its recipes. (This will be a major topic in chapters 8, 9, and 10.)

4. POSSIBILITY NATURALIZED

With the help of the Library of Mendel, we can now resolve—or at least unite under a single perspective—some of the nagging problems about "biological laws" and what is possible, impossible, and necessary in the world. Recall that we needed to get clear about these issues because, if we are to explain the way things *are*, it must be against a background of how things *might* have been, or *must* be, or *couldn't* be. We can now define a restricted concept of biological possibility:

> *x* is biologically possible if and only if *x* is an instantiation of an accessible genome or a feature of its phenotypic products.

Accessible from where? By what processes? Ah, there's the rub. We have to specify a starting point in the Library of Mendel, and a means of "travel." Suppose we were to start where we are today. Then we will be talking, first, about what is possible *now*—that is to say, in the near future, using whatever means of travel are currently available to us. We count as possible all *actual* contemporary species and all their features—including the features they have in virtue of their relations with other species and their features—plus anything that can be obtained by traveling from that broad front either just "in the course of nature"—without human manipulation—or with the help of such artificial cranes as the techniques of traditional animal-breeding (and, for that matter, surgery), or via the fancy new vehicles of genetic engineering. After all, we human beings and all our tricks are just another product of the contemporary biosphere. Thus it is biologically possible for you to have a fresh turkey dinner on Christmas Day, 2001, if and only if at least one instantiated turkey genome has produced the requisite phenotypic effects in time for dinner. It is biologically possible for you to ride a pteranodon before you die if and only if *Jurassic Park*-ish technology permits *that* sort of genome to get expressed in time.

No matter how we set these "travel" parameters, the resulting notion of biological possibility will have an important property: some things will be "more possible" than others—that is, nearer in the multidimensional search space, and more accessible, "easier" to get to. Things that would have been viewed as biological impossibilities just a few years ago—such as plants that glowed in the dark in virtue of having firefly genes in them—are now not only possible but actual. Are twenty-first-century dinosaurs possible? Well, the vehicles for *getting there* from here have been developed to the point where we can at least tell a cracking good story—one requiring remarkably little poetic license. ("There" is a portion of the Library of Mendel through which the Tree of Life stopped meandering about sixty million years ago.)

What rules govern travel through this space? What rules or laws constrain the relations between genomes and their phenotypic products? So far, all we have acknowledged are logical or mathematical necessities on the one hand and the laws of physics on the other. That is, we have proceeded as if we knew what both logical possibility and (mere) physical possibility were. These are difficult and controversial issues, but we may consider them *clamped:* we simply assume some fixed version of those varieties of possibility and necessity, and then develop our restricted notion of biological possibility in terms of it. The law of large numbers and the law of gravity, for instance, are both deemed to hold unreservedly and timelessly over the space. Clamping physical law lets us say flat out, for instance, that all the different genomes are physically possible—because chemistry says they are all stable, if encountered.

Keeping logic and physics and chemistry clamped, we could choose a different starting point. We could choose some moment on Earth five billion years ago, and consider what was biologically possible then. Not much, because before tigers could become possible (on Earth), eukaryotes, and then plants producing atmospheric oxygen in large quantities, and many other things had to become actual. With hindsight, we can say that tigers were in fact possible all along, if distant and extremely improbable. One of the virtues of this way of thinking of possibility is that it joins forces with probability, thus permitting us to trade in flat all-or-nothing claims about possibility for claims about relative distance, which is what matters for most purposes. (The all-or-nothing claims of biological possibility were all but impossible [hmm, that word again] to adjudicate, so this is no loss.) As we saw in our exploration of the Library of Babel, it doesn't make much difference what our verdict is about whether it is "possible in principle" to find some particular volume in that Vast space. What matters is what is practically possible, in one or another sense of "practical"—take your pick.

This is certainly not a standard definition of possibility, or even a standard *sort* of definition of possibility. The idea that some things might be "more possible" than others (or more possible from over here than from over there) is at odds with one standard understanding of the term, and some philosophical critics might say that this is simply not a definition of *possi-*

bility, whatever it is. Some other philosophers have defended views of comparative possibility (see especially Lewis 1986, pp. 10ff.), but I don't want to fight over it. If this is not an account of possibility, so be it. It is, then, a proposed *replacement* for a definition of possibility. Perhaps after all we don't need the concept of biological possibility (with its required all-or-nothing application) for any serious investigative purpose. Perhaps degree of accessibility in the space of the Library of Mendel is all we need, and is in fact a better concept than any all-or-nothing version could be. It would be nice, for instance, to have some way of *ranking* the following in terms of biological possibility: ten-pound tomatoes, aquatic dogs, flying horses, flying trees.

That will not be enough to satisfy many philosophers, and their objections are serious. Briefly considering them will at least make it clearer what I am claiming and what I am not claiming. First of all, isn't there something viciously circular about defining possibility in terms of access*ibility*? (Doesn't the latter term just reintroduce the former in its suffix, and still undefined?) Well, not quite. It does leave some definitely unfinished business, which I will simply acknowledge before moving on. We have supposed that we are holding some concept or other of *physical* possibility clamped for the time being; our idea of accessibility presupposes that this physical possibility, whatever it is, leaves us *some* elbow room—some openness of pathways (not just a single pathway) in the space. In other words, we are taking on the assumption that *nothing stops us* from going down any of the pathways that are open so far as physics is concerned.[11]

Quine's questions (at the head of this chapter) invited us to worry about

---

11. This idea of elbow room is something we need to presuppose in any case, for it is the minimal denial of actualism, the doctrine that only the actual is possible. David Hume, in *A Treatise of Human Nature* (1739), spoke of "a certain looseness" we want to exist in our world. This is the looseness that prevents the possible from shrinking tightly around the actual. This looseness is presupposed by *any* use of the word "can"—a word we can hardly do without! Some people have thought that, if determinism were true, actualism would be true—or, to turn it around, if actualism is *false*, *in*determinism must be true—but this is highly dubious. The implied argument against determinism would be disconcertingly simple: this oxygen atom has valence 2; therefore, it can unite with two hydrogen atoms to form a molecule of water (it *can* right now, whether or not it does); therefore, something is possible that isn't actual, so determinism is false. There are impressive arguments from physics that lead to the conclusion that determinism is false—but this isn't one of them. I am prepared to assume that actualism is false (and that this assumption is independent of the determinism/indeterminism question), even if I can't claim to prove it, if only because the alternative would be to give up and go play golf or something. But for a somewhat fuller discussion of actualism, see my book *Elbow Room* (1984), especially ch. 6, "Could Have Done Otherwise," from which material in this note is drawn. See also David Lewis' (1986, ch. 17) concurring opinion, about the related issue of the irrelevance of the issue of indeterminism to our sense that the future is "open."

whether we could count nonactual possible objects. One of the virtues of the proposed treatment of biological possibility is that, thanks to its "arbitrary" formal system—the system arbitrarily imposed on us by nature, at least in our neck of the woods—we can count the different nonactual possible genomes; they are Vast but finite in number, and no two are exactly alike. (By definition, genomes are distinct if they fail to share a nucleotide at any one of several billion loci.) In what sense are the nonactual genomes *really* possible? Only in this sense: *if* they were formed, they'd be stable. But whether or not any conspiracy of events could lead to their being formed is another matter, to be addressed in terms of accessibility from one location or another. Most of the genomes in this set of stable possibilities will *never* be formed, we can be sure, since the heat death of the universe will overtake the building process before it has made a sizable dent in the space.

Two other objections to this proposal about biological possibility cry out to be heard. First, isn't it outrageously "gene-centered," in anchoring *all* considerations of biological possibility to the accessibility of one genome or another in the Library of Mendel? Our proposed treatment of biological possibility flatly ignores (and hence implicitly rules impossible) "creatures" that are not end points of some branch of the Tree of Life that has already taken us as far as we are today. But that just *is* the grand unification of biology that Darwin discovered! Unless you harbor fantasies about spontaneous creation of new life forms by "Special Creation" or (the philosophers' secular version) "Cosmic Coincidence," you accept that every feature of the biosphere is one fruit or another of the Tree of Life (or, if not *our* Tree of Life, some other Tree of Life, with its own accessibility relations). No man is an island, John Donne proclaims, and Charles Darwin adds that neither is any clam or tulip—every *possible* living thing is connected by isthmuses of descent to all other living things. Notice that this doctrine rules *in* whatever marvels technology can produce in the future, provided—as we have already noted—that technologists themselves, and their tools and methods, are firmly located on the Tree of Life. It is a small further step to rule in life forms from outer space, provided they, too, are the products of a Tree of Life rooted, as ours is, in some nonmiraculous physical ground. (This topic will be explored in chapter 7.)

Second, why should we treat biological possibility so differently from physical possibility? If we assume that "laws of physics" fix the limits of physical possibility, why shouldn't we attempt to define biological possibility in terms of "laws of biology"? (We will turn to an examination of physical laws and physical necessity in chapter 7, but in the meantime, the difference appears large.) Many biologists and philosophers of science have maintained that there are biological laws. Doesn't the proposed definition rule them out? Or does it declare them superfluous? It doesn't rule them out. It permits someone to argue for the dominion of some law of biology over the space of the Library of Mendel, but it does put a difficult burden of proof on anyone who thinks that there are laws of biology *over and above* the laws of mathematics and physics. Consider the fate of "Dollo's Law," for instance.

> 'Dollo's Law' states that evolution is irreversible. . . . [But] There is no reason why general trends in evolution shouldn't be reversed. If there is a trend towards large antlers for a while in evolution, there can easily be a subsequent trend towards smaller antlers again. Dollo's Law is really just a statement about the statistical improbability of following exactly the same evolutionary trajectory twice (or indeed any *particular* trajectory), in either direction. A single mutational step can easily be reversed. But for larger numbers of mutational steps . . . the mathematical space of all possible trajectories is so vast that the chance of two trajectories ever arriving at the same point becomes vanishingly small. . . . There is nothing mysterious or mystical about Dollo's Law, nor is it something that we go out and 'test' in nature. It follows simply from the elementary laws of probability. [Dawkins 1986a, p. 94.]

There is no shortage of candidates for the role of "irreducible biological law." For instance, many have argued that there are "developmental laws" or "laws of form" that constrain the relation between genotype and phenotype. In due course we will consider their status, but already we can locate at least some of the most salient constraints on biological possibility as not "laws of biology" but just inescapable features of the geometry of design space, like Dollo's Law (or the Hardy-Weinberg Law of gene frequency, which is another application of probability theory, pure and simple).

Take the case of the horned birds. As Maynard Smith notes, there aren't any, and we don't know why. Might it be because they are ruled out by a biological *law*? Are horned birds flat impossible? Would they have to be inviable in any and all possible environments, or is there simply no path at all "from here to there" because of restrictions on the genome-reading process? As we have already noted, we should be impressed by the severe restrictions encountered by this process, but we should not be carried away. Those restrictions may not be a *universal* feature, but a temporally and spatially local feature, analogous to what Seymour Papert has dubbed the QWERTY phenomenon in the culture of computers and keyboards.

> The top row of alphabetic keys of the standard typewriter reads QWERTY. For me this symbolizes the way in which technology can all too often serve not as a force for progress but for keeping things stuck. The QWERTY arrangement has no rational explanation, only a historical one. It was introduced in response to a problem in the early days of the typewriter: The keys used to jam. The idea was to minimize the collision problem by separating those keys that followed one another frequently. . . . Once

adopted, it resulted in many millions of typewriters and ... the social cost of change ... mounted with the vested interest created by the fact that so many fingers now knew how to follow the QWERTY keyboard. QWERTY has stayed on despite the existence of other, more "rational" systems. [Papert 1980, p. 33.][12]

The imperious restrictions we encounter inside the Library of Mendel may look like universal laws of nature from our myopic perspective, but from a different perspective they may appear to count as merely local conditions, with historical explanations.[13] If so, then a restricted concept of biological possibility is the sort we want; the ideal of a universal concept of biological possibility will be misguided. But as I have already allowed, this does not rule out biological laws; it merely sets the burden of proof for those who want to propose any. And in the meantime, it gives us a framework for describing large and important classes of regularity we discover in the patterns in *our* biosphere.

CHAPTER 5: *Biological possibility is best seen in terms of accessibility (from some stipulated location) in the Library of Mendel, the logical space of all genomes. This concept of possibility treats the connectedness of the Tree of Life as a fundamental feature of biology, while leaving it open that there may also be biological laws that will also constrain accessibility.*

CHAPTER 6: *The R and D done by natural selection in the course of creating actual trajectories in the Vast space of possibilities can be measured to some extent. Among the important features of this search space are the solutions to problems that are perennially attractive and hence predictable, like forced moves in chess. This explains some of our intuitions about original- ity, discovery, and invention, and also clarifies the logic of Darwinian infer- ence about the past. There is a single, unified Design Space in which the processes of both biological and human creativity make their tracks, using similar methods.*

---

12. Others have exploited the QWERTY phenomenon to make similar points: David 1985, Gould 1991a.

13. George Williams (1985, p. 20) puts it this way: "I once insisted that '... the laws of physical science plus natural selection can furnish a complete explanation for any bio- logical phenomenon' [Williams 1966, pp. 6–7]. I wish now I had taken a less extreme view and merely identified natural selection as the only theory that a biologist needs in addition to those of the physical scientist. Both the biologist and the physical scientist need to reckon with historical legacies to explain any real-world phenomenon."

CHAPTER SIX

Threads of Actuality in Design Space

～ル～

1. DRIFTING AND LIFTING THROUGH DESIGN SPACE

The actual animals that have ever lived on Earth are a tiny subset of the theoretical animals that could exist. These real animals are the products of a very small number of evolutionary trajectories through genetic space. The vast majority of theoretical trajectories through animal space give rise to impossible monsters. Real animals are dotted around here and there among the hypothetical monsters, each perched in its own unique place in genetic hyperspace. Each real animal is surrounded by a little cluster of neighbours, most of whom have never existed, but a few of whom are its ancestors, its descendants and its cousins.

—RICHARD DAWKINS 1986a, p. 73

The actual genomes that have ever existed are a Vanishingly small subset of the combinatorially possible genomes, just as the actual books in the world's libraries are a Vanishingly small subset of the books in the imaginary Library of Babel. As we survey the Library of Babel, we may be struck by how hard it is to specify a *category* of books that isn't Vast in membership, however Vanishingly small it is in relation to the whole. The set of books composed entirely of grammatical English sentences is a Vast but Vanishing subset, and the set of readable, sense-making books is a Vast but Vanishing subset of it. Vanishingly hidden in that subset is the Vast set of books about people named Charles, and within that set (though Vanishingly hard to find) is the Vast set of books purporting to tell the truth about Charles Darwin, and a Vast but Vanishing subset of these consists of books composed entirely in limericks. So it goes. The number of *actual* books about Charles Darwin is a huge number, but not a Vast number, and we won't get down to that set (that set as of today, or as of the year 3000 A.D.) by just piling on the restricting adjectives in this fashion. To get to the actual books, we have to turn to the historical process that created them, in all its grubby particularity. The same is true of the actual organisms, or their actual genomes.

We don't need laws of biology to "prevent" most of the physical possibilities from becoming actualities; sheer absence of opportunity will account for most of them. The only "reason" *all* your nonactual aunts and uncles never came into existence is that your grandparents didn't have time or energy (to say nothing of the inclination) to create more than a few of the nearby genomes. Among the many nonactual possibles, some are—or were—"more possible" than others: that is, their appearance was more *probable* than the appearance of others, simply because they were *neighbors* of actual genomes, only a few choices away in the random zipping-up process that puts together the new DNA volume from the parent drafts, or only one or a few random typos away in the great copying process. Why didn't the near-misses happen? No reason; they just didn't happen to happen. And then, as the actual genomes that *did* happen to happen began to move away from the locations in Design Space of the near-misses, their probability of ever happening grew smaller. They were so close to becoming actual, and then their moment passed! Will they get another chance? It is possible, but Vastly improbable, given the Vast size of the space in which they reside.

But what forces, if any, bend the paths of actuality farther and farther away from their locations? The motion that occurs if there are no forces at all is called random genetic drift. You might think that drift, being random, would tend always to cancel itself out, bringing the path back to the same genomes again and again in the absence of any selective forces, but the very fact that there is only limited sampling in the huge space (which has a million dimensions, remember!) leads inevitably to the accumulation of "distance" between actual genomes (the upshot of "Dollo's Law").

Darwin's central claim is that when the force of natural selection is imposed on this random meandering, in addition to drifting there is lifting. Any motion in Design Space can be measured, but the motion of random drift is, intuitively, merely sideways; it doesn't get us anywhere important. Considered as R-and-D work, it is idle, leading to the accumulation of mere *typographical change*, but not to the accumulation of *design*. In fact, it is worse than that, for most mutations—typos—will be neutral, and most of the typos that aren't neutral will be deleterious. In the absence of natural selection, the drift is inexorably *downward* in Design Space. The situation in the Library of Mendel is thus precisely like the situation in the Library of Babel. Most typographical changes to *Moby Dick* can be supposed to be practically neutral—as good as invisible to most readers; of the few that make a difference, most will *do damage* to the text, making it a worse, less coherent, less comprehensible tale. Consider as an exercise, however, the version of Peter De Vries' game in which the object is to *improve* a text by a single typographical change. It is not impossible, but far from easy!

These intuitions about getting somewhere important, about design *improvement*, about *rising* in Design Space, are powerful and familiar, but are they reliable? Are they perhaps just a confusing legacy of the pre-Darwinian vision of Design coming down from a Handicrafter God? What is the relationship between the ideas of Design and Progress? There is no fixed agreement among evolutionary theorists about this. Some biologists are fastidious, going to great lengths to avoid allusions to design or function in their own work, while others build their whole careers around the functional analysis of this or that (an organ, patterns of food-gathering, reproductive "strategies," etc.). Some biologists think you can speak of design or function without committing yourself to any dubious doctrine about progress. Others are not so sure. Did Darwin deal a "death blow to Teleology," as Marx exclaimed, or did he show how "the rational meaning" of the natural sciences was to be empirically explained (as Marx went right on to exclaim), thereby making a safe home in science for functional or teleological discussion?

Is Design something that can be measured, even indirectly and imperfectly? Curiously enough, skepticism about this prospect actually undercuts the most potent source of skepticism about Darwinism. As I pointed out in chapter 3, the most powerful challenges to Darwinism have always taken this form: are Darwinian mechanisms powerful enough, or efficient enough, to have *done all that work* in the time available? All what work? If the question concerned mere sideways drifting in the typographical space of possible genomes, the answer would be obvious and uncontroversial: Yes, there has been *much* more than enough time. The speed at which random drift should accumulate mere typographical distance can be calculated, giving us a sort of posted speed limit, and both theory and observation agree that actual evolution happens much slower than that.[1] The "products" that are impressive to the skeptics are not the diverse DNA strings in themselves, but the amazingly intricate, complex, and *well-designed* organisms whose genomes those strings are.

---

1. See, for instance, the discussion in Dawkins 1986a, pp. 124–25, which concludes: "Conversely, strong 'selection pressure', we could be forgiven for thinking, might be expected to lead to rapid evolution. Instead, what we find is that natural selection exerts a braking effect on evolution. The baseline rate of evolution, in the absence of natural selection, is the maximum possible rate. That is synonymous with the mutation rate."

No analysis of the genomes in isolation of the organisms they create could yield the dimension we are looking for. It would be like trying to define the difference between a good novel and a great novel in terms of the relative frequencies of the alphabetical characters in them. We have to look at the whole organism, in its environment, to get any purchase on the issue. As William Paley saw, what is truly impressive is the bounty of astonishingly ingenious and smoothly functioning arrangements of matter that go to compose living things. And when we turn to examining the organism, we find again that no mere tabulation of the items composing it is going to give us what we want.

What could be the relationship between amounts of complexity and amounts of design? "Less is more," said the architect Ludwig Mies van der Rohe. Consider the famous British Seagull outboard motor, a triumph of simplicity, a design that honors the principle that what isn't there can't break. We want to be able to acknowledge—and even measure, if possible—the design excellence manifest in the right sort of simplicity. But what is the right sort? Or what is the right sort of *occasion* for simplicity? Not every occasion. Sometimes more *is* more, and of course what makes the British Seagull so wonderful is that it is such an elegant marriage of complexity and simplicity; nobody has quite such high regard, nor should they, for a paddle.

We can begin to get a clear view of this if we think about convergent evolution and the occasions on which it occurs. And, as is so often the case, choosing extreme—and imaginary—examples is a good way of focusing on what counts. In this instance, a favorite extreme case to consider is extra-terrestrial life, and of course it may someday soon be turned from fantasy into fact, if SETI, the ongoing Search for Extra-Terrestrial Intelligence, finds anything. If life on Earth is massively contingent—if its mere occurrence in any form at all is a happy accident—then what can we say, if anything, about life on other planets in the universe? We can lay down *some* conditions with confidence approaching certainty. These at first appear to fall into two contrasting groups: necessities and what we might call "obvious" optimalities.

Let's consider a necessity first. Life anywhere would consist of entities with autonomous metabolisms. Some people would say this is "true by definition." By defining life in this way, they can exclude the viruses as living forms, while keeping the bacteria in the charmed circle. There may be good reasons for such a definitional fiat, but I think we see more clearly the importance of autonomous metabolism if we see it as a deep if not utterly necessary condition for the sort of complexity that is needed to fend off the gnawing effects of the Second Law of Thermodynamics. All complex macromolecular structures tend to break down over time, so, unless a system is an *open* system, capable of taking in fresh materials and replenishing itself, it will tend to have a short career. The question "What does it live on?"

might get wildly different answers on different planets, but it does not betray a "geocentric"—let alone "anthropocentric"—assumption.

What about vision? We know that eyes have evolved independently many times, but vision is certainly not a necessity on Earth, since plants get along fine without it. A strong case can be made, however, that *if* an organism is going to further its metabolic projects by locomoting, and *if* the medium in which the locomoting takes place is transparent or translucent and amply supplied by ambient light, then *since* locomoting *works much better* (at furthering self-protective, metabolic, and reproductive aims) if the mover is guided by information about distal objects, and *since* such information can be garnered in a high-fidelity, low-cost fashion by vision, vision is a very good bet. So we would not be surprised to find that locomoting organisms on other planets (with transparent atmospheres) had eyes. Eyes are an obviously good solution to a very general problem that would often be encountered by moving metabolizers. Eyes may not always be "available," of course, for QWERTY reasons, but they are obviously rational solutions to this highly abstract design problem.

2. FORCED MOVES IN THE GAME OF DESIGN

Now that we have encountered this appeal to what is obviously rational under some general set of circumstances, we can look back and see that our case of necessity, having an autonomous metabolism, can be recast as simply the *only* acceptable solution to the *most general* design problem of life. If you wanna live, you gotta eat. In chess, when there is only one way of staving off disaster, it is called a *forced move*. Such a move is not forced by the rules of chess, and certainly not by the laws of physics (you can always kick the table over and run away), but by what Hume might call a "dictate of reason." It is simply dead obvious that there is one and only one solution, as anybody with an ounce of wit can plainly see. Any alternatives are immediately suicidal.

In addition to having an autonomous metabolism, any organism must also have a more or less definite boundary, distinguishing itself from everything else. This condition, too, has an obvious and compelling rationale: "As soon as something gets into the business of self-preservation, boundaries become important, for if you are setting out to preserve yourself, you don't want to squander effort trying to preserve the whole world: you draw the line" (Dennett 1991a, p. 174). We would also expect the locomoting organisms on an alien planet to have efficiently shaped boundaries, like those of organisms on Earth. Why? (Why on Earth?) If cost were no object, one might have no regard for streamlining in organisms that move through a relatively dense fluid, such as water. But cost is *always* an object—the Second Law of Thermodynamics guarantees that.

So at least some "biological necessities" may be recast as obvious solutions to most general problems, as *forced moves in Design Space*. These are cases in which, for one reason or another, there is only one way things can be done. But reasons can be deep or shallow. The deep reasons are the constraints of the laws of physics—such as the Second Law of Thermodynamics, or the laws of mathematics or logic.[2] The shallow reasons are just historical. There *used to be* two or more ways this problem might be solved, but now that some ancient historical accident has set us off down one particular path, only one way is remotely available; it has become a "virtual necessity," a necessity for all practical purposes, given the cards that have been dealt. The other options are no longer really options at all.

This marriage of chance and necessity is a hallmark of biological regularities. People often want to ask: "Is it merely a massively contingent fact that circumstances are as they are, or can we read some deep necessity into them?" The answer almost always is: Both. But note that the type of necessity that fits so well with the chance of random, blind generation is the necessity of *reason*. It is an inescapably teleological variety of necessity, the dictate of what Aristotle called *practical reasoning*, and what Kant called a *hypothetical imperative*:

> If you *want to achieve goal G*, then this is what you *must* do, given the circumstances.

The more universal the circumstances, the more universal the necessity. That is why we would not be surprised to find that the living things on other planets included locomotors with eyes, and why we would be more than surprised—utterly dumfounded—if we found things scurrying around on various projects but lacking any metabolic processes. But now let us consider the difference between the similarities that would surprise us and the similarities that would not. Suppose SETI struck it rich, and established communication with intelligent beings on another planet. We would not be surprised to find that they understood and used the same arithmetic that we do. Why not? Because arithmetic is *right*.

Might there not be different kinds of arithmetic-like systems, all equally good? Marvin Minsky, one of the founders of Artificial Intelligence, has

---

2. Are the constraints of pure logic deep or shallow? Some of each, I guess, depending on their obviousness. A delicious parody of adaptationist thinking is Norman Ellestrand's "Why are Juveniles Smaller Than Their Parents?" (1983), which explores with a heroically straight face a variety of "strategic" reasons for JSS (Juvenile Small Size). It ends with a brave look towards future research: "In particular, another juvenile character is even more widespread than JSS and deserves some thoughtful theoretical attention, the fact that juveniles *always* seem to be younger than their parents."

explored this curious question, and his ingeniously reasoned answer is No. In "Why Intelligent Aliens Will Be Intelligible," he offers grounds for believing in something he calls the

> Sparseness Principle: Whenever two relatively simple processes have products which are similar, those products are likely to be completely identical! [Minsky 1985a, p. 119, exclamation point in the original.]

Consider the set of *all possible processes*, which Minsky interprets *à la* the Library of Babel as all permutations of all possible computers. (Any computer can be identified, abstractly, as one "Turing machine" or another, and these can be given unique identifying numbers, and then put in numerical order, just like the alphabetical order in the Library of Babel.) Except for a Vanishing few, the Vast majority of these processes "do scarcely anything at all." So if you find "two" that do something similar (and worth noticing), they are almost bound to be one and the same process, at some level of analysis. Minsky (p. 122) applies the principle to arithmetic:

> From all this, I conclude that any entity who searches through the simplest processes will soon find fragments which do not merely resemble arithmetic but *are* arithmetic. It is not a matter of inventiveness or imagination, only a fact about the geography of the universe of computation, a world far more constrained than that of real things.

The point is clearly not restricted to arithmetic, but to all "necessary truths"—what philosophers since Plato have called *a priori* knowledge. As Minsky (p. 119) says, "We can expect certain '*a priori*' structures to appear, almost always, whenever a computation system evolves by selection from a universe of possible processes." It has often been pointed out that Plato's curious theory of reincarnation and reminiscence, which he offers as an explanation of the source of our *a priori* knowledge, bears a striking resemblance to Darwin's theory, and this resemblance is particularly striking from our current vantage point. Darwin himself famously noted the resemblance in a remark in one of his notebooks. Commenting on the claim that Plato thought our "necessary ideas" arise from the pre-existence of the soul, Darwin wrote: "read monkeys for preexistence" (Desmond and Moore 1991, p. 263).

We would not be surprised, then, to find that extra-terrestrials had the same unshakable grip on "2 + 2 = 4" and its kin that we do, but we would be surprised, wouldn't we, if we found them using the decimal system for expressing their truths of arithmetic. We are inclined to believe that our fondness for it is something of a historical accident, derived from counting on our two five-digit hands. But suppose they, too, have a pair of hands, each with five subunits. The "solution" of using-whatever-you've-got to count on is a fairly obvious one, if not quite in the forced-move category.[3] It would not be particularly surprising to find that our aliens had a *pair* of prehensile appendages, considering the good reasons there are for bodily symmetry, and the frequency of problems that require one thing to be manipulated relative to another. But that there should be five subunits on each appendage looks like a QWERTY phenomenon that has been deeply rooted for hundreds of millions of years—a mere historical happenstance that has restricted *our* options, but should not be expected to have restricted theirs. But perhaps we underestimate the rightness, the rationality, of having five subunits. For reasons we have not yet fathomed, it may be a Good Idea in general, and not merely something we are stuck with. Then it would not be amazing after all to find that our interlocutors from outer space had converged on the same Good Idea, and counted in tens, hundreds, and thousands.

We would be flabbergasted, however, to find them using the very symbols we use, the so-called arabic numerals: "1," "2," "3" . . . We know that right here on Earth there are perfectly fine alternatives, such as the Arabic numerals, "١," "٢," "٣," "٤" . . . as well as some not-so-viable alternatives, such as roman numerals, "*i*," "*ii*," "*iii*," "*iv*" . . . If we found the inhabitants of another planet using our arabic numerals, we would be quite sure that it was no coincidence—there *had* to be a historical connection. Why? Because the space of possible numeral shapes in which there is no *reason* for choosing one over the others is Vast; the likelihood of two independent "searches" ending up in the same place is Vanishing.

Students often have a hard time keeping clear about the distinction between numbers and numerals. Numbers are the abstract, "Platonic" objects that numerals are the names of. The arabic numeral "4" and the roman numeral "IV" are simply different *names* for one and the same thing—the *number* 4. (I can't talk about the number without naming it in one way or another, any more than I can talk about Clinton without using some word

---

3. Seymour Papert (1993, p. 90) describes observing a "learning disabled" boy in a classroom in which counting on your fingers was forbidden: "As he sat in the resource room I could see him itching to do finger manipulations. But he knew better. Then I saw him look around for something else to count with. Nothing was at hand. I could see his frustration grow. What could I do? . . . Inspiration came! I walked casually up to the boy and said out loud: 'Did you think about your teeth?' I knew instantly from his face that he got the point, and from the aide's face that she didn't. 'Learning disability indeed!' I said to myself. He did his sums with a half-concealed smile, obviously delighted with the subversive idea." (When considering using-whatever-you've-got as a possible forced move, it is worth recalling that not all peoples of our Earth have used the decimal system; the Mayans, for instance, used a base-20 system.)

or words that refer to him, but Clinton is a man, not a word, and numbers aren't symbols either—numerals are.) Here is a vivid way of seeing the importance of the distinction between numbers and numerals; we have just observed that it would *not* be surprising at all to find that extra-terrestrials used the same *numbers* we do, but simply incredible if they used the same *numerals*.

In a Vast space of possibilities, the odds of a similarity between two independently chosen elements is Vanishing *unless there is a reason*. There is for numbers (arithmetic is *true* and variations on arithmetic aren't) and there isn't for numerals (the symbol "§" would function exactly as well as the symbol "5" as a name for the number that follows 4).

Suppose we found the extra-terrestrials, like us, using the decimal system for most informal purposes, but converting to binary arithmetic when doing computation with the aid of mechanical prosthetic devices (computers). Their use of 0 and 1 in their computers (supposing they had invented computers!) would not surprise us, since there are good engineering *reasons* for adopting the binary system, and though these reasons are not dead obvious, they are probably within striking distance for average-type thinkers. "You don't have to be a rocket scientist" to appreciate the virtues of binary.

In general, we would expect them to have discovered many of the various ways things have of *being the right way*. Wherever there are many different ways of skinning a cat, and none is much better than any other, our surprise at their doing it *our* way will be proportional to how many different ways we think there are. Notice that even when we are contemplating some Vast number of *equivalent ways*, a value judgment is implicit. For us to recognize items as things falling in one of these Vast sets, they have to be seen as equally good ways, as ways of *performing the function x*. Function-alistic thinking is simply inescapable in this sort of inquiry; you can't even enumerate the possibilities without presupposing a concept of function. (Now we can see that even our deliberately antiseptic formalization of the Library of Mendel invoked functional presuppositions; we can't identify something as a *possible genome* without thinking of genomes as things that might serve a particular function within a reproductive system.)

So there turn out to be general principles of practical reasoning (including, in more modern dress, *cost-benefit analysis*) that can be relied upon to impose themselves on all life forms anywhere. We can argue about particular cases, but not about the applicability in general of the principles. Are such design features as bilateral symmetry in locomotors, or mouth-at-the-bow-end, to be explained as largely a matter of historical contingency, or largely a matter of practical wisdom? The only issues to debate or investigate are their relative contributions, and the historical order in which the contributions were made. (Recall that in the actual QWERTY phenomenon, there was a perfectly good engineering *reason* for the initial choice—it was just a reason whose supporting circumstances had long ago lapsed.)

Design work—lifting—can now be characterized as the work of discovering good ways of solving "problems that arise." Some problems are given at the outset, in all environments, under all conditions, to all species. Further problems are then created by the initial "attempts at solution" made by different species faced with the first problems. Some of these subsidiary problems are created by the other species of organisms (who must make a living, too), and other subsidiary problems are created by a species' own solutions to its own problems. For instance, now that one has decided—by flipping a coin, perhaps—to search for solutions in *this* area, one is stuck with problem B instead of problem A, which poses subproblems p, q, and r, instead of subproblems x, y, and z, and so forth. Should we personify a species in this way and treat it as an agent or practical reasoner (Schull 1990, Dennett 1990a)? Alternatively, we may choose to think of species as perfectly mindless nonagents, and put the rationale in the process of natural selection itself (perhaps jocularly personified as Mother Nature). Remember Francis Crick's quip about evolution's being cleverer than you are. Or we may choose to shrink from these vivid modes of expression altogether, but the analyses we do will have the same logic in any case.

This is what lies behind our intuition that design work is somehow intellectual work. Design work is discernible (in the otherwise uninterpretable typography of shifting genomes) only if we start imposing *reasons* on it. (In earlier work, I characterized these as "free-floating rationales," a term that has apparently induced terror or nausea in many otherwise well-disposed readers. Bear with me; I will soon provide some more palatable ways of making these points.)

So Paley was right in saying not just that Design was a wonderful thing to explain, but also that Design took Intelligence. All he missed—and Darwin provided—was the idea that this Intelligence could be broken into bits so tiny and stupid that they didn't count as intelligence at all, and then distributed through space and time in a gigantic, connected network of algorithmic process. The work must get done, but which work gets done is largely a matter of chance, since chance helps determine which problems (and subproblems and subsubproblems) get "addressed" by the machinery. Whenever we find a problem solved, we can ask: Who or what did the work? Where and when? Has a solution been worked out locally, or long ago, or was it somehow borrowed (or stolen) from some other branch of the tree? If it exhibits peculiarities that could only have arisen in the course of solving the subproblems in some apparently remote branch of the Tree that grows in Design Space, then barring a miracle or a coincidence too Cosmic to credit, there must be a copying event of some kind that moved that completed design work to its new location.

There is no single summit in Design Space, nor a single staircase or ladder with calibrated steps, so we cannot expect to find a scale for comparing amounts of design work across distant developing branches. Thanks to the vagaries and digressions of different "methods adopted," something that is in some sense just one problem can have both hard and easy solutions, requiring more or less work. There is a famous story about the mathematician and physicist (and coinventor of the computer) John von Neumann, who was legendary for his lightning capacity to do prodigious calculations in his head. (Like most famous stories, this one has many versions, of which I choose the one that best makes the point I am pursuing.) One day a colleague approached him with a puzzle that had two paths to solution, a laborious, complicated calculation and an elegant, Aha!-type solution. This colleague had a theory: in such a case, mathematicians work out the laborious solution while the (lazier, but smarter) physicists pause and find the quick and easy solution. Which solution would von Neumann find? You know the sort of puzzle: Two trains, 100 miles apart, are approaching each other on the same track, one going 30 miles per hour, the other going 20 miles per hour. A bird flying 120 miles per hour starts at train A (when they are 100 miles apart), flies to train B, turns around and flies back to the approaching train A, and so forth, until the trains collide. How far has the bird flown when the collision occurs? "Two hundred forty miles," Von Neumann answered almost instantly. "Darn," replied his colleague, "I predicted you'd do it the hard way." "Ay!" von Neumann cried in embarrassment, smiting his forehead. "There's an easy way!" (Hint: how long till the trains collide?)

Eyes are the standard example of a problem that has been solved many times, but eyes that may look just the same (and see just the same) may have been achieved by R-and-D projects that involved different amounts of work, thanks to the historical peculiarities of the difficulties encountered along the way. And the creatures that don't have eyes at all are neither better nor worse on any absolute scale of design; their lineage has just never been given this problem to solve. It is this same variability in *luck* in the various lineages that makes it impossible to define a single Archimedean point from which global progress could be measured. Is it progress when you have to work an extra job to pay for the high-priced mechanic you have to hire to fix your car when it breaks because it is too complex for you to fix in the way you used to fix your old clunker? Who is to say? Some lineages get trapped in (or are lucky enough to wander into—take your pick) a path in Design Space in which complexity begets complexity, in an arms race of competitive design. Others are fortunate enough (or unfortunate enough— take your pick) to have hit upon a relatively simple solution to life's problems at the outset and, having nailed it a billion years ago, have had nothing much to do in the way of design work ever since. We human beings, complicated creatures that we are, tend to appreciate complexity, but that may well be just an aesthetic preference that goes with our sort of lineage; other lineages may be as happy as clams with their ration of simplicity.

3. THE UNITY OF DESIGN SPACE

The formation of different languages and of distinct species, and the proofs that both have been developed through a gradual process, are curiously the same.

—CHARLES DARWIN 1871, p. 59

It will not have gone unnoticed that my examples in this chapter have wandered back and forth between the domain of organisms or biological design, on the one hand, and the domain of human artifacts—books, problems solved, and engineering triumphs on the other. This was by design, not accident, of course. It was to help set the stage for, and provide lots of ammunition for, a Central Salvo: *there is only one Design Space, and everything actual in it is united with everything else.* And I hardly need add that it was Darwin who taught us this, whether he quite realized it or not.

Now I want to go back over the ground we have covered, highlighting the evidence for this claim, and drawing out a few more implications of it and grounds for believing it. The similarities and continuities are of tremendous importance, I think, but in later chapters I will also point to some important dissimilarities between the human-made portions of the designed world and the portions that were created without benefit of the sort of locally concentrated, foresighted intelligence we human artificers bring to a problem.

We noted at the outset that the Library of Mendel (in the form of printed volumes of the letters A, C, G, T) is contained within the Library of Babel, but we should also note that at least a very large portion of the Library of Babel (What portion? See chapter 15) is in turn "contained" in the Library of Mendel, because *we* are in the Library of Mendel (our genomes are, and so are the genomes of all the life forms our lives depend on). The Library of Babel describes one aspect of our "extended phenotype" (Dawkins 1982). That is, in the same way that spiders make webs and beavers make dams, we make (among many other things) books. You can't assess the spider's genome for viability without a consideration of the web that is part of the normal equipment of the spider, and you can't assess the viability of our genomes (not any longer, you can't) without recognizing that we are a species with culture, a representative part of which is in the form of books. We are not just designed, we are designers, and all our talents as designers, and our products, must emerge non-miraculously from the blind, mechanical processes of Darwinian mechanisms *of one sort or another*. How many cranes-on-top-of-cranes does it take to get from the early design explorations of prokaryotic lineages to the mathematical investigations of Oxford dons? That is the question posed by Darwinian thinking. The resistance comes from those who think there must be some discontinuities somewhere, some skyhooks, or moments of Special Creation, or some other sort of miracles, between the prokaryotes and the finest treasures in our libraries.

There may be—that will be a question we will look at in many different ways in the rest of the book. But we have already seen a variety of deep parallels, instances in which the very same principles, the very same strategies of analysis or inference, apply in both domains. There are many more where they came from.

Consider, for instance, Darwin's pioneering use of a certain sort of historical inference. As Stephen Jay Gould has stressed (e.g., 1977a, 1980a), it is the imperfections, the curious fallings short of what would seem to be perfect design, that are the best evidence for a historical process of descent with modification; they are the best evidence of copying, instead of independent re-inventing, of the design in question. We can now see better why this is such good evidence. The odds against two independent processes' arriving at the same region of Design Space are Vast unless the design element in question is obviously right, a forced move in Design Space. Perfection will be independently hit upon again and again, especially if it is obvious. It is the idiosyncratic *versions* of near-perfection that are a dead giveaway of copying. In evolutionary theory, such traits are called *homologies*: traits that are similar not because they have to be for functional reasons, but because of copying. The biologist Mark Ridley observes, "Many of what are often presented as separate arguments for evolution reduce to the general form of the argument from homology," and he boils the argument down to its essence:

> The ear-bones of mammals are an example of a homology. They are homologous with some of the jaw-bones of reptiles. The ear-bones of mammals did not have to be formed from the same bones as form the jaw of reptiles; but in fact they are. . . . The fact that species share homologies is an argument for evolution, for if they had been created separately *there would be no reason why* [emphasis added] they should show homologous similarities. [Mark Ridley 1985, p. 9.]

This is how it is in the biosphere, and also how it is in the cultural sphere of plagiarism, industrial espionage, and the honest work of *recension of texts*.

Here is a curious historical coincidence: while Darwin was fighting his way clear to an understanding of this characteristically Darwinian mode of inference, some of his fellow Victorians, in England and especially in Germany, had already perfected the same bold, ingenious strategy of historical inference in the domain of *paleography* or *philology*. I have several times alluded to the works of Plato in this book, but it is "a miracle" that Plato's work survives for us to read today in any version at all. All the texts of his *Dialogues* were essentially lost for over a thousand years. When they re-emerged at the dawn of the Renaissance in the form of various tattered, dubious, partial copies of copies of copies from who knows where, this set in motion five hundred years of painstaking scholarship, intended to "purify the text" and establish a proper informational link with the original sources, which of course would have been in Plato's own hand, or the hand of the scribe to whom he dictated. The originals had presumably long since turned to dust. (Today there are some fragments of papyrus with Platonic text on them, and these bits of text may be roughly contemporaneous with Plato himself, but they have played no important role in the scholarship, having been uncovered quite recently.)

The task that faced the scholars was daunting. There were obviously many "corruptions" in the various nonextinct copies (called "witnesses"), and these corruptions or errors had to be fixed, but there were also many puzzling—or exciting—passages of dubious authenticity, and no way of asking the author which were which. How could they be properly distinguished? The corruptions could be more or less rank-ordered in obviousness: (1) typographical errors, (2) grammatical errors, (3) stupid or otherwise baffling expressions, or (4) bits that were just not stylistically or doctrinally like the rest of Plato. By Darwin's day, the philologists who devoted their entire professional lives to re-creating the genealogy of their witnesses had not only developed elaborate and—for their day—rigorous methods of comparison, but had succeeded in extrapolating whole lineages of copies of copies, and deduced many curious facts about the historical circumstances of their birth, reproduction, and eventual death. By an analysis of the patterns of shared and unshared errors in the existing documents (the carefully preserved parchment treasures in the Bodleian Library at Oxford, in Paris, in the Vienna Nationalbibliothek, in the Vatican, and elsewhere), they were able to deduce hypotheses about how many different copyings there had to have been, roughly when and where some of these must have been made, and which witnesses had relatively recent shared ancestors and which did not.

Sometimes the deductive boldness of their work is the equal of anything in Darwin: a particular group of manuscript errors, uncorrected and re-copied in all the descendants in a particular lineage, was almost certainly due to the fact that the scribe who took the dictation did not pronounce Greek the same way the reader did, and consequently misheard a particular phoneme on many occasions! Such clues, together with evidence from other sources on the history of the Greek language, might even suggest to the scholars which monastery, on which Greek island or mountaintop, in which century must have been the scene for the creation of this set of mutations—even though the actual parchment document created then and there has long since succumbed to the Second Law of Thermodynamics and turned to dust.[4]

Did Darwin ever learn anything from the philologists? Did any philologists recognize that Darwin had re-invented one of their wheels? Nietzsche was himself one of these stupendously erudite students of the ancient texts, and he was one of many German thinkers who were swept up in the Darwin boom, but, so far as I know, he never noticed a kinship between Darwin's method and that of his colleagues. Darwin himself was struck in later years by the curious similarity between his arguments and those of the philologists studying the genealogy of languages (not, as in the case of the Plato scholars, the genealogy of specific texts). In *The Descent of Man* (1871, p. 59) he pointed explicitly to their shared use of the distinction between homologies and analogies that could be due to convergent evolution: "We find in distinct languages striking homologies due to community of descent, and analogies due to a similar process of formation."

Imperfections or errors are just special cases of the variety of marks that speak loudly—and intuitively—of a shared history. The role of chance in twisting the paths taken in a bit of design work can create the same effect without creating an error. A case in point: In 1988, Otto Neugebauer, the great historian of astronomy, was sent a photograph of a fragment of Greek papyrus with a few numbers in a column on it. The sender, a classicist, had no clue about the meaning of this bit of papyrus, and wondered if Neugebauer had any ideas. The eighty-nine-year-old scholar recomputed the line-to-line differences between the numbers, found their maximum and minimum limits, and determined that this papyrus had to be a translation of part of "Column G" of a Babylonian cuneiform tablet on which was written a Babylonian "System B" lunar ephemeris! (An ephemeris is, like the *Nautical Almanac*, a tabular system for computing the location of a heavenly body for every time in a particular period.) How *could* Neugebauer make this Sherlock Holmes–ian deduction? Elementary: what was written in Greek (a sequence of sexagesimal—not decimal—numbers) was recognized by him to be part—column G!—of a highly accurate calculation of the moon's

---

4. Scholarship marches on. With the aid of computers, more recent researchers have shown "that the nineteenth-century model of the constitution and descent of our manuscripts of Plato was so oversimplified that it must be counted wrong. That model, in its original form, assumed that all the extant manuscripts were direct or indirect copies of one or more of the three oldest extant manuscripts, each a literal copy; variants in the more recent manuscripts were then to be explained either as scribal corruption or arbitrary emendation, growing cumulatively with each new copy. . . ." (Brumbaugh and Wells 1968, p. 2; the introduction provides a vivid picture of the fairly recent state of play.)

location that had been worked out by the Babylonians. There are lots of different ways of calculating an ephemeris, and Neugebauer knew that anyone working out their own ephemeris independently, using their own system, would not have come up with exactly the same numbers as anyone else, though the numbers might have been close. The Babylonian system B was excellent, so the design had been gratefully conserved, in translation, with all its fine-grained particularities. (Neugebauer 1989.)[5]

Neugebauer was a great scholar, but you can probably execute a parallel feat of deduction, following in his footsteps. Suppose you were sent a photocopy of the text below, and asked the same questions: What does it mean? Where might this be from?

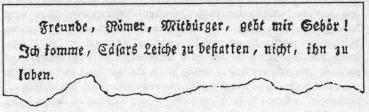

Freunde, Römer, Mitbürger, gebt mir Gehör! Ich komme, Cäsars Leiche zu bestatten, nicht, ihn zu loben.

FIGURE 6.1

Before reading on, try it. You can probably figure it out even if you don't really know how to read the old German *Fraktur* typeface—and even if you don't know German! Look again, closely. Did you get it? Impressive stunt! Neugebauer may have his Babylonian column G, but you quickly determined, didn't you, that this fragment must be part of a German translation of some lines from an Elizabethan tragedy (*Julius Caesar*, act III, scene ii, lines 79–80, to be exact). Once you think about it, you realize that it could hardly be anything else! The odds against *this* particular sequence of German letters' getting strung together under any other circumstances are Vast. Why? What is the particularity that marks such a string of symbols?

Nicholas Humphrey (1987) makes the question vivid by posing a more drastic version: if you were forced to "consign to oblivion" one of the following masterpieces, which would you choose: Newton's *Principia*, Chaucer's *Canterbury Tales*, Mozart's *Don Giovanni*, or Eiffel's Tower? "If the choice were forced," Humphrey answers,

> I'd have little doubt which it should be: the *Principia* would have to go.
> How so? Because, of all those works, Newton's was the only one that was

5. I am grateful to Noel Swerdlow, who told this story during the discussion following his talk "The Origin of Ptolemy's Planetary Theory," at the Tufts Philosophy Colloquium, October 1, 1993, and subsequently provided me with Neugebauer's paper and an explanation of its fine points.

replaceable. Quite simply: if Newton had not written it, then someone else would—probably within the space of a few years.... The *Principia* was a glorious monument to human intellect, the Eiffel Tower was a relatively minor feat of romantic engineering; yet the fact is that while Eiffel did it *his* way, Newton merely did it God's way.

Newton and Leibniz famously quarreled over who got to the calculus first, and one can readily imagine Newton having another quarrel with a contemporary over who should get priority on discovering the laws of gravitation. But had Shakespeare never lived, for example, no one else would ever have written his plays and poems. "C. P. Snow, in the *Two Cultures*, extolled the great discoveries of science as 'scientific Shakespeare'. But in one way he was fundamentally mistaken. Shakespeare's plays were Shakespeare's plays and no one else's. Scientific discoveries, by contrast, belong—ultimately—to no one in particular" (Humphrey 1987). Intuitively, the difference is the difference between discovery and creation, but we now have a better way of seeing it. On the one hand, there is design work that homes in on a best move or forced move which can be seen (in retrospect, at least) to be a uniquely favored location in Design Space accessible from many starting points by many different paths; on the other hand, there is design work the excellence of which is much more dependent on exploiting (and amplifying) the many contingencies of history that shape its trajectory, a trajectory about which the bus company's slogan is an understatement: getting there is much more than half the fun.

We saw in chapter 2 that even the long-division algorithm can avail itself of randomness or arbitrary idiosyncrasy—choose a digit at random (or your favorite digit) and check to see if it's the "right" one. But the actual idiosyncratic choices made as you go along cancel out, leaving no trace in the final answer, the right answer. Other algorithms can incorporate the random choices into the structure of their final products. Think of a poetry-writing algorithm—or a doggerel-writing algorithm, if you insist—that begins: "Choose a noun at random from the dictionary...." Such a design process can produce something that is definitely under quality control—selection pressure—but which nevertheless bears the unmistakable signs of its particular history of creation.

Humphrey's contrast is sharp, but his vivid way of drawing it might mislead. Science, unlike the arts, is engaged in journeys—sometimes races—with definite destinations: solutions to specific problems in Design Space. But scientists do care just as much as artists do about the routes taken, and hence would be appalled at the idea of discarding Newton's actual work and just saving his destination (which someone else would eventually have led us to in any case). They care about the actual trajectories because the methods used in them can often be used again, for other journeys; the good methods are cranes, which can be borrowed, with acknowledgment, and used to do lifting in other parts of Design Space. In the extreme case, the crane developed by a scientist may be of much more value than the particular lifting accomplished by it, the destination reached. For instance, a proof of a trivial result may nevertheless pioneer a new mathematical method of great value. Mathematicians put a high value on coming up with a simpler, more elegant proof of something they have already proved—a more efficient crane.

In this context, philosophy can be seen to lie about midway between science and the arts. Ludwig Wittgenstein famously stressed that in philosophy the process—the arguing and analyzing—is more important than the product—the conclusions reached, the theories defended. Though this is hotly (and correctly, in my opinion) disputed by many philosophers who aspire to solve real problems—and not just indulge in a sort of interminable logotherapy—even they would admit that we would never want to consign Descartes's famous "cogito ergo sum" thought experiment, for example, to oblivion, even though none would accept its conclusions; it is just too nifty an intuition pump, even if all it pumps is falsehoods (Dennett 1984, p. 18).

Why can't you copyright a successful multiplication of two numbers? Because anyone could do it. It's a forced move. The same is true of any simple fact that a genius isn't needed to discover. So how do the creators of tables or other routine (but labor-intensive) masses of printed data protect themselves from unscrupulous copiers? Sometimes they set traps. I am told, for instance, that the publishers of *Who's Who* have dealt with the problem of competitors' simply stealing all their hard-won facts and publishing their own biographical encyclopedias by quietly inserting a few entirely bogus entries. You can be sure that if one of those shows up on a competitor's pages, it was no coincidence!

In the larger perspective of the whole Design Space, the crime of plagiarism might be defined as *theft of crane*. Somebody or something has done some design work, thereby creating something that is useful in further design work and therefore may have value to anyone or anything embarked on a design project. In our world of culture, where the transmission of designs from agent to agent is enabled by many media of communication, the acquisition of designs developed in other "shops" is a common event, almost the defining mark of cultural evolution (which will be the topic of chapter 12). It has commonly been assumed by biologists that such transactions were impossible in the world of genetics (until the dawn of genetic engineering). You might say, in fact, that this has been the Official Dogma. Recent discoveries suggest otherwise—though only time will tell; no Dogma ever rolled over and died without a fight. For instance, Marilyn Houck (Houck et al. 1991) has found evidence that, about forty years ago, in either Florida or Central America, a tiny mite that feeds on fruit flies happened to puncture the egg of a fly of the *Drosophila willistoni* species, and in the process picked up some of that species' characteristic DNA, which it then inadvertently transmitted to the egg of a (wild) *Drosophila melanogaster* fly! This could explain the sudden explosion in the wild of a particular DNA element common in *D. willistoni* but previously unheard of in *D. melanogaster* populations. She might add: What else could explain it? It sure looks like species plagiarism.

Other researchers are looking at other possible vehicles for speedy design travel in the world of natural (as opposed to artificial) genetics. If they find them, they will be fascinating—but no doubt rare—exceptions to the orthodox pattern: genetic transmission of design by chains of direct descent only.[6] We are inclined, as just noted, to contrast this feature sharply with what we find in the freewheeling world of cultural evolution, but even there we can detect a powerful dependence on the combination of luck and copying.

Consider all the wonderful books in the Library of Babel that will never be written, even though the process that could create each of them involves no violation or abridgment of the laws of nature. Consider some book in the Library of Babel that you yourself might love to write—and that only you could write—for instance, the poetically expressed autobiographical tale of your childhood that would bring tears and laughter to all readers. We know that there are Vast numbers of books with just these features in the Library of Babel, and each is composable in only a million keystrokes. At the dawdling rate of five hundred strokes a day, the whole project shouldn't take you much longer than six years, with generous vacations. Well, what's stopping you? You have fingers that work, and all the keys on your word-processor can be depressed independently.

Nothing is stopping you. That is, there *needn't* be any identifiable forces, or laws of physics or biology or psychology, or salient disabilities brought on by identifiable circumstances (such as an ax embedded in your brain, or a gun pointed at you by a credible threatener). There are Vastly many books that you are never going to write "for no reason at all." Thanks to the myriad particular twists and turns of your life to date, you just don't happen to be well disposed to compose those sequences of keystrokes.

If we want to get some perspective—limited, to be sure—on what patterns go into creating your own authorial dispositions, we will have to consider the transmission of Design to you from the books you have read. The books that actually come to exist in the world's libraries are deeply

---

6. The genetic elements transferred in *Drosophila* are "intragenomic parasites" and probably have a negative effect on the adaptedness of their host organisms, so we shouldn't get our hopes up unduly. See Engels 1992.

dependent not just on their authors' biological inheritance, but on the books that have come before them. This dependence is conditioned by coincidences or accidents at every turning. Just look at my bibliography to discover the main lines of genealogy of this book. I have been reading and writing about evolution since I was an undergraduate, but if I had not been encouraged by Doug Hofstadter in 1980 to read Dawkins' *The Selfish Gene*, I probably would not have begun coalescing some of the interests and reading habits that have been major shapers of this book. And if Hofstadter had not been asked by *The New York Review of Books* to review my book *Brainstorms* (1978), he probably would never have hit upon the bright idea of proposing that we collaborate on a book, *The Mind's I* (1981), and then we would not have had the opportunity for mutual book-recommending that led me to Dawkins, and so forth. Even if I had read the same books and articles by following other paths, in a different order, I would not be conditioned in exactly the same way by that reading, and hence would have been unlikely to have composed (and edited, and re-edited) *just* the string of symbols you are now reading.

Can we measure this transmission of Design in culture? Are there units of cultural transmission analogous to the genes of biological evolution? Dawkins (1976) has proposed that there are, and has given them a name: *memes*. Like genes, memes are supposed to be replicators, in a different medium, but subject to much the same principles of evolution as genes. The idea that there might be a scientific theory, memetics, strongly parallel to genetics, strikes many thinkers as absurd, but at least a large part of their skepticism is based on misunderstanding. This is a controversial idea, which will get careful consideration in chapter 12, but in the meantime we can set aside the controversies and just use the term as a handy word for a salient (*memorable*) cultural item, something with enough Design to be worth saving—or stealing or replicating.

⇀⁓⁓

The Library of Mendel (or its twin, the Library of Babel—they are contained in each other, after all) is as good an approximate model of Universal Design Space as we could ever need to think about. For the last four billion years or so, the Tree of Life has been zigzagging through this Vast multidimensional space, branching and blooming with virtually unimaginable fecundity, but nevertheless managing to fill only a Vanishingly small portion of that space of the Possible with Actual designs.[7] According

---

7. "I confess that I believe the emptiness of phenotypic space is filled with red herrings.... Under the null hypothesis that no constraints at all exist, the branching pathways through space taken by this process constitute a random-branching walk in a

to Darwin's dangerous idea, all *possible* explorations of Design Space are connected. Not only all your children and your children's children, but all your brainchildren and your brainchildren's brainchildren must grow from the common stock of Design elements, genes and memes, that have so far been accumulated and conserved by the inexorable lifting algorithms, the ramps and cranes and cranes-atop-cranes of natural selection and its products.

If this is right, then all the achievements of human culture—language, art, religion, ethics, science itself—are themselves artifacts (of artifacts of artifacts . . .) of the same fundamental process that developed the bacteria, the mammals, and *Homo sapiens*. There is no Special Creation of language, and neither art nor religion has a literally divine inspiration. If there are no skyhooks needed to make a skylark, there are also no skyhooks needed to make an ode to a nightingale. No meme is an island.

Life and all its glories are thus united under a single perspective, but some people find this vision hateful, barren, odious. They want to cry out against it, and above all, they want to be magnificent exceptions to it. They, if not the rest, are made in God's image by God, or, if they are not religious, they want to be skyhooks themselves. They want somehow to be *intrinsic sources* of Intelligence or Design, not "mere" artifacts of the same processes that mindlessly produced the rest of the biosphere.

So a lot is at stake. Before we turn, in part III, to examine in detail the implications of the upward spread of universal acid through human culture, we need to secure the base camp, by considering a variety of deep challenges to Darwinian thinking within biology itself. In the process, our vision of the intricacy and power of the underlying ideas will be enhanced.

CHAPTER 6: *There is one Design Space, and in it the Tree of Life has grown a branch that has recently begun casting its own exploratory threads into that Space, in the form of human artifacts. Forced moves and other good ideas are like beacons in Design Space, discovered again and again, by the ultimately algorithmic search processes of natural selection and human investigation. As Darwin appreciated, we can retrospectively detect the historical fact of descent, anywhere in Design Space, when we find shared design features that would be Vastly unlikely to coexist unless there was a thread of descent between them. Historical reasoning about evolution thus depends on accepting Paley's premise: the world is full of good Design, which took work to create.*

This completes the introduction to Darwin's dangerous idea. Now we

---

high-dimensional space. The typical property of such a walk in a high-dimensional space is that most of the space is empty" (Kauffman 1993, p. 19).

must secure its base camp in biology, in part II, before looking at its power to transform our understanding of the human world, in part III.

CHAPTER 7: *How did the Tree of Life get started? Skeptics have thought a stroke of Special Creation—a skyhook—must be needed to get the evolutionary process going. There is a Darwinian answer to this challenge, however, which exhibits the power of Darwin's universal acid to work its way down through the lowest levels of the Cosmic Pyramid, showing how even the laws of physics might emerge from chaos or nothingness without recourse to a Special Creator, or even a Lawgiver. This dizzying prospect is one of the most feared aspects of Darwin's dangerous idea, but the fear is misguided.*

PART II

DARWINIAN THINKING IN BIOLOGY

Evolution is a change from a no-howish untalkaboutable all-alikeness by continuous sticktogetherations and somethingelsifications.

—WILLIAM JAMES 1880

Nothing in biology makes sense except in the light of evolution.

—THEODOSIUS DOBZHANSKY 1973

CHAPTER SEVEN

Priming Darwin's Pump

---

1. BACK BEYOND DARWIN'S FRONTIER

And God said, Let the earth bring forth grass, the herb yielding seed, and the fruit tree yielding fruit after his kind, whose seed is in itself, upon the earth: and it was so.

And the earth brought forth grass, and herb yielding seed after his kind, and the tree yielding fruit, whose seed was in itself, after his kind: and God saw that it was good.

—GENESIS 1:11–12

From what sort of seed could the Tree of Life get started? That all life on Earth has been produced by such a branching process of generation is now established beyond any reasonable doubt. It is as secure an example of a scientific fact as the roundness of the Earth, thanks in large part to Darwin. But how did the whole process get started in the first place? As we saw in chapter 3, Darwin not only started in the middle; he cautiously refrained from pushing his own published thinking back to the beginning—the ultimate origin of life and its preconditions. When pressed by correspondents, he had little more to say in private. In a famous letter, he surmised that it was quite possible that life began in "a warm little pond," but he had no details to offer about the likely recipe for this primeval preorganic soup. And in response to Asa Gray, as we saw (see page 67), he left wide open the possibility that the *laws* that would govern this Earth-shattering move were themselves designed—presumably by God.

His reticence on this score was wise on several counts. First, no one knew better than he the importance of anchoring a revolutionary theory in the bedrock of empirical facts, and he knew that he could only speculate, with scant hope in his own day of getting any substantive feedback. After all, as we have already seen, he didn't even have the Mendelian concept of the gene, let alone any of the molecular machinery underlying it. Darwin was an intrepid deducer, but he also knew when he didn't have enough premises to go on. Besides, there was his concern for his beloved wife, Emma, who desperately wanted to cling to her religious beliefs, and who could already see the threat looming in her husband's work. Yet his reluctance to push any farther into this dangerous territory, at least in public, went beyond his consideration for her feelings. There is a wider ethical consideration at stake, which Darwin certainly appreciated.

Much has been written about the moral dilemmas that scientists face when the discovery of a potentially dangerous fact puts their love of truth at odds with their concern for the welfare of others. Under what conditions, if any, would they be obliged to conceal the truth? These can be real dilemmas, with powerful and hard-to-plumb considerations on both sides. But there is no controversy at all about what a scientist's (or philosopher's) moral obligations should be regarding his or her speculations. Science doesn't often advance by the methodical piling up of demonstrable facts; the "cutting edge" is almost always composed of several rival edges, sharply competing and boldly speculative. Many of these speculations soon prove to be misbegotten, however compelling at the outset, and these necessary by-products of scientific investigation should be considered to be as potentially hazardous as any other laboratory wastes. One must consider their environmental impact. If their misapprehension by the public would be apt to cause suffering—by misleading people into dangerous courses of action, or by undercutting their allegiance to some socially desirable principle or creed—scientists should be particularly cautious about how they proceed, scrupulous about labeling speculations as such, and keeping the rhetoric of persuasion confined to its proper targets.

But ideas, unlike toxic fumes or chemical residues, are almost impossible to quarantine, particularly when they concern themes of abiding human curiosity, so, whereas there is no controversy at all about the principle of responsibility here, there has been scant agreement, then or now, about how to honor it. Darwin did the best he could: he kept his speculations pretty much to himself.

We can do better. The physics and chemistry of life are now understood in dazzling detail, so that much more can be deduced about the necessary and (perhaps) sufficient conditions for life. The answers to the big questions must still involve a large measure of speculation, but we can mark the speculations as such, and note how they could be confirmed or disconfirmed. There would be no point any more in trying to pursue Darwin's policy of reticence; too many very interesting cats are already out of the bag. We may not yet know exactly *how* to take all these ideas seriously, but thanks to Darwin's secure beachhead in biology, we know that we can and must.

It is small wonder that Darwin didn't hit upon a suitable mechanism of heredity. What do you suppose his attitude would have been to the speculation that within the nucleus of each of the cells in his body there was a copy of a set of instructions, written on huge macromolecules, in the form of double helixes tightly kinked into snarls to form a set of forty-six chromosomes? The DNA in your body, unsnarled and linked, would stretch to the sun and back several—ten or a hundred—times. Of course, Darwin is the man who painstakingly uncovered a host of jaw-dropping complexities in the lives and bodies of barnacles, orchids, and earthworms, and described them with obvious relish. Had he had a prophetic dream back in 1859 about the wonders of DNA, he would no doubt have reveled in it, but I wonder if he could have recounted it with a straight face. Even to those of us accustomed to the "engineering miracles" of the computer age, the facts are hard to encompass. Not only molecule-sized copying machines, but proofreading enzymes that correct mistakes, all at blinding speed, on a scale that supercomputers still cannot match. "Biological macromolecules have a storage capacity that exceeds that of the best present-day information stores by several orders of magnitude. For example, the 'information density' in the genome of *E. coli*, is about 10^{27} bits/m^3" (Küppers 1990, p. 180)."

In chapter 5, we arrived at a Darwinian definition of biological possibility in terms of accessibility within the Library of Mendel, but the precondition for that Library, as we noted, was the existence of genetic mechanisms of staggering complexity and efficiency. William Paley would have been transported with admiration and wonder at the atomic-level intricacies that make life possible at all. *How could they themselves have evolved if they are the precondition for Darwinian evolution?*

Skeptics about evolution have argued that this is the fatal flaw in Darwinism. As we have seen, the power of the Darwinian idea comes from the way it distributes the huge task of Design through vast amounts of time and space, preserving the partial products as it proceeds. In *Evolution: A Theory in Crisis*, Michael Denton puts it this way: the Darwinian assumes "that islands of function are common, easily found in the first place, and that it is easy to go from island to island through functional intermediates" (Denton 1985, p. 317). This is almost right, but not quite. Indeed, the central claim of Darwinism is that the Tree of Life spreads out its branches, connecting "islands of function" with isthmuses of intermediate cases, but nobody said the passage would be "easy" or the safe stopping places "common." There is only one strained sense of "easy" in which Darwinism is committed to these isthmus-crossings' being easy: since every living thing is a descendant of a living thing, it has a tremendous leg up; all but the tiniest fraction of its recipe is guaranteed to have time-tested viability. The lines of genealogy are lifelines indeed; according to Darwinism, the only hope of entering this cosmic maze of junk and staying alive is to stay on the isthmuses.

THE ORIGIN OF LIFE

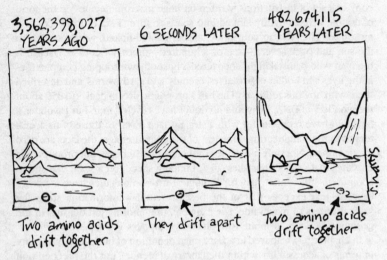

3,562,398,027 YEARS AGO

6 SECONDS LATER

482,674,115 YEARS LATER

Two amino acids drift together

They drift apart

Two amino acids drift together

FIGURE 7.1

But how could this process get started? Denton (p. 323) goes to some lengths to calculate the improbability of such a start-up, and arrives at a suitably mind-numbing number.

> To get a cell by chance would require at least one hundred functional proteins to appear simultaneously in one place. That is one hundred simultaneous events each of an independent probability which could hardly be more than 10^{-20} giving a maximum combined probability of 10^{-2000}

This probability is Vanishing indeed—next to impossible. And it looks at first as if the standard Darwinian response to such a challenge could not *as a matter of logic* avail us, since the very preconditions for its success—a system of replication with variation—are precisely what only its success would permit us to explain. Evolutionary theory appears to have dug itself into a deep pit, from which it cannot escape. Surely the only thing that could save it would be a skyhook! This was Asa Gray's fond hope, and the more we have learned about the intricacies of DNA replication, the more enticing this idea has become to those who are searching for a place to bail out science with some help from religion. One might say that it has appeared to many to be a godsend. Forget it, says Richard Dawkins:

Maybe, it is argued, the Creator does not control the day-to-day succession of evolutionary events, maybe he did not frame the tiger and the lamb, maybe he did not make a tree, but he *did* set up the original machinery of replication and replicator power, the original machinery of DNA and protein that made cumulative selection, and hence all of evolution, possible.

This is a transparently feeble argument, indeed it is obviously self-defeating. Organized complexity is the thing we are having difficulty explaining. Once we are allowed simply to *postulate* organized complexity, if only the organized complexity of the DNA/protein replicating engine, it is relatively easy to invoke it as a generator of yet more organized complexity. . . . But of course any God capable of intelligently designing something as complex as the DNA/protein replicating machine must have been at least as complex and organized as the machine itself. [Dawkins 1986a, p. 141.]

As Dawkins goes on to say (p. 316), "The one thing that makes evolution such a neat theory is that it explains how organized complexity can arise out of primeval simplicity." This is one of the key strengths of Darwin's idea, and the key weakness of the alternatives. In fact, I once argued, it is unlikely that any other theory could have this strength:

Darwin explains a world of final causes and teleological laws with a principle that is, to be sure, mechanistic but—more fundamentally—utterly independent of "meaning" or "purpose". It assumes a world that is *absurd* in the existentialist's sense of the term: not ludicrous but pointless, and this assumption is a necessary condition of any non-question-begging account of *purpose*. Whether we can imagine a *non*-mechanistic but also non-question-begging principle for explaining design in the biological world is doubtful; it is tempting to see the commitment to non-question-begging accounts here as tantamount to a commitment to mechanistic materialism, but the priority of these commitments is clear. . . . One argues: Darwin's materialistic theory may not be the only non-question-begging theory of these matters, but it is one such theory, and the only one we have found, which is quite a good reason for espousing materialism. [Dennett 1975, pp. 171–72.]

Is that a fair or even an appropriate criticism of the religious alternatives? One reader of an early draft of this chapter complained at this point, saying that by treating the hypothesis of God as just one more scientific hypothesis, to be evaluated by the standards of science in particular and rational thought in general, Dawkins and I are ignoring the very widespread claim by believers in God that their faith is quite beyond reason, not a matter to which such mundane methods of testing applies. It is not just unsympathetic, he claimed, but strictly unwarranted for me simply to assume that the scientific method continues to apply with full force in this domain of faith.

Very well, let's consider the objection. I doubt that the defender of religion will find it attractive, once we explore it carefully. The philosopher Ronald de Sousa once memorably described philosophical theology as "intellectual tennis without a net," and I readily allow that I have indeed been assuming without comment or question up to now that the net of rational judgment was up. But we can lower it if you really want to. It's your serve. Whatever you serve, suppose I return service rudely as follows: "What you say implies that God is a ham sandwich wrapped in tinfoil. That's not much of a God to worship!" If you then volley back, demanding to know how I can logically justify my claim that your serve has such a preposterous implication, I will reply: "Oh, do you want the net up for my returns, but not for your serves? Either the net stays up, or it stays down. If the net is down, there are no rules and anybody can say anything, a mug's game if there ever was one. I have been giving you the benefit of the assumption that you would not waste your own time or mine by playing with the net down."

Now if you want to *reason* about faith, and offer a reasoned (and reason-responsive) defense of faith as an extra category of belief worthy of special consideration, I'm eager to play. I certainly grant the existence of the phenomenon of faith; what I want to see is a reasoned ground for taking faith seriously as a *way of getting to the truth*, and not, say, just as a way people comfort themselves and each other (a worthy function that I do take seriously). But you must not expect me to go along with your defense of faith as a path to truth if at any point you appeal to the very dispensation you are supposedly trying to justify. Before you appeal to faith when reason has you backed into a corner, think about whether you really want to abandon reason when reason is on your side. You are sightseeing with a loved one in a foreign land, and your loved one is brutally murdered in front of your eyes. At the trial it turns out that in this land friends of the accused may be called as witnesses for the defense, testifying about their faith in his innocence. You watch the parade of his moist-eyed friends, obviously sincere, proudly proclaiming their undying faith in the innocence of the man you saw commit the terrible deed. The judge listens intently and respectfully, obviously more moved by this outpouring than by all the evidence presented by the prosecution. Is this not a nightmare? Would you be willing to live in such a land? Or would you be willing to be operated on by a surgeon who tells you that whenever a little voice in him tells him to disregard his medical training, he listens to the little voice? I know it passes in polite company to let people have it both ways, and under most circumstances I wholeheartedly cooperate with this benign arrangement. But we're seriously trying to get at the truth here, and if you think that this common but unspoken understanding about faith is anything better than socially useful obfuscation to avoid mutual embarrassment and loss of face, you have either seen much more deeply into this issue than any philosopher ever has (for none has ever come up with a good defense of this) or you are kidding yourself. (The ball is now in your court.)

Dawkins' retort to the theorist who would call on God to jump-start the evolution process is an unrebuttable refutation, as devastating today as when Philo used it to trounce Cleanthes in Hume's *Dialogues* two centuries earlier. A skyhook would at best simply postpone the solution to the problem, but Hume couldn't think of any cranes, so he caved in. Darwin came up with some magnificent cranes to do *middle*-level lifting, but can the principles that worked so well once be applied again to do the lifting required to get the booms of Darwin's cranes off the ground in the first place? Yes. Just when it might appear that the Darwinian idea has come to the end of its resources, it jumps niftily *down* a level and keeps right on going, not just one idea but many, multiplying like the brooms of the sorcerer's apprentice.

If you want to understand this trick, which at first glance seems unimaginable, you have to wrestle with some difficult ideas and a raft of details, both mathematical and molecular. This is not the book, and I am not the author, you should consult for those details, and nothing less could really secure your understanding, so what follows comes with a warning: although I will try to *acquaint* you with these ideas, you won't really know them unless you study them in the primary literature. (My own grasp on them is that of an amateur.) Imaginative theoretical and experimental explorations of the possibilities are now being conducted by so many different researchers that it practically constitutes a subdiscipline at the boundary between biology and physics. Since I cannot hope to demonstrate to you the validity of these ideas—and you shouldn't trust me if I claimed to do so—why am I presenting them? Because my purpose is philosophical: I wish to break down a prejudice, the conviction that a certain sort of theory couldn't *possibly* work. We have seen how Hume's philosophical trajectory got deflected by his inability to take seriously an opening in the wall that he dimly saw. He *thought he knew* that there was no point in heading any further in that direction, and, as Socrates never tired of pointing out, thinking you know when you don't is the main cause of philosophical paralysis. If I can show that it is *conceivable* that the Darwinian idea can carry through "all the way down," this will pre-empt a family of glib dismissals that is all too familiar, and open our minds to other possibilities.

2. Molecular Evolution

The smallest catalytically active protein molecules of the living cell consist of at least a hundred amino acids. For even such a short molecule, there exist $20^{100} \approx 10^{130}$ alternative arrangements of the twenty basic monomers. This shows that already on the lowest level of com-

*plexity, that of the biological macromolecules, an almost unlimited
variety of structures is possible.*

—BERND-OLAF KÜPPERS 1990, p. 11

*Our task is to find an algorithm, a natural law that leads to the origin of
information.*

—MANFRED EIGEN 1992, p. 12

In describing the power of the central claim of Darwinism in the previous
section, I helped myself to a slight (!) exaggeration: I said that every living
thing is the descendant of a living thing. This cannot be true, for it implies
an infinity of living things, a set with no first member. Since we know that
the total number of living things (on Earth, up till now) is large but finite,
we seem to be obliged, logically, to identify a first member—Adam the
Protobacterium, if you like. But how could such a first member come to
exist? A whole bacterium is much, much too complicated just to happen
into existence by cosmic accident. The DNA of a bacterium such as *E. coli*
has around four million nucleotides in it, almost all of them precisely in
order. It is quite clear, moreover, that a bacterium could not get by with
much less. So here is a quandary: since living things have existed for only a
finite time, there must have been a first one, but since all living things are
complex, there couldn't have been a first one!

There could only be one solution, and we know it well in outline: before
there were bacteria, with autonomous metabolisms, there were much simpler, quasi-living things, like viruses, but unlike them in not (yet) having any
living things to live off parasitically. From the chemist's point of view,
viruses are "just" huge, complex crystals, but thanks to their complexity,
they don't just sit there; they "do things." In particular, they reproduce or
self-replicate, with variations. A virus travels light, packing no metabolic
machinery, so it either stumbles upon the energy and materials required for
self-replication or self-repair, or eventually it succumbs to the Second Law
of Thermodynamics and falls apart. Nowadays, living cells provide concentrated storehouses for viruses, and viruses have evolved to exploit them, but
in the early days, they had to scrounge for less efficient ways of making more
copies of themselves. Viruses today don't all use double-stranded DNA;
some use an ancestral language, composed of single-stranded RNA (which
of course still plays a role in our own reproductive system, as an interme-
diary "messenger" system during "expression"). If we follow standard practice and reserve the term *virus* for a parasitic macromolecule, we need a
name for these earliest ancestors. Computer programmers call a cobbled-
together fragment of coded instructions that performs a particular task a
"macro," so I propose to call these pioneers *macros*, to stress that although
they are "just" huge macromolecules, they are also bits of *program* or

algorithm, bare, minimal self-reproducing mechanisms—remarkably like the computer viruses that have recently emerged to fascinate and plague us (Ray 1992, Dawkins 1993).[1] Since these pioneer macros reproduced, they met the necessary Darwinian conditions for evolution, and it is now clear that they spent the better part of a billion years evolving on Earth before there were any living things.

Even the simplest replicating macro is far from simple, however, a composition with thousands or millions of parts, depending on how we count the raw materials that go to make it. The alphabet letters *A*denine, *C*ytosine, *G*uanine, *T*hymine, and *U*racil are bases that are not too complex to arise in the normal course of prebiotic affairs. (RNA, which came before DNA, has Uracil, whereas DNA has Thymine.) Expert opinion differs, however, on whether these blocks could synthesize themselves by a series of coincidences into something as fancy as a self-replicator. The chemist Graham Cairns-Smith (1982, 1985) presents an updated version of Paley's argument, aimed at the molecular level: The process of synthesizing DNA fragments, even by the advanced methods of modern organic chemists, is highly elaborate; this shows that their chance creation is as improbable as Paley's watch in a windstorm. "Nucleotides are too expensive" (Cairns-Smith 1985, pp. 45–49). DNA exhibits too much design work to be a mere product of chance, Cairns-Smith argues, but he then proceeds to deduce an ingenious—if speculative and controversial—account of how that work might have been done. Whether or not Cairns-Smith's theory is eventually confirmed, it is well worth sharing simply because it so perfectly instantiates the fundamental Darwinian strategy.[2]

A good Darwinian, faced yet again with the problem of finding a needle in a haystack of Design Space, would cast about for a *still simpler* form of

---

1. Warning: biologists already use the term *macroevolution*, in contrast to microevolution, to refer to large-scale evolutionary phenomena—the patterns of speciation and extinction, for instance, in contrast to the refinement of wings or changes in resistance to toxins within a species. What I am calling the evolution of macros has nothing much to do with macroevolution in that established sense. The term *macro* is so apt for my purposes, however, that I have decided to stick with it, and try to offset its shortcomings with this patch—a tactic Mother Nature also often uses.

2. For just this reason, Richard Dawkins also presents a discussion and elaboration of Cairns-Smith's ideas in *The Blind Watchmaker* (1986a, pp. 148–58). Since Cairns-Smith's 1985 account and Dawkins' elaboration are such good reading for nonexperts, I will refer you to them for the delicious details, and provide just enough summary here to whet your appetite, adding the warning that there are problems with Cairns-Smith's hypotheses, and balancing the warning with the reassurance that even if his hypotheses are all ultimately rejected—an open question—there are other, less readily understandable, alternatives to take seriously next.

replicator that could somehow serve as a temporary scaffolding to hold the protein parts or nucleotide bases in place until the whole protein or macro could get assembled. Wondrous to say, there is a candidate with just the right properties, and more wondrous still, it is just what the Bible ordered: clay! Cairns-Smith shows that in addition to the carbon-based self-replicating crystals of DNA and RNA, there are also much simpler (he calls them "low-tech") silicon-based self-replicating crystals, and these silicates, as they are called, could themselves be the product of an evolutionary process. They form the ultra-fine particles of clay, of the sort that builds up just outside the strong currents and turbulent eddies in streams, and the individual crystals differ subtly at the level of molecular structure in ways that they pass on when they "seed" the processes of crystallization that achieve their self-replication.

Cairns-Smith develops intricate arguments to show how fragments of protein and RNA, which would be naturally attracted to the surfaces of these crystals like so many fleas, could eventually come to be used by the silicate crystals as "tools" in furthering their own replication processes. According to this hypothesis (which, like all really fertile ideas, has many neighboring variations, any one of which might prove to be the eventual winner), the building blocks of life began their careers as quasi-parasites of sorts, clinging to replicating clay particles and growing in complexity in the furtherance of the "needs" of the clay particles until they reached a point where they could fend for themselves. No skyhook—just a ladder that could be thrown away, as Wittgenstein once said in another context, once it had been climbed.

But this cannot be close to the whole story, even if it is all true. Suppose that short self-replicating strings of RNA got created by this low-tech process. Cairns-Smith calls these entirely self-involved replicators "naked genes," because they aren't *for* anything except their own replication, which they do without outside help. We are still left with a major problem: How did these naked genes ever come to be clothed? How did these solipsistic self-reproducers ever come to *specify* particular proteins, the tiny enzyme-machines that build the huge bodies that carry today's genes from generation to generation? But the problem is worse than that, for these proteins don't just build bodies; they are needed to assist in the very process of self-replication once a string of RNA or DNA gets long. Although short strings of RNA can replicate themselves without enzyme assistants, longer strings need a retinue of helpers, and specifying *them* requires a very long sequence—longer than could be replicated with high-enough fidelity until those very enzymes were already present. We seem to face paradox once again, in a vicious circle succinctly described by John Maynard Smith: "One cannot have accurate replication without a length of RNA of, say, 2000 base pairs, and one cannot have that much RNA without accurate replication" (Maynard Smith 1979, p. 445).

One of the leading researchers on this period of evolutionary history is Manfred Eigen. In his elegant little book, *Steps Towards Life* (1992)—a good place to continue your exploration of these ideas—he shows how the macros gradually built up what he calls the "molecular tool-kit" that living cells use to re-create themselves, while also building around themselves the sorts of structures that became, in due course, the protective membranes of the first prokaryotic cells. This long period of precellular evolution has left no fossil traces, but it has left plenty of clues of its history in the "texts" that have been transmitted to us through its descendants, including, of course, the viruses that swarm around us today. By studying the actual surviving texts, the specific sequences of A, C, G, and T in the DNA of higher organisms and the A, C, G, and U of their RNA counterparts, researchers can deduce a great deal about the actual identity of the earliest self-replicating texts, using refined versions of the same techniques the philologists used to reconstruct the words that Plato actually wrote. Some sequences in our own DNA are truly ancient, even traceable (via translation back into the earlier RNA language) to sequences that were composed in the earliest days of macro evolution!

Let's go back to the time when the nucleotide bases (A, C, G, T, and U) were occasionally present here and there in varying amounts, possibly congregated around some of Cairns-Smith's clay crystals. The twenty different amino acids, the building blocks for all proteins, also occur with some frequency under a wide range of nonbiotic conditions, so we can help ourselves to them as well. Moreover, it has been shown by Sidney Fox (Fox and Dose 1972) that individual amino acids can condense into "proteinoids," protein-like substances that have a very modest catalytic ability (Eigen 1992, p. 32). This is a small but important step up, since catalytic ability—the capacity to facilitate a chemical reaction—is the fundamental talent of any protein.

Now suppose some of the bases come to pair up, C with G, and A with U, and make smallish complementary sequences of RNA—less than a hundred pairs long—that can replicate, crudely, without enzymatic helpers. In terms of the Library of Babel, we would now have a printing press and a bookbindery, but the books would be too short to be good for anything except making more of themselves, with lots of misprints. And they would not be *about* anything. We may seem to be right back where we started—or even worse. When we bottom out at the level of molecular building blocks, we face a design problem that is more like construction out of Tinker Toy than gradual sculpting in modeling clay. Under the rigid rules of physics, either the atoms jump together into stable patterns or they don't.

Fortunately for us—indeed, fortunately for all living things—scattered in the Vast space of possible proteins there happen to be protein constructions that—if found—permit life to go forward. How might they get found? Somehow we have to get those proteins together with the protein-hunters, the fragments of self-replicating nucleotide strings that will *eventually* come to "specify" them in the macros they compose. Eigen shows how the vicious circle can turn friendly if it is expanded into a "hypercycle" with more than two elements (Eigen and Schuster 1977). This is a difficult technical concept, but the underlying idea is clear enough: imagine a circumstance in which fragments of type A can enhance the prospects of hunks of B, which in turn promote the well-being of bits of C, which, completing the loop, permit the replication of more fragments of A, and so forth, in a mutually reinforcing community of elements, until the point is reached where the whole process can take off, creating environments that normally serve to replicate longer and longer strings of genetic material. (Maynard Smith 1979 is a great help in understanding the idea of a hypercycle; see also Eigen 1983.)

But even if this is possible in principle, how could it get started? If all possible proteins and all possible nucleotide "texts" were truly equiprobable, then it would be hard to see how the process could ever get going. Somehow, the bland, mixed-up confetti of ingredients has to get some structure imposed on it, concentrating a few "likely-to-succeed" candidates and thereby making them *still more* likely to succeed. Remember the coin-tossing tournament in chapter 2? Somebody has to win, but the winner wins in virtue of no virtue, but simply in virtue of historical accident. The winner is not bigger or stronger or better than the other contestants, but is still the winner. It now seems that something similar happens in prebiotic molecular evolution, with a Darwinian twist: winners get to make extra copies of themselves for the next round, so that, without any selection "for cause" (as they say when dismissing potential jurors), dynasties of sheer replicative prowess begin to emerge. If we start with a purely random assortment of "contestants" drawn from the pool of self-replicating fragments, even if they are not initially distinguishable in terms of their replicative prowess, those that *happen* to win in the early rounds will occupy more of the slots in the subsequent rounds, flooding the space with trails of highly similar (short) texts, but still leaving vast hypervolumes of the space utterly empty and inaccessible for good. The initial threads of proto-life can emerge before there is any difference in skill, becoming the actuality from which the Tree of Life can then grow, thanks to tournaments of skill. As Eigen's colleague Bernd-Olaf Küppers (1990, p. 150) puts it, "The theory predicts *that* biological structures exist, but not *what* biological structures exist."[3] This is

---

3. Küppers (1990, pp. 137–46) borrows an example from Eigen (1976) to illustrate the underlying idea: a game of "non-Darwinian selection" you can play on a checkerboard with differently colored marbles. Start by randomly placing the marbles on all the squares, creating the initial confetti effect. Now throw two (eight-sided!) dice to determine a square (column 5, row 7, for instance) on which to act. Remove the marble on that

enough to build plenty of bias into the probability space from the outset.

So some of the possible macros, inevitably, are more probable—more likely to be stumbled upon in the Vast space of possibilities—than others. Which ones? The "fitter" ones? Not in any nontrivial sense, but just in the tautological sense of being identical to (or nearly identical to) previous "winners," who in turn tended to be almost identical to still earlier "winners." (In the million-dimension Library of Mendel, sequences that differ at a single locus are shelved "next to" each other in some dimension; the distance of any one volume from another is technically known as the Hamming distance. This process spreads "winners" out gradually—taking leaps of small Hamming distances—from any initial starting point in any and all directions in the Library.) This is the most rudimentary possible case of "the rich get richer," and since the success of the string has an explanation with no reference beyond the string itself and its resemblance as a string to its parent string, this is a purely *syntactic* definition of fitness, as opposed to a *semantic* definition of fitness (Küppers 1990, p. 141). That is, you don't have to consider what the string *means* in order to determine its fitness. We saw in chapter 6 that mere typographical change could never explain the Design that needs explaining, any more than you could explain the difference in quality between two books by comparing their relative frequencies of alphabetic characters, but before we can have the meaningful self-replicating codes that make this possible, we have to have self-replicating codes that don't mean a thing; their only "function" is to replicate themselves. As Eigen (1992, p. 15) puts it, "The structural stability of the molecule has no bearing upon the semantic information which it carries, and which is not expressed until the product of translation appears."

This is the birth of the ultimate QWERTY phenomenon, but, like the cultural case that gives it its name, it was not *entirely* without point even from the outset. Perfect equiprobability could have dissolved into a monopoly by a purely random process, as we have just seen, but perfect equiprobability is hard to come by in nature at any point, and at the very beginnings of this process of text generation, a bias was present. Of the four bases—A, C, G, and T—G and C are the most structurally stable: "Calculation of the necessary binding energies, along with experiments on binding

---

square. Throw the dice again; go to the square they name and check the color of the marble on this square and put a marble of that color on the just-vacated square ("reproduction" of that marble). Repeat the process, over and over. Eventually, it has the effect of unrandomizing the initial distribution of colors, so that one color ends up "winning" but for no reason at all—just historical luck. He calls this "non-Darwinian selection" because it is selection in the absence of a biasing cause; *selection without adaptation* would be the more familiar term. It is non-Darwinian only in the sense that Darwin didn't see the importance of allowing for it, not in the sense that Darwin (or Darwinism) cannot accommodate it. Manifestly it can.

and synthesis, show that sequences rich in G and C are best at self-replication by template instruction without the help of enzymes" (Eigen 1992, p. 34). This is, you might say, a natural or physical *spelling* bias. In English, "e" and "t" appear more frequently than, say, "u" or "j," but not because "e"s and "t"s are harder to erase, or easier to photocopy, or to write. (In fact, of course, the explanation runs the other way around; we tend to use the easiest-to-read-and-write symbols for the most frequently used letters; in Morse code, for example, "e" is assigned a single dot and "t" a single dash.) In RNA and DNA, this explanation is reversed: G and C are favored because they are the most stable in replication, not because they occur most frequently in genetic "words." This spelling bias is just "syntactic" at the outset, but it unites with a *semantic* bias:

> Examination of the genetic code [by the "philological methods"] ... indicates that its first codons were rich in G and C. The sequences GGC and GCC code respectively for the amino acids glycine and alanine, and because of their chemical simplicity these were formed in greater abundance ... [in the prebiotic world]. The assertion that the first code-words were *assigned* [emphasis added] to the most common amino acids is nothing if not plausible, and it underlines the fact that the logic of the coding scheme results from physical and chemical laws and their outworkings in Nature. [Eigen 1992, p. 34.]

These "outworkings" are *algorithmic sorting processes*, which take the probabilities or biases that are due to fundamental laws of physics and produce structures that would otherwise be wildly improbable. As Eigen says, the resulting scheme has a logic; it is not just two things coming together but an "assignment," a system that comes to make sense, and makes sense because—and only because—*it works*.

These very first "semantic" links are of course so utterly simple and local that they hardly count as semantic at all, but we can see a glimmer of *reference* in them nevertheless: there is a fortuitous wedding of a bit of nucleotide string with a protein fragment that helped directly or indirectly to reproduce *it*. The loop is closed; and once this "semantic" assignment system is in place, everything speeds up. Now a fragment of code-string can be the code *for* something—a protein. This creates a new dimension of evaluation, because some proteins are better than others at doing catalytic work, and particularly at assisting in the replication process.

This raises the stakes. Whereas at the outset, macro strings could differ only in their self-contained capacity to self-replicate, now they can magnify their differences by creating—and linking their fates to—other, larger, structures. Once this feedback loop is created, an arms race ensues: longer and longer macros compete for the available building blocks to build ever bigger, faster, more effective—but also more expensive—self-replicating systems. Our pointless coin-tossing tournament of luck has transformed itself into a tournament of skill. It has a point, for there is now something for the succession of winners to be better at than just, trivially, winning the coin-toss.

And does the new tournament ever work! There are tremendous "skill" differences between proteins, so there is plenty of room for improvement beyond the minuscule catalytic talents of the proteinoids. "In many cases, enzymic catalysis accelerates a reaction by a factor between one million and one thousand million. Wherever such a mechanism has been analysed quantitatively, the result has been the same: enzymes are optimal catalysts" (Eigen 1992, p. 22). Catalytic work done creates new jobs to be done, so the feedback cycles spread out to encompass more elaborate opportunities for improvement. "Whatever task a cell is adapted to, it carries out with optimal efficiency. The blue-green alga, a very early product of evolution, transforms light into chemical energy with an efficiency approaching perfection" (Eigen 1992, p. 16). Such optimality cannot be happenstance; it must be the result of a gradual homing-in process of improvement. So, from a set of tiny biases in the initial probabilities and competences of the building blocks, a process of snowballing self-improvement is initiated.

3. THE LAWS OF THE GAME OF LIFE

This most beautiful system of the sun, planets, and comets, could only proceed from the counsel and dominion of an Intelligent and Powerful Being.

—ISAAC NEWTON 1726 (passage translated in Ellegård 1956, p. 176)

The more I examine the universe and study the details of its architecture, the more evidence I find that the universe in some sense must have known that we were coming.

—FREEMAN DYSON 1979, p. 250

It is easy to imagine a world that, though ordered, nevertheless does not possess the right sort of forces or conditions for the emergence of significant depth.

—PAUL DAVIES 1992

Fortunately for us, the laws of physics vouchsafe that there are, in the Vast space of possible proteins, macromolecules of such breathtaking catalytic virtuosity that they can serve as the active building blocks of complex life. And, just as fortunately, the same laws of physics provide for just enough nonequilibrium in the world so that algorithmic processes can jump-start themselves, eventually discovering those macromolecules and turning them into tools for another wave of exploration and discovery. Thank God for those laws!

Well? Shouldn't we? If the laws were any different, we have just seen, the Tree of Life might never have sprung up. We may have figured out a way of excusing God from the task of designing the replication-machinery system (which can design itself automatically if any of the theories discussed in the previous section are right, or on the right track) but even if we concede that this is so, we still have the stupendous fact that the laws *do* permit this wonderful unfolding to happen, and that has been quite enough to inspire many people to surmise that the Intelligence of the Creator is the Wisdom of the Lawgiver, instead of the Ingenuity of the Engineer.

When Darwin entertains the idea that the laws of nature are designed by God, he has distinguished company, past and present. Newton insisted that the original arrangement of the universe was inexplicable by "meer natural causes" and could only be ascribed to "the Counsel and Contrivance of a Voluntary Agent." Einstein spoke of the laws of nature as the "secrets of the Old One" and famously expressed his disbelief in the role of chance in quantum mechanics by proclaiming "*Gott würfelt nicht*"—God does not play dice. More recently, the astronomer Fred Hoyle has said, "I do not believe that any scientist who examined the evidence would fail to draw the inference that the laws of nuclear physics have been deliberately designed with regard to the consequences they produce inside the stars" (quoted in Barrow and Tipler 1988, p. 22). The physicist and cosmologist Freeman Dyson puts the point much more cautiously: "I do not claim that the architecture of the universe proves the existence of God. I claim only that the architecture of the universe is consistent with the hypothesis that mind plays an essential role in its functioning" (Dyson 1979, p. 251). Darwin himself was prepared to propose an honorable truce at this point, but Darwinian thinking carries on, with a momentum created by the success of its earlier applications to the same issue in other contexts.

As more and more has been learned about the development of the universe since the Big Bang, about the conditions that permitted the formation of galaxies and stars and the heavy elements from which planets can be formed, physicists and cosmologists have been more and more struck by the exquisite sensitivity of the laws of nature. The speed of light is approximately 186,000 miles per second. What if it were only 185,000 miles per second, or 187,000 miles per second? Would that change much of anything? What if the force of gravity were 1 percent more or less than it is? The fundamental constants of physics—the speed of light, the constant of gravitational attraction, the weak and strong forces of subatomic interaction, Planck's constant—have values that of course permit the actual development of the universe as we know it to have happened. But it turns out that if in imagination we change any of these values by just the tiniest amount, we thereby posit a universe in which none of this could have happened, and indeed in which apparently nothing life-like could ever have emerged: no planets, no atmospheres, no solids at all, no elements except hydrogen and helium, or maybe not even that—just some boring plasma of hot, undifferentiated stuff, or an equally boring nothingness. So isn't it a wonderful fact that the laws are *just right* for us to exist? Indeed, one might want to add, we almost didn't make it!

Is this wonderful fact something that needs an explanation, and, if so, what kind of explanation might it receive? According to the Anthropic Principle, we are entitled to infer facts about the universe and its laws from the undisputed fact that we (we *anthropoi*, we human beings) are here to do the inferring and observing. The Anthropic Principle comes in several flavors. (Among the useful recent books is Barrow and Tipler 1988 and Breuer 1991. See also Pagels 1985, Gardner 1986.)

In the "weak form," it is a sound, harmless, and on occasion useful application of elementary logic: if x is a necessary condition for the existence of y, and y exists, then x exists. If consciousness depends on complex physical structures, and complex structures depend on large molecules composed of elements heavier than hydrogen and helium, then, since we are conscious, the world must contain such elements.

But notice that there is a loose cannon on the deck in the previous sentence: the wandering "must." I have followed the common practice in ordinary English of couching a claim of necessity in a technically incorrect way. As any student in logic class soon learns, what I really should have written is:

> *It must be the case that:* if consciousness depends ... then, since we are conscious, the world *contains* such elements.

The conclusion that can be validly drawn is only that the world *does* contain such elements, not that it *had to* contain such elements. It *has to* contain such elements *for us to exist*, we may grant, but it might not have contained such elements, and if that had been the case, we wouldn't be here to be dismayed. It's as simple as that.

Some attempts to define and defend a "strong form" of the Anthropic Principle strive to justify the late location of the "must" as not casual expression but a conclusion about the way the universe necessarily is. I admit that I find it hard to believe that so much confusion and controversy are actually generated by a simple mistake of logic, but the evidence is really quite strong that this is often the case, and not just in discussions of the Anthropic Principle. Consider the related confusions that surround Darwinian deduction in general. Darwin deduces that human beings *must have* evolved from a common ancestor of the chimpanzee, or that all life *must have* arisen from a single beginning, and some people, unaccountably, take these deductions as claims that human beings are somehow a necessary product of evolution, or that life is a necessary feature of our planet, but nothing of the kind follows from Darwin's deductions properly construed. What must be the case is not that we are here, but that *since* we are here, we evolved from primates. Suppose John is a bachelor. Then he *must* be single, right? (That's a truth of logic.) Poor John—he can never get married! The fallacy is obvious in this example, and it is worth keeping it in the back of your mind as a template to compare other arguments with.

Believers in any of the proposed strong versions of the Anthropic Principle think they can deduce something wonderful and surprising from the fact that we conscious observers are here—for instance, that in some sense the universe exists *for* us, or perhaps that we exist *so that* the universe as a whole can exist, or even that God created the universe the way He did so that we would be possible. Construed in this way, these proposals are attempts to restore Paley's Argument from Design, readdressing it to the Design of the universe's most general laws of physics, not the particular constructions those laws make possible. Here, once again, Darwinian countermoves are available.

These are deep waters, and most of the discussions of the issues wallow in technicalities, but the logical force of these Darwinian responses can be brought out vividly by considering a much simpler case. First, I must introduce you to the Game of Life, a nifty meme whose principal author is the mathematician John Horton Conway. (I will be putting this valuable thinking tool to several more uses, as we go along. This game does an excellent job of taking in a complicated issue and reflecting back only the dead-simple essence or skeleton of the issue, ready to be understood and appreciated.)

Life is played on a two-dimensional grid, such as a checkerboard, using simple counters, such as pebbles or pennies—or one could go high-tech and play it on a computer screen. It is not a game one plays to win; if it is a game at all, it is solitaire.[4] The grid divides space into square cells, and each cell

4. This description of Life is adapted from an earlier exposition of mine (1991b). Martin Gardner introduced the Game of Life to a wide audience in two of his "Mathematical Games" columns in *Scientific American*, in October 1970 and February 1971. Poundstone 1985 is an excellent exploration of the game and its philosophical implications.

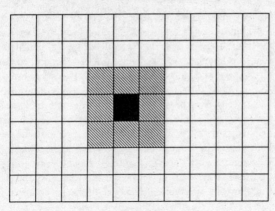

Figure 7.2

is either ON or OFF at each moment. (If it is ON, place a penny on the square; if it is OFF, leave the square empty.) Notice in figure 7.2 that each cell has eight neighbors: the four adjacent cells—north, south, east, and west—and the four diagonals—northeast, southeast, southwest, and northwest.

Time in the Life world is discrete, not continuous; it advances in ticks, and the state of the world changes between each two ticks according to the following rule:

> *Life Physics:* For each cell in the grid, count how many of its eight neighbors are ON at the present instant. If the answer is exactly two, the cell stays in its present state (ON or OFF) in the next instant. If the answer is exactly three, the cell is ON in the next instant whatever its current state. Under all other conditions, the cell is OFF.

That's it—that's the only rule of the game. You now know all there is to know about how to play Life. *The entire physics of the Life world is captured in that single, unexceptioned law.* Although this is the fundamental law of the "physics" of the Life world, it helps at first to conceive this curious physics in biological terms: think of cells going ON as births, cells going OFF as deaths, and succeeding instants as generations. Either overcrowding (more than three inhabited neighbors) or isolation (fewer than two inhabited neighbors) leads to death. Consider a few simple cases.

In the configuration in figure 7.3, only cells *d* and *f* each have exactly three neighbors ON, so they will be the only birth cells in the next generation. Cells *b* and *h* each have only one neighbor ON, so they die in the next generation. Cell *e* has two neighbors ON, so it stays on. Thus the next "instant" will be the configuration shown in figure 7.4.

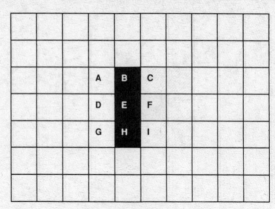

Figure 7.3

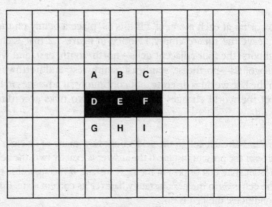

Figure 7.4

Obviously, the configuration will revert back in the next instant, and this little pattern will flip-flop back and forth indefinitely, unless some new on cells are brought into the picture somehow. It is called a *flasher* or traffic light. What will happen to the configuration in figure 7.5?

Nothing. Each on cell has three neighbors on, so it is reborn just as it is. No off cell has three neighbors on, so no other births happen. This configuration is called a *still life*. By the scrupulous application of our single law, one can predict with perfect accuracy the next instant of any configuration of on and off cells, and the instant after that, and so forth. In other words, the Life world is a toy world that perfectly instantiates the determinism made famous by Laplace: if we are given the state description of this world at an instant, we observers can perfectly predict the future instants by the simple

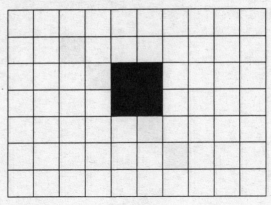

FIGURE 7.5

application of our one law of physics. Or, in the terms I have developed in earlier writings (1971, 1978, 1987b), when we *adopt the physical stance* towards a configuration in the Life world, our powers of prediction are perfect: there is no noise, no uncertainty, no probability less than one. Moreover, it follows from the two-dimensionality of the Life world that nothing is hidden from view. There is no backstage; there are no hidden variables; the unfolding of the physics of objects in the Life world is directly and completely visible.

If you find following the simple rule a tedious exercise, there are computer simulations of the Life world in which you can set up configurations on the screen and let the computer execute the algorithm for you, changing the configuration again and again according to the single rule. In the best simulations, one can change the scale of both time and space, alternating between close-up and bird's-eye view. A nice touch added to some color versions is that ON cells (often just called *pixels*) are color-coded by their age; they are born blue, let us say, and then change color each generation, moving through green to yellow to orange to red to brown to black and then staying black unless they die. This permits one to see at a glance how old certain patterns are, which cells are co-generational, where the birth action is, and so forth.[5]

One soon discovers that some simple configurations are more interesting than others. Consider a diagonal line segment, such as the one in figure 7.6.

---

5. Poundstone 1985 provides simple BASIC and IBM-PC assembly language simulations you can copy for your own home computer, and describes some of the interesting variations.

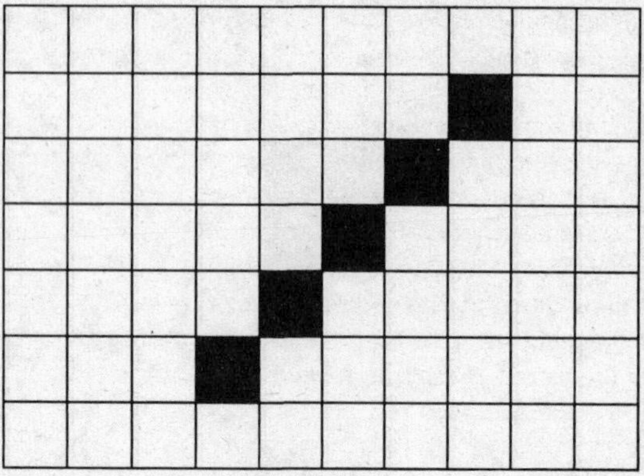

FIGURE 7.6

It is *not* a flasher; each generation, its two end ON cells die of isolation, and there are no birth cells. The whole segment soon evaporates. In addition to the configurations that never change—the still lifes—and those that evaporate entirely—such as the diagonal line segment—there are configurations with all manner of periodicity. The flasher, we saw, has a two-generation period that continues *ad infinitum*, unless some other configuration encroaches. Encroachment is what makes Life interesting: among the periodic configurations are some that swim, amoebalike, across the plane. The simplest is the *glider*, the five-pixel configuration shown taking a single stroke to the southeast in figure 7.7.

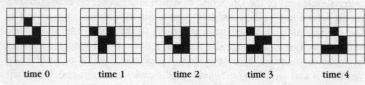

| time 0 | time 1 | time 2 | time 3 | time 4 |

FIGURE 7.7

Then there are the *eaters*, *puffer trains*, *space rakes*, and a host of other aptly named denizens of the Life world that emerge as recognizable objects at a new level. (This level is analogous to what in earlier work I have called the *design level*.) This level has its own language, a transparent foreshortening of the tedious descriptions one could give at the physical level. For instance:

An eater can eat a glider in four generations. Whatever is being consumed, the basic process is the same. A bridge forms between the eater and its prey. In the next generation, the bridge region dies from overpopulation, taking a bite out of both eater and prey. The eater then repairs itself. The prey usually cannot. If the remainder of the prey dies out as with the glider, the prey is consumed. [Poundstone 1985, p. 38.]

Notice that something curious happens to our "ontology"—our catalogue of what exists—as we move between levels. At the physical level there is no motion, just ON and OFF, and the only individual things that exist, cells, are defined by their fixed spatial location. At the design level we suddenly have the motion of persisting objects; it is one and the same glider (though composed each generation of different cells) that has moved southeast in figure 7.6, changing shape as it moves; and there is one less glider in the world after the eater has eaten it in figure 7.8.

time 0   time 1   time 2   time 3   time 4

FIGURE 7.8

Notice, too, that, whereas at the physical level there are absolutely no exceptions to the general law, at this level our generalizations have to be hedged: they require "usually" or "provided nothing encroaches" clauses. Stray bits of debris from earlier events can "break" or "kill" one of the objects in the ontology at this level. Their *salience as real things* is considerable, but not guaranteed. To say that their salience is considerable is to say that one can, with some small risk, ascend to this design level, adopt its ontology, and proceed to predict—sketchily and riskily—the behavior of larger configurations or systems of configurations, without bothering to compute the physical level. For instance, one can set oneself the task of designing some interesting supersystem out of the "parts" that the design level makes available.

This is just what Conway and his students set out to do, and they succeeded majestically. They designed, and proved the viability of the design of, a self-reproducing entity composed entirely of Life cells that was also (for good measure) a Universal Turing machine—it was a two-dimensional computer that in principle can compute any computable function! What on Earth inspired Conway and his students to create first this world and then this amazing denizen of that world? They were trying to answer at a very abstract level one of the central questions we have been considering in this chapter: what is the minimal complexity required for a self-reproducing thing? They were following up the brilliant early speculations of John von Neumann, who had been working on the question at the time of his death in 1957. Francis Crick and James Watson had discovered DNA in 1953, but how it worked was a mystery for many years. Von Neumann had imagined in some detail a sort of floating robot that picked up pieces of flotsam and jetsam that could be used to build a duplicate of itself that would then be able to repeat the process. His description (posthumously published, 1966) of how an automaton would read its own blueprint and then copy it into its new creation anticipated in impressive detail many of the later discoveries about the mechanisms of DNA expression and replication, but in order to make his proof of the possibility of a self-reproducing automaton mathematically rigorous and tractable, von Neumann had switched to simple, two-dimensional abstractions, now known as *cellular automata*. Conway's Life-world cells are a particularly agreeable example of cellular automata.

Conway and his students wanted to confirm von Neumann's proof in detail by actually constructing a two-dimensional world with a simple physics in which such a self-replicating construction would be a stable, *working* structure. Like von Neumann, they wanted their answer to be as general as possible, and hence as independent as possible of actual (Earthly? local?) physics and chemistry. They wanted something dead simple, easy to visualize and easy to calculate, so they not only dropped from three dimensions to two; they also "digitized" both space and time—all times and distances, as we saw, are in whole numbers of "instants" and "cells." It was von Neumann who had taken Alan Turing's abstract conception of a mechanical computer (now called a "Turing machine") and engineered it into the specification for a general-purpose stored-program serial-processing computer (now called a "von Neumann machine"); in his brilliant explorations of the spatial and structural requirements for such a computer, he had realized—and proved—that a Universal Turing machine (a Turing machine that can compute any computable function at all) could in principle be "built" in a two-dimensional world.[6] Conway and his students also set out to confirm this with their own exercise in two-dimensional engineering.[7]

It was far from easy, but they showed how they could "build" a working computer out of simpler Life forms. Glider streams can provide the input-output "tape," for instance, and the tape-reader can be some huge assembly of eaters, gliders, and other bits and pieces. What does this machine look like? Poundstone calculates that the whole construction would be on the order of 10^{13} cells or pixels.

---

6. See Dennett 1987b, ch. 9, for more on the theoretical implications of this trade-off in space and time.

7. For a completely different perspective on two-dimensional physics and engineering, see A. K. Dewdney's *The Planiverse* (1984), a vast improvement over Abbott's *Flatland* (1884).

Displaying a 10^{13}-pixel pattern would require a video screen about 3 million pixels across at least. Assume the pixels are 1 millimeter square (which is very high resolution by the standards of home computers). Then the screen would have to be 3 kilometers (about two miles) across. It would have an area about six times that of Monaco.

Perspective would shrink the pixels of a self-reproducing pattern to invisibility. If you got far enough away from the screen so that the entire pattern was comfortably in view, the pixels (and even the gliders, eaters and guns) would be too tiny to make out. A self-reproducing pattern would be a hazy glow, like a galaxy. [Poundstone 1985, pp. 227–28.]

In other words, by the time you have built up enough pieces into something that can reproduce itself (in a two-dimensional world), it is roughly as much larger than its smallest bits as an organism is larger than its atoms. You probably can't do it with anything much less complicated, though this has not been strictly proven. The hunch with which we began this chapter gets dramatic support: it takes a *lot* of design work (the work done by Conway and his students) to turn available bits and pieces into a self-replicating thing; self-replicators don't just fall together in cosmic coincidences; they are too large and expensive.

The Game of Life illustrates many important principles, and can be used to construct many different arguments or thought experiments, but I will content myself here with just two points that are particularly relevant to this stage in our argument, before turning to my main point. (For further reflections on Life and its implications, see Dennett 1991b.)

First, notice how the distinction between Order and Design gets blurred here, just as it did for Hume. Conway *designed* the whole Life world—that is, he set out to articulate an Order that would *function* in a certain way. But do gliders, for instance, count as designed things, or as just natural objects—like atoms or molecules? Surely the tape-reader Conway and his students cobbled together out of gliders and the like is a designed object, but the simplest glider would seem just to fall out of the basic physics of the Life world "automatically"—nobody had to design or invent the glider; it just was *discovered* to be implied by the physics of the Life world. But that, of course, is actually true of *everything* in the Life world. Nothing happens in the Life world that isn't strictly implied—logically deducible by straightforward theorem-proving—by the physics and the initial configuration of cells. Some of the things in the Life world are just more marvelous and unanticipated (by us, with our puny intellects) than others. There is a sense in which the Conway self-reproducing pixel-galaxy is "just" one more Life macromolecule with a very long and complicated periodicity in its behavior.

What if we set in motion a huge herd of these self-reproducers, and let them compete for resources. And suppose they then evolved—that is, their descendants were not exact duplicates of them. Would these descendants have a greater claim to having been designed? Perhaps, but there is no line to be drawn between merely ordered things and designed things. The engineer starts with some *objets trouvés*, found objects with properties that can be harnessed in larger constructions, but the differences between a designed and manufactured nail, a sawn plank, and a naturally occurring slab of slate are not "principled." Seagull wings are great lifters, hemoglobin macromolecules are superb transporting machines, glucose molecules are nifty energy-packets, and carbon atoms are outstanding all-purpose stickum-binders.

The second point is that Life is an excellent illustration of the power—and an attendant weakness—of computer simulations addressed to scientific questions. It used to be that the only way to persuade oneself of very abstract generalizations was to prove them rigorously from the fundamental principles or axioms of whatever theory one had: mathematics, physics, chemistry, economics. Earlier in this century, it was beginning to become clear that many of the theoretical calculations one would like to make in these sciences were simply beyond human capacity—"intractable." Then the computer came along to provide a new way of addressing such questions: massive simulations. Simulation of the weather is the example familiar to all of us from watching television meteorologists, but computer simulation is also revolutionizing how science is conducted in many other fields, probably the most important *epistemological* advance in scientific method since the invention of accurate timekeeping devices. In evolutionary theory, the new discipline of Artificial Life has recently sprung up to provide a name and an umbrella to cover a veritable Gold Rush of researchers at different levels, from the submolecular to the ecological. Even among those researchers who have not taken up the banner of Artificial Life, however, there is general acknowledgment that most of their theoretical research on evolution—most of the recent work discussed in this book, for instance—would have been simply unthinkable without computer simulations to test (to confirm *or* disconfirm) the intuitions of the theoreticians. Indeed, as we have seen, the very idea of evolution as an algorithmic process could not be properly formulated and evaluated until it was possible to test huge, complicated algorithmic models in place of the wildly oversimple models of earlier theorists.

Now, some scientific problems are not amenable to solution-by-simulation, and others are probably only amenable to solution-by-simulation, but in between there are problems that can in principle be addressed in two different ways, reminiscent of the two different ways of solving the train problem given to von Neumann—a "deep" way via theory, and a "shallow" way via brute-force simulation and inspection. It would be a shame if the many undeniable attractions of simulated worlds drowned out our aspirations to understand these phenomena in the deep ways of theory. I spoke with Conway once about the creation of the Game of Life, and he lamented the fact that explorations of the Life world were now almost exclusively by "empirical" methods—setting up all the variations of interest on a computer and letting her rip to see what happens. Not only did this usually shield one from even the opportunity of devising a strict proof of what one found, but, he noted, people using computer simulations are typically insufficiently patient; they try out combinations and watch them for fifteen or twenty minutes, and if nothing of interest has happened, they abandon them, marking them as avenues already explored and found barren. This myopic style of exploration risks closing off important avenues of research prematurely. It is an occupational hazard of all computer simulators, and it is simply their high-tech version of the philosopher's fundamental foible: mistaking a failure of imagination for an insight into necessity. A prosthetically enhanced imagination is still liable to failure, especially if it is not used with sufficient rigor.

But now it is time for the my main point. When Conway and his students first set out to create a two-dimensional world in which interesting things would happen, they found that nothing seemed to work. It took more than a year for this industrious and ingenious group of intelligent searchers to find the simple Life Physics rule in the Vast space of possible simple rules. All the obvious variations turned out to be hopeless. To get some sense of this, try altering the "constants" for birth and death—change the birth rule from three to four, for instance—and see what happens. The worlds these variations govern either freeze up solid in no time or evaporate into nothingness in no time. Conway and his students wanted a world in which growth was possible, but not too explosive; in which "things"—higher-order patterns of cells—could move, and change, but also retain their identity over time. And of course it had to be a world in which structures could "do things" of interest (like eat or make tracks or repel things). Of all the imaginable two-dimensional worlds, so far as Conway knows, there is only one that meets these *desiderata*: the Life world. In any event, the variations that have been checked in subsequent years have never come close to measuring up to Conway's in terms of interest, simplicity, fecundity, elegance. The Life world might indeed be the best of all possible (two-dimensional) worlds.

Now suppose that some self-reproducing Universal Turing machines in the Life world were to have a conversation with each other about the world as they found it, with its wonderfully simple physics—expressible in a single sentence and covering all eventualities.[8] They would be committing a log-

---

8. John McCarthy has for years been exploring the theoretical question of the minimal Life-world configuration that can learn the physics of its own world, and has tried to enlist

ical howler if they argued that since they existed, the Life world, with its particular physics, *had* to exist—for after all, Conway might have decided to be a plumber or play bridge instead of hunting for this world. But what if they deduced that their world was just too wonderful, with its elegant, Life-sustaining physics, to have come into existence without an Intelligent Creator? If they jumped to the conclusion that they owed their existence to the activities of a wise Lawgiver, they'd be right! There is a God and his name is Conway.

But they would be *jumping* to a conclusion. The existence of a universe obeying a set of laws even as elegant as the Life law (or the laws of our own physics) does not logically require an intelligent Lawgiver. Notice first how the actual history of the Game of Life divided the intellectual labor in two: on the one hand there was the initial exploratory work that led to the physical law promulgated by the Lawgiver, and on the other hand there was the engineering work of the law-exploiters, the Artificers. These *might* have happened in that temporal order—first Conway, in a stroke of inspired genius, promulgates the physics of the Life world, and then he and his students design and build the wonderful denizens of that world according to the law laid down. But in fact the two tasks were intermixed; many trial-and-error attempts to make things that were interesting provided the guidance for Conway's legislative search. Notice, second, that this postulated division of labor illustrates a fundamental Darwinian theme from the previous chapter. The task of the wise God required to put this world into motion is a task of discovery, not creation, a job for a Newton, not a Shakespeare. What Newton found—and what Conway found—are eternal Platonic fixed points that anybody else in principle could have discovered, not idiosyncratic creations that depend in any way on the particularities of the minds of their authors. If Conway had never turned his hand to designing cellular-automata worlds—if Conway had never even existed—some other mathematician might very well have hit upon *exactly* the Life world that Conway gets the credit for. So, as we follow the Darwinian down this path, God the Artificer turns first into God the Law*giver*, who now can be seen to merge with God the Law*finder*. God's hypothesized contribution is thereby becoming less personal—and hence more readily performable by something dogged and mindless!

Hume has already shown us how the argument runs, and now, bolstered by our experience with Darwinian thinking in more familiar terrain, we can

---

his friends and colleagues in this quest. I have always found the prospect of such a proof mouth-watering, but the paths to it are totally beyond me. So far as I know, nothing substantive has yet been published on this most interesting epistemological question, but I want to encourage others to address it. The same thought experiment is posed, independently, in Stewart and Golubitsky 1992, pp. 261–62.

extrapolate a positive Darwinian alternative to the hypothesis that our laws are a gift from God. What would the Darwinian alternative have to be? That there has been an evolution of worlds (in the sense of whole universes), and the world we find ourselves in is simply one among countless others that have existed through eternity. There are two quite different ways of thinking about the evolution of laws, one of them stronger, more "Darwinian," than the other in that it involves something like natural selection.

Might it be that there has been some sort of *differential reproduction* of universes, with some varieties having more "offspring" than others? Hume's Philo toyed with this idea, as we saw in chapter 1:

> And what surprise must we entertain, when we find him a stupid mechanic, who imitated others, and copied an art, which, through a long succession of ages, after multiplied trials, mistakes, corrections, deliberations, and controversies, had been gradually improving? Many worlds might have been botched and bungled, throughout an eternity, ere this system was struck out: Much labour lost: Many fruitless trials made: And a slow, but continued improvement carried on during infinite ages of world-making. [Pt. V.]

Hume imputes the "continued improvement" to the minimal selective bias of a "stupid mechanic," but we can replace the stupid mechanic with something even stupider without dissipating the lifting power: a purely algorithmic Darwinian process of world-trying. Though Hume obviously didn't think this was anything but an amusing philosophical fantasy, the idea has recently been developed in some detail by the physicist Lee Smolin (1992). The basic idea is that the singularities known as black holes are in effect the birthplaces of offspring universes, in which the fundamental physical constants would differ slightly, in random ways, from the physical constants in the parent universe. So, according to Smolin's hypothesis, we have both differential reproduction and mutation, the two essential features of any Darwinian selection algorithm. Those universes that just happened to have physical constants that encouraged the development of black holes would *ipso facto* have more offspring, which would have more offspring, and so forth—that's the selection step. Note that there is no grim reaper of universes in this scenario; they all live and "die" in due course, but some merely have more offspring. According to this idea, then, it is no mere interesting coincidence that we live in a universe in which there are black holes, nor is it an absolute logical necessity. It is, rather, the sort of conditional near-necessity you find in any evolutionary account. The link, Smolin claims, is carbon, which plays a role both in the collapse of gaseous clouds (or in other words, the birth of stars, a precursor to the birth of black holes) and, of course, in our molecular engineering.

Is the theory testable? Smolin offers some predictions that would, if dis-confirmed, pretty well eliminate his idea: it should be the case that all the "near" variations in physical constants from the values we enjoy should yield universes in which black holes are less probable or less frequent than in our own. In short, he thinks our universe should manifest at least a local, if not global, optimum in the black-hole-making competition. The trouble is that there are too few constraints, so far as I can see, on what should count as a "near" variation and why, but perhaps further elaboration on the theory will clarify this. Needless to say, it is hard to know what to make of this idea yet, but whatever the eventual verdict of scientists, the idea already serves to secure a philosophical point. Freeman Dyson and Fred Hoyle, among many others, think they see a wonderful pattern in the laws of physics; if they or anyone else were to make the tactical mistake of asking the rhetor-ical question "What else but God could *possibly* explain it?" Smolin would have a nicely deflating reply. (I advise my philosophy students to develop hypersensitivity for rhetorical questions in philosophy. They paper over whatever cracks there are in the arguments.)

But suppose, for the sake of argument, that Smolin's speculations are all flawed; suppose *selection* of universes doesn't work after all. There is a weaker, semi-Darwinian speculation that also answers the rhetorical ques-tion handily. Hume toyed with this weaker idea, too, as we already noted, in part VIII of his *Dialogues*:

> Instead of supposing matter infinite, as Epicurus did, let us suppose it finite. A finite number of particles is only susceptible of finite transpositions: And it must happen, in an eternal duration, that every possible order or position must be tried an infinite number of times. . . .
>
> Suppose . . . that matter were thrown into any position, by a blind, unguided force; it is evident that this first position must in all probability be the most confused and most disorderly imaginable, without any resem-blance to those works of human contrivance, which, along with a symme-try of parts, discover an adjustment of means to ends and a tendency to self-preservation. . . . Suppose, that the actuating force, whatever it be, still continues in matter. . . . Thus the universe goes on for many ages in a continued succession of chaos and disorder. But is it not possible that it may settle at last . . . ? May we not hope for such a position, or rather be assured of it, from the eternal revolutions of unguided matter, and may not this account for all the appearing wisdom and contrivance which is in the universe?

This idea exploits no version of selection at all, but simply draws atten-tion to the fact that we have eternity to play with. There is no five-billion-year deadline in this instance, the way there is for the evolution of life on Earth. As we saw in our consideration of the Libraries of Babel and Mendel, we need reproduction and selection if we are to traverse Vast spaces in non-Vast amounts of time, but when time is no longer a limiting consideration, selection is no longer a requirement. In the course of eternity, you can go *everywhere* in the Library of Babel or the Library of Mendel—or the Library of Einstein (all possible values of all the constants of physics)—as long as you keep moving. (Hume imagines an "actuating force" to keep the shuffling going, and this reminds us of Locke's argument about matter without motion, but it does not suppose that the actuating force has any intelligence at all.) In fact, if you shuffle through all the possibilities for eternity, you will pass through each possible place in these Vast (but finite) spaces not just once but an infinity of times!

Several versions of this speculation have been seriously considered by physicists and cosmologists in recent years. John Archibald Wheeler (1974), for instance, has proposed that the universe oscillates back and forth for eternity: a Big Bang is followed by expansion, which is followed by contraction into a Big Crunch, which is followed by another Big Bang, and so forth forever, with random variations in the constants and other crucial parameters occurring in each oscillation. Each possible setting is tried an infinity of times, and so every variation on every theme, both those that "make sense" and those that are absurd, spins itself out, not once but an infinity of times.

It is hard to believe that this idea is empirically testable in any meaningful way, but we should reserve judgment. Variations or elaborations on the theme just might have implications that could be confirmed or disconfirmed. In the meantime, it is worth noting that this family of hypotheses does have the virtue of extending the principles of explanation that work so well in testable domains all the way out. Consistency and simplicity are in its favor. And that, once again, is certainly enough to blunt the appeal of the traditional alternative.[9]

Anybody who won a coin-tossing tournament would be tempted to think he was blessed with magical powers, especially if he had no direct knowledge of the other players. Suppose you were to create a ten-round coin-tossing tournament without letting each of the 1,024 "contestants" realize he was entered in a tournament. You say to each one as you recruit him: "Congratulations, my friend. I am Mephistopheles, and I am going to bestow great powers on you. With me at your side, you are going to win ten

---

9. For a more detailed analysis of these issues, and a defense of a "neo-Platonist" middle ground, see J. Leslie 1989. (Like most middle grounds, this is not likely to appeal to either the devout or the skeptical, but it is at least an ingenious attempt at a compromise.) Van Inwagen (1993a, chh. 7 and 8) provides a clear and relentless analysis of the arguments—Leslie's, but also the arguments I have presented here—from a position of unusual neutrality. Anyone less than satisfied with my treatment should turn to this source first.

consecutive coin-tosses without a loss!" You then arrange for your dupes to meet, pairwise, until you have a final winner. (You never let the contestants discuss your relation to them, and you kiss off the 1,023 losers along the way with some *sotto voce* gibe to the effect that they were pretty gullible to believe your claim about being Mephistopheles!) The winner—and there must be one—will certainly have been given evidence of being a Chosen One, but if he falls for it, this is simply an illusion of what we might call retrospective myopia. The winner doesn't see that the situation was structured so that somebody simply had to be the lucky one—and he just happened to be *it*.

Now *if* the universe were structured in such a way that an infinity of different "laws of physics" got tried out in the fullness of time, we would be succumbing to the same temptation were we to draw any conclusions about the laws of nature being prepared especially for us. This is not an argument for the conclusion that the universe is, or must be, so structured, but just for the more modest conclusion that no feature of the observable "laws of nature" could be invulnerable to this alternative, deflationary interpretation.

Once these ever more speculative, ever more attenuated Darwinian hypotheses are formulated, they serve—in classic Darwinian fashion—to diminish by small steps the explanatory task facing us. All that is left over in need of explanation at this point is a certain perceived elegance or wonderfulness in the observed laws of physics. If you doubt that the hypothesis of an infinity of variant universes could actually explain this elegance, you should reflect that this has at least as much claim to being a non-question-begging explanation as any traditional alternative; by the time God has been depersonalized to the point of being some abstract and timeless principle of beauty or goodness, it is hard to see how the existence of God could explain anything. What would be asserted by the "explanation" that was not already given in the description of the wonderful phenomenon to be explained?

Darwin began his attack on the Cosmic Pyramid in the middle: Give me Order, and time, and I will explain Design. We have now seen how the downward path of universal acid flows: if we give his successors Chaos (in the old-fashioned sense of pure meaningless randomness), and eternity, they will explain Order—the very Order needed to account for the Design. Does utter Chaos in turn *need* an explanation? What is there left to explain? Some people think there is still one leftover "why" question: *Why is there something rather than nothing?* Opinions differ on whether the question makes any intelligible demand at all.[10] If it does, the answer "Because God

---

10. For an engaging examination of the question, see ch. 2 of Robert Nozick's *Philosophical Explanation*. Nozick offers several different candidate answers, all of them admittedly bizarre, but notes, disarmingly: "The question cuts so deep, however, that any

exists" is probably as good an answer as any, but look at its competition: "Why not?"

4. ETERNAL RECURRENCE—LIFE WITHOUT FOUNDATIONS?

Science is wonderful at destroying metaphysical answers, but incapable of providing substitute ones. Science takes away foundations without providing a replacement. Whether we want to be there or not, science has put us in a position of having to live without foundations. It was shocking when Nietzsche said this, but today it is commonplace; our historical position—and no end to it is in sight—is that of having to philosophize without 'foundations'.

—HILARY PUTNAM 1987, p. 29

The sense that the meaning of the universe had evaporated was what seemed to escape those who welcomed Darwin as a benefactor of mankind. Nietzsche considered that evolution presented a correct picture of the world, but that it was a disastrous picture. His philosophy was an attempt to produce a new world-picture which took Darwinism into account but was not nullified by it.

—R. J. HOLLINGDALE 1965, p. 90

In the wake of Darwin's publication of *Origin of Species*, Friedrich Nietzsche rediscovered what Hume had already toyed with: the idea that an eternal recurrence of blind, meaningless variation—chaotic, pointless shuffling of matter and law—would inevitably spew up worlds whose evolution through time would yield the *apparently* meaningful stories of our lives. This idea of eternal recurrence became a cornerstone of his nihilism, and thus part of the foundation of what became existentialism.

The idea that what is happening now has all happened before must be as old as the *déjà-vu* phenomenon that so often inspires superstitious versions of it. Cyclical cosmogonies are not uncommon in the catalogue of human cultures. But when Nietzsche hit upon a version of Hume's—and John Archibald Wheeler's—vision, he took it to be much more than an amusing thought experiment or an elaboration of ancient superstitions. He thought—at least for a while—he had stumbled upon a scientific proof of

---

approach that stands a chance of yielding an answer will look extremely weird. Someone who proposes a non-strange answer shows he didn't understand the question" (Nozick 1981, p. 116).

the greatest importance.[11] I suspect that Nietzsche was encouraged to take the idea more seriously than Hume had done by his dim appreciation of the tremendous power of Darwinian thinking.

Nietzsche's references to Darwin are almost all hostile, but there are quite a few, and that in itself supports Walter Kaufmann's argument (1950, preface) that Nietzsche "was not a Darwinist, but only aroused from his dogmatic slumber by Darwin, much as Kant was a century earlier by Hume." Nietzsche's references to Darwin also reveal that his acquaintance with Darwin's ideas was beset with common misrepresentations and misunderstandings, so perhaps he "knew" Darwin primarily through the enthusiastic appropriations of the many popularizers in Germany, and indeed throughout Europe. On the few points of specific criticism he ventures, he gets Darwin utterly wrong, complaining, for instance, that Darwin has ignored the possibility of "unconscious selection," when that was one of Darwin's most important bridging ideas in *Origin*. He refers to the "complete *bêtise* in the Englishmen, Darwin and Wallace," and complains, "At last, confusion goes so far that one regards Darwinism as philosophy: and now the scholars and scientists dominate" (Nietzsche 1901, p. 422). Others, however, regularly saw *him* as a Darwinian—"Other scholarly oxen have suspected me of Darwinism on this account" (Nietzsche 1889, III, i)—a label which he scoffed at, while proceeding to write, in his *Genealogy of Morals* (1887), one of the first and still subtlest of the Darwinian investigations of the evolution of ethics, a topic to which we will return in chapter 16.

Nietzsche viewed his argument for eternal recurrence as a proof of the absurdity or meaninglessness of life, a proof that no meaning was *given* to the universe from on high. And this is undoubtedly the root of the fear that many experience when encountering Darwin, so let us examine it in Nietzsche's version, as extreme as any we are apt to find. Why, exactly, would eternal recurrence make life meaningless? Isn't it obvious?

---

11. For a clear reconstruction of Nietzsche's uncharacteristically careful deduction of what he once described as "the most scientific of hypotheses," see Danto 1965, pp. 201–9. For a discussion and survey of this and other interpretations of Nietzsche's notorious idea of eternal recurrence, see Nehamas 1980, which argues that by "scientific" Nietzsche meant specifically "not-teleological." A recurring—but, so far, not eternally recurring—problem with the appreciation of Nietzsche's version of the eternal recurrence is that, unlike Wheeler, Nietzsche seems to think that *this* life will happen again not because it *and all possible variations on it* will happen over and over, but because there is only one possible variation—this one—and it will happen over and over. Nietzsche, in short, seems to have believed in actualism. I think that this is inessential to an appreciation of the moral implications Nietzsche thought he could or should draw from the idea, and perhaps to Nietzsche scholarship as well (but what do I know?).

What if a demon were to creep after you one day or night, in your loneliest loneness, and say: "This life which you live and have lived, must be lived again by you, and innumerable times more. And there will be nothing new in it, but every pain and every joy and every thought and every sigh— everything unspeakably small and great in your life—must come again to you, and in the same sequence and series. . . ." Would you not throw yourself down and curse the demon who spoke to you thus? Or have you once experienced a tremendous moment, in which you would answer him: "Thou art a god, and never have I heard anything more divine!" [*The Gay Science* (1882), p. 341 (passage translated in Danto 1965, p. 210).]

Is this message liberating, or horrifying? Nietzsche couldn't seem to make up his own mind, perhaps because he often chose to clothe the implications of his "most scientific of hypotheses" in these rather mystical trappings. We can get a little fresh air into the discussion by considering a delectable parody version, by the novelist Tom Robbins, in *Even Cowgirls Get the Blues*:

For Christmas that year, Julian gave Sissy a miniature Tyrolean village. The craftsmanship was remarkable.

There was a tiny cathedral whose stained-glass windows made fruit salad of sunlight. There was a plaza and *ein Biergarten*. The *Biergarten* got quite noisy on Saturday nights. There was a bakery that smelled always of hot bread and strudel. There was a town hall and a police station, with cutaway sections that revealed standard amounts of red tape and corruption. There were little Tyroleans in leather britches, intricately stitched, and, beneath the britches, genitalia of equally fine workmanship. There were ski shops and many other interesting things, including an orphanage. The orphanage was designed to catch fire and burn down every Christmas Eve. Orphans would dash into the snow with their nightgowns blazing. Terrible. Around the second week of January, a fire inspector would come and poke through the ruins, muttering, "If they had only listened to me, those children would be alive today." [Robbins 1976, pp. 191–92.]

The craftsmanship of this passage is itself remarkable. The repetition of the orphanage drama year after year seems to rob the little world of any real meaning. But why? Why exactly should it be the repetition of the fire inspector's lament that makes it sound so hollow? Perhaps if we looked closely at what that entails we would find the sleight of hand that makes the passage "work." Do the little Tyroleans rebuild the orphanage themselves, or is there a RESET button on this miniature village? What difference would that make? Well, where do the new orphans come from? Do the "dead" ones come back to life (Dennett 1984, pp. 9–10)? Notice that Robbins says that the orphanage *was designed* to catch fire and burn down every Christmas

Eve. The creator of this miniature world is clearly taunting us, ridiculing the seriousness with which we face our life problems. The moral seems clear: *if* the meaning of this drama must come from on high, from a Creator, it would be an obscene joke, a trivialization of the strivings of the individuals in that world. But what if the meaning is somehow the creation of the individuals themselves, arising anew in each incarnation rather than as a gift from on high? This might open up the possibility of meaning that was not threatened by repetition.

This is the defining theme of existentialism in its various species: the only meaning there can be is the meaning you (somehow) create for yourself. How *that* trick might be accomplished has always been something of a mystery among existentialists, but as we shall soon see, Darwinism does have some demystification to offer in its account of the process of meaning-creation. The key, once again, is the abandonment of John Locke's Mind-first vision, and its replacement with a vision in which *importance itself*, like everything else that we treasure, gradually evolves from nothingness.

We might pause, before turning to some of these details, to consider where our roundabout journey has brought us so far. We began with a somewhat childish vision of an anthropomorphic, Handicrafter God, and recognized that this idea, taken literally, was well on the road to extinction. When we looked through Darwin's eyes at the actual processes of design of which we and all the wonders of nature are the products to date, we found that Paley was right to see these effects as the result of a lot of design work, but we found a non-miraculous account of it: a massively parallel, and hence prodigiously wasteful, process of mindless, algorithmic design-trying, in which, however, the minimal increments of design have been thriftily husbanded, copied, and re-used over billions of years. The wonderful *particularity* or *individuality* of the creation was due, not to Shakespearean inventive genius, but to the incessant contributions of chance, a growing sequence of what Crick (1968) has called "frozen accidents."

That vision of the creative process still apparently left a role for God as Lawgiver, but this gave way in turn to the Newtonian role of Lawfinder, which also evaporated, as we have recently seen, leaving behind no Intelligent Agency in the process at all. What is left is what the process, shuffling through eternity, mindlessly finds (when it finds anything): a timeless Platonic possibility of order. That is indeed a thing of beauty, as mathematicians are forever exclaiming, but it is not itself something intelligent but, wonder of wonders, something intelligible. Being abstract and outside of time, it is nothing with an *initiation* or *origin* in need of explanation.[12]

---

12. Descartes had raised the question of whether God had created the truths of mathematics. His follower Nicolas Malebranche (1638–1715) firmly expressed the view that they needed no inception, being as eternal as anything could be.

What does need its origin explained is the concrete universe itself, and as Hume's Philo long ago asked: Why not stop at the material world? *It*, we have seen, does perform a version of the ultimate bootstrapping trick; it creates itself *ex nihilo*, or at any rate out of something that is well-nigh indistinguishable from nothing at all. Unlike the puzzlingly mysterious, time-less self-creation of God, this self-creation is a non-miraculous stunt that has left lots of traces. And, being not just concrete but the product of an ex-quisitely particular historical process, it is a creation of utter uniqueness—encompassing and dwarfing all the novels and paintings and symphonies of all the artists—occupying a position in the hyperspace of possibilities that differs from all others.

Benedict Spinoza, in the seventeenth century, *identified* God and Na-ture, arguing that scientific research was the true path of theology. For this heresy he was persecuted. There is a troubling (or, to some, enticing) Janus-faced quality to Spinoza's heretical vision of *Deus sive Natura* (God, or Nature): in proposing his scientific simplification, was he person-ifying Nature or depersonalizing God? Darwin's more generative vision pro-vides the structure in which we can see the intelligence of Mother Nature (or is it merely apparent intelligence?) as a non-miraculous and non-mysterious—and hence all the more wonderful—feature of this self-creating thing.

CHAPTER 7: *There must have been a first living thing, but there couldn't have been one—the simplest living thing is too complex, too designed, to spring into existence by sheer chance. This dilemma is solved not by a skyhook, but by a long series of Darwinian processes: self-replicating mac-ros, preceded or accompanied perhaps by self-replicating clay crystals, gradually advancing from tournaments of luck to tournaments of skill over a billion years. And the regularities of physics on which those cranes de-pend could themselves be the outcome of a blind, uncaring shuffle through Chaos. Thus, out of next to nothing, the world we know and love created itself.*

CHAPTER 8: *The work done by natural selection is R and D, so biology is fundamentally akin to engineering, a conclusion that has been deeply re-sisted out of misplaced fear for what it might imply. In fact, it sheds light on some of our deepest puzzles. Once we adopt the engineering perspective, the central biological concept of function and the central philosophical concept of meaning can be explained and united. Since our own capacity to respond to and create meaning—our intelligence—is grounded in our status as advanced products of Darwinian processes, the distinction be-tween real and artificial intelligence collapses. There are important differ-*

ences, however, between the products of human engineering and the products of evolution, because of differences in the processes that create them. We are just now beginning to get the grand processes of evolution into focus, by directing products of our own technology, computers, onto the outstanding questions.

CHAPTER EIGHT

Biology Is Engineering

～〜

1. THE SCIENCES OF THE ARTIFICIAL

Since World War II the discoveries that have changed the world were not made so much in lofty halls of theoretical physics as in the less-noticed labs of engineering and experimental physics. The roles of pure and applied science have been reversed; they are no longer what they were in the golden age of physics, in the age of Einstein, Schrödinger, Fermi and Dirac. . . . Historians of science have seen fit to ignore the history of the great discoveries in applied physics, engineering and computer science, where real scientific progress is nowadays to be found. Computer science in particular has changed and continues to change the face of the world more thoroughly and more drastically than did any of the great discoveries in theoretical physics.

—NICHOLAS METROPOLIS 1992

In this chapter I want to trace some of the overlooked and underappreciated implications of a central—I venture to say *the* central—feature of the Darwinian Revolution: the marriage, after Darwin, of biology and engineering. My goal in this chapter is to tell the positive side of the story of biology as engineering. Later chapters will deal with various assaults and challenges, but before they steal the limelight, I want to make out the case that the engineering perspective on biology is not merely occasionally useful, not merely a valuable option, but the obligatory organizer of all Darwinian thinking, and the primary source of its power. I expect a fair amount of emotional resistance to this claim. Be honest: doesn't this chapter's title provoke a negative reaction in you, along the lines of "Oh no, what a dreary, philistine, reductionist claim! Biology is *much* more than engineering!"?

The idea that a study of living forms is at least a close kin to engineering has been available since Aristotle's own pioneering investigations of organisms, and his analysis of teleology, the fourth of his causes, but only since Darwin has the idea begun to come into focus. It is quite explicit, of course, in the Argument from Design, which invites the observer to marvel at the cunning interplay of parts, the elegant planning and exquisite workmanship of the Artificer. But engineering has always had second-class status in the intellectual world. From Leonardo da Vinci to Charles Babbage to Thomas Edison, the engineering genius has always been acclaimed but nevertheless regarded with a certain measure of condescension by the mandarin elite of science and the arts. Aristotle did not help matters by proposing a distinction, adopted by the medievals, between what was *secundum naturam*, according to nature, and what was *contra naturam*, against nature, artificial. Mechanisms—but not organisms—were *contra naturam*. Then there were the things that were *praeter naturam,* or *un*natural (monsters and mutants), and the things that were *super naturam*—miracles (Gabbey 1993). How could the study of what was *against* nature shed much light on the glories—yea, even the monsters and miracles—of nature?

The fossil traces of this negative attitude are everywhere in our culture. For instance, in my own home discipline of philosophy, the subdiscipline known as philosophy of science has a long and respected history; many of ▬most eminent and influential philosophers these days are philosophers of science. There are excellent philosophers of physics, philosophers of biology, philosophers of mathematics, and even of social science. I have never even heard anybody in the field described as a philosopher of engineering—as if there couldn't possibly be enough conceptual material of interest in engineering for a philosopher to specialize in. But this is changing, as more and more philosophers come to recognize that engineering harbors some of the deepest, most beautiful, most important thinking ever done. (The title of this section is taken from Herbert Simon's seminal book [1969] on these topics.)

Darwin's great insight was that all the designs in the biosphere could be the products of a process that was as patient as it was mindless, an "automatic" and gradual lifter in Design Space. In retrospect, we can see that Darwin himself could hardly have imagined, let alone supported with evidence, the refinements and extensions of his idea that have permitted later Darwinians to go beyond his own cautious agnosticism about the origins of life itself, and even the "design" of the physical Order his idea presupposed. He was in no better position to characterize that Order than he was to describe the constraints and powers of the hereditary mechanism; he just knew there had to be such a mechanism, and it had to exploit the Order, whatever it was, that made "descent with modification" not only possible but fruitful.

The century-plus of subsequent focusing and extending of Darwin's great idea has been punctuated by controversy, amply illustrating, by the way, the reflexive extension of his idea to itself: the evolution of the Darwinian memes about evolution has been not just accompanied, but positively sped along, by competition between ideas. And as he hypothesized with regard to organisms, "competition will generally be most severe between those forms which are most nearly related to each other" (*Origin*, p. 121). Biologists themselves have not been immune to the heritage of negative attitudes towards engineering, of course. What is the hankering after skyhooks, after all, but the fond hope that a miracle will somehow come along to lift us above the cranes? Continued subliminal resistance to this feature of Darwin's fundamental idea has heightened controversy, impeded comprehension, and distorted expression—while at the same time propelling some of the most important challenges to Darwinism.

In response to these challenges, Darwin's idea has grown stronger. Today we can see that not only Aristotle's divisions but also other cherished compartmentalizations of science are threatened by its territorial expansion. The Germans divide learning into *Naturwissenschaften*, the natural sciences, and *Geisteswissenschaften*, the sciences of mind, meaning, and culture, but this sharp divide—cousin to C. P. Snow's Two Cultures (1963)—is threatened by the prospect that an engineering perspective will spread from biology up through the human sciences and arts. If there is just one Design Space, after all, in which the offspring of both our bodies and our minds are united under one commodious set of R-and-D processes, then these traditional walls may tumble.

Before proceeding, I want to confront a suspicion. Since I have just granted that Darwin himself didn't appreciate many of the issues that have to be dealt with if the theory of evolution by natural selection is to survive, isn't there something trivial or tautological about my claim that Darwin's idea survives all these challenges? No wonder it can keep on spreading, since it keeps on changing in response to new challenges! If my point were to crown Darwin as author and hero, there would be merit to this suspicion, but of course this is not primarily such an exercise of intellectual history. It doesn't really matter to my main thesis whether Darwin himself even existed! He could be, like the Average Taxpayer, a sort of mythical Virtual Author, for all I care. (Some authorities place Homer in that category.) The actual historical man does fascinate me; his curiosity, integrity, and stamina inspire me; his personal fears and flaws make him lovable. But he is, in a way, beside the point. He had the good fortune to be the midwife for an idea that has a life of its own, precisely because it *does* grow and change. Most ideas can't do that.

In fact, a great deal of rhetoric has been expended by partisans on both sides of the controversies about whether Darwin himself—St. Charles, you might call him—was a gradualist, an adaptationist, a catastrophist, a capitalist, a feminist. The answers to these questions are of considerable historical interest in their own right, and if they are carefully divorced from questions of ultimate justification, they can actually help us see what the scientific issues really are. What various thinkers *think* they are doing—saving the world from one ism or another, or finding room for God in science, or combating superstition—often turns out to be at right angles to the contribution their campaigns actually succeed in making. We have already seen instances of this, and more are in the offing. Probably no area of scientific research is driven by more hidden agendas than evolutionary theory, and it certainly will help to expose them, but nothing follows directly from the fact that some people are trying desperately—whether they realize it or not—to protect something evil or destroy something evil. People sometimes get it right in spite of having been driven by the most unpresentable hankerings. Darwin was who he was, and thought what he thought, warts and all. And now he is dead. Darwinism, on the other hand, has more than nine lives. It bids fair to being immortal.

2. DARWIN IS DEAD—LONG LIVE DARWIN!

I have taken the section title from the title of the "Résumé" with which Manfred Eigen ends his 1992 book. There is an unmistakable engineering flair to Eigen's thinking. His research is a sequence of biological construction problems posed and solved: how do the materials get amassed at the building site, and how does the design get determined, and in what order are the various parts assembled so that they don't fall apart before the whole structure is completed? His claim is that the ideas he presents are revolutionary, but that after the revolution, Darwinism is not only alive and well, but strengthened. I want to explore this theme in more detail, since we will see other versions of it that are nowhere near as clearcut as Eigen's.

What is supposed to be revolutionary about Eigen's work? In chapter 3 we looked at a fitness landscape with a single peak, and saw how the Baldwin Effect could turn a well-nigh-invisible telephone pole on a plain into Mount Fuji, with a steadily rising surrounding slope, so that no matter where in the space you started, you would eventually get to the summit if you simply followed the Local Rule:

Never step down; step up whenever possible.

The idea of a fitness landscape was introduced by Sewall Wright (1932), and it has become a standard imagination prosthesis for evolutionary theorists. It has proven its value in literally thousands of applications, including many outside of evolutionary theory. In Artificial Intelligence, economics, and other problem-solving domains, the model of problem-solving by incremental *hill-climbing* (or "gradient ascent") has been deservedly popular. It has even been popular enough to motivate theorists to calculate its limitations, which are severe. For certain classes of problems—or, in other words, in certain types of landscape—simple hill-climbing is quite impotent, for an intuitively obvious reason: the climbers get stuck on local second-rate summits instead of finding their way to the global summit, the Mount Everest of perfection. (The same limitations beset the method of simulated annealing.) The Local Rule is fundamental to Darwinism; it is equivalent to the requirement that there cannot be any intelligent (or "far-seeing") foresight in the design process, but only ultimately stupid opportunistic exploitation of whatever lucky lifting happens your way.

What Eigen has shown is that this simplest Darwinian model of steady improvement up a single slope of fitness to the optimal peak of perfection just doesn't work to describe what goes on in molecular or viral evolution. The rate of adaptation by viruses (and also of bacteria and other pathogens) is measurably faster than the "classical" models predict—so fast that it seems to involve illicit "look-ahead" by the climbers. So does this mean that Darwinism must be abandoned? Not at all, for what counts as local depends (not surprisingly) on the scale you use.

Eigen draws our attention to the fact that when viruses evolve, they don't go single-file; they travel in huge herds of almost identical variants, a fuzzy-edged cloud in the Library of Mendel that Eigen calls a "quasi-species." We already saw the unimaginably large cloud of *Moby Dick* variants in the Library of Babel, but any actual library is likely to have more than one or two variant editions of a book on its shelves, and in the case of a really popular book like *Moby Dick* it is also likely to have multiple copies of the same edition. Like actual *Moby Dick* collections, then, actual viral clouds include multiple identical copies but also multiple copies of minor typographical variants, and this fact has some implications, according to Eigen, that have been ignored by "classical" Darwinians. It is the *shape* of the cloud of variants that holds the key to the speed of molecular evolution.

A classical term among geneticists for the canonical version of a species (analogous to the canonical text of *Moby Dick*) is the *wild type*. It was often supposed by biologists that among the many different genotypes in a population, the pure wild type would predominate. Analogous would be the claim that in any library collection of copies of *Moby Dick*, most copies will be of the received or canonical edition—if there is one! But this doesn't have to be the case for organisms any more than for books in libraries. In fact, the wild type is really just an abstraction, like the Average Taxpayer, and a population may contain no individuals at all that have exactly "the" wild-type genome. (Of course, the same is true of books—scholars might debate for years over the purity of a particular word in a particular text, and until such debates were resolved, nobody could say exactly what the canonical or wild-type text of that work was, but the identity of the work would hardly be in jeopardy. James Joyce's *Ulysses* would be a good case in point.)

Eigen points out that this distribution of the "essence" over a variety of nearly identical vehicles turns out to make that essence much more movable, much more adaptable, especially in "rugged" fitness landscapes, with multiple peaks and few smooth slopes. It permits the essence to send out efficient scouting parties into the neighboring hills and ridges, ignoring wasteful exploration of the valleys, and thereby vastly (not Vastly, but enough to make a huge difference) enhancing its capacity to find higher peaks, better optima, at some distance from its center, where the (virtual) wild type sits.[1]

The reasons it works are summarized by Eigen as follows:

Functionally competent mutants, whose selection values come close to that of the wild type (though remaining below it), reach far higher population numbers than those that are functionally ineffective. An asymmetric spectrum of mutants builds up, in which mutants far removed from the wild type arise successively from intermediates. The population in such a chain of mutants is influenced decisively by the structure of the value landscape. The value landscape consists of connected plains, hills, and mountain ranges. In the mountain ranges, the mutant spectrum is widely scattered, and along ridges even distant relatives of the wild type appear with finite [that is, not infinitesimal] frequency. It is precisely in the mountainous regions that further selectively superior mutants can be expected. As soon as one of these turns up on the periphery of a mutation spectrum the established ensemble collapses. A new ensemble builds up around the superior mutant, which thus takes over the role of the wild type.... This causal chain results in a kind of 'mass action', *by which the superior mutants are tested with much higher probability* than inferior mutants, even if the latter are an equal distance away from the wild type. [Eigen 1992, p. 25.]

So there is a tight interaction between the shape of the fitness landscape and the population that occupies it, creating a series of feedback loops,

---

1. The similarity between these themes and the themes I develop in *Consciousness Explained* (1991a) about the need to break up the Cartesian Theater, with its Central Meaner, and distribute its intelligence work around to a variety of peripheral agents, is of course no accident. It is, however, mainly a case of convergent evolution, so far as I can determine. I had not read any of Eigen's work at the time I was writing my book, though it certainly would have inspired me if I had. A useful bridge between Eigen on molecules and me on consciousness is Schull 1990 on the intelligence of species, and my commentary, Dennett 1990a.

leading—usually—from one temporarily stable problem-setting to another. No sooner do you climb a peak than the whole landscape pitches and billows into a new mountain range and you start climbing all over again. In fact, the landscape is constantly shifting under your feet (if you are a quasi-species of viruses).

Now, this is really not as revolutionary as Eigen claims. Sewall Wright himself, in his "shifting balance theory," tried to explain how multiple peaks and shifting landscapes would be traversable not by individual "wild-type" exemplars, but by various-sized populations of variants, and Ernst Mayr has stressed for many years that "population thinking" is at the heart of Darwinism, something overlooked by geneticists at their peril. So Eigen has really not revolutionized Darwinism but, rather—no small contribution—created some theoretical innovations that clarify and strengthen underappreciated and imperfectly formulated ideas that had been around for years. When Eigen (1992, p. 125) says, "The (quantitative) acceleration of evolution that this brings about is so great that it appears to the biologist as a surprising new *quality*, an apparent ability of selection to 'see ahead', something that would be viewed by classical Darwinians as the purest heresy!" he is indulging in a familiar form of overdramatization, ignoring the many biologists who at least anticipated, and perhaps even fomented, his "revolution."

After all, when traditional Darwinian theorists postulate fitness landscapes and then randomly sprinkle genotypes on them in order to calculate what theory says would happen to them, they know that, in nature, genotypes don't just get thrown randomly into pre-existing parts of the world. Every model of a time-consuming process has to start at some arbitrary "moment"; the curtain rises and the model then plots what happens next. If we look at such a model and see that at the "outset" it shows a bunch of candidates down in the valleys, we can be pretty sure that the theorist recognizes that they weren't "always" down there—whatever that would mean! Wherever on the fitness landscape there are candidates at one time, there were peaks before, or those candidates wouldn't be there, so these must be relatively *new* valleys these candidates are occupying, a new predicament that evolution has placed before them. Only that assumption could justify locating the candidates in the valleys in the first place. Eigen's contribution reinforces the appreciation that we have to add these complications to the models if we want them actually to do the work that Darwinians have always supposed that their simpler models could do.

It is certainly no accident that our appreciation of the need for these much more complicated models coincides in time (almost down to the month, and certainly to the year) with our capacity to build and explore such models on existing computers. No sooner do more powerful computers become available than we discover with their help that more complex models of evolution are not only possible but positively required if we are really to explain what Darwinism has always claimed it can explain. Darwin's idea that evolution is an algorithmic process is now becoming an ever more enriched family of hypotheses, undergoing its own population explosion thanks to the opening up of new environments for it to live in.

In Artificial Intelligence, a prized strategy is to work on deliberately simplified versions of the phenomena of interest. These are engagingly called "toy problems." In the Tinker Toy world of molecular biology, we get to see the simplest versions of the fundamental Darwinian phenomena in action, but these are *real* toy problems! We can take advantage of the *relative* simplicity and purity of this lowest-level Darwinian theory to introduce and illustrate some of the themes that we will trace through the higher levels of evolution in later chapters.

Evolutionists have always helped themselves to claims about fitness and optimality and the growth of complexity, for instance, and these claims have been recognized by claimant and critic alike to be serious oversimplifications at best. In the world of molecular evolution, no such apologies are required. When Eigen speaks of optimality, he has a crisp definition of what he means, and experimental measurements to back him up and keep him honest. His fitness landscapes and measures of success are neither subjective nor *ad hoc*. Molecular complexity can be measured in several mutually supporting and objective ways, and there is no poetic license at all in Eigen's use of the term "algorithm." When we envision a proofreading enzyme, for instance, chugging along a pair of DNA strands, checking and fixing and copying and then moving one step along and repeating the process, we can hardly doubt that we are watching a microscopic automaton at work, and the best simulations match the observed facts so closely that we can be very sure there are no magical helper-elves, no skyhooks, lurking in these quarters. In the world of molecules, the application of Darwinian thinking is particularly pure and unadulterated. Indeed, when we adopt this vantage point, it can seem something of a marvel that Darwinian theory, which works so beautifully on molecules, applies at all to such ungainly—galactic-sized—conglomerations of cells as birds and orchids and mammals. (We don't expect the periodic table to enlighten us about corporations or nations, so why would we expect Darwinian evolutionary theory to work on such complexities as ecosystems or mammalian lineages!?)

In *macro*scopic biology—the biology of everyday-sized organisms such as ants and elephants and redwood trees—everything is untidy. Mutation and selection can usually only be indirectly and imperfectly inferred, thanks to a mind-boggling array of circumstantial complications. In the molecular world, mutation and selection events can be directly measured and manipulated, and the generation time for viruses is so short that huge Darwinian effects can be studied. For instance, it is the horrifying capacity of toxic viruses to mutate in deadly combat with modern medicine that spurs on and funds much of this research. (The AIDS virus has undergone so much mutation in the last decade that its history over that period exhibits more genetic diversity—measured in codon revisions—than is to be found in the entire history of primate evolution!)

The research of Eigen and hundreds of others has definite practical applications for all of us. It is fitting to observe, then, that this important work is an instance of Darwinism triumphant, reductionism triumphant, mechanism triumphant, materialism triumphant. It is also, however, the farthest thing from *greedy* reductionism. It is a breathtaking cascade of levels upon levels upon levels, with new principles of explanation, new phenomena appearing at each level, forever revealing that the fond hope of explaining "everything" at some one lower level is misguided. Here is Eigen's own summary of what his survey shows; you will note that it is written in terms that should be congenial to the most ardent critic of reductionism:

> Selection is more like a particularly subtle demon that has operated on the different steps up to life, and operates today at the different levels of life, with a set of highly original tricks. Above all, it is highly active, driven by an internal feedback mechanism that searches in a very discriminating manner for the best route to optimal performance, not because it possesses an inherent drive towards any predestined goal, but simply by virtue of its inherent non-linear mechanism, which gives the appearance of goal-directedness. [Eigen 1992, p. 123.]

3. FUNCTION AND SPECIFICATION

Shape is destiny in the world of macromolecules. A one-dimensional sequence of amino acids (or of the nucleotide codons that code for them) determines the identity of a protein, but the sequence only partially constrains the way this one-dimensional protein string folds itself up. It typically springs into just one of many possible shapes, an idiosyncratically shaped snarl that its sequence type almost always prefers. This three-dimensional shape is the source of its power, its capacity as a catalyst—as a builder of structures or a fighter of antigens or a regulator of development, for instance. It is a machine, and what it does is a very strict function of the shape of its parts. Its overall three-dimensional shape is much more important, functionally, than the one-dimensional sequence that is responsible for it. The important protein lysozyme, for instance, is a particular-shaped molecular machine that is produced in many different versions—more than a hundred different amino-acid sequences have been found in nature that fold into the same functional shape—and of course differences in these amino-acid sequences can be used as "philological" clues in re-creating the evolutionary history of the production and use of lysozyme.

And here is a puzzle, first noted by Walter Elsasser (1958, 1966), but quite conclusively solved by Jacques Monod (1971). Considered very abstractly, the fact that a one-dimensional code can be "for" a three-dimensional structure shows that information is added. Indeed, *value* is added. The individual amino acids have value (by contributing to the functional prowess of a protein) not just in virtue of their location in the one-dimensional sequence that forms the string, but in virtue of their location in three-dimensional space once the string is folded up.

> Thus there is a seeming contradiction between the statement that the genome 'entirely defines' the function of a protein and the fact that this function is linked to a three-dimensional structure whose data content is *richer* than the direct contribution made to the structure by the genome. [Monod 1971, p. 94.]

As Küppers (1990, p. 120) points out, Monod's solution is straightforward: "The seemingly irreducible, or excess, information is contained in the specific conditions of the protein's environment, and only together with these can the genetic information determine unambiguously the structure and thus the function of the protein molecule." Monod (1971, p. 94) puts it this way:

> ... of all the structures possible only one is actually realized. Initial conditions hence enter among the items of information finally enclosed within the ... structure. Without specifying it, they contribute to the realization of a unique shape by eliminating all alternative structures, in this way proposing—or rather imposing—an unequivocal interpretation of a potentially equivocal message.[2]

What does this mean? It means—not surprisingly—that the language of DNA and the "readers" of that language have to evolve together; neither can work on its own. When the deconstructionists say that the reader brings something to the text, they are saying something that applies just as surely to DNA as to poetry; the something that the reader brings can be charac-

---

2. Philosophers will recognize, I trust, that Monod thus both posed and solved Putnam's (1975) problem of Twin Earth, at least in the context of the "toy problem" of molecular evolution. Meaning "ain't in the head," as Putnam famously observed, and it ain't (all) in the DNA either. Twin Earth, otherwise known as the problem of broad versus narrow content, will get exhumed briefly in chapter 14, so I can give it its proper Darwinian funeral.

terized most generally and abstractly as information, and only the combination of information from the code and the code-reading environment suffices to create an organism.[3] As we noted in chapter 5, some critics have fastened on this fact as if it were somehow the refutation of "gene centrism," the doctrine that the DNA is the sole information store for inheritance, but that idea was always only a handy oversimplification. Though libraries are commonly allowed to be storehouses of information, of course it is really only libraries-*plus-readers* that preserve and store the information. Since libraries have not—up till now, at any rate—contained among their volumes the information needed to create more readers, their capacity to store information (effectively) has been dependent on there being *another* information-storage system—the human genetic system, of which DNA is the principle medium. When we apply the same reasoning to DNA itself, we see that it, too, requires a continuing supply of "readers" that it does not itself entirely specify. Where does the rest of the information come from to specify these readers? The short answer is that it comes from the very continuities of the environment—the persistence in the environment of the necessary raw (and partially constructed) materials, and the conditions in which they can be exploited. Every time you make sure that your dishrag gets properly dry in between uses, you break the chain of environmental continuity (e.g., lots of moisture) that is part of the informational background presupposed by the DNA of the bacteria in the dishrag whose demise you seek.

We see here a special case of a very general principle: any *functioning* structure carries *implicit* information about the environment in which its function "works." The wings of a seagull magnificently embody principles of aerodynamic design, and thereby also imply that the creature whose wings these are is excellently adapted for flight in a medium having the specific density and viscosity of the atmosphere within a thousand meters or so of the surface of the Earth. Recall the example in chapter 5 of sending the score of Beethoven's Fifth Symphony to "Martians." Suppose we carefully preserved the body of a seagull and sent it off into space (without any accompanying explanation), to be discovered by these Martians. If they

---

3. David Haig (personal communication) has drawn my attention to a fascinating new wrinkle in this unfolding story about folding proteins: molecular chaperones. "Chaperones are molecular cranes *par excellence*. They are proteins with which an amino acid chain associates while it is folding that allows the chain to adopt a conformation that would be unavailable in the absence of the chaperone. The chaperone is then discarded by the folded protein. Chaperones are highly conserved. . . . Molecular chaperones were named by analogy to the functions of chaperones at a debutante ball: their role was to encourage some interactions and to discourage others." For recent details, see Martin et al. 1993, Ellis and van der Vies 1991.

made the fundamental assumption that the wings were functional, and that their function was flight (which might not be as obvious to them as we, who have seen them do it, think), they could use this assumption to "read off" the implicit information about an environment for which these wings would be well designed. Suppose they then asked themselves how all this aerodynamic theory came to be implicit in the structure, or, in other words: How did all this information get into these wings? The answer *must* be: By an interaction between the environment and the seagull's ancestors. (Dawkins 1983a explores these issues in more detail.)

The same principle applies at the most basic level, where the function is specification itself, *the function on which all other functions depend*. When we wonder, with Monod, how the three-dimensional shape of the proteins gets fixed, given that the information in the genome must underspecify them, we see that only a pruning of the nonfunctional (or less functional) could explain it. So the acquisition of a *particular* shape by a molecule involves a mixture of historical accident on the one hand and the "discovery" of important truths on the other.

From the outset, the process of the design of molecular "machines" exhibits these two features of human engineering. Eigen (1992, p. 34) provides a good instance of this in his reflections on the structure of the DNA code. "One might well ask why Nature has used four symbols, when she might just as well have made do with two." Why indeed? Notice how naturally and inevitably a "why" question arises at this point, and notice that it calls for an "engineering" answer. Either the answer is that there is no reason—it is historical accident, pure and simple—or there *is* a reason: a condition was or is present that makes this the *right* way or *best* way for the coding system to get designed, given the conditions that obtained.[4]

All the deepest features of molecular design may be considered from the engineering perspective. On the one hand, consider the fact that macromolecules come in two basic shape categories: symmetrical and *chiral* (with left-handed and right-handed versions). There is a reason why so many should be symmetrical:

> The selective advantage in a symmetrical complex is enjoyed by all the subunits, while in an asymmetric complex the advantage is only effective in the subunit in which the mutation arises. It is for this reason that we find so many symmetric structures in biology, "because they were able to make the most effective use of their advantage, and thus—*a posteriori*—won the

---

4. Eigen suggests that there is a reason why there are four letters, not two, but I am not going to pass it on. Perhaps you can figure out for yourself what it might be before seeing what Eigen says. You already have at your fingertips the relevant principles of engineering to give it a good shot.

selection competition; this was not, however, because symmetry is—*a priori*—an indispensable requirement for the fulfillment of a functional purpose." [Küppers 1990, p. 119, incorporating a quotation from Eigen and Winkler-Oswatitsch 1975.]

But what about the asymmetric or chiral shapes? Is there a reason why they should be one way—left-handed, say—rather than the other? No, probably not, but: "Even if there is no *a priori* physical explanation for the decision, even if it was just a brief fluctuation that gave one or the other equivalent possibility a momentary advantage, the self-reinforcing character of selection would turn the random decision into a major and permanent breach of symmetry. The cause would be a purely 'historical' one" (Eigen 1992, p. 35).[5]

The shared chirality of organic molecules (in our part of the universe) was thus probably another pure QWERTY phenomenon, or what Crick (1968) has called a "frozen accident." But even in the case of such a QWERTY phenomenon, if the conditions are just right and the opportunities and hence pressures are great enough, the tables might be turned and a new standard established. This is apparently just what happened when the DNA language displaced the RNA language as the *lingua franca* of encoding for complex organisms. The *reasons* for its preferability are clear: by being double-stranded, the DNA language permitted a system of error-correcting or proofreading enzymes, which could repair copying errors in one strand by reference to its mate. This made the creation of longer, more complicated genomes feasible (Eigen 1992, p. 36).

Note that this reasoning does *not* yield the conclusion that double-stranded DNA *must* develop, for Mother Nature had no advance intention to create multicellular life. It just reveals that *if* double-stranded DNA happens to begin to develop, it opens up opportunities that are dependent on it. Hence it becomes a necessity for those exemplars in the space of all possible life forms that avail themselves of it, and if those life forms prevail

---

5. Danny Hillis, the creator of the Connection Machine, once told me a story about some computer scientists who designed an electronic component for a military application (I think it was part of a guidance system in airplanes). Their prototype had two circuit boards, and the top one kept sagging, so, casting about for a quick fix, they spotted a brass doorknob in the lab which had just the right thickness. They took it off its door and jammed it into place between the two circuit boards on the prototype. Sometime later, one of these engineers was called in to look at a problem the military was having with the actual manufactured systems, and found to his amazement that between the circuit boards in each unit was a very precisely milled brass duplicate of the original doorknob. This is an Ur-story that has many well-known variations in engineering circles and among evolutionary biologists. For instance, see Primo Levi's amusing account of the mystery of the varnish additive in *The Periodic Table* (1984).

over those that do not avail themselves of it, that yields a retroactive endorsement of this *raison d'être* of the DNA language. This is the way evolution always discovers reasons—by retroactive endorsement.

4. ORIGINAL SIN AND THE BIRTH OF MEANING

The road to wisdom?
Well, it's plain and simple to express:
Err and err and err again
but less and less and less.

—PIET HEIN

The solution to the problem of life is seen in the vanishing of this problem.

—LUDWIG WITTGENSTEIN 1922, prop. 6.521

Once upon a time, there was no mind, and no meaning, and no error, and no function, and no reasons, and no life. Now all these wonderful things exist. It has to be possible to tell the story of how they all came to exist, and that story must pass, by subtle increments, from elements that manifestly lack the marvelous properties to elements that manifestly have them. There will have to be isthmuses of dubious or controversial or just plain unclassifiable intermediates. All these wonderful properties must have come into existence gradually, by steps that are barely discernible *even in retrospect*.

Recall that in the previous chapter it seemed to be obvious, maybe even a truth of logic, that either there had to be a First Living Thing or there had to be an infinite regress of Living Things. Neither horn of the dilemma would do, of course, and the standard Darwinian solution, which we will see over and over again, was this: in its place we described a *finite* regress, in which the sought-for marvelous property (life, in this case) was acquired by slight, perhaps even imperceptible, amendments or increments.

Here is the most general form of the schema of Darwinian explanation. The task of getting from the early time when there wasn't any *x* to the later time when there is lots of *x* is completed by a finite series of steps in which it becomes less and less clear that "there still isn't any *x* here, not really," through a series of "debatable" steps until we eventually find ourselves on steps where it is really quite obvious that "of course there is *x*, lots of *x*." We never draw any lines.

Notice what happens in the particular case of the origin of life if we try to draw the line. There are a slew of truths—no doubt largely unknowable in detail by us—any one of which we could "in principle" identify, if we wished, as the truth that confirms the identify of Adam the Protobacterium. We can sharpen up the conditions on being the First Living Thing however we like, but when we then get in our time machine and go back to witness the moment, we find that Adam the Protobacterium, no matter how we have defined it, is probably as undistinguished as Mitochondrial Eve. We know as a matter of logic that there was at least one start that has us as its continuation, but there were probably many false starts that differed *in no interesting way at all* from the one that initiated the winning series. The title of Adam is, once again, a retrospective honor, and we make a fundamental mistake of reasoning if we ask, *In virtue of what essential difference* is this the beginning of life? There need be no difference at all between Adam and Badam, an atom-for-atom duplicate of Adam who just happened not to have founded anything of note. This is not a *problem* for Darwinian theory; this is a source of its power. As Küppers puts it (1990, p. 133), "The fact that we obviously are not in a position to give a comprehensive definition of the phenomenon 'life' speaks not against but indeed for the possibility of a completely physical explanation of life phenomena."

Exactly the same gratuitous predicament faces anyone who, despairing of defining something as complicated as life, decides to define the apparently simpler notion of *function* or *teleology*. At exactly what point does function make its appearance? Did the very first nucleotides have functions, or did they just have causal powers? Did Cairns-Smith's clay crystals exhibit *genuine* teleological properties, or just "*as if*" teleological properties? Do gliders in the Life world have the *function* of locomotion, or do they just move? It doesn't make any difference how you legislate the answer; the interesting world of functioning mechanisms has to start with mechanisms that "straddle the line," and, however far back you place the line, there will be precursors that differ in arguably nonessential ways from the anointed ones.[6]

Nothing complicated enough to be really interesting could have an essence.[7] This anti-essentialist theme was recognized by Darwin as a truly

---

6. See Bedau 1991 for an exploration of this point that arrives at a somewhat different destination, and Unger 1990 for arguments that go directly counter to it. Unger insists we have conventions such that there must be (on logical grounds) "straddle pairs" in such circumstances, such that one is the last item in the series to lack x and the other is the first in the series to have x. But as van Inwagen (1993b) observes, a more inviting conclusion is: so much the worse for those conventions.

7. These are fighting words for some philosophers. For a clear attempt to salvage a formal logic of essences that specifically addresses the problems raised by the complexity of artifacts and organisms, see Forbes 1983, 1984. The conclusion I draw from Forbes' work is that it constructs what may be a Pyrrhic victory over Quine's staunch skepticism about essences, but in the process it confirms his underlying warning: contrary to what you might think, there is nothing natural about essentialist thinking; seeing the world through essentialist glasses does not at all make your life easy.

202 BIOLOGY IS ENGINEERING

revolutionary epistemological or metaphysical accompaniment to his science; we should not be surprised by how hard it is for people to swallow. Ever since Socrates taught Plato (and all the rest of us) how to play the game of asking for necessary and sufficient conditions, we have seen the task of "defining your terms" as a proper preamble to all serious investigations, and this has sent us off on interminable bouts of essence-mongering.[8] We *want* to draw lines; we often *need* to draw lines—just so we can terminate or forestall sterile explorations in a timely fashion. Our perceptual systems are even genetically designed to force straddling candidates for perception into one classification or another (Jackendoff 1993), a Good Trick but not a forced move. Darwin shows us that evolution does not need what we need; the real world can get along just fine with the *de facto* divergences that emerge over time, leaving lots of emptiness between clusters of actuality.

We have just glanced briefly at a particularly important instance of this characteristic Darwinian explanatory schema, and we should pause to confirm the effect. Through the microscope of molecular biology, we get to witness the birth of *agency*, in the first macromolecules that have enough complexity to "do things." This is not florid agency—*echt* intentional action, with the representation of reasons, deliberation, reflection, and conscious decision—but it is the only possible ground from which the seeds of intentional action could grow. There is something alien and vaguely repellent about the quasi-agency we discover at this level—all that purposive hustle and bustle, and yet *there's nobody home*. The molecular machines perform their amazing stunts, obviously exquisitely designed, and just as obviously none the wiser about what they are doing. Consider this account of the activity of an RNA phage, a replicating virus:

> First of all, the virus needs a material in which to pack and protect its own genetic information. Secondly, it needs a means of introducing its information into the host cell. Thirdly, it requires a mechanism for the specific replication of its information in the presence of a vast excess of host cell RNA. Finally, it must arrange for the proliferation of its information, a process that usually leads to the destruction of the host cell.... The virus

---

8. One of the major themes of the German philosopher Martin Heidegger was that Socrates is to blame for much of what is wrong with philosophy, because he taught us all to demand necessary and sufficient conditions. It cannot be often that Darwin and Heidegger support each other, so the occasion is worth noting. Hubert Dreyfus has long maintained (e.g., 1972, 1979) that Artificial Intelligence is based on a failure to appreciate Heidegger's critique of Socrates, and though that may be true of some strands of AI, it is not true of the field in general, which is firmly with Darwin, a claim that I will defend later in this chapter, and in greater detail in chapters 13 to 15.

even gets the cell to carry out its replication; its only contribution is one protein factor, specially adapted for the viral RNA. This enzyme does not become active until a 'password' on the viral RNA is shown. When it sees this, it reproduces the viral RNA with great efficiency, while ignoring the very much greater number of RNA molecules of the host cell. Consequently the cell is soon flooded with viral RNA. This is packed into the virus' coat protein, which is also synthesized in large quantities, and finally the cell bursts and releases a multitude of progeny virus particles. All this is a programme that runs automatically and is rehearsed down to the smallest detail. [Eigen 1992, p. 40.]

Love it or hate it, phenomena like this exhibit the heart of the power of the Darwinian idea. An impersonal, unreflective, robotic, mindless little scrap of molecular machinery is the ultimate basis of all the agency, and hence meaning, and hence consciousness, in the universe.

Right from the beginning, the cost of *doing something* is running the risk of doing it *wrong*, of making a mistake. Our slogan could be: No taking without mistaking. The first error that ever was made was a typographical error, a copying mistake that then became the opportunity for creating a new task environment (or fitness landscape) with a new criterion of right and wrong, better and worse. A copying error "counts" as an error here only because there is a cost to getting it wrong: termination of the reproductive line at worst, or a diminution in the capacity to replicate. These are all objective matters, differences that are there whether or not we look at them, or care about them, but they bring in their train a new perspective. Before that moment, no opportunity for error existed. However things went, they went neither right nor wrong. Before that moment, there was no stable, predictive way of exercising the option of adopting the perspective from which errors might be discerned, and every mistake anybody or anything has ever made since is dependent on that original error-making process. In fact, there is strong selection pressure for making the genetic copying process as high-fidelity as possible, minimizing the likelihood of error. Fortunately, it cannot quite achieve perfection, for if it did, evolution would grind to a halt. This is Original Sin, in scientifically respectable guise. Like the Biblical version, it purports to explain something: the emergence of a new level of phenomena with special characteristics (meaners in one case, sinners in the other). Unlike the Biblical version, it provides an explanation that makes sense; it does not proclaim itself to be a mysterious fact that one has to take on faith, and it has testable implications.

Notice that one of the first fruits of the perspective from which error is discernible is a clarification of the concept of a species. When we consider all the actual genomic texts that get created in the process of copying, copying, copying—with occasional mutations—nothing *intrinsically* counts as a canonical version of anything. That is, although we can identify mutations by simply comparing the "before" sequence with the "after" sequence, there is no intrinsic way of telling which of the uncorrected typographical *errors* might more fruitfully be viewed as editorial *improvements*.[9] Most mutations are what engineers would call "don't-cares," variations that make no discernible difference to viability, but as selection gradually takes its toll, the better versions begin to cluster. It is only relative to a "wild type" (a center of gravity, in effect, of such a cluster) that we can identify a particular version as a mistaken version, and even then there is the possibility, remote in practice but omnipresent in principle, that what seems a mistake from the perspective of one wild type is a brilliant improvement from the perspective of a wild-type-in-the-making. And as new wild types emerge as the foci or summits of fitness landscapes, the direction of the steady pressure of error *correction* can reverse in any particular neighborhood of Design Space. Once a particular family of similar texts is no longer subject to "correction" relative to a receding or lapsed norm, it is free to wander into the attractive basin of a new norm.[10] Reproductive isolation is thus both a cause and an effect of the clumpiness of phenotypic space. Wherever there are competing error-correcting regimes, one regime or the other will win out, and hence the isthmus between the competitors will tend to dissolve, leaving empty space between occupied zones of Design Space. Thus, just as norms of pronunciation and word use reinforce clustering in speech communities (a theoretically important point made by Quine 1960 in his discussion of error and the emergence of norms in language), so norms of genomic expression are the ultimate basis of speciation.

Through the same molecular-level microscope we see the birth of *meaning*, in the acquisition of "semantics" by the nucleotide sequences, which at first are mere syntactic objects. This is a crucial step in the Darwinian campaign to overthrow John Locke's Mind-first vision of the cosmos. Philosophers commonly agree, for good reason, that meaning and mind can

9. Note the parallel here with my discussion of the false dichotomy between Orwellian and Stalinesque models of consciousness in *Consciousness Explained* (1991a). In that case as well, there is no *intrinsic* mark of the canonical.

10. Once again we see the tolerance for topsy-turvy imagery. Some theorists speak of *basins of attraction*, guided by the metaphor of balls rolling blindly *down*hill to the local *minimum* instead of climbing blindly *up*hill to the local *maximum*. Just turn an adaptive landscape inside out and the mountains become basins, the ridges become canyons, and "gravity" provides the analogue of selection pressure. It doesn't make any difference whether you choose "up" or "down" as the favored direction, just so long as you are consistent. Here I have slipped, momentarily, into the rival perspective, just to make this point.

never be pulled apart, that there could never be meaning where there was no mind, or mind where there was no meaning. *Intentionality* is the philosopher's technical term for this meaning; it is the "aboutness" that can relate one thing to another—a name to its bearer, an alarm call to the danger that triggered it, a word to its referent, a thought to its object.[11] Only some things in the universe manifest intentionality. A book or a painting can be about a mountain, but a mountain itself is not about anything. A map or a sign or a dream or a song can be about Paris, but Paris is not about anything. Intentionality is widely regarded by philosophers as *the* mark of the mental. Where does intentionality come from? It comes from minds, of course.

But that idea, perfectly good in its own way, becomes a source of mystery and confusion when it is used as a metaphysical principle, rather than a fact of recent natural history. Aristotle called God the Unmoved Mover, the source of all motion in the universe, and Locke's version of Aristotelian doctrine, as we have seen, identifies this God as Mind, turning the Unmoved Mover into the Unmeant Meaner, the source of all Intentionality. Locke took himself to be proving deductively what the tradition already took to be obvious: original intentionality springs from the Mind of God; we are God's creatures, and derive our intentionality from Him.

Darwin turned this doctrine upside down: intentionality doesn't come from on high; it percolates up from below, from the initially mindless and pointless algorithmic processes that gradually acquire meaning and intelligence as they develop. And, perfectly following the pattern of all Darwinian thinking, we see that the first meaning is not full-fledged meaning; it certainly fails to manifest all the "essential" properties of *real* meaning (whatever you may take those properties to be). It is mere quasi-meaning, or semi-semantics. It is what John Searle (1980, 1985, 1992) has disparaged as mere "*as if* intentionality" as opposed to what he calls "Original Intentionality." But you have to start somewhere, and the fact that the first step in the right direction is just barely discernible as a step towards meaning at all is just what we should expect.

There are two paths to intentionality. The Darwinian path is diachronic, or historical, and concerns the gradual accretion, over billions of years, of the sorts of Design—of functionality and purposiveness—that can support an intentional interpretation of the activities of organisms (the "doings" of "agents"). Before intentionality can be fully *fledged*, it must go through its awkward, ugly period of featherless pseudo-intentionality. The synchronic

---

11. The topic of intentionality has been written about extensively by philosophers of many different traditions in recent years. For an overview and a general definition, see my article "Intentionality" (co-authored with John Haugeland) in Gregory 1987. For more detailed analyses, see my earlier books (1969, 1978, 1987b).

path is the path of Artificial Intelligence: in an organism with genuine intentionality—such as yourself—there are, right now, many parts, and some
of these parts exhibit a sort of semi-intentionality, or mere *as if* intentionality, or pseudo-intentionality—call it what you like—and your own genuine, fully fledged intentionality is in fact the product (with no further miracle
ingredients) of the activities of all the semi-minded and mindless bits that
make you up (this is the central thesis defended in Dennett 1987b, 1991a).
That is what a mind *is*—not a miracle-machine, but a huge, semi-designed,
self-redesigning amalgam of smaller machines, each with its own design
history, each playing its own role in the "economy of the soul." (Plato was
right, as usual, when he saw a deep analogy between a republic and a
person—but of course he had much too simple a vision of what this might
mean.)

There is a deep affinity between the synchronic and diachronic paths to
intentionality. One way of dramatizing it is to parody an ancient anti-
Darwinian sentiment: the monkey's uncle. Would you want your daughter
to marry a robot? Well, if Darwin is right, your great-great- . . . grandmother
was a robot! A macro, in fact. That is the unavoidable conclusion of the
previous chapters. Not only are you descended from macros; you are composed of them. Your hemoglobin molecules, your antibodies, your neurons,
your vestibular-ocular reflex machinery—at every level of analysis we find
machinery that dumbly does a wonderful, elegantly designed job. We have
ceased to shudder, perhaps, at the scientific vision of viruses and bacteria
busily and mindlessly executing their subversive projects—horrid little automata doing their evil deeds. But we should not think that we can take
comfort in the thought that *they* are alien invaders, so unlike the more
congenial tissues that make up *us*. We are made of the same sorts of automata that invade us—no halos of *élan vital* distinguish your antibodies
from the antigens they combat; they simply belong to the club that is you,
so they fight on your behalf.

Can it be that if you put enough of these dumb homunculi together you
make a real conscious person? The Darwinian says there could be no other
way of making one. Now, it certainly does not follow from the fact that you
are descended from robots that you are a robot. After all, you are also a
direct descendant of some fish, and you are not a fish; you are a direct
descendant of some bacteria, and you are not a bacterium. But unless dualism or vitalism is true (in which case you have some extra, secret ingredient in you), you are *made of* robots—or what comes to the same thing,
a collection of trillions of macromolecular machines. And all of these are
ultimately descended from the original macros. So something made of robots *can* exhibit genuine consciousness, or genuine intentionality, because
you do if anything does.

No wonder, then, that there should be so much antagonism to both

Darwinian thinking and Artificial Intelligence. Together they strike a fundamental blow at the last refuge to which people have retreated in the face of the Copernican Revolution: the mind as an inner sanctum that science cannot reach. (See Mazlish 1993.) It is a long and winding road from molecules to minds, with many diverting spectacles along the way—and we will tarry over the most interesting of these in subsequent chapters—but now is the time to look more closely than usual at the Darwinian beginnings of Artificial Intelligence.

5. THE COMPUTER THAT LEARNED TO PLAY CHECKERS

Alan Turing and John von Neumann were two of the greatest scientists of the century. If anybody could be said to have invented the computer, they did, and their brainchild has come to be recognized as both a triumph of engineering and an intellectual vehicle for exploring the most abstract realms of pure science. Both thinkers were at one and the same time awesome theorists and deeply practical, epitomizing an intellectual style that has been playing a growing role in science since the Second World War. In addition to creating the computer, both Turing and von Neumann made fundamental contributions to theoretical biology. Von Neumann, as we have already noted, applied his brilliant mind to the abstract problem of self-replication, and Turing (1952) did pioneering work on the most basic theoretical problems of embryology or morphogenesis: how can the complex topology—the shape—of an organism arise from the simple topology of the single fertilized cell from which it grows? The process begins, as every high-school student knows, with an event of quite symmetrical division. (As François Jacob has said, the dream of every cell is to become two cells.) Two cells become four, and four become eight, and eight become sixteen; how do hearts and livers and legs and brains get started under such a regime?[12] Turing saw the continuity between such molecular-level problems and the problem of how a poet writes a sonnet, and from the earliest days of computers, the ambition of those who saw what Turing saw has

---

12. Two highly accessible accounts of Turing's work on morphogenesis are Hodges (1983, ch. 7), and Stewart and Golubitsky (1992), which also discusses their relation to more recent theoretical explorations in the field. Beautiful as Turing's ideas are, they probably have at best a very attenuated application to real biological systems. John Maynard Smith (personal communication) recalls being entranced by Turing's 1952 paper (which his supervisor, J. B. S. Haldane, had shown him), and for years he was convinced that "my fingers must be Turing waves; my vertebrae must be Turing waves"—but he eventually came to realize, reluctantly, that it could not be that simple and beautiful.

been to use his wonderful machine to explore the mysteries of thought.[13]

Turing published his prophetic essay, "Computing Machinery and Intelligence," in the philosophical journal *Mind* in 1950, surely one of the most frequently cited articles ever to appear in that journal. At the time he wrote it, there were no Artificial Intelligence programs—there were really only two operating computers in the world—but within a few years, there were enough machines up and running twenty-four hours a day so that Arthur Samuel, a research scientist at IBM, could fill the otherwise idle late-night time on one of the early giants with the activities of a program that is as good a candidate as any for the retrospective title of AI-Adam. Samuel's program played checkers, and it got better and better by playing against itself through the small hours of the night, redesigning itself by throwing out earlier versions that had not fared well in the nightly tournament and trying out new mutations that were mindlessly generated. It eventually became a much better checkers-player than Samuel himself, providing one of the first clear counterexamples to the somewhat hysterical myth that "a computer can really only do what its programmer tells it to do." From our perspective we can see that this familiar but mistaken idea is nothing but an expression of Locke's hunch that only Minds can Design, an exploitation of *ex nihilo nihil fit* that Darwin had clearly discredited. The way Samuel's program transcended its creator, moreover, was by a strikingly classical process of Darwinian evolution.

Samuel's legendary program is thus not only the progenitor of the intel-

----

13. In fact, the bridge between computers and evolution goes back even farther, to Charles Babbage, whose 1834 conception of the "Difference Engine" is generally credited with inaugurating the prehistory of the computer. Babbage's notorious *Ninth Bridgewater Treatise* (1838) exploited his theoretical model of a computing engine to offer a mathematical proof that God had in effect programmed nature to generate the species! "On Babbage's smart machine any sequence of numbers could be programmed to cut in, however long another series had been running. By analogy, God at Creation had appointed new sets of animals and plants to appear like clockwork throughout history—he had created the laws which produced them, rather than creating them direct" (Desmond and Moore 1991, p. 213). Darwin knew Babbage and his *Treatise*, and even attended his parties in London. Desmond and Moore (pp. 212–18) offer some tantalizing glimpses into the traffic of ideas that may have crossed this bridge.

More than a century later, another London society of like-minded thinkers, the Ratio Club, served as the hotbed for more recent ideas. Jonathan Miller drew my attention to the Ratio Club, and urged me to research its history in the course of writing this book, but I have not made much progress to date. I am tantalized, however, by the 1951 photograph of its membership that graces the front of A. M. Uttley's *Information Transmission in the Nervous System* (1979): Alan Turing is seated on the lawn, along with the neurobiologist Horace Barlow (a direct descendant of Darwin, by the way); standing behind are Ross Ashby, Donald MacKay, and other major figures of the earliest days of what has become cognitive science. It's a small world.

lectual species, AI, but also of its more recent offshoot, AL, Artificial Life. Legendary though it is, few people today are familiar with its remarkable details, many of which deserve to be more widely known.[14] Samuel's first checkers program was written in 1952, for the IBM 701, but the learning version wasn't finished until 1955, and ran on an IBM 704; a later version ran on the IBM 7090. Samuel found some elegant ways of coding any state of a checkers game into four thirty-six-bit "words" and any move into a simple arithmetical operation on those words. (Compared with today's prodigiously wasteful computer programs which run on for megabytes, Samuel's basic program was microscopic in size—a "low-tech" genome indeed, with fewer than six thousand lines of code—but, then, he had to write it in machine code; this was before the days of computer programming languages.) Once he'd solved the problem of representing the basic process of legal checkers play, he had to face the truly hard part of the problem: getting the computer program to *evaluate* the moves, so it could *select* the best move (or at least one of the better moves) whenever possible.

What would a good evaluation function look like? Some trivial games, like tic-tac-toe, have feasible algorithmic solutions. There is a guaranteed win or draw for one player, and this best strategy can be computed in realistic amounts of time. Checkers is not such a game. Samuel (p. 72) pointed out that the space of possible checkers games has on the order of 10^{40} choice-points, "which, at 3 choices per millimicrosecond, would still take 10^{21} centuries to consider." Although today's computers are millions of times faster than the lumbering giants of Samuel's day, they still couldn't make a dent on the problem using the brute-force approach of exhaustive search. The search space is Vast, so the method of search must be "heuristic"—the branching tree of all possible moves has to be ruthlessly pruned by semi-intelligent, myopic demons, leading to a risky, chance-ridden exploration of a tiny subportion of the whole space.

Heuristic search is one of the foundational ideas of Artificial Intelligence. One might even define the task of the field of AI as the creation and inves-

---

14. Samuel's 1959 paper was reprinted in the first anthology of Artificial Intelligence, Feigenbaum and Feldman's classic, *Computers and Thought* (1964). Although I had read that paper in Feigenbaum and Feldman when it first came out, I had, like most readers, passed over most of the details and savored the punch line: a 1962 match between the "adult" program and Robert Nealey, a checkers champion. Nealey was gracious in defeat: "In the matter of the end game, I have not had such competition from any human being since 1954, when I lost my last game." It took a superb lecture by my colleague George Smith in an introductory course in computer science that we cotaught at Tufts to rekindle my interest in the details of Samuel's article, in which I find something new and valuable every time I reread it.

tigation of heuristic algorithms. But there is also a tradition within computer science and mathematics of *contrasting* heuristic methods with algorithmic methods: heuristic methods are risky, not guaranteed to yield results, whereas algorithms come with a guarantee. How do we resolve this "contradiction"? There is no contradiction at all. Heuristic algorithms are, like all algorithms, mechanical procedures that are guaranteed to do what they do, but what they do is engage in risky search! They are not guaranteed to *find* anything—or at least they are not guaranteed to find the specific thing sought in the amount of time available. But, like well-run tournaments of skill, good heuristic algorithms *tend* to yield highly interesting, reliable results in reasonable amounts of time. They are risky, but the good ones are good risks indeed. You can bet your life on them (Dennett 1989b). Failure to appreciate the fact that algorithms can be heuristic procedures has misled more than a few critics of Artificial Intelligence. In particular, it has misled Roger Penrose, whose views will be the topic of chapter 15.

Samuel saw that the Vast space of checkers could only be *feasibly* explored by a process that riskily pruned the search tree, but how do you go about constructing the pruning and choosing demons to do this job? What readily programmable stop-looking-now rules or evaluation functions would have a better-than-chance power to grow a search tree in wise directions? Samuel was searching for a good algorithmic searching method. He proceeded empirically, beginning by devising ways of mechanizing whatever obvious rules of thumb he could think of. Look before you leap, of course, and learn from your mistakes, so the system should have a memory in which to store past experience. "Rote learning" carried the prototype quite far, by simply storing thousands of positions it had already encountered and seen the fruits of. But rote learning can only take you so far; Samuel's program confronted rapidly diminishing returns when it had stored in the neighborhood of a million words of description of past experience and began to be overcome with indexing and retrieval problems. When higher or more versatile performance is required, a different strategy of design has to kick in: *generalization*.

Instead of trying to find the search procedure himself, Samuel tried to get the computer to find it. He wanted the computer to design its own evaluation function, a mathematical formula—a polynomial—that would yield a number, positive or negative, for every move it considered, such that, in general, the higher the number, the better the move. The polynomial was to be concocted of lots of pieces, each contributing positively or negatively, multiplied by one coefficient or another, and adjusted to various other circumstances, but Samuel had no idea what sort of concoction would work well. He made some thirty-eight different chunks—"terms"—and threw them into a "pool." Some of the terms were intuitively valuable, such as those giving points for increased mobility or potential captures, but others were more or less off the wall—such as "DYKE: the parameter is credited with

1 for each string of passive pieces that occupy three adjacent diagonal squares." At any one time, sixteen of the terms were thrown together into the working genome of the active polynomial and the rest were idle. By a lot of inspired guesswork and even more inspired tuning and tinkering, Samuel devised rules for elimination from the tournament, and found ways of keeping the brew stirred up, so that the trial-and-error process was likely to hit upon good combinations of terms and coefficients and recognize them when it did. The program was divided into Alpha, a rapidly mutating pioneer, and Beta, a conservative opponent that played the version that had won the most recent game. "Alpha generalizes on its experience after each move by adjusting the coefficients in its evaluation polynomial and by replacing terms which appear to be unimportant by new parameters drawn from the reserve list" (Samuel 1964, p. 83).

> At the start an arbitrary selection of 16 terms was chosen and all terms were assigned equal weights. . . . During [the early rounds] a total of 29 different terms was discarded and replaced, the majority of these on two different occasions. . . . The quality of the play was extremely poor. During the next seven games there were at least eight changes made in the top listing involving five different terms. . . . Quality of play improved steadily but the machine still played rather badly. . . . Some fairly good amateur players who played the machine during this period [after seven more games] agreed that it was 'tricky but beatable'. [Samuel 1964, p. 89.]

Samuel noted (p. 89) that, although the learning at this early stage was surprisingly fast, it was "quite erratic and none too stable." He was discovering that the problem space being explored was a rugged fitness landscape in which a program using simple hill-climbing techniques tended to fall into traps, instabilities, and obsessive loops from which the program could not recover without a helping nudge or two from its designer. He was able to recognize the "defects" in his system responsible for these instabilities and patch them. The final system—the one that beat Nealey—was a Rube Goldberg amalgam of rote learning, kludges,[15] and products of self-design that were quite inscrutable to Samuel himself.

---

15. Pronounced to rhyme with "stooge," a kludge is an *ad hoc* or jury-rigged patch or software repair. Purists spell this slang word "kluge," drawing attention to its (likely) etymology in the deliberate mispronunciation of the German word *klug(e)*, meaning "clever"; but according to *The New Hacker's Dictionary* (Raymond 1993), the term may have an earlier ancestor, deriving from the Kluge paper feeder, an "adjunct to mechanical printing presses" in use as early as 1935. In its earlier use, it named "a complex and puzzling artifact with a trivial function." The mixture of esteem and contempt hackers exhibit for kluges ("How could anything so dumb be so smart!") perfectly reproduces the attitude of biologists when they marvel at the perversely intricate solutions Mother Nature so often discovers.

It is not surprising that Samuel's program caused a tremendous sensation, and greatly encouraged the early visionaries of AI, but the enthusiasm for such learning algorithms soon faded. The more people looked into the attempt to extend his methods to more complex problems—chess, for instance, to say nothing of real-world, non-toy problems—the more the success of Samuel's Darwinian learner seemed to be attributable to the relative simplicity of checkers rather than to the power of the underlying learning capacity. Was this, then, the end of Darwinian AI? Of course not. It just had to hibernate for a while until computers and computer scientists could advance a few more levels of complexity.

Today, the offspring of Samuel's program are multiplying so fast that at least three new journals have been founded in the last year or two to provide a forum: *Evolutionary Computation*, *Artificial Life*, and *Adaptive Behavior*. The first of these emphasizes traditional engineering concerns: using simulated evolution as a method to expand the practical design powers of programmers or software engineers. The "genetic algorithms" devised by John Holland (who worked with Art Samuel at IBM on his checkers program) have demonstrated their power in the no-nonsense world of software development and have mutated into a phylum of algorithmic variations. The other two journals concentrate on more biologically flavored research, in which the simulations of evolutionary processes permit us, really for the first time, to *study* the biological design process itself by *manipulating* it—or, rather, by manipulating a large-scale simulation of it. As Holland has said, Artificial Life programs *do* permit us to "rewind the tape of life" and replay it, again and again, in many variations.

6. ARTIFACT HERMENEUTICS, OR REVERSE ENGINEERING

The strategy of interpreting organisms as if they were artifacts has a lot in common with the strategy known to engineers as *reverse engineering* (Dennett 1990b). When Raytheon wants to make an electronic widget to compete with General Electric's widget, they buy several of GE's widgets and proceed to analyze them: that's reverse engineering. They run them, benchmark them, X-ray them, take them apart, and subject every part of them to interpretive analysis: Why did GE make these wires so heavy? What are these extra ROM registers for? Is this a double layer of insulation, and, if so, why did they bother with it? Notice that the reigning assumption is that all these "why" questions have answers. Everything has a *raison d'être*; GE did nothing in vain.

Of course, if the wisdom of the reverse engineers includes a healthy helping of self-knowledge, they will recognize that this default assumption of optimality is too strong: sometimes engineers put stupid, pointless things in their designs, sometimes they forget to remove things that no longer have a function, sometimes they overlook retrospectively obvious shortcuts. Still, optimality must be the default assumption; if the reverse engineers can't assume that there is a good rationale for the features they observe, they can't even begin their analysis.[16]

Darwin's revolution does not discard the idea of reverse engineering but, rather, permits it to be reformulated. Instead of trying to figure out what God intended, we try to figure out what reason, if any, "Mother Nature"—the process of evolution by natural selection itself—"discerned" or "discriminated" for doing things one way rather than another. Some biologists and philosophers are very uncomfortable with any such talk about Mother Nature's reasons. They think it is a step backwards, an unmotivated concession to pre-Darwinian habits of thought, at best a treacherous metaphor. So they are inclined to agree with the recent critic of Darwinism, Tom Bethell, in thinking there is something fishy about this double standard (see page 73). I claim that it is not just well motivated; it is extremely fruitful and, in fact, unavoidable. As we have already seen, even at the molecular level you just can't do biology without doing reverse engineering, and you can't do reverse engineering without asking what reasons there are for whatever it is you are studying. You have to ask "why" questions. Darwin didn't show us that we don't have to ask them; he showed us how to answer them (Kitcher 1985a).

Since the next chapter will be devoted to defending this claim by demonstrating the ways in which the process of evolution by natural selection is *like* a clever engineer, it is important that we first establish two important ways in which it is *not* like a clever engineer.

When we human beings design a new machine, we usually start with a

---

16. This fact has been exploited by counter-reverse-engineers. I discuss an example in Dennett 1978 (p. 279):

> There is a book on how to detect fake antiques (which is also, inevitably, a book on how to *make* fake antiques) which offers this sly advice to those who want to fool the "expert" buyer: once you have completed your table or whatever (having utilized all the usual means of simulating age and wear) take a modern electric drill and drill a hole right through the piece in some conspicuous but puzzling place. The would-be buyer will argue: no one would drill such a disfiguring hole without a reason (it can't be supposed to look "authentic" in any way) so it must have served a purpose, which means this table must have been in use in someone's home; since it was in use in someone's home, it was not made expressly for sale in this antique shop. . . . Therefore it is authentic. Even if this "conclusion" left room for lingering doubts, the buyer will be so preoccupied dreaming up uses for that hole it will be months before the doubts can surface.

It has been claimed, with what plausibility I do not know, that Bobby Fischer has used the same strategy to defeat opponents in chess, especially when time is running out: make a deliberately "off-the-wall" move and watch your opponent waste precious time trying to make sense of it.

pretty-good version of the machine on hand, either an earlier model, or a "mockup" or scale model that we have built. We examine it carefully, and try out various alterations: "If we just bend this jaw up a little bit like so, and move this zipper-bit over a tad like so, it would work even better." But that is not the way evolution works. This comes out especially clearly at the molecular level. A particular molecule is the shape it is, and won't tolerate much bending or reshaping. What evolution has to do when it improves molecular design is to make *another* molecule—one that is almost like the one that doesn't work very well—and simply discard the old one.

People are advised *never* to switch horses in midstream, but evolution *always* switches horses. It can't *fix* anything, except by selecting and discarding. So in every evolutionary process—and hence in every true evolutionary explanation—there is always a faint but disconcerting odor of something dicey. I will call this phenomenon *bait-and-switch*, after the shady practice of attracting customers by advertising something at a bargain price and then, when you've lured them to the store, trying to sell them a substitute. Unlike that practice, evolutionary bait-and-switch is not really nefarious; it just seems to be, because it doesn't explain what at first you thought you wanted explained. It subtly changes the topic.

We saw the ominous shadow of bait-and-switch in its purest form in chapter 2, in the weird wager that I can produce somebody who wins ten consecutive coin-tosses without a loss. I don't know in advance who that somebody is going to be; I just know that the mantle will pass—has to pass, as a matter of algorithmic necessity—to *somebody or other* so long as I execute the algorithm. If you overlook this possibility and take my sucker bet, it is because you are too used to the human practice of tracking individuals and building projects around identified individuals and their future prospects. And if the winner of the tournament thinks there has to be an explanation of why *he* won, he is mistaken: there is no reason at all why *he* won; there is only a very good reason why *somebody* won. But, being human, the winner will no doubt think there *ought* to be a reason why he won: "If your 'evolutionary account' can't explain it, then you are leaving out something important!" To which the evolutionist must calmly reply: "Sir, I know that is what you came in here wanting, but let me try to interest you in something a little more affordable, a little less presumptuous, a little more defensible."

Has it ever occurred to you how lucky you are to be alive? More than 99 percent of all the creatures that have ever lived have died without progeny, but not a single one of your ancestors falls into that group! What a royal lineage of winners you come from! (Of course, the same thing is true of every barnacle, every blade of grass, every housefly.) But it's even eerier than that. We have learned, have we not, that evolution works by weeding out the unfit? Thanks to their design defects, these losers have a "pathetic but praiseworthy tendency to die before reproducing their kind" (Quine

1969, p. 126). This is the very engine of Darwinian evolution. If, however, we look back with tunnel vision at your family tree, we will find many different organisms, with a wide variety of strengths and weaknesses, but, curiously enough, *their* weaknesses never led a single one of them to a premature demise! So it looks from this angle as if evolution can't explain even a *single feature* that you inherited from your ancestors! Suppose we look back at the fan-out of your ancestors. Notice first that eventually it stops fanning out and begins to double up; you share *multiple* ancestors with everybody else alive today, and are multiply related to many of your own ancestors. When we look at the whole tree over time, we see that the later, more recent ancestors have improvements that the earlier ones lacked, but all the crucial events—all the selection events—happen offstage: not a single one of your ancestors, all the way back to the bacteria, succumbed to predation before reproducing, or lost out to the competition for a mate.

Of course, evolution *does* explain all the features that you inherited from your ancestors, but not by explaining why *you* are lucky enough to have them. It explains why today's winners have the features they do, but not why *these individuals* have the features they do.[17] Consider: You order a new car, and specify that it be green. On the appointed day, you go to the dealership and there it sits, green and new. Which is the right question to ask: "Why is this car green?" or "Why is this (green) car here?" (In later chapters we will look further at the implications of bait-and-switch.)

The second important difference between the processes—and hence the products—of natural selection and human engineering concerns the feature of natural selection that strikes many people as most paradoxical: its utter lack of foresight. When human engineers design something (forward engineering), they must guard against a notorious problem: unforeseen side effects. When two or more systems, well designed in isolation, are put into a supersystem, this often produces interactions that were not only not part of the intended design, but positively harmful; the activity of one system inadvertently clobbers the activity of the other. The only practical way to guard against unforeseen side effects, since by their very nature they are unforeseeable by those whose gaze is perforce restricted to just one of the subsystems being designed, is to design all subsystems to have relatively impenetrable boundaries that coincide with the epistemic boundaries of their creators. Human engineers typically attempt to insulate subsystems from each other, and insist on an overall design in which each subsystem has a single, well-defined function within the whole.

The set of supersystems having this fundamental abstract architecture is

17. "But this is not to explain why, e.g., contractile vacuoles occur in certain protozoans; it is to explain why the sort of protozoan incorporating contractile vacuoles occurs" (Cummins 1975, in Sober 1984b, pp. 394–95).

vast and interesting, of course, but it does not include very many of the systems designed by natural selection! The process of evolution is notoriously lacking in foresight. Since it has no foresight at all, unforeseen or unforeseeable side effects are nothing to it; it proceeds, unlike human engineers, via the profligate process of creating vast numbers of relatively *un*insulated designs, most of which are hopelessly flawed because of self-defeating side effects, but a few of which, by dumb luck, are spared that ignominious fate. Moreover, this apparently inefficient design philosophy carries a tremendous bonus that is relatively unavailable to the more efficient, top-down process of human engineers: thanks to its having no bias against unexamined side effects, it can take advantage of the rare cases where beneficial *serendipitous* side effects emerge. Sometimes, that is, designs emerge in which systems interact to produce more than was aimed at. In particular (but not exclusively), one gets elements in such systems that have multiple functions.

Elements with multiple functions are not unknown to human engineering, of course, but their relative rarity is signaled by the delight we are apt to feel when we encounter a new one. (A favorite of mine is found in the Diconix portable printer. This optimally tiny printer runs on largish rechargeable batteries, which have to be stored somewhere; they fit snugly inside the platen, or roller.) On reflection, we can see that such instances of multiple function are epistemically accessible to engineers under various salubrious circumstances, but we can also see that by and large such solutions to design problems must be exceptions against a background of strict isolation of functional elements. In biology, we encounter quite crisp anatomical isolation of functions (the kidney is entirely distinct from the heart; nerves and blood vessels are separate conduits strung through the body, etc.), and without this readily discernible isolation, reverse engineering in biology would no doubt be humanly impossible. But we also see superimposition of functions that apparently goes "all the way down." It is very, very hard to think about entities in which the elements have multiple overlapping roles in superimposed subsystems, and, moreover, in which some of the most salient effects observable in the interaction of these elements may not be functions at all, but merely by-products of the multiple functions being served.[18]

Until recently, biologists who wanted to be reverse engineers had to concentrate on figuring out the designed features of "finished products"— organisms. These they could collect by the hundreds or thousands, study the variations thereof, take apart, and manipulate *ad lib*. It was much more difficult to get any epistemic purchase on the *developmental* or *building*

---

18. The preceding three paragraphs are drawn, with revisions, from Dennett 1994a.

process by which a genotype gets "expressed" in a fully formed phenotype. And the *design* processes that shaped the developmental processes that shape the "finished products" were largely inaccessible to the sorts of intrusive observation and manipulation that good science (or good reverse engineering) thrives on. You could look at the sketchy historical record, and run it in fast-forward (like "elapsed-time" photography of plants growing, weather developing, etc.—always a nifty way to make the patterns visible), but you couldn't "rewind the tape" and run variations on the initial conditions. Now, thanks to computer simulations, it is possible to *study* the hypotheses about the design process that have always lain at the heart of the Darwinian vision. Not surprisingly, they turn out to be more complicated, and themselves more intricately designed, than we had thought.

Once the processes of R and D and construction begin to come into focus, we can see that an affliction of shortsightedness that has often misled interpreters of human artifacts has multiple parallels in biology. When we engage in artifact hermeneutics, trying to decipher the design of items uncovered by archeologists, or trying to recover a proper interpretation of the ancient monuments that we have grown up with, there is a tendency to overlook the possibility that some of the features that puzzle us have no function at all in the finished product, but played a crucial functional role in the process that created the product.

Cathedrals, for instance, have many curious architectural features that have provoked functional fantasies and fierce debates among art historians. The functions of some of these features are fairly obvious. The many "vises" or circular stairways that twist their way up inside the piers and walls are useful ways for custodians to gain access to remote parts of the building: to the roof, say, and to the space between the vault and the roof where the machinery is hidden that lowers the chandeliers to the floor so that candles may be replaced. But many of the vises would be there even if no such later access had been anticipated by the builders; it was simply the best, or maybe the only, way for the builders to get the building crew and materials where they needed to be during construction. Other passageways leading nowhere inside the walls are probably there in order to get fresh air into the interior of the walls (Fitchen 1961). Medieval mortar took a long time—years, in some cases—to cure, and as it cured it shrank, so care was taken to keep wall thickness minimized so that distortion was minimized as the building cured. (Thus those passageways have a similar function to the heat-dissipation "fins" on automobile-engine housings, except that their functions lapsed once the buildings reached maturity.)

Moreover, much that appears unremarkable when you look at a cathedral as simply a finished product seems deeply puzzling when you start asking how it could have been built. Chicken-egg problems abound. If you build the flying buttresses before you build the central vault, they will push the walls in; if you build the vault first, it will spread the walls before the buttresses can be installed; if you try to build them both at once, it seems likely that the staging for one would get in the way of the staging for the other. It is surely a problem that has a solution—probably many different ones—but thinking them up and then looking for the evidence to confirm or disconfirm them is a challenging exercise. One strategy that recurs is one we have already seen in action in Cairns-Smith's clay-crystal hypothesis: there must have been scaffolding members that have disappeared, that functioned only during the building process. Such structures often leave clues of their former presence. Plugged "putlog holes" are the most obvious. Heavy timbers called "putlogs" were temporarily fixed in the walls to bear the scaffolding above them.

Many of the decorative elements of Gothic architecture, such as the elaborate patterns of ribs in the vaults, are really structurally functional members—but only during the construction phase. They had to be erected before the "web courses" of the vaults could be filled in between them. They stiffened the relatively delicate wooden "centering" scaffolding, which would otherwise have tended to buckle and deform under the temporarily uneven weight of partially built vaults. There were severe limits on the strength of scaffolding that could be constructed and held securely in place at great heights using medieval materials and methods. These limits dictated many of the "ornamental" details of the finished church. Another way of making the same point: many readily *conceivable* finished products were simply impossible to erect, given the constraints on the building process, and many of the apparently non-functional features of existing buildings are in fact enabling design features without which the finished product could not exist. The invention of cranes (real cranes) and their kin opened up regions of the space of architectural possibility that were previously inaccessible.[19]

The point is simple, but casts a long shadow: When you ask functional questions about *anything*—organism or artifact—you must remember that it has to come into its current or final form by a process that has its own requirements, and these are exactly as amenable to functional analysis as any features of the end state. No bell rings to mark the end of building and the beginning of functioning (cf. Fodor 1987, p. 103). The requirement that an organism be a going concern at every stage of its life places iron constraints on its later features.

---

19. Four classic explorations of these issues are John Fitchen's *The Construction of Gothic Cathedrals*, which reads like a detective story, Fitchen's *Building Construction Before Mechanization* (1986), William Barclay Parsons' *Engineers and Engineering in the Renaissance* (1939, republished by MIT Press, 1967) and Bertrand Gille's *Engineers of the Renaissance* (1966).

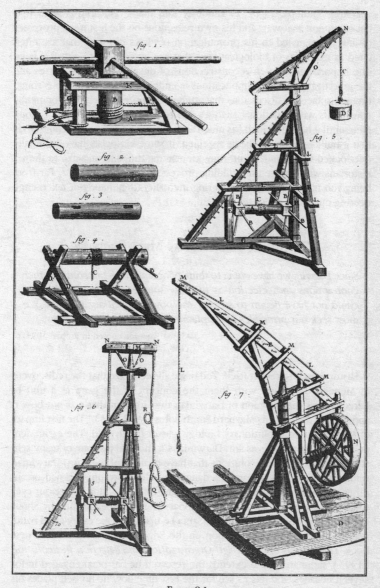

FIGURE 8.1.
Early rotating cranes and other devices for raising or moving loads. (From Diderot and d'Alembert, *Encyclopédie* [1751–1772], reproduced in Fitchen 1986.)

D'Arcy Thompson (1917) famously said that everything is what it is because it got that way, and his own reflections on the historical processes of development led to his promulgation of "laws of form" that are often cited as examples of biological laws that are irreducible to physical laws. The importance of such reconstructions of developmental processes and the investigation of their implications is undeniable, but this issue is sometimes misplaced in discussions that attempt to *contrast* such developmental constraints with functional analyses. No sound functional analysis is complete until it has confirmed (as much as these points ever can be confirmed) that a building path has been specified. If some biologists have habitually overlooked this requirement, they are making the same mistake as the art historians who ignore the building process of their monuments. Far from being too taken with an engineering mentality, they have not taken engineering questions seriously enough.

7. STUART KAUFFMAN AS META-ENGINEER

Since Darwin, we have come to think of organisms as tinkered-together contraptions and selection as the sole source of order. Yet Darwin could not have begun to suspect the power of self-organization. We must seek our principles of adaptation in complex systems anew.

—STUART KAUFFMAN, quoted in Ruthen 1993, p. 138

History tends to repeat itself. Today we all recognize that the rediscovery of Mendel's laws, and with them the concept of the gene as a unit of heredity, was the salvation of Darwinian thinking, but that was not how it appeared at the time. As Maynard Smith notes (1982, p. 3), "The first impact of Mendelism on evolutionary biology was distinctly odd. The early Mendelians saw themselves as anti-Darwinians." This was just one of many self-styled anti-Darwinian revolutions that have turned out to be pro-Darwinian reformations, dragging Darwin's dangerous idea from one sickbed or another and putting it back to work. Another that is unfolding before our eyes today is the new direction in evolutionary thinking spearheaded by Stuart Kauffman and his colleagues at the Santa Fe Institute. Like every good bandwagon, it has a slogan: "Evolution on the Edge of Chaos." Kauffman's new book, *The Origins of Order: Self-Organization and Selection in Evolution* (1993), summarizes and extends the research he has been engaged in for several decades, and lets us see for the first time how he himself places his ideas in the context of the history of the field.

Many have heralded him as a Darwin-slayer, finally driving that oppressive presence from the scene, and doing it, moreover, with the flashing blade of brand-new science: chaos theory and complexity theory, strange attractors and fractals. He himself has been tempted by that view in the past (Lewin 1992, pp. 40–43), but his book bristles with warnings, fending off the embrace of the anti-Darwinians. He begins the preface of his book (p. vii) by describing it as "an attempt to include Darwinism in a broader context":

> Yet our task is not only to explore the sources of order which may lie available to evolution. We must also integrate such knowledge with the basic insight offered by Darwin. Natural selection, whatever our doubt in detailed cases, is surely a preeminent force in evolution. Therefore, to combine the themes of self-organization *and* selection, we must expand evolutionary theory so that it stands on a broader foundation and then raise a new edifice. [Kauffman 1993, p. xiv.]

I go to such lengths to quote Kauffman himself on this point since I have also felt the strong wind of anti-Darwinian sentiment among my own readers and critics, and know that they will be strongly motivated to suspect that I am merely reworking Kauffman's ideas to fit my own biased view! No, he himself—for whatever that is worth—now sees his work as a deepening of Darwinism, not an overthrow. But, then, what can be his point about "spontaneous self-organization" as a source of "order" if not a flat denial that selection is the ultimate source of order?

Now that it is possible to build truly complex evolutionary scenarios on computers, rewinding the tape over and over, we can see patterns that eluded earlier Darwinian theorists. What we see, Kauffman claims, is that order "shines through" *in spite of* selection, not because of it. Instead of witnessing the gradual accrual of organization under the steady pressure of cumulative selection, we witness the *inability* of selective pressure (which can be carefully manipulated and monitored in the simulations) to overcome an inherent tendency of the populations in question to resolve themselves into ordered patterns. So this seems at first to be a striking demonstration that natural selection cannot be the source of organization and order after all—which would indeed be the downfall of the Darwinian idea.

But there is another way of looking at it, as we have seen. What conditions have to be in effect for evolution by natural selection to occur? The words I put into Darwin's mouth were simple: Give me Order, and time, and I will give you Design. But what we have subsequently learned is that not every variety of Order is sufficient for evolvability. As we saw illustrated by Conway's Game of Life, you have to have just the right sort of Order, with just the right mix of freedom and constraint, growth and decay, rigidity and fluidity, for good things to happen at all. You only get evolution, as the Santa Fe motto proclaims, on the edge of chaos, in the regions of possible law that form the hybrid zone between stifling order and destructive chaos. Fortunately, our portion of the universe is poised in just such a zone, in which the conditions for evolvability are tuned just right. And where did those salubrious conditions come from? They could "in principle" have come from the wisdom and foresight of a designer like Conway, *or* they could have come from a *prior* evolutionary process, either one with selection or one without. In fact—and this, I think, is the heart of Kauffman's vision—evolvability itself not only must evolve (for us to be here), but is *likely* to evolve, is almost sure to evolve, because it is a forced move in the game of Design.[20] Either you find the path that leads to evolvability or you don't go anywhere, but finding the path to evolvability is not such a big deal; it's "obvious." The principles of design that make biological evolution possible will be found, again and again and again, no matter how many times we rerun the tape. "Contrary to all our expectations, the answer, I think, is that it may be surprisingly *easy*" (Kauffman 1993, p. xvi).

When we considered forced moves in Design Space in chapter 6, we were thinking about features of the final products that were so obviously "right" that we would not be surprised to find them independently appearing—arithmetic among the alien intelligences, eyes wherever there is locomotion through a transparent medium. But what about features of the process of creating those products? If there are fundamental rules about how things have to be designed, about the order in which design innovations can be created, the strategies of design that are bound to work or fail, then these should be homed in on by evolution just as surely as the features of the finished products. What Kauffman has discovered, I submit, is not so much *laws of form* as *rules for designing*: the imperatives of meta-engineering. Kauffman has many telling observations to make about just such principles of meta-engineering that govern the process by which new designs could, in practice, be created. We can consider them to be features of the whole phenomenon of evolution that have *already* been discovered, have *already* gone to fixation, in effect, in our part of the universe. (We will not be surprised to find them everywhere else in the universe where there are designed things, because this is the only way to design things.)

Adaptive evolution is a search process—driven by mutation, recombination, and selection—on fixed or deforming fitness landscapes. An adapting population flows over the landscape under these forces. The structure of

---

20. The evolution of evolvability is a (retrospectively!) obvious recursive move for Darwinians to promulgate—a likely source of cranes, you might say—and it has been discussed by many thinkers. For an early discussion, see Wimsatt 1981. For a different slant on the issue, see Dawkins 1989b.

such landscapes, smooth or rugged, governs both the evolvability of populations and the sustained fitness of their members. The structure of fitness landscapes inevitably imposes limitations on adaptive search. [Kauffman 1993, p. 118.]

Notice that this is all pure Darwinism—every bit acceptable and nonrevolutionary, but with a major shift of emphasis to the role of the topology of the fitness landscape, which, Kauffman argues, has a profound effect on the *rate* at which design innovations can be found, and the *order* in which design chances can accumulate. If you have ever tried to write a sonnet, you have confronted the basic design problem that Kauffman's models examine: "epistasis," or the interactions between genes. As the budding poet soon discovers, writing a sonnet isn't easy! Saying something meaningful—let alone beautiful—within the rigid constraints of the sonnet form is a frustrating exercise. No sooner do you tentatively fix one line than you have to revise many of the other lines, and that forces you to abandon some hard-won excellences, and so forth, round and round in circles, searching for an overall good fit—or, we might say, searching for overall good fitness. The mathematician Stanislaw Ulam saw that the constraints of poetry could be a source of creativity, not a hindrance. The idea may apply to the creativity of evolution, for just the same reason:

When I was a boy I felt that the role of rhyme in poetry was to compel one to find the unobvious because of the necessity of finding a word which rhymes. This forces novel associations and almost guarantees deviations from routine chains or trains of thought. It becomes paradoxically a sort of automatic mechanism of originality. [Ulam 1976, p. 180.]

Before Kauffman, biologists tended to ignore the prospect that evolution would confront the same sort of pervasive interactions, because they had no clear way of studying it. His work shows that making a viable genome is more like writing a good poem than simply jotting down a shopping list. *Since* the structure of fitness landscapes is more important than we had thought (with our simpler, Mount Fuji models of hill-climbing), there are constraints on design-improvement *methods* that keep engineering projects channeled into more narrow paths to success than we had imagined.

Evolvability, the capacity to search a reasonable fraction of the space, may be optimized when landscape structure, mutation rate, and population size are adjusted so that populations just begin to 'melt' from local regions of the space. [Kauffman 1993, p. 95.]

One ubiquitous feature in biological evolution that Kauffman concentrates on is the principle that "local rules generate global order." This is not a principle that governs human engineering. Pyramids are always built from the bottom up, of course, but the organization of the building process, since the days of the pharaohs, has been top-down, under the control of a single autocrat who had a clear and literally commanding vision of the whole, but probably was a bit vague about how the local details would be accomplished. "Global" direction from on high puts in motion a hierarchical cascade of "local" projects. This is such a common feature of large-scale human projects that we have a hard time imagining alternatives (Papert 1993, Dennett 1993a). Since we don't recognize the principle Kauffman discerns as one that is familiar from human engineering, we are not apt to see it as a principle of engineering at all, but I suggest that it is. Reformulated slightly, we could put it as follows. Until you manage to evolve communicating organisms that can form large engineering organizations, you are bound by the following Preliminary Design Principle: all global order must be generated by local rules. So all the early products of design, up to the creation of something with some of the organizational talents of *Homo sapiens*, must obey whatever constraints follow from the "management decision" that all order must be accomplished by local rules. Any "attempts" to create living forms that violate this precept will end in immediate failure—or, more accurately, will not even get started sufficiently to be discernible as attempts.

If no bell rings, as I have said, to mark the moment when the R-and-D process ends and the life of the "finished product" begins, it should at least sometimes be hard to tell whether a design principle in question is a principle of engineering or of meta-engineering. A case in point is Kauffman's (1993, pp. 75ff.) proposed rederivation of "von Baer's laws" of embryology. One of the most striking patterns in the embryos of animals is the fact that they all start out so much the same.

> Thus early fish, frog, chick and human embryos are remarkably similar. . . . The familiar explanation for these laws is that mutants [I think he means "mutations"] affecting early ontogeny are more disruptive than mutants affecting late ontogeny. Thus mutants altering early development are less likely to accumulate, and early embryos remain more similar from one order of organisms to another than do late embryos. Is this plausible argument actually so plausible? [Kauffman 1993, p. 75.]

The traditional Darwinian, on Kauffman's reading, places the responsibility for von Baer's laws in a "special mechanism," built right into organisms. Why don't we see many finished products with strikingly different early embryos? Well, since change-orders that affect early parts of the process tend to be more disastrous in their effect on the finished product than change-orders that affect later parts of the process, Mother Nature has designed a special developmental mechanism to protect against such experimentation. (This would be analogous to IBM's forbidding its computer scientists to investigate alternative architectures for its CPU or central-processing-unit chip—*designed* resistance to change.)

And what is Kauffman's contrasting explanation? It starts with the same point and takes it in a rather different direction:

> ... a locking-in of early development, and hence von Baer's laws, do not represent a special mechanism of developmental canalization, the usual sense of which is a buffering of the phenotype against genetic alteration. ... Instead, locking-in of early development is a direct reflection of the fact that the number of ways to improve organisms by altering early ontogeny has dwindled faster than the number of ways to improve by altering late development. [Kauffman 1993, p. 77. See also Wimsatt 1986.]

Think of the issue from the point of view of human engineering for a moment. Why is it that the foundations of churches are more alike than their upper stories? Well, says the traditional Darwinian, they have to be built first, and any wise contractor will tell you that if you *must* tinker with design elements, work on the steeple ornament first, or the windows. You are less apt to have a disastrous crash than if you try to come up with a new way of preparing the foundation. So it is not so surprising that churches all start out looking more or less alike, with the big differences emerging in the later elaborations of the building process. Actually, says Kauffman, there really just aren't as many different *possible* solutions to the foundation problem as there are to later building problems. Even stupid contractors who butted their heads against this fact for eons would not come up with a wide variety of foundation designs. This difference of emphasis may look small, but it has some important implications. Kauffman says we don't need to look for a canalization *mechanism* to explain this fact; it will take care of itself. But there is also an underlying agreement between Kauffman and the tradition he wants to supplant: there are only so many good ways of building things, given the starting constraints, and evolution finds them again and again.

It is the *non-optionality* of these "choices" that Kauffman wants to stress, and so he and his colleague Brian Goodwin (e.g., 1986) are particularly eager to discredit the powerful image, first made popular by the great French biologists Jacques Monod and François Jacob, of Mother Nature as a "tinker," engaging in the sort of tinkering the French call *bricolage*. The term was first made salient by the anthropologist Claude Lévi-Strauss (1966). A tinker or *bricoleur* is an opportunistic maker of gadgets, a "satisficer" (Simon 1957) who is always ready to settle for mediocrity if it is cheap enough. A tinker is not a deep thinker. The two elements of classical

Darwinism that Monod and Jacob concentrate on are chance on the one hand and, on the other, the utter directionlessness and myopia (or blindness) of the watchmaker. But, says Kauffman, "Evolution is not just 'chance caught on the wing.' It is not just a tinkering of the ad hoc, of bricolage, of contraption. It is emergent order honored and honed by selection" (Kauffman 1993, p. 644).

Is he saying the watchmaker *isn't* blind? Of course not. But then what is he saying? He is saying that there are principles of order that govern the design process, and that force the tinker's hand. Fine. Even a blind tinker will find the forced moves; it doesn't take a rocket scientist, as one says. A tinker who can't find the forced moves is not worth a tinker's damn, and won't design a thing. Kauffman and his colleagues have made an interesting set of discoveries, but the attack on the image of the tinker is to a large extent, I think, misplaced. The tinker, says Lévi-Strauss, is willing to be guided by the nature of the material, whereas the engineer wants the material to be perfectly malleable—like the concrete so beloved by the Bauhaus architects. So the tinker is a deep thinker after all, complying with constraints, not fighting them. The truly wise engineer works not *contra naturam* but *secundum naturam*.

One of the virtues in Kauffman's attack is that it draws attention to an underappreciated possibility, one that we can make vivid with the help of an imaginary example from human engineering. Suppose that the Acme Hammer Company discovers that the new hammers made by its rival, Bulldog Hammer, Inc., have plastic handles with exactly the same intricate pattern of colored whorls on them as is sported by the new Acme Model Zeta. "Theft!" scream their legal representatives. "You copied our design!" Maybe, but then again, maybe not. It just might be that there is only one way of making plastic handles with any strength, and that is to stir up the plastic somehow as it sets. The result is inevitably a distinctive pattern of whorls. It would be almost impossible to make a serviceable plastic hammer handle that *didn't* have those whorls in it, and the discovery of this fact might be one that would be eventually imposed on just about anybody who tried to make a plastic hammer-handle. This could explain the otherwise suspicious similarity without any hypothesis of "descent" or copying. Now, maybe the Bulldog people did copy Acme's design, *but they would have found it in any case*, sooner or later. Kauffman points out that biologists tend to overlook this sort of possibility when they draw their inferences about descent, and he draws attention to many compelling cases in the biological world in which similarity of pattern has nothing to do with descent. (The most striking cases he discusses are illuminated by Turing's 1952 work on the mathematical analysis of the creation of spatial patterning in morphogenesis.)

In a world with no discoverable principles of design, all similarities are suspicious—likely to be due to copying (plagiarism or descent).

We have come to think of selection as essentially the only source of order in the biological world. If 'only' is an overstatement, then surely it is accurate to state that selection is viewed as the overwhelming source of order in the biological world. It follows that, in our current view, organisms are largely ad hoc solutions to design problems cobbled together by selection. It follows that most properties which are widespread in organisms are widespread by virtue of common descent from a tinkered-together ancestor, with selective maintenance of the useful tinkerings. It follows that we see organisms as overwhelmingly contingent historical accidents, abetted by design. [Kauffman 1993, p. 26.]

Kauffman wants to stress that the biological world is much more a world of Newtonian discoveries (such as Turing's) than Shakespearean creations, and he has certainly found some excellent demonstrations to back up his claim. But I fear that his attack on the metaphor of the tinker feeds the yearning of those who don't appreciate Darwin's dangerous idea; it gives them a false hope that they are seeing not the forced hand of the tinker but the divine hand of God in the workings of nature.

Kauffman himself has called what he is doing the quest for "the physics of biology" (Lewin 1992, p. 43), and that is not really in conflict with what I am calling it: meta-engineering. It is the investigation of the most general constraints on the processes that can lead to the creation and reproduction of designed things. But when he declares this a quest for "laws," he feeds the antiengineering prejudice (or you might call it "physics envy") that distorts so much philosophical thinking about biology.

Does anyone suppose that there are *laws* of nutrition? Laws of locomotion? There are all sorts of highly imperturbable boundary conditions on nutrition and locomotion, owing to fundamental laws of physics, and there are plenty of regularities, rules of thumb, trade-offs, and the like that are encountered by any nutritional or locomotive mechanisms. But these are not laws. They are like the highly robust regularities of automotive engineering. Consider the regularity that (*ceteris paribus*) ignition is accomplished only by or after the use of a key. There is a reason for this, of course, and it has to do with the perceived value of automobiles, their susceptibility to theft, the cost-effective (but not foolproof) options provided by pre-existing locksmith technology, and so forth. When one understands the myriad cost-benefit trade-offs of the design decisions that go into creating automobiles, one appreciates this regularity. It is not any kind of law; it is a regularity that tends to settle out of a complex set of competing *desiderata* (otherwise known as norms). These highly reliable, norm-tracking generalizations are not laws of automotive engineering, nor are their biological counterparts laws of locomotion or nutrition. The location of the mouth at the bow rather than the stern end of the locomoting organism (*ceteris paribus*— there are exceptions!) is a deep regularity, but why call it a law? We understand *why* it should be so, because we see what mouths—or locks and keys—are *for*, and why certain ways are the best ways of accomplishing those ends.

CHAPTER 8: *Biology is not just like engineering; it is engineering. It is the study of functional mechanisms, their design, construction, and operation. From this vantage point, we can explain the gradual birth of function, and the concomitant birth of meaning or intentionality. Achievements that at first seem either literally miraculous (e.g., the creation of recipe-readers where none were before) or at least intrinsically Mind-dependent (learning to play winning checkers) can be broken down into the ever smaller achievements of ever smaller and stupider mechanisms. We have now begun to pay close attention to the design process itself, not just its products, and this new research direction is deepening Darwin's dangerous idea, not overthrowing it.*

CHAPTER 9: *The task of reverse engineering in biology is an exercise in figuring out "what Mother Nature had in mind." This strategy, known as adaptationism, has been an amazingly powerful method, generating many spectacular leaps of inference that have been confirmed—along with some that have not, of course. The famous critique of adaptationism by Stephen Jay Gould and Richard Lewontin focuses attention on the suspicions people have harbored about adaptationism, but is largely misdirected. The applications of game theory in adaptationism have been particularly fruitful, but one must be cautious: there may be more hidden constraints than theorists often assume.*

CHAPTER NINE

Searching for Quality

~~~

1. THE POWER OF ADAPTATIONIST THINKING

'Naked as Nature intended' was a persuasive slogan of the early Naturist movement. But Nature's original intention was that the skin of all primates should be un-naked.

—ELAINE MORGAN 1990, p. 66

Judging a poem is like judging a pudding or a machine. One demands that it work. It is only because an artifact works that we infer the intention of an artificer.

—W. WIMSATT and M. BEARDSLEY 1954, p. 4

If you know something about the design of an artifact, you can predict its behavior without worrying yourself about the underlying physics of its parts. Even small children can readily learn to manipulate such complicated objects as VCRs without having a clue as to how they work; they know just what will happen when they press a sequence of buttons, because they know what is designed to happen. They are operating from what I call the *design stance.* The VCR repairer knows a great deal more about the design of the VCR, and knows, roughly, how all the interior parts interact to produce both proper functioning and pathological functioning, but may also be quite oblivious of the underlying physics of the processes. Only the designers of the VCR had to understand the physics; they are the ones who must descend to what I call the *physical stance* in order to figure out what sorts of design revisions might enhance picture quality, or diminish wear and tear on the tape, or reduce the electricity consumption of the product. But when they engage in *reverse* engineering—of some other manufacturer's VCR, for instance—they avail themselves not only of the physical stance, but also of what I call the *intentional stance*—they try to figure out *what the designers*

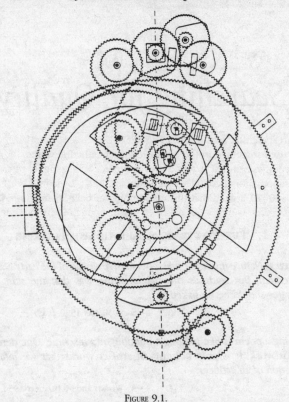

FIGURE 9.1.

Figure diagramming the wheel-work of the Antikythera mechanism by Derek de-Solla Price (Yale University).

had in mind. They treat the artifact under examination as a product of a process of *reasoned* design development, a series of *choices* among alternatives, in which the *decisions* reached were those *deemed best* by the designers. Thinking about the postulated functions of the parts is making assumptions about the *reasons* for their presence, and this often permits one to make giant leaps of inference that finesse one's ignorance of the underlying physics, or the lower-level design elements of the object.

Archeologists and historians sometimes encounter artifacts whose meaning—whose function or purpose—is particularly obscure. It is instructive to look briefly at a few examples of such *artifact hermeneutics* to see how one reasons in such cases.[1]

The Antikythera mechanism, discovered in 1900 in a shipwreck, and

1. For an expanded analysis of these issues, see Dennett 1990b.

dating from ancient Greece, is an astonishingly complex assembly of bronze gears. What was it for? Was it a clock? Was it the machinery for moving an automaton statue, like Vaucanson's marvels of the eighteenth century? It was—almost certainly—an orrery or a planetarium, and the proof is that it would be a *good* orrery. That is, calculations of the periods of rotation of its wheels led to an interpretation that would have made it an accurate (Ptolemaic) representation of what was then known about the motions of the planets.

The great architectural historian Viollet-le-Duc described an object called a *cerce*, used somehow in the construction of cathedral vaults.

He hypothesized that it was a movable piece of staging, used as a temporary support for incomplete web-courses, but a later interpreter, John Fitchen (1961), argued that this could not have been its function. For one thing, the cerce would not have been strong enough in its extended position, and, as figure 9.2 shows, its use would have created irregularities in the vault webbing which are not to be found. Fitchen's extended and elaborate

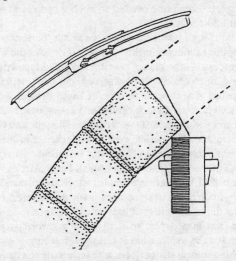

Viollet-le-Duc's cerce device as support for each web course during the erection of the vault. The smaller-scale drawing shows a cerce, based on Viollet-le-Duc's representation and description. Its extended position clearly indicates how one slotted board laps the other. Hung vertically as support for the stones of a web course, it is seen (in the detailed section) that the stones of any given course cannot line up throughout: those that lean against the far board (shown in outline) tilt much more than those that lean against the near board (shown hatched). As no such break does occur in the alignment of the web stone-coursing, it is obvious that the cerce device was not used in this fashion, in spite of Viollet-le-Duc's assertion that it was. [Fitchen 1961, p. 101.]

FIGURE 9.2

argument concludes that the cerce was no more than an adjustable template, a conclusion he supports by coming up with a much more elegant and versatile solution to the problem of temporary support of web courses.

The important feature in these arguments is the reliance on optimality considerations; it counts against the hypothesis that something is a cherry-pitter, for instance, if it would have been a demonstrably inferior cherry-pitter. Occasionally, an artifact loses its original function and takes on a new one. People buy old-fashioned sadirons not to iron their clothes with, but to use as bookends or doorstops; a handsome jam pot can become a pencil-holder, and lobster traps get recycled as outdoor planters. The fact is that sadirons are much better as bookends than they are at ironing clothes—when compared with the competition today. And a Dec-10 mainframe computer today makes a nifty heavy-duty anchor for a large boat-mooring. No artifact is immune from such appropriation, and however clearly its *original* purpose may be read from its current form, its new purpose may be related to that original purpose by mere historic accident—the fellow who owned the obsolete mainframe needed an anchor badly, and opportunistically pressed it into service.

The clues about such historical processes would be simply unreadable without assumptions about optimality of design. Consider the so-called dedicated word-processor—the cheap, portable, glorified typewriter that uses disk storage and an electronic display screen, but can't be used as an all-purpose computer. If you open up one of these devices, you find it is governed by an all-purpose CPU or central processing unit, such as an 8088 chip—a full-power computer vastly more powerful, swift, and versatile than the biggest computer Alan Turing ever saw—locked into menial service, performing a minuscule fraction of the tasks it *could* be harnessed to perform. Why is all this excess functionality found here? Martian reverse engineers might be baffled, but there is a simple historical explanation, of course: the genealogy of computer development gradually lowered costs of chip manufacture to the point where it was much cheaper to install a whole computer-on-a-chip in a device than to build a special-purpose control circuit. Notice that the explanation is historical but also, inescapably, proceeds from the intentional stance. It became *wise* to design dedicated word-processors this way, when the cost-benefit analysis showed that this was the *best, cheapest* way to *solve the problem.*

What is amazing is how powerful the intentional stance can be in reverse engineering, not only of human artifacts, but also of organisms. In chapter 6, we saw the role of practical reasoning—cost-benefit analysis in particular—in distinguishing the forced moves from what we might call the *ad lib* moves, and we saw how Mother Nature could be predicted to "discover" the forced moves again and again. The idea that we can impute such "free-floating rationales" to the mindless process of natural selection is dizzying, but there is no denying the fruits of the strategy. In chapters 7 and 8, we saw how the engineering perspective informs research at every level from the molecules on up, and how this perspective *always* involves distinguishing the better from the worse, and the reasons Mother Nature has found for the distinction. The intentional stance is thus the crucial lever in all attempts to reconstruct the biological past. Did *Archaeopteryx*, the extinct birdlike creature that some have called a winged dinosaur, ever really get off the ground? Nothing could be more ephemeral, less likely to leave a fossil trace, than a flight through the air, but if you do an engineering analysis of its claws, they turn out to be excellent adaptations for *perching on branches*, not for *running*. An analysis of the claw curvature, supplemented by aerodynamic analysis of the archaeopteryx wing structure, makes it quite plain that the creature was *well designed* for flight (Feduccia 1993). So it almost certainly flew—or had ancestors that flew (we mustn't forget the possibility of excess functionality persisting, like the computer in the word-processor). The hypothesis that the archaeopteryx flew has not yet been fully confirmed to every expert's satisfaction, but it suggests many further questions to address to the fossil record, and when those questions are pursued, either the evidence will mount in favor of the hypothesis or it won't. The hypothesis is testable.

The lever of reverse engineering is not just for prying out secrets of history; it is even more spectacular as a predictor of unimagined secrets of the present. Why are there colors? Color-coding is generally viewed as a recent engineering innovation, but it is not. Mother Nature discovered it much earlier (for the details, see the section on why there are colors in Dennett 1991a, pp. 375–83). We know this thanks to lines of research opened up by Karl von Frisch, and, as Richard Dawkins points out, von Frisch used a bold exercise in reverse engineering to make the initial move.

Von Frisch (1967), in defiance of the prestigious orthodoxy of von Hess, conclusively demonstrated colour vision in fish and in honeybees by controlled experiments. He was driven to undertake those experiments by his refusal to believe that, for example, the colours of flowers were there for no reason, or simply to delight men's eyes. [Dawkins 1982, p. 31.]

A similar inference led to the discovery of the endorphins, the morphine-like substances that we produce in our own bodies when we are put under enough stress or pain—creating the "runner's high," for instance. The reasoning was the reverse of von Frisch's. Scientists found receptors in the brain that are highly specific for morphine, which has a powerful painkilling effect. Reverse engineering insists that wherever there is a highly particular lock, there must be a highly particular key to fit it. *Why are these receptors here?* (Mother Nature could not have foreseen the development of morphine!) There must be some molecules produced internally under some conditions, the original keys that these locks were designed to receive. Seek a molecule that fits this receptor and is produced under circumstances in which a shot of morphine might be beneficial. Eureka! Endogenously created morphine—endorphin—was discovered.

Even more devious Sherlock-Holmesian leaps of deduction have been executed. Here, for instance, is a general mystery: "Why do some genes change their pattern of expression depending on whether they are maternally or paternally inherited?" (Haig and Graham 1991, p. 1045). This phenomenon—in which the genome-reading machinery *pays more attention*, in effect, to either the paternal text or the maternal text—is known as *genomic imprinting* (for a general account, see Haig 1992), and has been confirmed to occur in special cases. What do the special cases have in common? Haig and Westoby (1989) developed a model that purports to solve the general mystery by *predicting* that genomic imprinting would be found only in organisms "in which females carry offspring by more than one male during their life span and a system of parental care in which offspring receive most of their postfertilization nutrients from one parent (usually the mother) and thus compete with offspring fathered by other males." In such circumstances, they reasoned, there should be a conflict between maternal and paternal genes—paternal genes will tend to favor exploiting the mother's body as much as possible, but maternal genes would "view" this as almost suicidal—and the result should be that the relevant genes will in effect choose sides in a tug-of-war, and genomic imprinting will result (Haig and Graham 1991, p. 1046).

See the model at work. There is a protein, "Insulin-like Growth Factor II" (IGF-II), which is, as its name suggests, a growth-enhancer. Not surprisingly, the genetic recipes of many species order the creation of large quantities of IGF-II during embryonic development. But, like all functioning machines, IGF-II needs the right supportive environment to do its work, and in this case it needs helper molecules known as "type 1 receptors." So far, our story is just like the endorphin story: we have a type of key (IGF-II) and a kind of lock (type 1 receptors) in which it fits and performs an obviously important role. But in mice, for instance, there is another kind of lock (type 2 receptors) in which it also fits. What are these secondary locks for? For nothing, apparently; they are descendants of molecules that in other species (toads, for instance) play a role in cells' "garbage-disposal" systems, but this is not what they do when they bind to IGF-II in mice. Then why are they there? Because they are "ordered" by the genetic recipe for making a mouse, of course, but here is the telltale twist: whereas both the maternal and paternal contributions to the chromosome contain recipe instructions for making them, these instructions are *preferentially expressed* from the maternal chromosome. Why? To counteract the instruction in the recipe that calls for too much growth-enhancer. The type 2 receptors are just there to soak up—to "capture and degrade"—all the excess growth-enhancer that the paternal chromosome would pump into the fetus if it had its way. Since mice are a species in which females tend to mate with more than one male, males in effect compete to exploit the resources of each female, a competition from which females must protect themselves (and their own genetic contributions).

Haig and Westoby's model predicts that genes would evolve in mice to protect females from this exploitation, and this imprinting has been confirmed. Moreover, their model predicts that type 2 receptors shouldn't work this way in species in which genetic conflict of this sort can't arise. They shouldn't work this way in chickens, because offspring can't influence how much yolk their eggs receive, so the tug-of-war can never get started. Sure enough, the type 2 receptors in chickens don't bind to IGF-II. Bertrand Russell once slyly described a certain form of illicit argument as having all the advantages of theft over honest toil, and one can sympathize with the hardworking molecular biologist who reacts with a certain envy when somebody like Haig swoops in, saying, in effect, "Go look under that rock—I bet you'll find a treasure of the following shape!"

But that is what Haig was able to do: he predicted what Mother Nature's move would be in the hundred-million-year game of mammal design. Of all the possible moves available, he saw that there was a good reason for this move, so this is what would be discovered. We can get a sense of the magnitude of the leap that such an inference takes by comparing it with a parallel leap that we can make in the Game of Life. Recall that one of the possible denizens of the Life world is a Universal Turing machine composed of trillions of pixels. Since a Universal Turing machine can compute any computable function, it can play chess—simply by mimicking the program of any chess-playing computer you like. Suppose, then, that such an entity occupies the Life plane, playing chess against itself, in the fashion of Samuel's computer playing checkers against itself. Looking at the configuration of dots that accomplishes this marvel would almost certainly be unilluminating to anyone who had no clue that a configuration with such powers could exist. But from the perspective of someone who *had the hypothesis* that this huge array of black dots was a chess-playing computer, enormously efficient ways of predicting the future of that configuration are made available.

Consider the savings you could achieve. At first you would be confronted by a screen on which trillions of pixels flash on and off. Since you know the single rule of Life Physics, you could laboriously calculate the behavior of each spot on the screen if you wanted, but it would take eons. As a first cost-cutting step, you could shift from thinking about individual pixels to thinking about gliders and eaters and still lifes, and so forth. Whenever you saw a glider approaching an eater, you would just predict "consumption in four generations" without bothering with the pixel-level calculations. As a second step, you could move to thinking of the gliders as symbols on the "tape" of a gigantic Turing machine, and then, adopting this higher design stance towards the configuration, predict its future *as* a Turing machine. At this level you would be "hand-simulating" the "machine language" of a computer program that plays chess, still a tedious way of making predictions, but orders of magnitude more efficient than working out the physics. As a third and still more efficient step, you could ignore the details of the chess-playing program itself and just assume that, whatever they are, they are *good*! That is, you could assume that the chess-playing program running on the Turing machine made of gliders and eaters played not just legal chess but good legal chess—it had been well designed (perhaps it has designed itself, in the manner of Samuel's checkers program) to find the good moves. This permits you to shift to thinking about chessboard positions, possible chess moves, and the grounds for evaluating them—to shift to reasoning about reasons.

Adopting the intentional stance towards the configuration, you could predict its future *as* a chess-player performing intentional actions—making chess moves and trying to achieve checkmate. First you would have to figure out the interpretation scheme that permits you to say which configurations of pixels count as which symbols: which glider pattern spells out "QxBch" (Queen takes Bishop; check) and the other symbols for chess moves. But then you could use the interpretation scheme to predict, for instance, that the next configuration to emerge from the galaxy would be such-and-such a glider stream—say, the symbols for "RxQ" (Rook takes Queen). There is risk involved, because the chess program being run on the Turing machine may be far from perfectly rational, and, at a different level, debris may wander onto the scene and "break" the Turing-machine configuration before it finishes the game. But if all goes well, as it normally will, if you have the right interpretation, you can astonish your friends by saying something like "I predict that the next stream of gliders to emerge in location L in this Life galaxy will have the following pattern: a singleton, followed by a group of three, followed by another singleton ..." How on Earth were you able to predict that that particular "molecular" pattern would appear then?[2]

In other words, real but (potentially) noisy patterns abound in such a configuration of the Life world, there for the picking up if only you are lucky or clever enough to hit on the right perspective. They are not *vi-*

2. In case you wondered, I imagined "RxQ" to be spelled out in Morse code, and "R" in Morse is dot-dash-dot—the group of three gliders counts as a dash.

sual patterns but, you might say, *intellectual* patterns. Squinting or twisting your head in front of the computer screen is not apt to help, whereas posing fanciful interpretations (or what Quine would call "analytical hypotheses") may uncover a gold mine. The opportunity confronting the observer of such a Life world is analogous to the opportunity confronting the cryptographer staring at a new patch of cipher text, or the opportunity confronting the Martian peering through a telescope at the Super-bowl Game. If the Martian hits on the intentional stance—otherwise known as folk psychology[3]—as the right level to look for pattern, shapes will readily emerge through the noisy jostling of people-particles and team-molecules.

The scale of compression when one adopts the intentional stance towards the two-dimensional chess-playing computer galaxy is stupendous: it is the difference between figuring out in your head what White's most likely (best) chess move is versus calculating the state of a few trillion pixels through a few hundred thousand generations. But the scale of the savings is really no greater in the Life world than in our own. Predicting that someone will duck if you throw a brick at him is easy from the intentional or folk-psychological stance; it is and will always be intractable if you have to trace the photons from brick to eyeball, the neurotransmitters from optic nerve to motor nerve, and so forth.

For such vast computational leverage one might be prepared to pay quite a steep price in errors, but in fact the intentional stance, used correctly, provides a description system that permits extremely reliable prediction of not only intelligent human behavior, but also the "intelligent behavior" of the process that designed organisms. All this would warm William Paley's heart. We can put the burden of proof on the skeptics with a simple challenge argument: if there weren't design in the biosphere, how come the intentional stance *works*? We can even get a rough measure of the design in the biosphere by comparing the cost of making predictions from the lowest-level physical stance (which assumes no design—well, almost no design, depending on how we treat the evolution of universes) with the cost of making predictions from the higher stances: the design stance and the intentional stance. The added leverage of prediction, the diminution of uncertainty, the shrinkage of the huge search space to a few optimal or

---

3. I introduced the term "folk psychology" in 1978 (Dennett 1981, 1987b) as the name for the natural, perhaps even partly innate, talent human beings have for adopting the intentional stance. See Baron-Cohen 1995 for a fascinating contribution to the current state of play. There is more agreement among philosophers and psychologists about the existence of the talent than there is about my analysis of it. See, for instance, the recent anthologies on the topic—Greenwood 1991, and Christensen and Turner 1993. See Dennett 1987b, 1990b, and 1991b for my account.

near-optimal paths, is a measure of the design that is observable in the world.

The biologists' name for this style of reasoning is *adaptationism*. It is defined by one of its most eminent critics as the "growing tendency in evolutionary biology to reconstruct or predict evolutionary events by *assuming* that all characters are established in evolution by direct natural selection of the most adapted state, that is, the state that is an optimum 'solution' to a 'problem' posed by the environment" (Lewontin 1983). These critics claim that, although adaptationism plays *some* important role in biology, it is not really all that central or ubiquitous—and, indeed, we should try to balance it with other ways of thinking. I have been showing, however, that it plays a crucial role in the analysis of every biological event at every scale from the creation of the first self-replicating macromolecule on up. If we gave up adaptationist reasoning, for instance, we would have to give up the best textbook argument for the very occurrence of evolution (I quoted Mark Ridley's version of it on page 136): the widespread existence of homologies, those suspicious similarities of design that are *not* functionally necessary.

Adaptationist reasoning is not optional; it is the heart and soul of evolutionary biology. Although it may be supplemented, and its flaws repaired, to think of *displacing* it from central position in biology is to imagine not just the downfall of Darwinism but the collapse of modern biochemistry and all the life sciences and medicine. So it is a bit surprising to discover that this is precisely the interpretation that many readers have placed on the most famous and influential critique of adaptationism, Stephen Jay Gould and Richard Lewontin's oft-cited, oft-reprinted, but massively misread classic, "The Spandrels of San Marco and the Panglossian Paradigm: A Critique of the Adaptationist Programme" (1979).

2. THE LEIBNIZIAN PARADIGM

If, among all the possible worlds, none had been better than the rest, then God would never have created one.

—GOTTFRIED WILHELM LEIBNIZ 1710

The study of adaptation is not an optional preoccupation with fascinating fragments of natural history, it is the core of biological study.

—COLIN PITTENDRIGH 1958, p. 395

Leibniz, notoriously, said that this was the best of all possible worlds, a striking suggestion that might seem preposterous from a distance, but turns out, as we have seen, to throw an interesting light on the deep questions of what it is to be a possible world, and on what we can infer about the actual world from the fact of its actuality. In *Candide*, Voltaire created a famous caricature of Leibniz, Dr. Pangloss, the learned fool who could rationalize any calamity or deformity—from the Lisbon earthquake to venereal disease—and show how, no doubt, it was all for the best. Nothing *in principle* could prove that this was not the best of all possible worlds.

Gould and Lewontin memorably dubbed the *excesses* of adaptationism the "Panglossian Paradigm," and strove to ridicule it off the stage of serious science. They were not the first to use "Panglossian" as a term of criticism in evolutionary theory. The evolutionary biologist J. B. S. Haldane had a famous list of three "theorems" of bad scientific argument: the Bellman's Theorem ("What I tell you three times is true"; from "The Hunting of the Snark" by Lewis Carroll), Aunt Jobisca's Theorem ("It's a fact the whole world knows"; from Edward Lear, "The Pobble Who Had No Toes"), and Pangloss's Theorem ("All is for the best in this best of all possible worlds"; from *Candide*). John Maynard Smith then used the last of these more particularly to name "the old Panglossian fallacy that natural selection favours adaptations that are good for the species as a whole, rather than acting at the level of the individual." As he later commented, "It is ironic that the phrase 'Pangloss's theorem' was first used in the debate about evolution (in print, I think, by myself, but borrowed from a remark of Haldane's), not as a criticism of adaptive explanations, but specifically as a criticism of 'group-selectionist', mean-fitness-maximising arguments" (Maynard Smith 1988, p. 88). But Maynard Smith is wrong, apparently. Gould has recently drawn attention to a still earlier use of the term by a biologist, William Bateson (1909), of which he, Gould, had been unaware when he chose to use the term. As Gould (1993a, p. 312) says, "The convergence is hardly surprising, as Dr. Pangloss is a standard synecdoche for this form of ridicule." As we saw in chapter 6, the more apt or fitting a brainchild is, the more likely it is to be born (or borrowed) independently in more than one brain.

Voltaire created Pangloss as a parody of Leibniz, and it is exaggerated and unfair to Leibniz—as all good parody is. Gould and Lewontin similarly caricatured adaptationism in their article attacking it, so parity of reasoning suggests that, if we wanted to undo the damage of that caricature, and describe adaptationism in an accurate and constructive way, we would have a title ready-made: we could call adaptationism, fairly considered, the "Leibnizian Paradigm."

The Gould and Lewontin article has had a curious effect on the academic world. It is widely regarded by philosophers and other humanists who have heard of it or even read it as some sort of *refutation of adaptationism*. Indeed, I first learned of it from the philosopher/psychologist Jerry Fodor, a lifelong critic of my account of the intentional stance, who pointed out that what I was saying was pure adaptationism (he was right about that), and went on to let me in on what the *cognoscenti* all knew: Gould and Lewontin's article had shown adaptationism "to be completely bankrupt." (For an instance of Fodor's view in print, see Fodor 1990, p. 70.) When I looked into it, I found out otherwise. In 1983, I published a paper in *Behavioral and Brain Sciences*, "Intentional Systems in Cognitive Ethology," and since it was unabashedly adaptationist in its reasoning, I included a coda, "The 'Panglossian Paradigm' Defended," which criticized both Gould and Lewontin's paper and—more particularly—the bizarre myth that had grown up around it.

The results were fascinating. Every article that appears in *BBS* is accompanied by several dozen commentaries by experts in the relevant fields, and my piece drew fire from evolutionary biologists, psychologists, ethologists, and philosophers, most of it friendly but some remarkably hostile. One thing was clear: it was not just some philosophers and psychologists who were uncomfortable with adaptationist reasoning. In addition to the evolutionary theorists who weighed in enthusiastically on my side (Dawkins 1983b, Maynard Smith 1983), and those who fought back (Lewontin 1983), there were those who, though they agreed with me that Gould and Lewontin had not refuted adaptationism, were eager to downplay the standard use of optimality assumptions that I claimed to be an essential ingredient in all evolutionary thinking.

Niles Eldredge (1983, p. 361) discussed the reverse engineering of functional morphologists: "You will find sober analyses of fulcra, force vectors and so forth: the understanding of anatomy as a living machine. Some of this stuff is very good. Some of it is absolutely dreadful." He went on to cite, as an example of good reverse engineering, the work of Dan Fisher (1975) comparing modern horseshoe crabs with their Jurassic ancestors:

> Assuming only that Jurassic horseshoe crabs also swam on their backs, Fisher showed they must have swum at an angle of 0–10 degrees (flat on their backs) and at the somewhat greater speed of 15–20 cm/sec. Thus the 'adaptive significance' of the slight differences in anatomy between modern horseshoe crabs and their 150-million-year-old relatives is translated into an understanding of their slightly different swimming capabilities. (In all honesty, I must also report that Fisher does use optimality in his arguments: He sees the differences between the two species as a sort of trade-off, where the slightly more efficient Jurassic swimmers appear to have used the same pieces of anatomy to burrow somewhat less efficiently than their modern-day relatives). In any case, Fisher's work stands as a really good example of functional morphological analysis. The notion of adaptation is naught but conceptual filigree—one that may have played a role in motivating the research, but one that was not vital to the research itself. [Eldredge 1983, p. 362.]

But in fact the role of optimality assumptions in Fisher's work—beyond the explicit role that Eldredge conceded—is so "vital" and indeed omnipresent that Eldredge entirely overlooked it. For instance, Fisher's inference that the Jurassic crabs swam at 15–20 cm/sec has as a tacit premise that those crabs *swam at the optimal speed for their design*. (How does he know they swam at all? Perhaps they just lay there, oblivious of the excess functionality of their body shapes.) Without this tacit (and, of course, dead obvious) premise, no conclusion at all could be drawn about what the *actual* swimming speed of the Jurassic variety was.

Michael Ghiselin (1983, p. 363) was even more forthright in denying this unobvious obvious dependence:

> Panglossianism is bad because it asks the wrong question, namely, What is good? . . . The alternative is to reject such teleology altogether. Instead of asking, What is good? we ask, What has happened? The new question does everything we could expect the old one to do, and a lot more besides.

He was fooling himself. There is hardly a single answer to the question "What has happened (in the biosphere)?" that doesn't depend crucially on assumptions about what is good.[4] As we just noted, you can't even avail yourself of the concept of a homology without taking on adaptationism, without taking the intentional stance.

So now what is the problem? It is the problem of how to tell good— irreplaceable—adaptationism from bad adaptationism, how to tell Leibniz from Pangloss.[5] Surely one reason for the extraordinary influence of Gould

---

4. Doesn't my assertion fly in the face of the claims of those cladists who purport to deduce history from a statistical analysis of shared and unshared "characters"? (For a philosophical survey and discussion, see Sober 1988.) Yes, I guess it does, and my review of their arguments (largely via Sober's analyses) shows me that the difficulties they create for themselves are largely if not entirely due to their trying so hard to find non-adaptationist ways of drawing the sound inferences that are dead obvious to adaptationists. For instance, those cladists who abstain from adaptation talk cannot just help themselves to the obvious fact that having webbed feet is a pretty good "character" and having dirty feet (when examined) is not. Like the behaviorists who pretended to be able to explain and predict "behavior" defined in the starkly uninterpreted language of geographical trajectory of body parts, instead of using the richly functionalistic language of searching, eating, hiding, chasing, and so forth, the abstemious cladists create majestic edifices of intricate theory, which is amazing, considering they do it with one hand tied behind their backs, but strange, considering that they wouldn't have to do it at all if they didn't insist on tying one hand behind their backs. (See also Dawkins 1986a, ch. 10, and Mark Ridley 1985, ch. 6.)

5. The myth that the point of the Gould and Lewontin paper was to destroy adaptationism, not correct its excesses, was fostered by the paper's rhetoric, but in some quarters it backfired on Gould and Lewontin, since adaptationists themselves tended to pay more

Calvin and Hobbes

by Bill Watterson

FIGURE 9.3

and Lewontin's paper (among nonevolutionists) is that it expressed, with many fine rhetorical flourishes, what Eldredge called the "backlash" against the concept of adaptationism among biologists. What were they reacting against? In the main, they were reacting against a certain sort of laziness: the adaptationist who hits upon a truly nifty explanation for why a particular circumstance should prevail, and then never bothers to test it—because it is too good a story, presumably, not to be true. Adopting another literary label, this time from Rudyard Kipling (1912), Gould and Lewontin call such explanations "Just So Stories." It is an enticing historical curiosity that Kipling wrote his *Just So Stories* at a time when this objection to Darwinian explanation had already been swirling around for decades;[6] forms of it were raised by some of Darwin's earliest critics (Kitcher 1985a, p. 156). Was Kipling inspired by the controversy? In any case, calling the adaptationists' flights of imagination "Just So Stories" hardly does them credit; as delightful as I have always found Kipling's fantasies about how the elephant got its trunk, and the leopard got its spots, they are quite simple and unsurprising tales compared with the amazing hypotheses that have been concocted by adaptationists.

Consider the greater honey guide, *Indicator indicator*, an African bird that owes its name to its talent for leading human beings to wild beehives hidden in the forest. When the Boran people of Kenya want to find honey, they call for the bird by blowing on whistles made of sculpted snail shells. When a bird arrives, it flies around them, singing a special song—its "follow-

---

attention to the rhetoric than the arguments: "The critique by Gould and Lewontin has had little impact on practitioners, perhaps because they were seen as hostile to the whole enterprise, and not merely to careless practise of it" (Maynard Smith 1988, p. 89).

6. Kipling began publishing the individual stories in 1897.

me" call. They follow as the bird darts ahead and waits for them to catch up, always making sure they can see where it's heading. When the bird reaches the hive, it changes its tune, giving the "here-we-are" call. When the Boran locate the beehive in the tree and break into it, they take the honey, leaving wax and larvae for the honey guide. Now, don't you ache to believe that this wonderful partnership actually exists, and has the clever functional properties described? Don't you want to believe that such a marvel could have evolved under some imagined series of selection pressures and opportunities? I certainly do. And, happily, in this case, the follow-up research is confirming the story, and even adding nifty touches as it does so. Recent controlled tests, for instance, showed that the Boran honey-hunters took much longer to find hives without the help of the birds, and 96 percent of the 186 hives found during the study were encased in trees in ways that would have made them inaccessible to the birds without human assistance (Isack and Reyer 1989).

Another fascinating story, which strikes closer to home, is the hypothesis that our species, *Homo sapiens*, descended from earlier primates via an intermediate species that was aquatic (Hardy 1960, Morgan 1982, 1990)! These aquatic apes purportedly lived on the shores of an island formed by the flooding of the area that is now in Ethiopia, during the late Miocene, about seven million years ago. Cut off by the flooding from their cousins on the African continent, and challenged by a relatively sudden change in their climate and food sources, they developed a taste for shellfish, and over a period of a million years or so they began the evolutionary process of returning to the sea that we know was undergone earlier by whales, dolphins, seals, and otters, for instance. The process was well under way, leading to the fixation of many curious characteristics that are otherwise found *only* in aquatic mammals—not in any other primate, for example—when circumstances changed once again, and these semi-seagoing apes returned to a life on the land (but typically on the shore of sea, lake, or river). There, they found that many of the adaptations they had developed for good reasons in their shell-diving days were not only not valuable but a positive hindrance. They soon turned these handicaps to good uses, however, or at least made compensations for them: their upright, bipedal posture, their subcutaneous layer of fat, their hairlessness, perspiration, tears, inability to respond to salt deprivation in standard mammalian ways, and, of course, the diving reflex—which permits even newborn human infants to survive sudden submersion in water for long periods with no ill effects. The details—and there are many, many more—are so ingenious, and the whole aquatic-ape theory is so shockingly antiestablishment, that I for one would *love* to see it vindicated. That does not make it true, of course.

The fact that its principal exponent these days is not only a woman, Elaine Morgan, but an amateur, a science writer without proper official credentials in spite of her substantial researches, makes the prospect of vindication all the more enticing.[7] The establishment has responded quite ferociously to her challenges, mostly treating them as beneath notice, but occasionally subjecting them to withering rebuttal.[8] This is not necessarily a pathological reaction. Most uncredentialed proponents of scientific "revolutions" are kooks who really are not worth paying any attention to. There really are a lot of them besieging us, and life is too short to give each uninvited hypothesis its proper day in court. But in this case, I wonder; many of the counterarguments seem awfully thin and *ad hoc*. During the last few years, when I have found myself in the company of distinguished biologists, evolutionary theorists, paleo-anthropologists, and other experts, I have often asked them just to tell me, please, exactly why Elaine Morgan must be wrong about the aquatic-ape theory. I haven't yet had a reply worth mentioning, aside from those who admit, with a twinkle in their eyes, that they have often wondered the same thing. There seems to be nothing *inherently* impossible about the idea; other mammals have made the plunge, after all. Why couldn't our ancestors have started back into the ocean and then retreated, bearing some telltale scars of this history?

Morgan may be "accused" of telling a good story—she certainly has—but not of declining to try to test it. On the contrary, she has used the story as leverage to coax a host of surprising predictions out of a variety of fields, and has been willing to adjust her theory when the results have demanded it. Otherwise, she has stuck to her guns and, in fact, invited attack on her views through the vehemence of her partisanship. As so often happens in such a confrontation, the intransigence and defensiveness, on both sides, have begun to take their toll, creating one of those spectacles that then discourage anyone who just wants to know the truth from having anything more to do with the subject. Morgan's latest book on the topic (1990)

---

7. Sir Alister Hardy, the Linacre Professor of Zoology at Oxford, who originally proposed the theory, could hardly have been a more secure member of the scientific establishment, however.

8. For instance, there is no mention at all of the aquatic-ape theory, not even to dismiss it, in two recent coffee-table books that include chapters on human evolution. Philip Whitfield's *From So Simple a Beginning: The Book of Evolution* (1993) offers a few paragraphs on the standard savanna theory of bipedalism. "The Primates' Progress," by Peter Andrews and Christopher Stringer, is a much longer essay on hominid evolution, in *The Book of Life* (Stephen Jay Gould, ed., 1993b), but it, too, ignores the aquatic-ape theory—the AAT. And, adding insult to oblivion, there has also been a wickedly funny parody of it by Donald Symons (1983), exploring the radical hypothesis that our ancestors used to *fly*—"The *fly*ing on *a*ir *t*heory—FLOAT, as it is acronymously (acrimoniously, among the reactionary human evolution 'establishment')." For an overview of the reactions, see G. Richards 1991.

responded with admirable clarity, however, to the objections that had been lodged to date, and usefully contrasted the strengths and weaknesses of the aquatic-ape theory to those of the establishment's history. And, more recently still, a book has appeared that collects essays by a variety of experts, for and against the aquatic-ape theory: Roede et al. 1991. The tentative verdict of the organizers of the 1987 conference from which that book sprang (p. 324) is that, "while there are a number of arguments favoring the AAT, they are not sufficiently convincing to counteract the arguments against it." That judicious note of mild disparagement helps ensure that the argument will continue, perhaps even with less rancor; it will be interesting to see where it all comes out.

My point in raising the aquatic-ape theory is not to defend it against the establishment view, but to use it as an illustration of a deeper worry. Many biologists would like to say, "A pox on both your houses!" Morgan (1990) deftly exposes the hand-waving and wishful thinking that have gone into the establishment's tale about how—and *why*—*Homo sapiens* developed bipedalism, sweating, and hairlessness on the savanna, not the seashore. Their stories may not be literally as fishy as hers, but some of them are pretty farfetched; they are every bit as speculative, and (I venture to say) no better confirmed. What they mainly have going for them, so far as I can see, is that they occupied the high ground in the textbooks before Hardy and Morgan tried to dislodge them. Both sides are indulging in adaptationist Just So Stories, and since *some story or other* must be true, we must not conclude we have found *the* story just because we have come up with *a* story that seems to fit the facts. To the extent that adaptationists have been less than energetic in seeking further confirmation (or dreaded disconfirmation) of their stories, this is certainly an excess that deserves criticism.[9]

But before leaving it at that, I want to point out that there are many adaptationist stories that *everybody* is happy to accept even though they

---

9. The geneticist Steve Jones (1993, p. 20) gives us another case in point: There are more than three hundred strikingly different species of cichlid fish in Lake Victoria. They are so different; how did they get there? "The conventional view is that Lake Victoria must once have dried up into many small lakes to allow each species to evolve. Apart from the fish themselves, there is no evidence that this ever happened." Adaptationist stories *do* get disconfirmed and abandoned, however. My favorite example is the now-discredited explanation of why certain sea turtles migrate all the way across the Atlantic between Africa and South America, spawning on one side, feeding on the other. According to this all-too-reasonable story, the habit started when Africa and South America were first beginning to split apart; at that time, the turtles were just going across the bay to spawn; the distance grew imperceptibly longer over the eons, until their descendants dutifully cross an ocean to get to where their instinct still tells them to spawn. I gather that the timing of the breakup of Gondwanaland turns out not to match the evolutionary timetable for the turtles, sad to say, but wasn't it a cute idea?

have never been "properly tested," just because they are too obviously true to be worth further testing. Does anybody seriously doubt that eyelids evolved to protect the eye? But that very obviousness may hide good research questions from us. George Williams points out that concealed behind such obvious facts may lie others that are well worth further investigation:

> A human eye blink takes about 50 milliseconds. That means that we are blind about 5% of the time when we are using our eyes normally. Many events of importance can happen in 50 milliseconds, so that we might miss them entirely. A rock or spear thrown by a powerful adversary can travel more than a meter in 50 milliseconds, and it could be important to perceive such motion as accurately as possible. Why do we blink with both eyes simultaneously? Why not alternate and replace 95% visual attentiveness with 100%? I can imagine an answer in some sort of trade-off balance. A blink mechanism for both eyes at once may be much simpler and cheaper than one that regularly alternates. [G. Williams 1992, pp. 152–53.]

Williams has not himself yet attempted to confirm or disconfirm any hypothesis growing out of this exemplary piece of adaptationist problem-setting, but he has called for the research by asking the question. It would be as pure an exercise in reverse engineering as can be imagined.

> Serious consideration of why natural selection permits simultaneous blinking might yield otherwise elusive insights. What change in the machinery would be needed to produce the first step towards my envisioned adaptive alternation or simple independent timing? How might the change be achieved developmentally? What other changes would be expected from a mutation that produced a slight lag in the blinking of one eye? How would selection act on such a mutation? [G. Williams 1992, p. 153.]

Gould himself has endorsed some of the most daring and delicious of adaptationist Just So Stories, such as the argument by Lloyd and Dybas (1966) explaining why cicadas (such as "seventeen-year locusts") have reproductive cycles that are prime-numbered years long—thirteen years, or seventeen, but never fifteen or sixteen, for instance. "As evolutionists," Gould says, "we seek answers to the question, why. Why, in particular, should such striking synchroneity evolve, and why should the period between episodes of sexual reproduction be so long?" (Gould 1977a, p. 99).[10]

---

10. Gould has recently (1993a, p. 318) described his antiadaptationism as the "zeal of the convert," and elsewhere (1991b, p. 13) confesses, "I sometimes wish that all copies of *Ever Since Darwin* would self-destruct," so perhaps he would recant these words today, which would be a pity, since they eloquently express the rationale of adaptationism.

The answer—which makes beautiful sense, in retrospect—is that, by having a large prime number of years between appearances, the cicadas minimize the likelihood of being discovered and later tracked as a predictable feast by predators who themselves show up every two years, or three years, or five years. If the cicadas had a periodicity of, say, sixteen years, then they would be a rare treat for predators who showed up every year, but a more reliable source of food for predators who showed up every two or four years, and an even-money gamble for predators that got in phase with them on an eight-year schedule. If their period is not a multiple of any lower number, however, they are a rare treat—not worth "trying" to track—for any species that isn't lucky enough to have exactly their periodicity (or some multiple of it—the mythical Thirty-four-Year Locust-Muncher would be in fat city). I don't know whether Lloyd and Dybas' Just So Story has been properly confirmed yet, but I don't think Gould is guilty of Panglossianism in treating it as established until proven otherwise. And if he really wants to ask and answer "why" questions, he has no choice but to be an adaptationist.

The problem he and Lewontin perceive is that there are no standards for when a particular bit of adaptationist reasoning is too much of a good thing. How serious, really, is this problem even if it has no principled "solution"? Darwin has taught us not to look for essences, for dividing lines between *genuine* function or *genuine* intentionality and mere *on-its-way-to-being* function or intentionality. We commit a fundamental error if we think that if we want to indulge in adaptationist thinking we need a license and the only license could be the possession of a strict definition of or criterion for a genuine adaptation. There are good rules of thumb to be followed by the prospective reverse engineer, made explicit years ago by George Williams (1966). (1) Don't invoke adaptation when other, lower-level, explanations are available (such as physics). We don't have to ask what advantage accrues to maple trees that explains the tendency of their leaves to fall *down*, any more than the reverse engineers at Raytheon need to hunt for a reason why GE made their widgets so that they would melt readily in blast furnaces. (2) Don't invoke adaptation when a feature is the outcome of some general developmental requirement. We don't need a special reason of increased fitness to explain the fact that heads are attached to bodies, or limbs come in pairs, any more than the people at Raytheon need to explain why the parts in GE's widget have so many edges and corners with right angles. (3) Don't invoke adaptation when a feature is a by-product of another adaptation. We don't need to give an adaptationist explanation of the capacity of a bird's beak to groom its feathers (since the features of the

---

Gould's attitude towards adaptationism is not so easily discerned, however. *The Book of Life* (1993b) is packed with adaptationist reasoning that made it past his red pencil, and thus presumably has his endorsement.

bird's beak are there for more pressing reasons), any more than we need a special explanation of the capacity of the GE widget's casing to shield the innards from ultraviolet rays.

But you will already have noticed that in each case these rules of thumb can be overridden by a more ambitious inquiry. Suppose someone marveling at the brilliant autumn foliage in New England asks *why* the maple leaves are so vividly colored in October. Isn't this adaptationism run amok? Shades of Dr. Pangloss! The leaves are the colors they are simply because once the summer energy-harvest season is over, the chlorophyll vanishes from the leaves, and the residual molecules have reflective properties that happen to determine the bright colors—an explanation at the level of chemistry or physics, not biological purpose. But wait. Although this may have been the only explanation that was true up until now, today it is true that human beings so prize the autumn foliage (it brings millions of tourist dollars to northern New England each year) that they protect the trees that are brightest in autumn. You can be sure that if you are a tree competing for life in New England, there is now a selective advantage to having bright autumn foliage. It may be tiny, and in the long run it may never amount to much (in the long run, there may be no trees at all in New England, for one reason or another), but this is how all adaptations get their start, after all, as fortuitous effects that get opportunistically picked up by selective forces in the environment. And of course there is also an adaptationist explanation for why right angles predominate in manufactured goods, and why symmetry predominates in organic limb-manufacturing. These may become utterly fixed traditions, which would be almost impossible to dislodge by innovation, but the reasons why *these* are the traditions are not hard to find, or controversial.

Adaptationist research always leaves unanswered questions open for the next round. Consider the leatherback sea turtle and her eggs:

> Near the end of egg laying, a variable number of small, sometimes misshapen eggs, containing neither embryo nor yolk (just albumin) are deposited. Their purpose is not well understood, but they become desiccated over the course of incubation and may moderate humidity or air volume in the incubation chamber. (It is also possible that they have no function or are a vestige of some past mechanisms not apparent to us today.) [Eckert 1992, p. 30.]

But where does it all end? Such open-endedness of adaptationist curiosity is unnerving to many theorists, apparently, who wish there could be stricter codes of conduct for this part of science. Many who have hoped to contribute to clearing up the controversy over adaptationism and its backlash have despaired of finding such codes, after much energy has been expended in drawing up and criticizing various legislative regimes. They are just not being Darwinian enough in their thinking. Better adaptationist thinking soon drives out its rivals by normal channels, just as second-rate reverse engineering betrays itself sooner or later.

> The eskimo face, once depicted as 'cold engineered' (Coon *et al.*, 1950) becomes an adaptation to generate and withstand large masticatory forces (Shea, 1977). We do not attack these newer interpretations; they may all be right. We do wonder, though, whether the failure of one adaptive explanation should always simply inspire a search for another of the same general form, rather than a consideration of alternatives to the proposition that each part is 'for' some specific purpose. [Gould and Lewontin 1979, p. 152.]

Is the rise and fall of successive adaptive explanations of various things a sign of healthy science constantly improving its vision, or is it like the pathological story-shifting of the compulsive fibber? If Gould and Lewontin had a serious alternative to adaptationism to offer, their case for the latter verdict would be more persuasive, but although they and others have hunted around energetically, and promoted their alternatives boldly, none has yet taken root.

> Adaptationism, the paradigm that views organisms as complex adaptive machines whose parts have adaptive functions subsidiary to the fitness-promoting function of the whole, is today about as basic to biology as the atomic theory is to chemistry. And about as controversial. Explicitly adaptationist approaches are ascendant in the sciences of ecology, ethology, and evolution because they have proven essential to discovery; if you doubt this claim, look at the journals. Gould and Lewontin's call for an alternative paradigm has failed to impress practicing biologists both because adaptationism is successful and well-founded, and because its critics have no alternative research program to offer. Each year sees the establishment of such new journals as *Functional Biology* and *Behavioral Ecology*. Sufficient research to fill a first issue of *Dialectical Biology* has yet to materialize. [Daly 1991, p. 219.]

What particularly infuriates Gould and Lewontin, as the passage about the Eskimo face suggests, is the blithe confidence with which adaptationists go about their reverse engineering, always sure that sooner or later they will find *the reason* why things are as they are, even if it so far eludes them. Here is an instance, drawn from Richard Dawkins' discussion of the curious case of the flatfish (flounders and soles, for instance) who when they are born are vertical fish, like herring or sunfish, but whose skulls undergo a weird twisting transformation, moving one eye to the other side, which then becomes the top of the bottom-dwelling fish. Why didn't they evolve like those other bottom-dwellers, skates, which are not on their side but on their belly, "like sharks that have passed under a steam roller" (Dawkins 1986a, p. 91)? Dawkins *imagines* a scenario (pp. 92–93):

> ... even though the skate way of being a flat fish might *ultimately* have been the best design for bony fish too, the would-be intermediates that set out along this evolutionary pathway apparently did less well in the short term than their rivals lying on their side. The rivals lying on their side were so much better, in the short term, at hugging the bottom. In genetic hyperspace, there is a smooth trajectory connecting free-swimming ancestral bony fish to flatfish lying on their side with twisted skulls. There is not a smooth trajectory connecting these bony fish ancestors to flatfish lying on their belly. There is such a trajectory in theory, but it passes through intermediates that would have been—in the short term, which is all that matters—unsuccessful if they had ever been called into existence.

Does Dawkins *know* this? Does he know that the postulated intermediates were less fit? Not because he has seen any data drawn from the fossil record. This is a purely theory-driven explanation, argued *a priori* from the assumption that natural selection tells us the true story—some true story or other—about every curious feature of the biosphere. Is that objectionable? It does "beg the question"—but what a question it begs! It assumes that Darwinism is basically on the right track. (Is it objectionable when meteorologists say, begging the question against supernatural forces, that there must be a purely physical explanation for the birth of hurricanes, even if many of the details so far elude them?) Notice that in this instance, Dawkins' explanation is almost certainly right—there is nothing especially daring about that particular speculation. Moreover, it is, of course, exactly the sort of thinking a good reverse engineer should do. "It seems so obvious that this General Electric widget casing ought to be made of two pieces, not three, but it's made of three pieces, which is wasteful and more apt to leak, so we can be damn sure that three pieces was seen as better than two in somebody's eyes, shortsighted though they may have been. Keep looking!" The philosopher of biology Kim Sterelny, in a review of *The Blind Watchmaker*, made the point this way:

> Dawkins is admittedly giving only scenarios: showing that it's *conceivable* that (e.g.) wings could evolve gradually under natural selection. Even so, one could quibble. Is it really true that natural selection is so fine-grained that, for a protostick insect, looking 5% like a stick is better than looking 4% like one? (pp. 82–83). A worry like this is especially pressing because

Dawkins' adaptive scenarios make no mention of the costs of allegedly adaptive changes. Mimicry might deceive potential mates as well as potential predators.... Still, I do think this objection is something of a quibble because essentially I agree that natural selection is the only possible explanation of complex adaptation. So something like Dawkins' stories have got to be right. [Sterelny 1988, p. 424.][11]

3. PLAYING WITH CONSTRAINTS

It is just as foolish to complain that people are selfish and treacherous as it is to complain that the magnetic field does not increase unless the electric field has a curl.

—JOHN VON NEUMANN, quoted in William Poundstone 1992, p. 235

As a general rule today a biologist seeing one animal doing something to benefit another assumes either that it is manipulated by the other individual or that it is being subtly selfish.

—GEORGE WILLIAMS 1988, p. 391

One may nevertheless be reasonably nervous about the size of the role of sheer, unfettered imagination in adaptationist thinking. What about butterflies with tiny machine guns for self-protection? This fantastic example is often cited as the sort of option that can be dismissed without detailed analysis by adaptationists seeking to describe the ensemble of possible butterfly adaptations from which Mother Nature has chosen the best, all things considered. It is just too distant a possibility in design space to be taken seriously. But as Richard Lewontin (1987, p. 156) aptly notes, "My guess is that if fungus-gardening ants had never been seen, the suggestion that this was a reasonable possibility for ant evolution would have been regarded as silly." Adaptationists are masters of the retrospective rationale, like the

---

11. Dawkins is not content to rest with Sterelny's dismissal of his own objections as "quibbles" since, he points out (personal communication), they raise an important point often misunderstood: "It is not up to individual humans like Sterelny to express their own commonsense scepticism of the proposition that 5% like a stick is significantly better than 4%. It is an easy rhetorical point to make: 'Come on, are you really trying to tell me that 5% like a stick really matters when compared to 4%?' This rhetoric will often convince laymen, but the population genetic calculations (e.g. by Haldane) belie common sense in a fascinating and illuminating way: because natural selection works on genes distributed over many individuals and over many millions of years, human actuarial intuitions are over-ruled."

chess-player who only notices *after* he's made the move that it forces check-mate in two moves. "How brilliant—and I almost thought of it!" But before we decide that this is a *flaw* in adaptationist character or method, we should remind ourselves that this retrospective endorsement of brilliance is the way Mother Nature herself always operates. Adaptationists should hardly be faulted for being unable to predict the brilliant moves that Mother Nature herself was oblivious of until she'd stumbled upon them.

The perspective of game-playing is ubiquitous in adaptationism, where mathematical *game theory* has played a growing role ever since its introduction into evolutionary theory by John Maynard Smith (1972, 1974).[12] Game theory is yet one more fundamental contribution to twentieth-century thinking from John von Neumann.[13] Von Neumann created game theory in collaboration with the economist Oskar Morgenstern, and it grew out of their realization that *agents* make a fundamental difference to the complexity of the world.[14] Whereas a lone "Robinson Crusoe" agent can view all problems as seeking stable maxima—hill-climbing on Mount Fuji, if you like—as soon as other (maxima-seeking) agents are included in the environment, strikingly different methods of analysis are required:

> A guiding principle cannot be formulated by the requirement of maximiz-ing two (or more) functions at once. . . . One would be mistaken to believe that it can be obviated . . . by a mere recourse to the devices of the theory of probability. Every participant can determine the variables which de-

---

12. Maynard Smith built his game-theory applications to evolution on the foundations already laid by R. A. Fisher (1930). One of Maynard Smith's many more recent contributions was showing Stuart Kauffman that he was, after all, a Darwinian, not an anti-Darwinian (see Lewin 1992, pp. 42–43).

13. I sometimes wonder if there is any important advance in thinking in the second half of this century that von Neumann is *not* the father of. The computer, the model of self-replication, game theory—and if that weren't enough, von Neumann also made major contributions to quantum physics. For what it is worth, however, I suspect that his formulation of the measurement problem in quantum mechanics is his one bad idea, a sleight-of-hand endorsement of a fundamentally Cartesian model of conscious observa-tion that has bedeviled quantum mechanics ever since. My student Turhan Canli first opened this door in his (undergraduate!) term paper for me on the problem of Schrö-dinger's cat, in which he developed the sketch of an alternative formulation of quantum physics in which time is quantized. If I ever master the physics (a very remote prospect, sad to say), I will tackle this hunch, which might extend in wildly ambitious ways my theory of consciousness (1991a); more likely, however, is the prospect that I will be a semi-comprehending but enthusiastic spectator of this development, wherever it leads.

14. For a fascinating account of the history of game theory and its relation to nuclear disarmament, see William Poundstone's 1992 book, *Prisoner's Dilemma: John von Neu-mann, Game Theory, and the Puzzle of the Bomb.*

scribe his own actions but not those of the others. Nevertheless those 'alien' variables cannot, from his point of view, be described by statistical assumptions. This is because the others are guided, just as he himself, by rational principles—whatever that may mean—and no *modus procedendi* can be correct which does not attempt to understand those principles and the interactions of the conflicting interests of all participants. [Von Neumann and Morgenstern 1944, p. 11.]

The fundamental insight that unites game theory and evolutionary theory is that the "rational principles—whatever that may mean" that "guide" agents in competition can exert their influence even on such unconscious, unreflective semi-agents as viruses, trees, and insects, because the stakes and payoff possibilities of competition determine which lines of play cannot help winning or losing if adopted, however mindlessly they are adopted.

The best-known example in game theory is the Prisoner's Dilemma, a simple two-person "game" which casts shadows, both obvious and surprising, into many different circumstances in our world. Here it is in basic outline (excellent detailed discussions of it are found in Poundstone 1992 and Dawkins 1989a). You and another person have been imprisoned pending trial (on a trumped-up charge, let's say), and the prosecutor offers each of you, separately, the same deal: if you both hang tough, neither confessing nor implicating the other, you will each get a short sentence (the state's evidence is not that strong); if you confess and implicate the other and he hangs tough, you go scot free and he gets life in prison; if you both confess-and-implicate, you both get medium-length sentences. Of course, if you hang tough and the other person confesses, he goes free and you get life. What should you do?

If you both could hang tough, defying the prosecutor, this would be much better for the two of you than if you both confess, so couldn't you just promise each other to hang tough? (In the standard jargon of the Prisoner's Dilemma, the hang-tough option is called *cooperating*.) You could promise, but you would each then feel the temptation—whether or not you acted on it—to *defect*, since then you would go scot free, leaving the *sucker*, sad to say, in deep trouble. Since the game is symmetrical, the other person will be just as tempted, of course, to make a sucker of you by defecting. Can you risk life in prison on the other person's keeping his promise? Probably safer to defect, isn't it? That way, you definitely avoid the worst outcome of all, and might even go free. Of course, the other fellow will figure this out, too, if it's such a bright idea, so he'll probably play it safe and defect, too, in which case you *must* defect to avoid calamity—unless you are so saintly that you don't mind spending your life in prison to save a promise-breaker!—so you'll both wind up with medium-length sentences. If only you could overcome this reasoning and cooperate!

The logical structure of the game is what matters, not this particular setting, which is a usefully vivid imagination-driver. We can replace the prison sentences with positive outcomes (it's a chance to win different amounts of cash—or, say, descendants) just so long as the payoffs are symmetrical, and ordered so that lone defection pays more than mutual cooperation, which pays each more than mutual defection does, which in turn pays more than the sucker payoff one gets when the other is a lone defector. (And in formal settings we set a further condition: the average of the sucker and mutual-defection payoffs must not be greater than the mutual-cooperation payoff.) Whenever this structure is instantiated in the world, there is a Prisoner's Dilemma.

Game-theoretic explorations have been undertaken in many fields, from philosophy and psychology to economics and biology. The most influential of the many applications of game-theoretic thinking to evolutionary theory is Maynard Smith's concept of an *evolutionarily stable strategy*, or ESS, a strategy that may not be "best" from any Olympian (or Fujian!) standpoint, but is unimprovable-upon and unsubvertible under the circumstances. Maynard Smith (1988, especially chh. 21 and 22) is an excellent introductory account of game theory in evolution. The revised edition of Richard Dawkins' *The Selfish Gene* (1989a) has a particularly good account of the development of ESS thinking in biology during the last decade or so, when large-scale computer simulations of various game-theoretic models revealed complications that had been overlooked by the earlier, less realistic versions.

> I now like to express the essential idea of an ESS in the following more economical way. An ESS is a strategy that does well against copies of itself. The rationale for this is as follows. A successful strategy is one that dominates the population. Therefore it will tend to encounter copies of itself. Therefore it won't stay successful unless it does well against copies of itself. This definition is not so mathematically precise as Maynard Smith's, and it cannot replace his definition because it is actually incomplete. But it does have the virtue of encapsulating, intuitively, the basic ESS idea. [Dawkins 1989a, p. 282.]

There can be no doubt that game-theoretic analyses work in evolutionary theory. Why, for instance, are the trees in the forest so tall? For the very same reason that huge arrays of garish signs compete for our attention along commercial strips in every region of the country! Each tree is looking out for itself, and trying to get as much sunlight as possible.

If only those redwoods could get together and agree on some sensible zoning restrictions and stop competing with each other for sunlight, they could avoid the trouble of building those ridiculous and expensive trunks, stay low and thrifty shrubs, and get just as much sunlight as before! [Dennett 1990b, p. 132.]

But they can't get together; under these circumstances, defection from any cooperative "agreement" is bound to pay off if ever or whenever it occurs, so trees would be stuck with the "tragedy of the commons" (Hardin 1968) if there weren't an essentially inexhaustible supply of sunshine. The tragedy of the commons occurs when there is a finite "public" or shared resource that individuals will be selfishly tempted to take more of than their fair share—such as the edible fish in the oceans. Unless very specific and enforceable agreements can be reached, the result will tend to be the destruction of the resource. Many species, in many regards, face various sorts of Prisoner's Dilemmas. And we human beings face them both consciously and unconsciously—sometimes in ways that we might never have imagined without the aid of adaptationist thinking.

Homo sapiens is not exempt from the sort of genetic conflict David Haig postulates to explain genomic imprinting; in an important new article (1993) he analyzes a variety of conflicts that exist between the genes of a pregnant woman and the genes of her embryo. It is in the embryo's interests, of course, that the mother bearing it stay strong and healthy, for its own survival depends on her not only completing her term of pregnancy but tending for her newborn. However, if the mother, in her attempt to stay healthy under trying circumstances—famine, for instance, which must have been a common circumstance in most generations of human existence—should cut down on the nutrition she provides her embryo, at some point this becomes more of a threat to the embryo's survival than the alternative, a weakened mother.

If the embryo were "given a choice" between being spontaneously aborted early in the pregnancy or being stillborn or of low birth weight on the one hand, versus being born at normal weight of a weak or even dying mother on the other, what would (selfish) reason dictate? It would dictate taking whatever steps are available to try to ensure that the mother does not cut her losses (she can always try to have another child later, when the famine is over), and this is just what the embryo does. Both embryo and mother can be entirely oblivious of this conflict—as oblivious as the trees rising competitively in the forest. The conflict plays out in the genes and their control of hormones, not in the brains of mother and embryo; it is the same sort of conflict we saw between maternal and paternal genes in the mouse. There is a flood of hormones; the embryo produces a hormone that will enhance its own growth at the expense of the mother's nutritional needs; her body responds with an antagonist hormone that attempts to undo the effect of the first; and so on, in an escalation that can produce hormone levels many times higher than normal. This tug-of-war usually ends in a mutually semi-satisfactory standoff, but it produces a host of by-products that would be utterly baffling and senseless were they not the predictable effects of such conflict. Haig concludes with an application of the fundamental game-theoretic insight: "Maternal and fetal genes would both benefit if a given transfer of resources was achieved with a lesser production of . . . hormones and less maternal resistance, but such an agreement is evolutionarily unenforceable" (Haig 1993, p. 518).

This is not, in many regards, welcome news. Von Neumann's all-too-casual remark on the inevitability of human selfishness epitomizes the Darwinian mind-set that many people view with loathing, and it is not hard to see why. They fear that Darwinian "survival of the fittest" would *entail* that people are nasty and selfish. Isn't that just what von Neumann is saying? No. Not quite. He is saying that it is indeed entailed by Darwinism that such virtues as cooperation should be *in general* "evolutionarily unenforceable" and hence hard to come by. If cooperation and the other unselfish virtues are to exist, *they must be designed*—they do not come for free. They *can* be designed under special circumstances. (See, for instance, Eshel 1984, 1985, and Haig and Grafen 1991.) After all, the eukaryotic revolution that made multicelled organisms possible was a revolution that began when an enforceable truce was somehow engineered between certain prokaryotic cells and their bacterial invaders. They found a way of joining forces and submerging their selfish interests.

Cooperation and the other virtues are, in general, rare and special properties that can only emerge under very particular and complex R-and-D circumstances. We might contrast the Panglossian Paradigm, then, with the Pollyannian Paradigm, which cheerfully assumes, with Pollyanna, that Mother Nature is Nice.[15] In general, she isn't—but that isn't the end of the world. Even in the present case, we can see that there are other perspectives to adopt. Aren't we really rather fortunate, for instance, that trees are so insuperably selfish? The beautiful forests—to say nothing of the beautiful wooden sailing ships and the clean white paper on which we write our poetry—could not exist if trees weren't selfish.

There can be no doubt, as I say, that game-theoretic analyses work in evolutionary theory, but do they *always* work? Under what conditions do they apply, and how can we tell when we are overstepping? Game-theory calculations always assume that there is a certain range of "possible" moves, from which the selfish-by-definition contestants make their choices. But how realistic is this *in general?* Just because a move in a particular circumstance is the move that *reason dictates*, is it the move nature will always

---

15. For a powerful antidote to the Pollyannian Paradigm, see G. Williams 1988.

take? Isn't this Panglossian optimism? (As we have just seen, this sometimes looks more like Panglossian *pessimism*: "Darn—organisms are 'too smart' to cooperate!"[16])

The standard assumption of game theory is that there will always be mutations that have the "right" phenotypic effects to rise to the occasion, but what if the right move just doesn't "occur to Mother Nature"? Is this ever or often very likely? We certainly know of cases in which Mother Nature *does* take the move—to make the forests, for instance. Are there perhaps just as many (or more) cases in which some sort of hidden constraint prevents this from happening? There may well be, but in every such case, adaptationists will want to persist by asking the next question: And *is there a reason* in this case why Mother Nature doesn't take the move, or is it just a brute, unthinking constraint on Mother Nature's rational gamesmanship?

Gould has suggested that a fundamental flaw of adaptationist reasoning is the assumption that in every fitness landscape, the way is always shown as clear to the tops of the various summits, but there might well be hidden constraints, rather like railroad tracks lying across the landscape. "The constraints of inherited form and developmental pathways may so channel any change that even though selection induces motion down permitted paths, the channel itself represents the primary determinant of evolutionary direction" (Gould 1982a, p. 383). Populations, then, do not get to spread *ad lib* across the terrain, but are forced to stay on the tracks, as in figure 9.4.

Suppose this is true. Now, how do we locate the hidden constraints? It is all very well for Gould and Lewontin to point to the possibility of hidden constraints—every adaptationist already acknowledges this as an omnipresent possibility—but we need to consider what methodology might be best for discovering them. Consider a curious variation on a standard practice in chess.

When a stronger player plays a weaker opponent in friendly matches, the stronger player often volunteers to take on a handicap, to make the game more evenly matched and exciting. The standard handicap is to give up a piece or two—to play with only one bishop or one rook, or, in a really extreme case, to play without a queen. But here is another handicapping system that might have interesting results. Before the match, the stronger player writes down on a piece of paper a hidden constraint (or constraints)

---

16. The Panglossian pessimist says, "Isn't it a shame that this is, after all, the best of all possible worlds!" Imagine a beer commercial: As the sun sets over the mountains, one of the hunks lounging around the campfire intones, "It doesn't get any better than this!"—at which point his beautiful companion bursts into tears: "Oh no! Is that really true?" It wouldn't sell much beer.

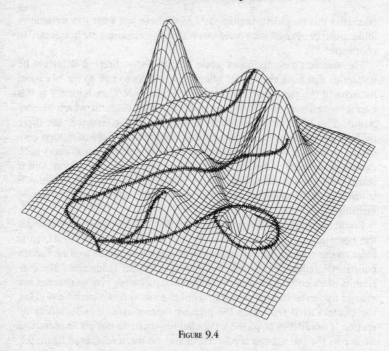

FIGURE 9.4

that she will undertake to play under, and hides the paper under the board. What is the difference between a constraint and a forced move? Reason dictates a forced move—and will always dictate it, again and again—whereas some frozen bit of history dictates a constraint, whether or not there was a reason for its birth, and whether or not there is a reason for or against it now. Here are a few of the possible constraints:

Unless I am forced by the rules to do so (because I am in check, and am obliged to play whatever legal move escapes check),

(1) I may never move the same piece on two consecutive turns.
(2) I may not castle.
(3) I may capture with pawns only three times in the whole game.
(4) My queen must move only in rook fashion, never diagonally.

Now imagine the epistemological predicament of the weaker player, who *knows* his opponent is playing with hidden constraints but doesn't know what they are. How should he proceed? The answer is quite obvious: he should play *as if* all the *apparently* possible moves—all the legal moves—

are available to her, and adjust his strategy only when evidence begins to mount that she is actually bound not to take what otherwise would be the obviously best move.

Such evidence is not at all easy to gather. If you think your opponent cannot move her queen diagonally, you might test that hypothesis by the risky tactic of offering a free capture to that queen on the diagonal. If the queen declines, that counts in favor of your hypothesis—*unless* there is a deeper reason of strategy (unimagined as yet by you) for declining the capture. (Remember Orgel's Second Rule: Evolution is cleverer than you are.)

Of course, another way of learning the hidden constraints at the chessboard is to peek at the paper, and one might think that what Gould and Lewontin are recommending is that adaptationists simply abandon their game-playing and go for the truth via a more direct examination of the molecular evidence. Unfortunately, this analogy is mistaken. You are certainly entitled to use whatever data-gathering tricks are available in the game of science, but when you peek at the molecules, all you find there is more machinery, more design (or apparent design) in need of reverse engineering. Nowhere are Mother Nature's hidden constraints *written down* in a way that can be read without the help of the interpretive rules of artifact hermeneutics (Dennett 1990b). The descent to the deeper level of the DNA, for instance, is indeed a valuable way of vastly improving one's investigative acuity—though usually at the intolerable cost of drowning in too much data—but in any case it is not an alternative to adaptationism; it is an extension of it.

The example of playing chess with hidden constraints lets us see a profound difference between Mother Nature and human chess-players that does have implications, I think, for a widespread foible in adaptationist thinking. If *you* were playing chess under hidden constraints, you would adjust your strategy accordingly. Knowing that you had secretly promised not to move your queen diagonally, you would probably forgo any campaign that put your queen at risk of capture thanks to her unusual limitation—although of course you could take a chance, hoping your weak opponent wouldn't notice the possibility. But you have knowledge of the hidden constraints, and foresight. Mother Nature does not. Mother Nature has no reason to avoid high-risk gambits; she takes them all, and shrugs when most of them lose.

Here is how the idea applies in evolutionary thinking. Suppose we notice that a particular butterfly has protective coloration on its wings that uncannily mimics the pattern of colors on the forest floor where it lives. We chalk that up as a fine adaptation, camouflage, which it undoubtedly is. This butterfly does better than its cousins *because* its coloration so perfectly reproduces the coloration of the forest floor. But there is a temptation, routinely succumbed to, to add, implicitly or explicitly: "And what's more, if the forest floor had any other color pattern on it, the butterfly would look like *that* pattern instead!" That is uncalled for. It may well not be true. It could even be, in the limit, that this is the *only* sort of forest floor that this lineage of butterfly could mimic with much success; if the forest floor were much different, this lineage would just not be here—never forget about the importance in evolution of bait-and-switch. If the forest floor changes, what will happen? Will the butterfly automatically adapt? All we can say is that either it will adapt by changing its camouflage or it won't! If it doesn't, then either it will find some other adaptation in its limited kit of available moves, or it will soon disappear.

The limiting case, in which exactly one path was ever open to explore, is an instance of our old nemesis actualism: only the actual was possible. Such straitjacketed explorations of the space of (apparent) possibility are not ruled out, I am saying, but they must be the exception, not the rule. If they were the rule, Darwinism would be defunct, utterly incapable of explaining any of the (apparent) design in the biosphere. It would be as if you wrote a chess-playing computer program that could just play one game by rote (say, Alekhine's moves in the famous Flamberg-Alekhine match in Mannheim in 1914) and, *mirabile dictu*, it regularly won against all competition! This would be a "pre-established harmony" of miraculous proportions, and would make a mockery of the Darwinian claim to have an explanation of how the "winning" moves have been found.

But our dismissal of actualism should not tempt us to err in the other direction, supposing that the space of real possibilities is much more densely populated than it actually is. The temptation, when we think about phenotypic variation, is to adopt a sort of Identikit tactic of assuming that all the minor variations we can imagine on the themes we find in actuality are truly available. Carried to extremes, this tactic will always vastly—Vastly—overestimate what is actually possible. If the *actual* Tree of Life occupies Vanishingly narrow threads through the Library of Mendel, the *actually possible* Tree of Life is itself some rather bushier but still far from dense partial filling of the *apparently possible*. We have already seen that the Vast space of all imaginable phenotypes—Identikit Space, we might call it—no doubt includes huge regions for which there are no recipes in the Library of Mendel. But even along the paths through which the Tree of Life wanders, we are not guaranteed that the neighboring regions of Identikit Space are actually all accessible.[17]

---

17. Gould is fond of pointing out the mistake of looking back in time and seeing "lineages" where we should be seeing "bushes"—including all the failures that have left no descendants. I am pointing out a contrary sort of mistake: imagining dense (or even

If hidden constraints guarantee that there is a largely invisible set of maze walls—or channels or railroad tracks—in the space of apparent possibility, then "you can't get there from here" is true much more often than we might imagine. Even if this is so, we still can do no better in our exploration of this possibility than to play out our reverse-engineering strategies at every opportunity, at every level. It is important not to overestimate the actual possibilities, but it is even more important not to underestimate them, an equally common foible, though not one that adaptationists typically manifest. Many adaptationist arguments are of the if-it's-possible-it-will-happen variety: cheats will emerge to invade the saints; or an arms race will ensue until such-and-such a first-order adaptive stability is achieved, etc. These arguments presuppose that enough of the space of possibilities is "habitable" to ensure that the process approximates the game-theory model used. But are these assumptions always appropriate? Will these bacteria mutate into a form that is resistant to our new vaccine? Not if we're lucky, but we're better off assuming the worst—namely, that there are, in the space actually accessible to these bacteria, countermoves in the arms race our medical innovation has set in motion (Williams and Nesse 1991).

CHAPTER 9: *Adaptationism is both ubiquitous and powerful in biology. Like any other idea, it can be misused, but it is not a mistaken idea; it is in fact the irreplaceable core of Darwinian thinking. Gould and Lewontin's fabled refutation of adaptationism is an illusion, but they have raised everybody's consciousness about the risks of incautious thinking. Good adaptationistic thinking is always on the lookout for hidden constraints, and in fact is the best method for uncovering them.*

CHAPTER 10: *The view of Darwinian thinking presented so far in this book has been challenged, repeatedly, by Stephen Jay Gould, whose influential writings have contributed to a seriously distorted picture of evolutionary biology among both lay people and philosophers and scientists in other fields. Gould has announced several different "revolutionary" abridgments of orthodox Darwinism, but they all turn out to be false alarms. There is a pattern to be discerned in these campaigns: Gould, like eminent evolutionary thinkers before him, has been searching for skyhooks to limit the power of Darwin's dangerous idea.*

---

continuous) bushes of unactualized possibility where in fact there may be rather sparse twigs creating paths to relatively isolated outposts in the huge space of apparent possibilities.

CHAPTER TEN

Bully for Brontosaurus

---

1. THE BOY WHO CRIED WOLF?

Scientists have power by virtue of the respect commanded by the discipline. We may therefore be sorely tempted to misuse that power in furthering a personal prejudice or social goal—why not provide that extra oomph by extending the umbrella of science over a personal preference in ethics or politics? But we cannot, lest we lose the very respect that tempted us in the first place.

—STEPHEN JAY GOULD 1991b, pp. 429–30

Many years ago, I saw a program on British television in which young children were interviewed about Queen Elizabeth II. Their confident answers were charming: the Queen, it seems, spends a large part of the day vacuum-cleaning Buckingham Palace—while wearing her crown, of course. She pulls the throne up to the telly when she is not occupied with affairs of state, and wears an apron over her ermine robes when she does the washing up. I realized then that the largely imaginary Queen Elizabeth II of these young children (what philosophers would call their *intentional object*) was in some regards a more potent and interesting object in the world than the actual woman. Intentional objects are the creatures of beliefs, and hence they play a more direct role in guiding (or misguiding) people's behavior than do the real objects they purport to be identical to. The gold in Fort Knox, for example, is less important than what is believed about it, and the Albert Einstein of myth is, like Santa Claus, much better known than the relatively dimly remembered historical fellow who was the primary source for the myth.

This chapter is about another myth—Stephen Jay Gould, Refuter of Orthodox Darwinism. Over the years, Gould has mounted a series of attacks on aspects of contemporary neo-Darwinism, and although none of these attacks has proven to be more than a mild corrective to orthodoxy at best, their rhetorical impact on the outside world has been immense and distorting. This presents me with a problem that I cannot ignore or postpone. In my own work over the years, I have often appealed to evolutionary considerations, and have almost as often run into a curious current of resistance: my appeals to Darwinian reasoning have been bluntly rejected as discredited, out-of-date science by philosophers, psychologists, linguists, anthropologists, and others who have blithely informed me that I have got my biology all wrong—I haven't been doing my homework, because Steve Gould has shown that Darwinism isn't in such good shape after all. Indeed, it is close to extinction.

That is a myth, but a very influential myth, even in the halls of science. I have tried in this book to present an accurate account of evolutionary thinking, deflecting the reader from common misunderstandings, and defending the theory against ill-grounded objections. I have had a lot of expert help and advice, and so I am confident that I have succeeded. But the view of Darwinian thinking I have presented is quite at odds with the view made familiar to many by Gould. Surely, then, my view must be mistaken? After all, who knows better about Darwin and Darwinism than Gould?

Americans are notoriously ill-informed about evolution. A recent Gallup poll (June 1993) discovered that 47 percent of adult Americans believe that *Homo sapiens* is a species created by God less than ten thousand years ago. But insofar as they know anything at all about the subject, it is probably due more to Gould than to anyone else. In the battle over the teaching of "creation science" in the schools, he has been a key witness for the defense of evolution in the court cases that continue to plague American education. For twenty years, his monthly column, "This View of Life," in *Natural History*, has provided professional and amateur biologists with a steady stream of arresting insights, fascinating facts, and well-needed correctives to their thinking. In addition to his collections of these essays, in such volumes as *Ever Since Darwin* (1977a), *The Panda's Thumb* (1980a), *Hen's Teeth and Horse's Toes* (1983b), *The Flamingo's Smile* (1985), *Bully for Brontosaurus* (1991b), and *Eight Little Piggies* (1993d), and his technical publications on snails and paleontology, he has written a major theoretical book, *Ontogeny and Phylogeny* (1977b); an attack on IQ testing, *The Mismeasure of Man* (1981); a book on the reinterpretation of the fauna of the Burgess Shale, *Wonderful Life* (1989a); and numerous other articles on topics ranging from Bach to baseball, from the nature of time to the compromises of *Jurassic Park*. Most of this is simply wonderful: astonishingly erudite, the very model of a scientist who recognizes, as my high-school physics teacher once said, that science, done right, *is* one of the humanities.

The title of Gould's monthly column comes from Darwin, the closing sentence of *Origin of Species*.

There is grandeur in this view of life, with its several powers, having been originally breathed into a few forms or into one; and that, whilst this planet has gone cycling on according to the fixed laws of gravity, from so simple a beginning endless forms most beautiful and most wonderful have been, and are being, evolved.

Anybody as prolific and energetic as Gould would surely have an agenda beyond that of simply educating and delighting his fellow human beings about the Darwinian view of life. In fact, he has had numerous agendas. He has fought hard against prejudice, and particularly against the abuse of scientific research (and scientific prestige) by those who would clothe their political ideologies in the potent mantle of scientific respectability. It is important to recognize that Darwinism has always had an unfortunate power to attract the most unwelcome enthusiasts—demagogues and psychopaths and misanthropes and other abusers of Darwin's dangerous idea. Gould has laid this sad story bare in dozens of tales, about the Social Darwinists, about unspeakable racists, and most poignantly about basically good people who got confused—seduced and abandoned, you might say—by one Darwinian siren or another. It is all too easy to run off half cocked with some poorly understood version of Darwinian thinking, and Gould has made it a major part of his life's work to protect his hero from this sort of abuse.

The irony is that his own strenuous efforts to protect Darwinism have sometimes backfired. Gould has been a defender of his own brand of Darwinism, but an ardent opponent of what he has called "ultra-Darwinism" or "hyper-Darwinism." What is the difference? The uncompromising "no-skyhooks-allowed" Darwinism I have presented is, by Gould's lights, hyper-Darwinism, an extremist view that needs overthrowing. Since in fact it is, as I have said, quite orthodox neo-Darwinism, Gould's campaigns have had to take the form of calls for revolution. Time and again, Gould has announced from his bully-pulpit to a fascinated world of onlookers that neo-Darwinism is dead, supplanted by a revolutionary new vision—still Darwinian, but overthrowing the establishment view. It hasn't happened. As Simon Conway Morris, one of the heroes of Gould's *Wonderful Life*, has said, "His views have done much to stir the established orthodoxies, even if, when the dust settles, the edifice of evolutionary theory still looks little changed" (Conway Morris 1991, p. 6).

Gould is not the only evolutionist to succumb to the urge of overdramatization. Manfred Eigen and Stuart Kauffman—and there are others we haven't considered—have also styled themselves at first as radical heretics. Who wouldn't prefer one's contributions to be truly revolutionary? But whereas Eigen and Kauffman, as we have seen, have moderated their rhetoric in due course, Gould has gone from revolution to revolution. So far, his declarations of revolution have all been false alarms, but he has kept on trying, defying the moral of Aesop's fable about the boy who cried wolf. This has earned him not just a credibility problem (among scientists), but also the animosity of some of his colleagues, who have felt the sting of what they consider to be undeserved public condemnation in the face of his influential campaigns. As Robert Wright (1990, p. 30) puts it, Gould is "America's evolutionist laureate. If he has been systematically misleading Americans about what evolution is and what it means, that amounts to a lot of intellectual damage."

Has he done this? Consider the following. If you believe:

(1) that adaptationism has been refuted or relegated to a minor role in evolutionary biology, or

(2) that since adaptationism is "the central intellectual flaw of sociobiology" (Gould 1993a, p. 319), sociobiology has been utterly discredited as a scientific discipline, or

(3) that Gould and Eldredge's hypothesis of punctuated equilibrium overthrew orthodox neo-Darwinism, or

(4) that Gould has shown that the fact of mass extinction refutes the "extrapolationism" that is the Achilles' heel of orthodox neo-Darwinism,

then what you believe is a falsehood. If you believe any of these propositions, you are, however, in very good company—both numerous and intellectually distinguished company. Quine once said of a misguided critic of his work, "He reads with a broad brush." We are all apt to do this, especially when we try to construe in simple terms the take-home message of work outside our own field. We tend to read, with bold brushstrokes, what we want to find. Each of these four propositions expresses a verdict that is rather more decisive and radical than Gould may have intended, but together they compose a message that is out there, in many quarters. I beg to differ, so it falls to me to dismantle the myth. Not an easy job, since I must painstakingly separate the rhetoric from the reality, all the while fending off—by explaining away—the entirely reasonable presumption that an evolutionist of Gould's stature couldn't be *that* wrong in his verdicts, could he? Yes and no. The real Gould has made major contributions to evolutionary thinking, correcting a variety of serious and widespread misapprehensions, but the mythical Gould has been created out of the yearnings of many Darwin-dreaders, feeding on Gould's highly charged words, and this has encouraged, in turn, his own aspirations to bring down "ultra-Darwinism," leading him into some misbegotten claims.

If Gould has kept crying wolf, why has he done this? The hypothesis I shall defend is that Gould is following in a long tradition of eminent thinkers who have been seeking skyhooks—and coming up with cranes. Since evolutionary theory has made great progress in recent years, the task of making room for a skyhook has become more difficult, raising the bar for any thinker who wants to find some blessed exemption. By following the repetition of theme and variation in Gould's work, I will uncover a pattern: each failed attempt defines a small portion of the shadow of his quarry, until eventually the source of Gould's driving discomfort will be clearly outlined. Gould's ultimate target is Darwin's dangerous idea itself; he is opposed to the very idea that evolution is, in the end, just an algorithmic process.

It would be interesting to ask the further question of why Gould is so set against this idea, but that is really a task for another occasion, and perhaps for another writer. Gould himself has shown how to execute such a task. He has examined the underlying assumptions, fears, and hopes of earlier scientists, from Darwin himself through Alfred Binet, the inventor of IQ testing, to Charles Walcott, the (mis)classifier of the Burgess Shale fauna, to name just three of his best-known case histories. What hidden agendas—moral, political, religious—have driven Gould himself? Fascinating though this question is, I am going to resist the temptation to try to answer it, though in due course I will briefly consider, as I must, the rival hypotheses that have been suggested. I have enough to do just defending the admittedly startling claim that the pattern in Gould's failed revolutions reveals that America's evolutionist laureate has always been uncomfortable with the fundamental core of Darwinism.

For years I was genuinely baffled by the ill-defined hostility to Darwinism that I encountered among many of my fellow academics, and although they cited Gould as their authority, I figured they were just wishfully misreading him, with a little help from the mass media, always eager to obliterate subtlety and fan the flames of every minor controversy. It really didn't occur to me that Gould was often fighting on the other side. He himself has been victimized so often by this hostility. Maynard Smith mentions just one example:

> One cannot spend a lifetime working on evolutionary theory without becoming aware that most people who do not work in the field, and some who do, have a strong wish to believe that the Darwinian theory is false. This was most recently brought home to me when my friend Stephen Gould, who is as convinced a Darwinist as I am, found himself the occasion of an editorial in the *Guardian* announcing the death of Darwinism, followed by an extensive correspondence on the same theme, merely because he had pointed out some difficulties the theory still faces. [Maynard Smith 1981, p. 221, as reprinted in Maynard Smith 1988.]

Why should such a "convinced Darwinist" as Gould keep getting himself in trouble by contributing to the public misconception that Darwinism is dead? There is no more committed or brilliant adaptationist than John Maynard Smith, but here I think we see the master napping: he doesn't ask himself this "why" question. After I began to notice that many of the most important contributions to evolutionary theory have been made by thinkers who were fundamentally ill-at-ease with Darwin's great insight, I could begin to take seriously the hypothesis that Gould himself is one of these. Making the case for this hypothesis will take patience and hard work, but there's no avoiding it. The mythology about what Gould has shown and hasn't shown is so widespread that it will befog all the other issues before us if I don't do what I can to disperse it first.

2. THE SPANDREL'S THUMB

I think I can see what is breaking down in evolutionary theory—the strict construction of the modern synthesis with its belief in pervasive adaptation, gradualism and extrapolation by smooth continuity from causes of change in local populations to major trends and transitions in the history of life.

—STEPHEN JAY GOULD 1980b

At issue is not the general idea that natural selection can act as a creative force; the basic argument, in principle, is sound. Primary doubt centers on the subsidiary claims—gradualism and the adaptationist program.

—STEPHEN JAY GOULD 1982a

Gould has done much to bring a central theme of Darwinism, that supposed perfection in design is a jury-rigged compromise adopting some improbable pieces of anatomy, to general notice. But some of these essays contain hints that somehow the Darwinian explanation is only partly correct. But is this a serious attack? Not on a closer reading.

—SIMON CONWAY MORRIS 1991

Gould (1980b, 1982a) sees two main problem elements in the modern synthesis: "pervasive adaptation" and "gradualism." And he sees them as related. How? He has given somewhat different answers over the years. We can begin with "pervasive adaptation." To see what the issue is, we should return to the Gould and Lewontin paper of 1979. The title is a good place to start: "The Spandrels of San Marco and the Panglossian Paradigm: A Critique of the Adaptationist Programme." In addition to their redefining of "Panglossian," they introduced another term, "spandrel," which has proven to be a highly successful coinage in one sense: it has spread through evolutionary biology and beyond. In a recent retrospective essay, Gould put it this way:

> Ten years later, my friend Dave Raup ... said to me, "We have all been spandrelized." When your example becomes both generic and a different part of speech, you have won. Call those San Marco spandrels "Kleenex," "Jell-O," and a most emphatically non-metaphorical "Band-Aid." [Gould 1993a, p. 325.]

Ever since Gould and Lewontin, evolutionists (and many others) have spoken of spandrels, thinking that they knew what they were talking about. What are spandrels? A good question. Gould wants to convince us that adaptation is not "pervasive," so he needs to have a term for the (presumably many) biological features that are not adaptations. They are to be called "spandrels." Spandrels are, um, things that *aren't* adaptations, whatever they are. Gould and Lewontin have shown us, haven't they, that spandrels are ubiquitous in the biosphere? Not so. Once we clear away the confusions about what the term might mean, we will see that either spandrels are not ubiquitous after all, or they are *the normal basis* for adaptations, and hence no abridgment at all of "pervasive adaptation."

Gould and Lewontin's paper begins with two famous architectural examples, and since a crucial misstep is made at the outset, we must look closely at the text. (One of the effects of classic texts is that people misremember them, having read them hurriedly once. Even if you are familiar with this oft-reprinted beginning, I urge you to read it again, slowly, to see how the misstep happens, right before your eyes.)

> The great dome of St Mark's Cathedral in Venice presents in its mosaic design a detailed iconography expressing the mainstays of Christian faith. Three circles of figures radiate out from a central image of Christ: angels, disciples, and virtues. Each circle is divided into quadrants, even though the dome itself is radially symmetrical in structure. Each quadrant meets one of the four spandrels in the arches below the dome. Spandrels—the tapering triangular spaces formed by the intersection of two rounded arches at right angle (figure [10.]1)—are necessary architectural by-products of mounting a dome on rounded arches. Each spandrel contains a design admirably fitted into its tapering space.... The design is so elaborate, harmonious and purposeful that we are tempted to view it as the starting point of any analysis, as the cause in some sense of the surrounding architecture. But this would invert the proper path of analysis. The system begins with an architectural constraint: the necessary four spandrels and their tapering triangular form. They provide a space in which the mosaicists worked; they set the quadripartite symmetry of the dome above.... Every fan vaulted ceiling must have a series of open spaces along the

© Alinari/Art Resources, N.Y.

FIGURE 10.1.
One of the spandrels of San Marco.

mid-line of the vault, where the sides of the fans intersect between the pillars (figure [10.]2). Since the spaces must exist, they are often used for ingenious ornamental effect. In King's College Chapel in Cambridge, for example, the spaces contain bosses alternately embellished with the Tudor rose and portcullis. In a sense, this design represents an 'adaptation', but the architectural constraint is clearly primary. The spaces arise as a necessary by-product of fan vaulting; their appropriate use is a secondary effect. Anyone who tried to argue that the structure exists because the alternation of rose and portcullis makes so much sense in a Tudor chapel would be inviting the same ridicule that Voltaire heaped on Dr Pangloss. . . . Yet evolutionary biologists, in their tendency to focus exclusively on immediate adaptation to local conditions, do tend to ignore architectural constraints and perform just such an inversion of explanation. [Gould 1993a, pp. 147–49.]

First, we should notice that from the outset Gould and Lewontin invite us to *contrast* adaptationism with a concern for architectural "necessity" or

FIGURE 10.2.
The ceiling of King's College Chapel.

"constraint"—as if the discovery of such constraints weren't an integral part of (good) adaptationist reasoning, as I have argued in the last two chapters. Now, perhaps we should stop right here and consider the possibility that Gould and Lewontin have been massively misunderstood, thanks to the misfiring rhetoric of this opening passage, rhetoric which they even correct somewhat, in the last sentence quoted above. Perhaps what Gould and Lewontin showed, in 1979, is that we must all be *better* adaptationists; we should *expand* our reverse-engineering perspective back onto the processes of R and D, and embryological development, instead of focusing "exclusively on *immediate* adaptation to *local* conditions." That, after all, is one of the main lessons of the last two chapters, and Gould and Lewontin could share the credit for drawing it to the attention of evolutionists. But almost everything else that Gould and Lewontin have said militates against this interpretation; they mean to oppose adaptationism, not enlarge it. They call for a "pluralism" in evolutionary biology of which adaptationism is to be just one element, its influence diminished by the other elements, if not utterly suppressed.

The spandrels of San Marco, we are told, "are necessary architectural by-products of mounting a dome on rounded arches." In what sense necessary? The standard assumption among biologists I have asked is that this is somehow a *geometric* necessity, and hence has nothing whatever to do with adaptationist cost-benefit calculations, since there is simply no choice to be made! As Gould and Lewontin (p. 161) put it, "Spandrels must exist once a blueprint specifies that a dome shall rest on rounded arches." But is that true? It might appear at first as if there were no alternatives to smooth, tapering triangular surfaces in between the dome and the four rounded arches, but there are in fact indefinitely many ways that those spaces could be filled with masonry, all of them about equal in structural soundness and ease of building. Here is the San Marco scheme (on the left) and two variations. The variations are both, in a word, ugly (I deliberately made them so), but that does not make them *impossible*.

Here there is a terminological confusion that seriously impedes discus-

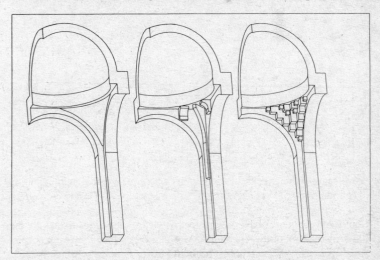

FIGURE 10.3

sion. Does figure 10.3 display three different sorts of spandrels, or does it display a spandrel on the left, and two ugly alternatives to spandrels? Like other specialists, art historians often indulge in both strict and loose usages for their terms. Strictly speaking, the tapering, roughly spherical surface illustrated in figure 10.1, the sort of surface illustrated on the left in figure 10.3, is called a *pendentive*, not a *spandrel*. Strictly speaking, spandrels are what remains of a wall once you punch an arch through it, as in figure 10.4. (But even that definition leaves room for confusion. In figure 10.4, are we shown spandrels on the left, and something else on the right, or do "pierced spandrels" count as spandrels, strictly speaking? I don't know.)

Speaking more loosely, spandrels are places-to-be-dealt-with, and in that looser sense, the three variations in figure 10.3 all count as spandrel varieties. Another variety of spandrel (in that sense) would be a *squinch*, shown in figure 10.5.

But sometimes art historians speak of spandrels when they are talking specifically about pendentives, the variety shown on the left in figure 10.3. In that sense, squinches are not types of spandrels, but rivals to spandrels.

Now, why does all this matter? Because, when Gould and Lewontin say that spandrels are "necessary architectural by-products," what they say is false, if they are using "spandrel" in the narrow sense (synonymous with "pendentive") and true only if we understand the term in the loose, all-inclusive sense. But in that sense of the term, spandrels are design *problems*, not *features* that might either be designed (adaptations) or not. Spandrels in the loose sense are indeed "geometrically necessary" in one regard: if you

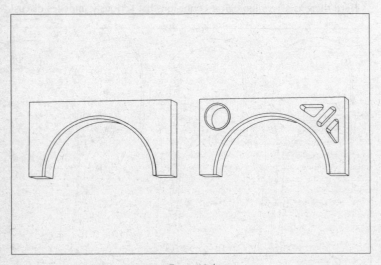

FIGURE 10.4

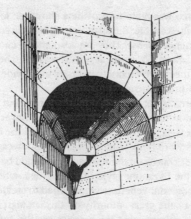

Squinch. A corbelling, usually a small arch or half-cornical niche, which is placed across the corners of a square bay in order to form an octagon suitable for carrying an octagonal cloister-vault or a dome. [Krautheimer 1981.]

FIGURE 10.5

place a dome over four arches, you have what you might call an *obligatory design opportunity*: you have to put something there to hold up the dome—some shape or other, you decide which. But if we interpret spandrels as obligatory places for one adaptation or another, they are hardly a challenge to adaptationism.

But is there nevertheless some other way in which spandrels in the narrow sense—pendentives—truly are nonoptional features of San Marco? That is what Gould and Lewontin seem to be asserting, but if so, they are wrong. Not only were the pendentives just one among many *imaginable* options; they were just one among the readily *available* options. Squinches had been a well-known solution to the problem of a dome over arches in Byzantine architecture since about the seventh century.[1]

What the actual design of the San Marco spandrels—that is, pendentives—has going for it are mainly two things. First, it is (approximately) the "minimal-energy" surface (what you would get if you stretched a soap film in a wire model of the corner), and hence it is close to the minimal surface area (and hence might well be viewed as the optimal solution if, say, the number of costly mosaic tiles was to be minimized!). Second, this smooth surface is *ideal* for the mounting of mosaic images—and that is why the

---

1. "Whatever the origin of the dome on squinches, however, the importance of the question, it seems to me, has been vastly overplayed. Squinches are an element of construction which can be incorporated into almost any kind of architecture." (Krautheimer 1981, p. 359.)

Basilica of San Marco was built: to provide a showcase for mosaic images. The conclusion is inescapable: the spandrels of San Marco aren't spandrels even in Gould's extended sense. They are adaptations, chosen from a set of equipossible alternatives for largely aesthetic reasons. They were *designed* to have the shape they have precisely in order to provide suitable surfaces for the display of Christian iconography.

After all, San Marco is not a granary; it is a church (but not a cathedral). The primary function of its domes and vaults was never to keep out the rain—there were less expensive ways of doing that in the eleventh century, when these domes were built—but to provide a showcase for symbols of the creed. An earlier church on the site had burned and been rebuilt in 976, but subsequently the Byzantine style of mosaic decoration had provoked the admiration of powerful Venetians who wanted to create a local example. Otto Demus (1984), the great authority on the San Marco mosaics, shows in four magnificent volumes that the mosaics are the *raison d'être* of San Marco, and hence of many of its architectural details. In other words, there wouldn't be any such pendentives in Venice if the "environmental problem" of how to display Byzantine mosaic images of Christian iconography had not been posed and this solution found. If you look closely at the pendentives (this is detectable in figure 10.1, but unmistakable if you look at the actual pendentives, as I did on a recent visit to Venice), you will see that care has been taken to round off the transition between the pendentive proper and the arches it connects, the better to provide a continuous surface for the application of mosaics.

Gould and Lewontin's other example from architecture was also ill-chosen, as it turns out, since we simply don't know whether the King's College bosses alternating rose and portcullis are the *raison d'être* of the fan vaulting or vice versa. We do know that fan vaulting was *not* part of the original design of that chapel, but a later revision, a change order introduced years after the construction had begun, for reasons unknown (Fitchen 1961, p. 248). The very heavy (and heavily carved) keystones at the intersections of the ribs of earlier Gothic vaults had been a sort of forced move for builders, as I noted in chapter 8, since they needed the extra weight of this keystone to counteract the rising tendency of the pointed arches, especially during the construction phase, when deformation of partially completed structures was a major problem to be solved. But in late fan vaulting of the King's College type, the purpose of the bosses is probably entirely to provide focal points for ornament. Did the bosses have to be there anyway? No. From an engineering point of view, there could have been neat round holes there, "lanterns" letting in daylight from above if it weren't for the roof. Maybe fan vaulting *was* chosen by the builders so that the ceiling could carry the Tudor symbols!

So the fabled spandrels of San Marco are not spandrels but adaptations after all.[2] That is curious, you may think, but not theoretically important, because, as Gould himself has often reminded us, one of Darwin's fundamental messages is that artifacts get recycled with new functions—"exapted," to use Gould and Vrba's coinage (1981). The panda's thumb is not really a thumb, but it is pretty good at doing what it does. Isn't the Gould-Lewontin concept of a spandrel a valuable tool in evolutionary thinking even if its birth was, to exapt yet another famous phrase, a frozen accident of history? Well, what *is* the function of the term "spandrel" in evolutionary thinking? So far as I know, Gould has never given the term (in application to biology) an official definition, and since the examples he has relied on to exhibit his intended meaning are at best misleading, we are left to our own devices: we should try to find the best, most charitable, interpretation of his texts. When we turn to that task, one point emerges from context with clarity: whatever a spandrel is, it is supposed to be a non-adaptation.

What would be a *good* architectural example of a spandrel (*sensu* Gould)? If adaptations are examples of (good, cunning) design, then perhaps a spandrel is a "no-brainer"—a feature exhibiting no design cunning at all. The existence of a doorway—just a rough opening—in a building might seem to be an example, since we would not be particularly impressed by the wisdom of the builder who included such a feature in his house. But there is, after all, a very good reason why dwellings should have doorways. If spandrels are just dead-obvious good solutions to design problems that tend therefore to become part of a relatively unthinking tradition of building, then spandrels abound. In that case, however, they would not be alternatives to adaptation, but examples *par excellence* of adaptation—either forced moves or, in any event, moves you'd be foolish not to consider. A better sort of example, then, might be what engineers sometimes call a "don't-care": something that has to be one way or another, but that nothing makes better one way than another. If we put a door in the doorway, the

---

2. I am not the first, I have recently discovered, to note these minor errors in Gould's excursion in art history. Some years ago, two evolutionary biologists were there before me: Alasdair Houston (1990) drew attention to the point about spandrels, pendentives, and squinches, and Tim Clutton-Brock, in a lecture at Harvard, questioned Gould's interpretation of the fan vaulting of King's College Chapel.

It is interesting that these points were overlooked by all the deconstructionists and rhetoricians who contributed essays to a recent book (Selzer 1993) devoted in its entirety to an analysis of the rhetoric of Gould and Lewontin's essay. You might suppose that someone among this group of sixteen humanists would have noticed the factual problems in the fundamental rhetorical device of the essay, but it must be remembered that these sophisticates are interested in "deconstructing knowledge"—which means that they have transcended the stodgy, old-fashioned dichotomy between fact and fiction, and hence are not professionally curious about whether what they read is the truth!

door will need hinges, but should they go on the left or the right? Perhaps nobody cares, so a coin is flipped, and hinges on the left get installed. If other builders copy the result unthinkingly, establishing a local tradition (reinforced by the latchmakers, who make latches for left-hinged doors only), this might be a spandrel masquerading as an adaptation. "Why are all the doors in this village hinged on the left?" would be a classic adaptationist question, to which the answer would be: "No reason. Just historical accident." So is that a good architectural example of a spandrel? Perhaps, but, as the example of the autumn leaves in the preceding chapter showed, it is never a mistake to *ask* the adaptationist's "why" question, even when the true answer is that there is no reason. Are there many features in the biosphere that exist for no reason? It all depends on what counts as a feature. Trivially, there are indefinitely many properties (e.g., the elephant's property of having more legs than eyes, the daisy's property of buoyancy) that are not themselves adaptations, but no adaptationist would deny this. Presumably, there is a more interesting doctrine that Gould and Lewontin are urging us to abandon.

What is the doctrine of "pervasive adaptation," then, that Gould supposes such an admission of widespread spandrels would overthrow? Let us consider the most extreme form of Panglossian adaptationism imaginable—the view that *every designed thing* is *optimally* designed. A sidelong glance at human engineering will show that even this view not only permits but requires the existence of plenty of undesigned stuff. Imagine, if you can, some masterpiece of human engineering—the perfectly designed widget-factory, energy-efficient, maximally productive, minimally expensive to operate, maximally humane to its workers, simply unimprovable in any dimension. The waste-paper collection system, for instance, makes recycling by type of wastepaper maximally convenient and agreeable to the staff, at minimal energy costs, and so forth. A Panglossian triumph, it seems. But wait—what is the *wastepaper* for? It's not for anything. It's a by-product of the other processes, and the wastepaper collection system is for dealing with it. You can't give an adaptationist explanation of why the disposal/recycling system is optimal without presupposing that the wastepaper itself is just ... waste! Of course, you can go on and ask whether the clerical operations could be made "paperless" by better use of computers, but if that happens not to be the case for one reason or another, there will still be wastepaper to deal with, and other wastes and by-products as well in any case, so there will always be plenty of undesigned features in a system that is maximally well designed. No adaptationist could be such a "pervasive" adaptationist as to deny it. The thesis that every property of every feature of everything in the living world is an adaptation is not a thesis anybody has ever taken seriously, or implied by what anybody has taken seriously, so far as I know. If I am wrong, there are some serious loonies out there, but Gould has never shown us one.

Sometimes, however, it does seem that he thinks this is the view to attack. He characterizes adaptationism as "pure adaptationism" and "panadaptationism"—apparently the view that every feature of every organism is to be explained as an adaptation selected for. In her recent book, *The Ant and the Peacock*, the philosopher of biology Helena Cronin is particularly acute in diagnosing this view (Cronin, pp. 66–110). She catches Gould in the act of sliding into exactly this misconstrual:

> . . . Stephen Gould talks about 'what may be the most fundamental question in evolutionary theory' and then, significantly, spells out not one question but two: 'How *exclusive* is natural selection as an agent of evolutionary change? Must *all* features of organisms be viewed as adaptations?' (Gould 1980[a], p. 49; my emphasis). But natural selection could be the only true begetter of adaptations without having begot all characteristics; one can hold that all adaptive characteristics are the result of natural selection without holding that all characteristics are, indeed, adaptive. [Cronin 1991, p. 86.]

Natural selection could still be the "exclusive agent" of evolutionary change even though many features of organisms were not adaptations. Adaptationists are—and should be—*always* on the lookout for adaptive explanations of whatever feature captures their attention, but this strategy falls short of committing anybody to the caricature that Gould calls "panadaptationism."

Perhaps what Gould opposes will become clearer if we look at what he recommends in its place. What alternatives to adaptationism did Gould and Lewontin suggest, as components of their recommended pluralism? Chief among them was the idea of a *Bauplan*, a German architectural term that had been adopted by certain continental biologists. The term would usually be translated in English as "ground plan" or "floor plan"—the basic outline of the structure as seen from above. It is curious that an architectural term should be highlighted in a *counter*adaptationist campaign, but it makes a certain daft sense when you see how the original *Bauplan* theorists pushed it. Adaptation, they said, could explain *superficial* modifications of the design of organisms to fit the environment, but not the fundamental features of living things: "The important steps in evolution, the construction of the *Bauplan* itself and the transition between *Baupläne*, must involve some other unknown, and perhaps 'internal' mechanism" (Gould and Lewontin 1979, p. 159). The floor plan is not designed by evolution, but just somehow given? Sounds a bit fishy, doesn't it? Were Gould and Lewontin buying this radical idea from the continent? Not for a moment. They quickly (p. 159) granted that English biologists had been right "in rejecting this strong form as close to an appeal to mysticism."

But once the mystical version of *Baupläne* is shunned, what is left? Our old friend: the claim that good reverse engineering takes the building process into account. As Gould and Lewontin put it (p. 160), their view of matters "does not deny that change, when it occurs, may be mediated by natural selection, but it holds that constraints restrict possible paths and modes of change so strongly that the constraints themselves become much the most interesting aspect of evolution." Whether or not they are the *most* interesting aspect, they are certainly important, as we have seen. Perhaps adaptationists (like art historians) need to have this point repeatedly drawn to their attention. When Dawkins, an arch-adaptationist if there ever was one, says, "There are some shapes that certain kinds of embryology seem incapable of growing" (Dawkins 1989b, p. 216), he is expressing a version of this point about the constraint of the *Bauplan*, and it was something of a revelation to him, he says. It was forcefully brought home to him by his own computer simulations of evolution, not by the Gould and Lewontin paper, but we might let them chime in: "We told you so!"

Gould and Lewontin also discuss other alternatives to adaptation, and these, too, are themes we have already encountered in orthodox Darwinism: random fixation of genes (the role of historical accident and its amplification), developmental constraints due to the way genes get expressed, and the problems of getting around in a fitness landscape with "multiple adaptive peaks." These are all real phenomena; as usual, the debate among evolutionists is not about whether they exist, but about how important they are. Theories that incorporate them have indeed played a significant role within the growing sophistication of the neo-Darwinian synthesis, but they are reforms or complications, not revolutions.

So some evolutionists have accepted Gould and Lewontin's pluralism in an irenic spirit, as a call not to abandon but, rather, to improve adaptationism. As Maynard Smith (1991, p. 6) has put it, "The effect of the Gould-Lewontin paper has been considerable, and on the whole welcome. I doubt if many people have stopped trying to tell adaptive stories. Certainly I have not done so myself." Gould and Lewontin's paper has had a welcome effect, then, but one of its by-products has not been so welcome. The inflammatory rhetoric suggesting that these somewhat neglected themes constituted a major alternative to adaptationism opened the floodgates to a lot of wishful thinking by Darwin-dreaders who would prefer that there *not* be an adaptationist explanation of one precious phenomenon or another. What would their dimly imagined alternative be? Either the "internal necessity" that Gould and Lewontin themselves dismiss as an appeal to mysticism, or utter cosmic coincidence—an equally mystical nonstarter. Neither Gould nor Lewontin explicitly endorsed either wild alternative to adaptation, but this was overlooked by those who wanted to be dazzled by the authority of these eminent Darwin-doubters.

Moreover, Gould, in spite of the appeal to pluralism in the co-authored paper, has persisted in describing it as laying waste to adaptationism (e.g., 1993a), and has held out for a "non-Darwinian" interpretation of its central concept, spandrels. It may have occurred to you that I have overlooked an obvious interpretation of spandrels: perhaps spandrels are just QWERTY phenomena. QWERTY phenomena, you recall, are constraints, but constraints with an adaptive history and hence an adaptationist explanation.[3] Gould himself briefly considered this alternative (1982a, p. 383): "If the channels [that constrain current options] are set by past adaptations, then selection remains preeminent, for all major structures are either expressions of immediate selection, or channeled by a phylogenetic heritage of previous selection." Nicely put, but he promptly rejected it, calling it Darwinian (which it certainly is), and recommending an alternative "non-Darwinian version" which he described as "not widely appreciated but potentially fundamental." Spandrels, he then suggested (p. 383), aren't the frozen constraints created by earlier adaptations; they are *exaptations*. What contrast was he trying to draw?

I think he saw the difference between the exploitation of something previously designed, and the exploitation of something originally undesigned, and was claiming that it was an important difference. Perhaps. Here is some indirect textual evidence for that reading. A recent article in the *Boston Globe* quotes the linguist Samuel Jay Keyser of MIT:

> "Language may well be a spandrel of the mind," Keyser says, and then waits patiently while his questioner looks "spandrel" up in the dictionary.... The first builder who supported domes with arches created spandrels *by accident* [emphasis added], and at first builders paid no attention to spandrels and decorated only the arches, Keyser says. But after a couple of centuries, builders began focussing on and decorating the spandrels. In the

---

3. In his own discussion of the original QWERTY phenomenon (1991a), Gould makes a useful point (1991a, p. 71), but does not develop it further, so far as I know: because of the curious historical sequence of events that led to the general adoption of the standard QWERTY typewriter keyboard, "An array of competitions that would have tested QWERTY were never held." That is, it is simply *irrelevant* to ask whether QWERTY is a better design than alternatives X, Y, and Z, since those alternatives were never pitted against QWERTY in the marketplace or the design workshop. They just never came up at a time when, it seems, they could have made a difference. Adaptationists should be alert to the fact that, even though whatever we see in nature has been "tested against all comers" and not found wanting, only a Vanishingly small (and biased) subset of all the imaginable competitions has ever been held. The inevitable parochiality of all actual tournaments means that one must be cautious in characterizing the virtues of the winners. An old Downeast joke makes the same point more succinctly: "Mornin', Edna." "Mornin', Bessie. How's yer husband?" "Compared to what?"

same way, Keyser says language—that is, the ability to convey information by speech—may have been a thinking and communicating "spandrel" accidentally created by the development of some cultural "arch." . . . "Language is very likely an accidental artifact of some evolutionary quirk of mind." [Robb 1991.]

Perhaps Keyser has been misquoted—I am always cautious about accepting any journalist's account of someone's words, having been burned badly myself—but if the quotation is accurate, then for Keyser spandrels are originally accidents, not necessities, don't-cares, or QWERTY phenomena. Once, when I was working at a bronze-casting foundry in Rome, we had an explosion in a cast as we were filling it; molten bronze went splashing all over the floor. One of the splashes hardened into a fantastic lacy shape that I promptly appropriated and turned into a sculpture. Was I exapting a spandrel? (The Dadaist artist Marcel Duchamp, in contrast, would not have been exapting a *spandrel* when he appropriated a urinal as his *objet trouvé* and called it a sculpture, since the urinal had a function in its earlier life.)

Gould himself (1993a, p. 31) has quoted this newspaper story with approval, not noticing that Keyser has the art history wrong, and not expressing any disagreement with Keyser's definition of a spandrel as an accident. So perhaps Keyser is right about the meaning of the term: spandrels are just accidents available for exaptation. Gould introduced "exaptation" in an article he co-authored with Elizabeth Vrba in 1982, "Exaptation: A Missing Term in the Science of Form." Their intent was to contrast exaptation to adaptation. Their chief stalking horse, however, was an astonishingly ill-favored term that had gained some currency in textbooks on evolution: *preadaptation*.

Preadaption seems to imply that the proto-wing, while doing something else in its incipient stages, knew where it was going—predestined for a later conversion to flight. Textbooks usually introduce the word and then quickly disclaim any odor of foreordination. (But a name is obviously ill-chosen if it cannot be used without denying its literal meaning.) [Gould 1991b, p. 144n.]

"Preadaptation" was a terrible term, for exactly the reasons Gould gives, but notice that he is not claiming that the targets of his criticism committed the major mistake of granting foresight to natural selection—he admits that they "quickly disclaimed" this heresy in the very act of introducing the term. They were making the minor mistake of choosing a usage perversely likely to foster this confusion. Switching from "preadaptation" to "exaptation" might well be seen, then, as a wise reform of usage, better suited to drive home the orthodox view of adaptationists. Gould, however, resisted this reformist interpretation. He wanted exaptation, and spandrels, to present a "potentially fundamental" and "non-Darwinian" alternative.

Elizabeth Vrba and I have proposed that the restrictive and confusing word "preadaptation" be dropped in favor of the more inclusive term "exaptation"—for any organ not evolved under natural selection for its current use—either because it performed a different function in ancestors (classical preadaptation) or because it represented a nonfunctional part available for later co-optation. [Gould 1991b, p. 144n.]

But, according to orthodox Darwinism, every adaptation is one sort of exaptation or the other—this is trivial, since no function is eternal; if you go back far enough, you will find that every adaptation has developed out of predecessor structures each of which either had some other use or no use at all. The only phenomena that Gould's exaptation revolution would rule out are the phenomena that orthodox adaptationists "quickly" disavowed in any case: planned-for preadaptations.

The spandrel revolution (against panadaptationism) and the exaptation revolution (against preadaptationism) evaporate on closer inspection, since both panadaptationism and preadaptationism have been routinely shunned by Darwinians ever since Darwin himself. These nonrevolutions not only do not challenge any orthodox Darwinian tenet; the coinages they introduce are as likely to confuse as the coinages they were supposed to replace.

It is hard to be a revolutionary if the establishment keeps co-opting you. Gould has often complained that his target, neo-Darwinism, recognizes the very exceptions he wants to turn into objections, "and this imposes a great frustration upon anyone who would characterize the modern synthesis in order to criticize it" (Gould 1980b, p. 130).

The modern synthesis has sometimes been so broadly construed, usually by defenders who wish to see it as fully adequate to meet and encompass current critiques, that it loses all meaning by including everything.... Stebbins and Ayala [two eminent defenders] have tried to win an argument by redefinition. The essence of the modern synthesis must be its Darwinian core. [Gould 1982a, p. 382.]

It is surprising to see a Darwinian give anything an essence, but we can take Gould's point, if not his language: there is *something* about the modern synthesis that he wants to overthrow, and before you can overthrow something you must pin it down. He has sometimes claimed (e.g., 1983a) he could see the modern synthesis doing his work for him, "hardening" into a brittle orthodoxy that would be easier to attack. If only! In fact, no sooner has he gone into battle than the modern synthesis has shown its flexibility, readily absorbing his punches, to his frustration. I think he is right, however, that the modern synthesis has a "Darwinian core," and I think he is right that it is his target; he just hasn't yet put his finger on it himself.

If the case against "pervasive adaptation" has vanished, then, what about the case against gradualism, the other main element in the modern synthesis that Gould sees "breaking down"? Gould's attempted revolution against gradualism was actually his first; it opened with a salvo in 1972 which introduced yet another familiar coinage to the vocabulary of evolutionists and onlookers alike: *punctuated equilibrium*.

3. PUNCTUATED EQUILIBRIUM: A HOPEFUL MONSTER

Punctuated equilibrium has finally obtained an unambiguous major-ity—that is, our theory is now 21 years old. We also, with parental pride (and therefore, potential bias), believe that primary controversy has ceded to general comprehension and that punctuated equilibrium has been accepted by most of our colleagues (a more conventional sort of majority) as a valuable addition to evolutionary theory.

—STEPHEN JAY GOULD and NILES ELDREDGE 1993,
p. 223

What needs to be said now, loud and clear, is the truth: that the theory of punctuated equilibrium lies firmly within the neo-Darwinian synthe-sis. It always did. It will take time to undo the damage wrought by the overblown rhetoric, but it will be undone.

—RICHARD DAWKINS 1986a, p. 251

Niles Eldredge and Gould co-authored the paper that introduced the term, "Punctuated Equilibria: An Alternative to Phyletic Gradualism" (1972). Whereas orthodox Darwinians, according to them, tended to envision all ev-olutionary change as gradual, they argued that, on the contrary, it proceeded by jerks: long periods of changelessness or *stasis*—equilibrium—interrupted by sudden and dramatic brief periods of rapid change—punctuations. The basic idea is often illustrated by contrasting a pair of trees of life (figure 10.6).

We can suppose that the horizontal dimension registers some one aspect of phenotypic variation or body design—we'd need a multidimensional space to represent it all, of course. The orthodox view on the left is pictured as showing that all motion through design space (that is, to the left or right in the diagram) is at a more or less steady pace. Punctuated equilibrium, in contrast, shows long periods of unchanged design (the vertical line seg-ments) interrupted by "instantaneous" sideways leaps in design space (the horizontal segments). To see the central claim of their theory, trace the evolutionary history of the species at K in each picture. The orthodox picture shows a more or less steady rightward trend from the diagram's Adam species, A. Their proposed alternative agrees that K is a descendant of

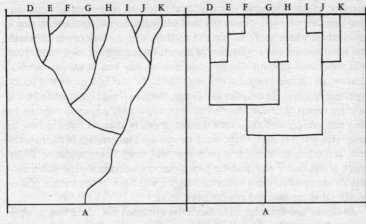

FIGURE 10.6

A, and that it accomplished the same rightward shift in Design Space in the same amount of time, but by fits and starts, not a steady climb. (These diagrams can be tricky to think about; the difference between a ramp and a staircase is the point of the contrast, but the giant steps are the *sideways* moves, not the vertical bits, which are the boring periods of "motion" through time only, with no motion through design space.)

There is a familiar trio of reactions by scientists to a purportedly radical hypothesis: (a) "You must be out of your mind!", (b) "What else is new? Everybody knows *that*!", and, later—if the hypothesis is still standing—(c) "Hmm. You *might* be on to something!" Sometimes these phases take years to unfold, one after another, but I have seen all three emerge in near synchrony in the course of a half-hour's heated discussion following a conference paper. In the case of the hypothesis of punctuated equilibrium, the phases are particularly pronounced, in large part because Gould has several times changed his mind about just what he and Eldredge were claiming. In its first appearance, the thesis of punctuated equilibrium was presented not as a revolutionary challenge at all, but as a conservative correction of an illusion to which orthodox Darwinians had succumbed: paleontologists were simply mistaken in thinking that Darwinian natural selection should leave a fossil record showing lots of intermediate forms.[4] There was no

---

4. "During the past thirty years, the allopatric theory [of speciation] has grown in popularity to become, for the vast majority of biologists, *the* theory of speciation" (Eldredge and Gould 1972, p. 92). This orthodox theory has some striking implications: "The theory of allopatric (or geographic) speciation suggests a different interpretation of

mention in the first paper of any radical theory of speciation or mutation. But later, about 1980, Gould decided that punctuated equilibrium was a revolutionary idea after all—not an explanation of the lack of gradualism in the fossil record, but a refutation of Darwinian gradualism itself. This claim was advertised as revolutionary—and now it truly was. It was too revolutionary, and it was hooted down with the same sort of ferocity the establishment reserves for heretics like Elaine Morgan. Gould backpedaled hard, offering repeated denials that he had ever meant anything so outrageous. In that case, responded the establishment, there is after all nothing new in what you say. But wait. Might there be still another reading of the hypothesis, according to which it is both true and new? There might be. Phase three is still under way, and the jury is out, considering several different— but all nonrevolutionary—alternatives. We will have to retrace the phases to see what the hue and cry has been about.

As Gould and Eldredge have themselves pointed out, there was an obvious problem of scale in such diagrams as figure 10.6. What if we zoomed way in on the orthodox picture and found that, once we enlarged it sufficiently, it looked like this:

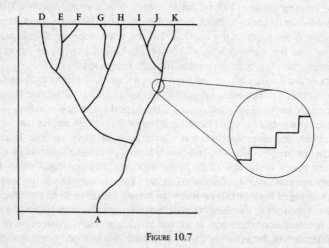

FIGURE 10.7

At *some* level of magnification, any evolutionary ramp must look like a staircase. Is figure 10.7 a picture of punctuated equilibrium? If it is, then orthodox Darwinism was already a theory of punctuated equilibrium. Even

---

paleontological data. If new species arise very rapidly in small, peripherally isolated local populations, then the great expectation of insensibly graded fossil sequences is a chimera" (Eldredge and Gould 1972, p. 82).

the most extreme gradualist can allow that evolution could take a breather for a while, letting the vertical lines extend indefinitely through time until some new selection pressure somehow arose. During this period of stasis, selection pressure would be conservative, keeping the design roughly constant by swiftly eliminating any experimental alternatives that arose. As the old mechanic said, "Don't fix what isn't broke." Whenever a new selection pressure arose, we'd see a "sudden" response of heightened evolution, a punctuation interrupting the equilibrium. Was there really a revolutionary point of disagreement being offered by Eldredge and Gould here, or were they merely offering an interesting observation about the variability in tempo of evolutionary processes and its predictable effects on the fossil record?

Punctuationists typically draw the punctuation parts of their revolutionary diagrams absolutely horizontal (to make strikingly clear that they are presenting a true alternative to the rampant ramp-view of orthodoxy). This makes it look as if each of the design revisions illustrated takes place in a twinkling, in no time at all. But that is just a misleading artifact of the huge vertical scale adopted, which shows millions of years to the inch. The sideways motion is not really instantaneous. It is only "geologically instantaneous."

> An isolated population may take a thousand years to speciate, and its transformation would therefore appear glacially slow if measured by the irrelevant scale of our personal lives. But a thousand years, appropriately recorded in geological time, is only an unresolvable moment, usually preserved on a single bedding plane [in fossil-bearing rock], in a lifetime of species that often live for several million years in stasis. [Gould 1992a, pp. 12–14.]

So suppose we zoom in on one of these thousand-year instants, changing the vertical scale of the time dimension by a few orders of magnitude to see what might actually be going on (figure 10.8). The horizontal step taken between time t and time t′ will have to be stretched out somehow, and we must turn it into relatively big steps or little steps or tiny steps, or some combination thereof.

Were any of the possibilities revolutionary? What exactly were Eldredge and Gould maintaining? Here their respective views diverged somewhat, at least for a while. The view was revolutionary, Gould claimed, because it maintained that the punctuations were not just business-as-usual evolution, not just *gradual* changes. Remember the old joke about the drunk who falls down the elevator shaft and says, on rising, "Look out for the first step—it's a doozy!"? For a while, Gould was proposing that the first step in the establishment of any new species was a doozy—a non-Darwinian *saltation* ("somersault" and "sauté" come from the same Latin root):

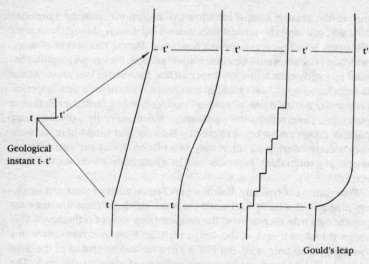

Gould's leap

FIGURE 10.8

Speciation is not always an extension of gradual, adaptive allelic substitution to greater effect, but may represent, as Goldschmidt argued, a different style of genetic change—rapid reorganization of the genome, perhaps nonadaptive. [Gould 1980b, p. 119.]

Speciation itself, in this view, is not an effect of accumulated adaptations gradually driving populations apart but, rather, a cause with its own, non-Darwinian explanation:

But in saltational, chromosomal speciation, reproductive isolation comes first and cannot be considered as an adaptation at all. . . . We can, in fact, reverse the conventional view and argue that speciation, by forming new entities stochastically, provides raw material for selection. [Gould 1980b, p. 124.]

This suggestion, which I call Gould's leap, is represented in the right-hand graph in figure 10.8. Only part of the punctuation process, the gradual, cleaning-up process at the end, is "Darwinian," Gould claimed:

If new *Baupläne* often arise in an adaptive cascade following the saltational origin of a key feature, then part of the process is sequential and adaptive, and therefore Darwinian; but the initial step is not, since selection does not play a creative role in building the key feature. [Gould 1982a, p. 383.]

It is this "creative role" of something other than selection that caught the skeptical attention of Gould's colleagues. To get clear about what caused the furor, we need to note that our diagram in figure 10.8 is really unable to distinguish several crucially different hypotheses. The trouble with the diagram is that it needs more dimensions, so we can compare the steps in *genotype space* (the typographical steps in the Library of Mendel) to the steps in *phenotype space* (the design innovations in Design Space) and then evaluate these differences on a *fitness landscape*. As we have seen, the relations between recipe and result are complex, and many possibilities might be illustrated. We saw in chapter 5 that a small typographical change in the genome could in principle have a large effect on the phenotype expressed. We also saw, in chapter 8, that some typographical changes in the genome can have no effect at all on the phenotype—there are over a hundred different ways of "spelling" lysozyme, for instance, and hence more than a hundred equivalent ways of spelling the order for lysozyme in DNA codons. We know, then, that at one extreme there can be organisms so similar in design as to be indistinguishable that nevertheless have large differences in their DNA—for instance, you and whoever that person is for whom you are often mistaken (your *Doppelgänger*—no philosophy book would be complete with mentioning doppelgängers). At the other extreme, there can be organisms that are bizarrely different in appearance, but almost identical genetically. A single mutation in just the wrong place can produce a monster—the medical term for such deformed offspring is *terata*, Greek (and Latin) for "monsters." And there can also be organisms that are almost identical in appearance and structure, and almost identical in DNA, but dramatically different in fitness—for instance, fraternal twins one of whom happens to have a gene that gives it either immunity or susceptibility to some disease.

A large leap in any of these three spaces, or a saltation, may also be called a *macromutation* (meaning a big mutation, not just a mutation in what I have called a macro—a macromolecular subsystem).[5] As Ernst Mayr (1960) has observed, there are three different reasons we could call a mutation big: it is a big step in the Library of Mendel; it produces a radical difference in phenotype (a monster); it produces (one way or another) a big increase in fitness—a lot of *lifting*, in our metaphor for *good work done* by design changes.

It is possible for the molecular replicating machinery to take large steps in the Library of Mendel—there are cases in which whole chunks of text get transposed, inverted, or deleted in a single copying "mistake." It is also

5. For an introduction to the term, see Dietrich's essay "Macromutation" in the excellent new sourcebook, *Keywords in Evolutionary Biology*, edited by Keller and Lloyd (1992).

possible for typographical differences to accumulate slowly (and, in general, randomly) over a long time in the large portion of DNA that never gets expressed, and if these accumulated changes suddenly got expressed, thanks to some transposing error, a huge phenotypic effect would be expected. But it is only when we turn to the third sense of macromutation—large differences in fitness—that we can get clear about what seemed to be radical in Gould's proposal. The terms "saltation" and "macromutation" have tended to be used to describe a successful move, a *creative* move, in which offspring in a single generation shift from one region of Design Space to another *and prosper as a result*. The idea had been promoted by Richard Goldschmidt (1933, 1940), and made unforgettable by his catchphrase: "hopeful monsters." What made his work notorious was that he claimed that such leaps were necessary for speciation to occur.

This suggestion had been roundly rejected by neo-Darwinian orthodoxy, for the reasons we have already considered. Even before Darwin, the received wisdom of biologists was, as Linnaeus said in his classic work of taxonomy (1751), "*Natura non facit saltus*"—nature does not make leaps—and this was one maxim that Darwin didn't just leave untouched; he provided enormous support for it. Large leaps sideways *in a fitness landscape* will almost never be to your benefit; wherever you currently find yourself, you are where you are because this has been a good region of Design Space for your ancestors—you are near the top of some peak or other in the space—so, the bigger the step you take (jumping randomly, of course), the more likely you are to jump off a cliff—into the low country, in any case (Dawkins 1986a, ch. 9). According to this standard reasoning, it is no accident that monsters are virtually always hopeless. That is what made Goldschmidt's views so heretical; he knew and accepted that this was true in general, but proposed that nevertheless the extremely rare exceptions to this rule were the main lifters of evolution.

Gould is a famous defender of underdogs and outcasts, and he deplored the "ritualistic ridicule" (Gould 1982b, p. xv) to which Goldschmidt had been subjected by the orthodox. Was Gould going to try to rehabilitate Goldschmidt? Yes and no. In "Return of the Hopeful Monster" (in Gould 1980a, p. 188), Gould complained that "defenders of the synthetic theory made a caricature of Goldschmidt's ideas in establishing their whipping boy." So it seemed to many biologists that Gould was arguing that punctuated equilibrium was a theory of Goldschmidtian speciation through macromutation. To them it seemed that Gould was trying to wave his wonderful historian's wand over the tarnished reputation of Goldschmidt, and bring his ideas back into favor. Here the mythic Gould, Refuter of Orthodoxy, seriously got in the way of the real Gould, so that even his colleagues succumbed to the temptation to read what he wrote with a broad brush. They scoffed in disbelief, and then, when he denied that he was endorsing—

had ever endorsed—Goldschmidt's saltationism, they scoffed all the more derisively. They knew what he'd said.

But did they? I must admit that I thought they did until Steve Gould insisted to me that I should check *all* his various publications, and see for myself that his opponents were foisting a caricature on him. He struck a nerve; no one knows better than I how frustrating it is to have the skeptics hang a crude but convenient label on one's subtle view. (I'm the guy who reputedly denies that people experience colors or pains, and thinks that thermostats think—just ask my critics.) So I checked. He chose to dub his denial of gradualism "the Goldschmidt break" (Gould 1980b), and recommended for serious consideration—without endorsing—some radical Goldschmidtian views, but in the same paper he was careful to say, "We do not now accept all his arguments about the nature of variation." In 1982, he made it clear that the only feature of Goldschmidt's view he was endorsing was the idea of "small genetic changes producing large effects by altering rates of development" (Gould 1982d, p. 338), and in his introduction to the reprinting of Goldschmidt's notorious book, he expanded on this point:

> Darwinians, with their traditional preferences for gradualism and continuity, might not shout hosannas for large phenotypic shifts induced rapidly by small genetic changes that affect early development; but nothing in Darwinian theory precludes such events, for the underlying continuity of small genetic changes remains. [Gould 1982b, p. xix.]

Nothing revolutionary, in other words:

> One may be excused for retorting: "so what else is new?" Has any biologist ever denied it? But ... progress in science often demands the recovery of ancient truths and their rendering in novel ways. [Gould 1982d, pp. 343–44.]

Still, he could not resist the urge to describe this possibly underappreciated fact about development as a *non*-Darwinian creative force in evolution, "because the constraints that it imposes upon the nature of phenotypic change guarantee that small and continuous Darwinian variation is not the raw material of all evolution," for it "relegates selection to a negative role (eliminating the unfit) and assigns the major creative aspect of evolution to variation itself" (Gould 1982d, p. 340).

It is still not clear how much importance to assign to this possibility in principle, but, in any event, Gould has not pursued it further: "Punctuated equilibrium is not a theory of macromutation" (Gould 1982c, p. 88). Confusion on this score still abounds, however, and Gould has had to keep issuing his disclaimers: "Our theory entails no new or violent mechanism, but only represents the proper scaling of ordinary events into the vastness of geological time" (Gould 1992b, p. 12).

So this was a false-alarm revolution that was largely if not entirely in the eyes of the beholders. But in that case, the view we find Gould and Eldredge maintaining once we zoom in beyond the misleading time-squashing of their geological diagrams, is not the rightmost doozy in figure 10.8 after all, but one of the other, gradual, nonviolent paths illustrated. As Dawkins has noted, the way in which Eldredge and Gould challenged "gradualism" was not, in the end, by positing some exciting new *non*gradualism, but by saying that evolution, when it occurred, was indeed gradual—but most of the time it was *not even* gradual; it was at a dead stop. The lefthand diagram in figure 10.6 is supposed to represent orthodoxy, but the feature of it that their theory challenged was not its gradualism—once we get the scale right, they are gradualists themselves. The feature they were challenging was what Dawkins (1986a, p. 244) calls "constant speedism."

Now, has neo-Darwinian orthodoxy ever been committed to constant speedism? In their original paper, Eldredge and Gould claimed that paleontologists were mistaken in thinking that orthodoxy required constant speedism. Was Darwin himself a constant speedist? Darwin often, and correctly, harped on the claim that evolution could *only* be gradual (at best, you might say). As Dawkins (1986a, p. 145) says, "For Darwin, any evolution that had to be helped over the jumps by God was not evolution at all. It made a nonsense of the central point of evolution. In the light of this, it is easy to see why Darwin constantly reiterated the *gradualness* of evolution." But documentary evidence in support of the claim that he was committed to *constant speedism* is not just hard to find; there is a famous passage in which Darwin clearly expresses the opposite view, the view that could be called—in two words—punctuated equilibrium:

> Many species once formed never undergo any further change ...; and the periods, during which species have undergone modification, though long as measured by years, have probably been short in comparison with the periods during which they retain the same form. [*Origin*, 4th and subsequent eds; see Peckham 1959, p. 727.]

Ironically, however, Darwin put just one diagram in *Origin*, and it happens to show steadily sloping ramps. Steven Stanley, another major exponent of punctuated equilibrium, reprints this diagram in his book (Stanley 1981, p. 36) and makes the inference explicit in his caption.

One effect of such claims is that today there is undoubtedly a tradition imputing constant speedism either to Darwin himself or to neo-Darwinian orthodoxy. For instance, Colin Tudge, a good science journalist, writing about Elizabeth Vrba's recent claims concerning the pulse of evolution,

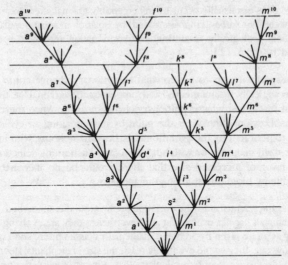

The tree of life published by Darwin in the *Origin* (1859, p. 117). The tree depicts a gradualistic pattern of evolution. Each fanlike pattern represents the slow evolutionary divergence of populations. Darwin believed that new species, and eventually new genera and families, formed by this kind of slow divergence. [Stanley 1981.]

FIGURE 10.9

points to the presumed implications for orthodoxy of current research on the evolution of impalas and leopards:

> Traditional Darwinism would predict a steady modification of the impala over 3 million years, even without climatic change, because it still needs to outrun leopards. But, in fact, neither impalas nor leopards have changed very much. They are both too versatile to be worried by climatic change, and competition between them and with their own kind does not—as Darwin supposed—provide sufficient selective pressure to cause them to alter. [Tudge 1993, p. 35.]

Tudge's presumption that the discovery of three million years of stasis in the impala and leopard would confound Darwin is a familiar one, but it is an artifact, direct or indirect, of a particular forced reading of the "ramps" in Darwin's (and other orthodox) diagrams.

Gould has proclaimed the death of gradualism, but is he himself, then, a gradualist (but not a constant speedist) after all? His denial that his theory proposes any "violent mechanism" suggests that he is, but it is hard to tell, for on the very same page he says that, according to the theory of punctuated equilibrium,

change does not usually occur by imperceptibly gradual alteration of entire species but *rather* [emphasis added] by isolation of small populations
and their geologically instantaneous transformation into new species.
[Gould 1992a, p. 12.]

This passage invites us to believe that evolutionary change *could not be*
both "geologically instantaneous" and "imperceptibly gradual" at the same
time. But that is just what evolutionary change *must* be when there are no
saltations. Dawkins dramatizes the point by passing along an eye-opening
thought experiment by the evolutionist G. Ledyard Stebbins, who imagines
a mouse-sized mammal for which he postulates such a tiny selection pressure in favor of increased size that there would be no increase in size
measurable by biologists studying the animal:

> As far as the scientist studying evolution on the ground is concerned, then,
> these animals are not evolving at all. Nevertheless they are evolving, very
> slowly at a rate given by Stebbins' mathematical assumption, and even at
> this slow rate, they would eventually reach the size of elephants. How long
> would this take? ... Stebbins calculates that at his assumed very slow rate
> of evolution, it would take about 12,000 generations.... Assuming a gen-
> eration time of five years, which is longer than that of a mouse but shorter
> than that of an elephant, 12,000 generations would occupy about 60,000
> years. 60,000 years is too *short* to be measured by ordinary geological
> methods of dating the fossil record. As Stebbins says, 'The origin of a new
> kind of animal in 100,000 years or less is regarded by paleontologists as
> "sudden" or "instantaneous".' [Dawkins 1986a, p. 242.]

Certainly Gould would not call such a locally imperceptible mouse-to-
elephant change a violation of gradualism, but in that case his own opposition
to gradualism is left with no support at all from the fossil record. In fact, he
grants this (1982a, p. 383)—the only evidence that his own field of paleontology is able to provide in opposition to gradualism goes in the wrong direction. Gould may hanker for evidence of a revolutionary speed-up of one
kind or another, but the fossil record could only show periods of stasis that
suggest that evolution is often not even gradual.

But perhaps *this* awkward fact can be turned to good use: perhaps the
challenge to orthodoxy should be, not that it can't account for the punctuations, but that it can't account for the equilibria! Perhaps Gould's challenge to the modern synthesis should be that it *is* committed to constant
speedism after all: that, although Darwin didn't positively deny equilibrium
(indeed, he asserted that it occurs), he can't actually explain equilibrium
when it does occur, and such equilibrium or stasis, it might be argued, is a
major pattern in the world in need of explanation. This is in fact the next
direction Gould turned in his attack on the modern synthesis.

How can we claim to understand evolution if we only study the percent or two of phenomena that construct life's directional history and leave the vast field of straight-growing bushes—the story of most lineages most of the time—in the limbo of conceptual oblivion? [Gould 1993c, p. 16.]

But this path has problems. First, we must be careful not to make the error that is the mirror image of Gould's error about *pan*adaptationism. We must not make the error of "panequilibriumism." However striking or "pervasive" the pattern of stasis turns out to be, we know in advance that most *lineages* do not exhibit stasis. Far from it. Remember our difficulties in coloring in Lulu and her conspecifics in chapter 4? Most lineages soon die out, never having time to establish stasis; we will only "see" a species where there is something salient and stable in the record. The "discovery" that all species exhibit stasis much of the time is like the discovery that all droughts last longer than a week. We wouldn't notice that there *was* a drought if it wasn't a long-lasting phenomenon. So, since a modicum of stasis is a precondition for the identification of a species, the fact that all species exhibit some degree of stasis is merely true by definition.

Nevertheless, the phenomenon of stasis might be a real one in need of explanation. We should ask not why species exhibit stability (something true by definition) but why there *are* salient, identifiable species—that is, why lineages go stable at all. But even here, neo-Darwinism has several obvious adaptationist explanations for why stasis should often occur in a lineage. We have already seen the primary one several times: every species is—must be—a going concern, and going concerns must be conservative; most deviations from the time-tested tradition will be quickly punished by extinction. Eldredge himself (1989) has suggested that a major reason for stasis is "habitat tracking." Sterelny (1992, p. 45) describes it this way:

As the environment changes, organisms may react by tracking their old habitat. They might move north as the climate cools, rather than by evolving adaptations to the cold. [This is not a mistake—Sterelny is a *Southern Hemisphere* philosopher of biology!] Selection will usually drive tracking. For migrants that follow the habitat (personally or by reproductive dispersion) will typically be fitter than the population fragment that fails to move, for the residual fragment will be less well adapted to the new environment and will be faced with new competition from other migrants tracking their old habitat.

Note that habitat tracking is as much a "strategy" of plants as of animals. Indeed, some of the clearest cases of speciation invoke this phenomenon. As the icecap recedes after an ice age, the range of some Northern Asian plant spreads to the north year after year, "following" the ice, and spreading east and west as it goes, crossing in the Bering Strait region, and perhaps even encircling the globe like the herring gulls. Then, as the ice advances south during the *next* ice age, it sheers off the connections between the Asian and North American parts of the family, creating two isolated ranges that then naturally diverge into distinct species, but as they both move southward in their respective hemispheres, they continue to look much the same, *because* they track their favored climatic conditions, instead of staying put and going in for further winter adaptations.[6]

Another possible explanation of punctuated equilibrium is purely theoretical. Stuart Kauffman and his colleagues have produced computer models that exhibit behavior in which relatively long periods of stasis are interrupted by brief periods of change not triggered by any "outside" interference, so this pattern seems to be an endogenous or internal feature of the operation of particular sorts of evolutionary algorithms. (For a recent discussion, see Bak, Flyvbjerg, and Sneppen 1994.)

It is quite clear, then, that equilibrium is no more a problem for the neo-Darwinian than punctuation; it can be accounted for, and even predicted. But Gould has seen yet another revolution lurking in punctuated equilibrium. Maybe the horizontal steps of punctuation are not just (relatively) rapid steps in Design Space; maybe what is important about them is that they are steps of *speciation*. How could this make a difference? Look at figure 10.10.

In both cases, the lineage at K got where it got by exactly the same sequence of punctuations and equilibria, but the case illustrated on the left shows a *single species* undergoing rapid periods of change followed by long periods of stasis. Such change without speciation is known as *anagenesis*. The case illustrated on the right is an instance of *cladogenesis*, change via speciation. Gould claims that the rightward trend in the two cases would have a different explanation. But how could this be true? Recall what we

---

6. George Williams (1992, p. 130) disputes the importance of habitat tracking in stasis, noting that parasites, "seasonal amplitude of insolation" (amount of sunshine), and many other environmental factors would always be different after a geographical move, so that populations would never be able to stay in exactly the same selective environment, and hence would be subjected to selection pressure in spite of moving. But it seems to me that much if not all of the adjustment to these selection pressures could be invisible to paleontology, which can only see in the fossil record the preserved changes in hard-part design. Habitat tracking could be responsible for much of the *paleontologically observable* stasis (and what other stasis do we know about?), even if Williams is right that this body-plan stasis would have to mask concurrent nonstasis at most if not all other design levels in response to the many environmental changes that would have to accompany any long-range habitat-tracking moves. And unless many species *moved in unison* in their habitat tracking, there couldn't be habitat tracking at all, since other species are such crucial elements in any species' selective environment.

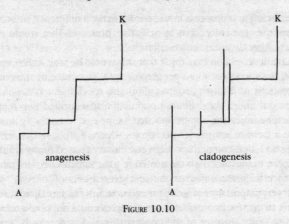

anagenesis   cladogenesis

A   A

FIGURE 10.10

learned in chapter 4: speciation is an event that can only be retrospectively identified. Nothing that happens *during the sideways move* could distinguish an anagenetic process from a cladogenetic process. There has been speciation only if there is a *later* flourishing of separate branches that survive long enough to be identifiable as separate species.

Couldn't there be special processes of what we might call *hopeful* speciation—or *incipient* speciation? Consider a case in which speciation does occur. Parent-species A splits into daughter-species B and C.

Now wind back the tape just far enough in time to drop a bomb (an asteroid, a tidal wave, a drought, poison) on the earliest members of the B species, as in the middle diagram. Doing this turns what had been a case of speciation into something indistinguishable from a case of anagenesis (on the right). The fact that the bomb prevents those whose offspring it kills

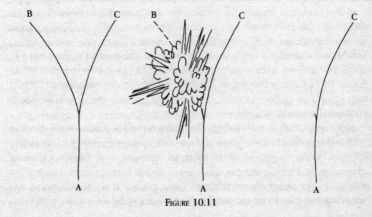

B   C   B   C   C

A   A   A

FIGURE 10.11

from ever being grandparents could hardly make a difference to how their contemporaries got sorted out by selective pressure. That would require some sort of backward-in-time causation.

Is this really true? You may think that this would be true if the event that kicked off the speciation was a geographic split, guaranteeing the complete causal isolation of the two groups (allopatric speciation), but what if the speciation got under way within a population that formed two reproductively incommunicative subgroups that competed directly against each other (in a form of sympatric speciation)? Darwin proposed, as we noted (see page 43), that competition between closely related forms would be a driving force in speciation, so the presence—the nonabsence—of *what can be retrospectively identified as* the first generations of a "rival" species might be very important indeed for speciation, but the fact that these rivals are "going to be" the founders of a new species could not play a role in the intensity or other features of the competition and hence in the speed or direction of the horizontal motion in Design Space.

We may well suppose that relatively rapid morphological change (sideways movement) is a normally necessary precondition for speciation. Rapidity of change is crucially affected by the size of the gene pool; large gene pools are conservative and tend to absorb innovation attempts without a trace. One way of making a large gene pool small is dividing it in two, and this may in fact be the most common sort of downsizing, but thereafter it makes no difference whether or not nature discards one of the halves (as in the middle graph in figure 10.11). It is the bottleneck of a diminished gene pool that permits the rapid motion, not the presence of two or more different bottlenecks. If there is speciation, then *two whole species* pass through their respective bottlenecks; if there is no speciation, then *one whole species* is pressed through a single bottleneck. So cladogenesis *cannot* involve a process during a punctuation period different from the process that occurs in anagenesis, because the difference between cladogenesis and anagenesis is definable only in terms of postpunctuation *sequelae*. Gould sometimes speaks as if speciation does make a difference. For instance, Gould and Eldredge (1993, p. 225) speak of "the crucial requirement of ancestral survival after punctuated branching" (as shown on the left in figure 10.11), but according to Eldredge (personal communication) this is only a crucial *epistemological* requirement for the theorist, who needs "ancestral survival" as *evidence* of descent.

His explanation is interesting. The fossil record is loaded with cases in which one form abruptly stops and another, quite different, form abruptly appears "in its place." Which of these are cases of swift sideways leaps of evolution, and which are cases of simple displacement due to sudden immigration of a rather distant relative? You can't tell. It is only when you can see what you take to be the parent species coexisting for a while with what you take to be its offspring that you can be quite sure that there is a direct path from the earlier form to the later form. As an epistemological point, this completely undercuts the claim that Gould has wanted to make: that most swift evolutionary change has been accomplished by speciation. For if, as Eldredge says, the fossil record *usually* shows abrupt shifts *without* any "ancestral survival after punctuated branching," and if there is no telling which of these are cases of punctuated anagenetic change (as opposed to immigration phenomena), then there is no way of telling from the fossil record whether speciation is a very frequent or very rare accompaniment of rapid morphological change.[7]

There might still be another way of making sense of Gould's insistence that it is speciation, not mere adaptation, that makes the big difference in evolution. What if it turned out that some lineages go in for a lot of punctuation (and, in the process, produce lots of daughter species) and other lineages do not, and those that do not do so tend to die out? Neo-Darwinians usually assume that adaptations occur by the gradual transformation of the organisms in particular lineages, but "if lineages do not change by transformation, then long term trends in lineages can hardly be the result of their slow transformation" (Sterelny 1992, p. 48). This has long been considered an interesting possibility (in their original article, Eldredge and Gould discuss it very briefly, and credit Sewall Wright 1967 as one of its sources). Gould's version of the idea (e.g., 1982a) is that whole species don't get revised by the piecemeal redesign of their individual members; species are rather brittle, unchanging things; the shifts in Design Space happen (largely? often? always?) because of species *extinction* and species *birth*. This idea is what Gould and Eldredge (1993, p. 224) call "higher level sorting." It is sometimes known as species selection, or *clade* selection. It is hard to get clear about, but we have the equipment already at hand to clarify its central point. Remember bait-and-switch? Gould is in effect proposing a new application of this fundamental Darwinian idea: don't think that evolution *makes adjustments* in existing lineages; evolution *throws away* whole lineages, and lets other, different, lineages prosper. It looks as if there are adjustments to lineages over time, but what is really going on is bait-and-switch at the species level. The *right level* at which to look for evolutionary trends, he could then claim, is not the level of the gene, or the organism, but the whole species or clade. Instead of looking at the loss of particular genes from gene pools, or the differential death of particular genotypes within a

---

7. For a similar criticism of Gould, see Ayala 1982. See also G. Williams 1992, pp. 53–54; Williams, who defines cladogenesis as the isolation, however short-lived, of any gene pool, also points out the triviality to evolutionary theory of short-term cladogenesis (pp. 98–100).

population, look at the differential extinction rate of whole species and the differential "birth" rate of species—the rate at which a lineage can speciate into daughter species.

This is an interesting idea, but it is not, as it first appears to be, a denial of the orthodox claim that whole species undergo transformation via "phyletic gradualism." Let it be true, as Gould proposes, that some lineages spawn lots of daughter species and others don't, and that the former tend to survive longer than the latter. Look at the trajectory through Design Space of each surviving species. It, the whole species, is at any period of time either in stasis or undergoing punctuated change, but that change itself is a "slow transformation of a lineage," after all. It may be true that the best way of seeing the long-term macroevolutionary pattern is to look for differences in "lineage fecundity" instead of looking at the transformations in the individual lineages. This is a powerful proposal worth taking seriously, but it neither refutes nor supplants gradualism; it builds on it.[8]

(The level shift Gould proposes reminds me of the level shift between hardware and software in computer science; the software level is the right level at which to answer certain large-scale questions, but it does not cast any doubt on the truth of the explanations of the same phenomena at the hardware level. You would be foolish to try to explain the visible differences between WordPerfect and Microsoft Word at the hardware level, and perhaps you would be foolish to try to explain some of the visible patterns of diversity in the biosphere by concentrating on the slow transformations of the various lineages, but that does not mean that they did not undergo slow transformations at various punctuation marks in their history.)

The relative importance of species selection of the sort Gould is now proposing has not yet been determined. And it is clear that however large a role species selection comes to play in the latest versions of neo-Darwinism, it is no skyhook. After all, the way new lineages come onto the scene as candidates for species selection is by standard gradualistic micromutation—unless Gould does want to embrace hopeful monsters. So Gould may have helped discover a new crane, if that is what it turns out to be: a heretofore unrecognized or unappreciated mechanism of design innovation, built out of the standard, orthodox mechanisms. Since my diagnosis, however, is that he has all along been hoping for skyhooks, not cranes, I must predict that he will keep on looking. Could there perhaps be something *else* about speciation that is so special that it cannot be handled by

---

8. Gould's ideas about "higher level species sorting" must be distinguished from some lookalike neighbors: the ideas about group selection or population selection currently under intense and controversial scrutiny among evolutionists. Those ideas will be discussed in the next chapter.

neo-Darwinism? Darwin's account of speciation, as we have just recalled, invoked competition between close relatives.

> New species usually win an address by driving out others in overt competition (a process that Darwin often described in his notebooks as "wedging"). This constant battle and conquest provides a rationale for progress, since victors, on average, may secure their success by general superiority in design. [Gould 1989b, p. 8.]

Gould does not like this image of the wedge. What is wrong with it? Well, it invites (he claims) a belief in progress, but this invitation, we have already seen, is as easily declined by neo-Darwinism as it was by Darwin himself. Global, long-term progress, amounting to the view that things in the biosphere are, in general, getting better and better and better, was denied by Darwin, and although it is often imagined by onlookers to be an implication of evolution, it is simply a mistake—a mistake no orthodox Darwinians fall for. What else might be wrong with the image of the wedge? Gould speaks in the same article (p. 15) of "the plodding predictability of the wedge," and I suggest that this is exactly what offends him in the image: like the ramp of gradualism, it suggests a sort of *predictable, mindless trudge* up the slopes of Design Space (see, e.g., Gould 1993d, ch. 21). The trouble with a wedge is simple: it is not a skyhook.

4. Tinker to Evers to Chance: The Burgess Shale Double-Play Mystery[9]

Even today a good many distinguished minds seem unable to accept or even understand that from a source of noise natural selection alone and unaided could have drawn all the music of the biosphere. In effect natural selection operates upon the products of chance and can feed nowhere else; but it operates in a domain of very demanding conditions, and from this domain chance is barred.

—Jacques Monod 1971, p. 118

---

9. "Tinker to Evers to Chance" is a baseball meme, immortalizing the double-play combination of three Hall of Fame infielders, Joe Tinker, Johnny (the Crab) Evers, and Frank Chance, who played together for Chicago in the National League from 1903 to 1912. In 1980, Richard Stern, a freshman in my introductory philosophy course, wrote an excellent essay for me, an update on Hume's *Dialogues*, this time between a Darwinian (Tinker, of course) and a believer in God (Evers, of course), ending up, appropriately, with Chance. The serendipitous multiple convergences of that title, given Gould's own encyclopedic knowledge and love of baseball, is simply irresistible.

> *But modern punctuationalism—especially in its application to the va-*
> *garies of human history—emphasizes the concept of contingency: the*
> *unpredictability of the nature of future stability, and the power of*
> *contemporary events and personalities to shape and direct the actual*
> *path taken among myriad possibilities.*
>
> —STEPHEN JAY GOULD 1992b, p. 21

Gould speaks here not just of unpredictability but of the power of contemporary events *and personalities* to "shape and direct the actual path" of
evolution. This echoes exactly the hope that drove James Mark Baldwin to
discover the effect now named for him: *somehow* we have to get personalities—consciousness, intelligence, agency—back in the driver's seat. If we
can just have contingency—radical contingency—this will give the *mind*
some elbow room, so it can *act*, and be *responsible* for its own destiny,
instead of being the mere effect of a mindless cascade of mechanical processes! This conclusion, I suggest, is Gould's ultimate destination, revealed
in the paths he has most recently explored.

I mentioned in chapter 2 that the main conclusion of Gould's *Wonderful
Life: The Burgess Shale and the Nature of History* (1989a) is that if the tape
of life were rewound and played again and again, chances are mighty slim
that *we* would ever appear again. There are three things about this conclusion that have baffled reviewers. First, why does he think it is so important?
(According to the dust jacket, "In this masterwork Gould explains why the
diversity of the Burgess Shale is important in understanding this tape of our
past and in shaping the way we ponder the riddle of existence and the
awesome improbability of human evolution.") Second, exactly what *is* his
conclusion—in effect, who does he mean by "we"? And, third, how does he
think this conclusion (whichever one it is) follows from his fascinating
discussion of the Burgess Shale, to which it seems almost entirely unrelated?
We will work our way through these questions from third to second to
first.[10]

Thanks to Gould's book, the Burgess Shale, a mountainside quarry in
British Columbia, has now been elevated from being a site famous among
paleontologists to the status of an international shrine of science, the birthplace of . . . well, Something Really Important. The fossils found there date
from the period known as the Cambrian Explosion, a time around six hundred million years ago when the multicellular organisms really took off,
creating the palm branches of the Tree of Life of figure 4.1. Formed under
peculiarly felicitous conditions, the fossils immortalized in the Burgess Shale

---

10. Yes, I know, Joe Tinker played shortstop, not third base, but cut me a little slack,
please!

are much more complete and three-dimensional than fossils usually are, and their classification by Charles Walcott early in this century was guided by his literal dissection of some of the fossils. He shoehorned the varieties he found into traditional phyla, and so matters stood (roughly) until the brilliant reinterpretations in the 1970s and 1980s by Harry Whittington, Derek Briggs, and Simon Conway Morris, who claimed that many of these creatures—and they are an astonishingly alien and extravagant lot—had been misclassified; they actually belonged to phyla that have no modern descendants at all, phyla never before imagined.

That is fascinating, but is it revolutionary? Gould (1989a, p. 136) certainly thinks so: "I believe that Whittington's reconstruction of *Opabinia* in 1975 will stand as one of the great documents in the history of human knowledge." His trio of heroes didn't put it that way (see, e.g., Conway Morris 1989), and their caution has proven to be prophetic; subsequent analyses have tempered some of their most radical reclassificatory claims after all (Briggs et al. 1989, Foote 1992, Gee 1992, Conway Morris 1992). If it weren't for the pedestal Gould had placed his heroes on, they wouldn't now seem to have fallen so far—the first step was a doozy, and they didn't even get to take the step for themselves.

But in any case, what was the revolutionary point that Gould thought was established by what we may have learned about these Cambrian creatures? The Burgess fauna appeared suddenly (and remember what that means to a geologist), and most of them vanished just as suddenly. This nongradual entrance and exit, Gould claims, demonstrates the fallacy of what he calls "the cone of increasing diversity," and he illustrates his claim with a remarkable pair of trees of life.

A picture is worth a thousand words, and Gould emphasizes again and again, with many illustrations, the power of iconography to mislead even the expert. Figure 10.12 is another example, and it is his own. On the top, he tells us, is the old false view, the cone of increasing diversity; on the bottom, the improved view of decimation and diversification. But notice that you can turn the bottom picture into a cone of increasing diversification by simply *stretching the vertical scale*. (Alternatively, you can turn the top picture into a new and approved icon of the bottom sort by squeezing the vertical scale down, in the style of standard punctuated-equilibrium diagrams—e.g., on the right in figure 10.6.) Since the vertical scale is arbitrary, Gould's diagrams don't illustrate any difference at all. The bottom half of his lower diagram perfectly illustrates a "cone of increasing diversity," and who knows whether the very next phase of activity in the top diagram would be a decimation that turned it into a replica of the bottom diagram.

The cone of increasing diversity is obviously not a fallacy, if we measure diversity by the number of different species. Before there were a hundred species there were ten, and before there were ten there were two, and so

The Cone of Increasing Diversity

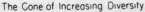
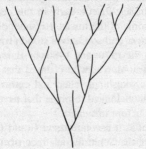

Decimation and Diversification

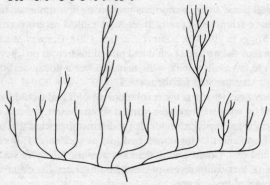

The cone of increasing diversity. The false but still conventional iconography of the cone of increasing diversity, and the revised model of diversification and decimation, suggested by the proper reconstruction of the Burgess Shale. [Gould 1989a, p. 46.]

FIGURE 10.12

it must be, on every branch of the Tree of Life. Species go extinct all the time, and perhaps 99 percent of all the species that have ever existed are now extinct, so we must have plenty of decimation to balance off the diversification. The Burgess Shale's flourishing and demise may have been less gradual than that of other fauna, before or since, but that does not demonstrate anything radical about the shape of the Tree of Life.

Some say this misses Gould's point: "What is special about the spectacular diversity of the Burgess Shale fauna is that these weren't just new *species*, but *whole new phyla*! These were *radically* novel designs!" I trust this was never Gould's point, because if it was, it was an embarrassing fallacy of retrospective coronation; as we have already seen, *all* new phyla—indeed, new kingdoms!—have to start out as mere new subvarieties and then become new species. The fact that from today's vantage point they appear to be early members of new *phyla* does not in itself make them special at all. They *might* be special, however, not because they were "going to be" the founders of new phyla, but because they were *morphologically* diverse in striking ways. The way for Gould to test this hypothesis would be, as Dawkins (1990) has said, to "take his ruler to the animals themselves, unprejudiced by modern preconceptions about 'fundamental body plans' and classification. The true index of how unalike two animals are is how unalike they actually are!" Such studies as have been done to date suggest, however, that in fact the Burgess Shale fauna, for all their peculiarity, exhibit no inexplicable or revolutionary morphological diversity after all (e.g., Conway Morris 1992, Gee 1992, McShea 1993).

The Burgess Shale fauna were, let us suppose (it is not really known), wiped out in one of the periodic mass extinctions that have visited the Earth. The dinosaurs, as we all know, succumbed to a later one, the Cretaceous Extinction (otherwise known as the extinction at the K-T boundary), probably triggered about sixty-five million years ago by the impact of a huge asteroid. Mass extinction strikes Gould as very important, and as a challenge to neo-Darwinism: "If punctuated equilibrium upset traditional expectations (and did it ever!), mass extinction is far worse" (Gould 1985, p. 242). Why? According to Gould, orthodoxy requires "extrapolationism," the doctrine that all evolutionary change is gradual and predictable: "But if mass extinctions are true breaks in continuity, if the slow building of adaptation in normal times does not extend into predicted success across mass extinction boundaries, then extrapolationism fails and adaptationism succumbs" (Gould 1992a, p. 53). This is just false, as I have pointed out:

> I cannot see why any adaptationist would be so foolish as to endorse anything like "extrapolationism" in a form so "pure" as to deny the possibility or even likelihood that mass extinction would play a major role in pruning the tree of life, as Gould puts it. It has always been obvious that the most perfect dinosaur will succumb if a comet strikes its homeland with a force hundreds of times greater than all the hydrogen bombs ever made. [Dennett 1993b, p. 43.]

Gould responded (1993e) by quoting a passage from Darwin himself, clearly expressing extrapolationist views. So is adaptationism (today) committed to this hopeless implication? Here is one instance when Charles Darwin himself has to count as a straw man, now that neo-Darwinism has moved on. It is true that Darwin tended to insist, shortsightedly, on the gradual nature of all extinctions, but it has long been recognized by neo-Darwinians that this was due to his eagerness to distinguish his view from the varieties of Catastrophism that stood in the way of acceptance of the theory of evolution by natural selection. We must remember that in Darwin's day miracles and calamities such as the Biblical Flood were the chief rival to Darwinian thinking. Small wonder he tended to shun anything that seemed suspiciously swift and convenient.

The fact (if it is one) that the Burgess fauna were decimated in a mass extinction is in any case less important to Gould than another conclusion he wants to draw about their fate: their decimation, he claims (1989a, p. 47n.), was *random*. According to the orthodox view, "Survivors won for cause," but, Gould opines (p. 48), "Perhaps the grim reaper of anatomical designs was only Lady Luck in disguise." Was it truly *just* a lottery that fixed *all* their fates? That would be an amazing—and definitely revolutionary—claim, especially if Gould then extended it as a generalization, but he has no evidence for such a strong claim, and backs away from it (p. 50):

> I am willing to grant that some groups may have enjoyed an edge (though we have no idea how to identify or define them), but I suspect that [the Lady Luck hypothesis] grasps a central truth about evolution. The Burgess Shale, in making this ... interpretation intelligible by the hypothetical experiment of the tape, promotes a radical view of evolutionary pathways and predictability.

Gould's suggestion, then, is not that he can prove the Lady Luck hypothesis, but that the Burgess Shale makes it at least intelligible. As Darwin insisted from the beginning, however, all it takes is "some groups" with an "edge" to put the wedge of competition into action. So is Gould just saying that *most* of the competition (or the competition with the largest, most important effects) was a true lottery? That is what he "suspects."

What is his evidence for this suspicion? He offers none at all. What he offers is the fact that he, looking at these amazing creatures, can't imagine why some would be better designed than others. They all seem about equally bizarre and ungainly to him. That is not good evidence that they didn't in fact differ dramatically in engineering quality, given their respective predicaments. If you don't even try to engage in reverse engineering, you are not in a good position to conclude that there *is no* reverse-engineering explanation to be discovered. He does offer a wager (p. 188): "I challenge any paleontologist to argue that he could have gone back to the Burgess seas and, without the benefit of hindsight, picked out *Naroia, Canadaspis, Aysheaia,* and *Sanctaris* for success, while identifying *Marrella, Odaraia, Sidneyia,* and *Leonchoilia* as ripe for the grim reaper." That's a pretty safe sucker bet, since all such a paleontologist would have to go on is the outlines of organs visible in fossil traces. But it could be lost. Some really ingenious reverse engineer might someday be able to tell an awfully convincing story about why the winners won and the losers lost. Who knows? One thing we all know: you can't make a scientific revolution out of an almost untestable hunch. (See also Gould 1989a, pp. 238–39, and Dawkins 1990 for further observations on this score.)

So we are *still* stuck with a mystery about what Gould thinks is so special about the unique flourishing and demise of these amazing creatures. They inspire a suspicion in him, but why? Here's a clue, from a talk Gould gave at the Edinburgh International Festival of Science and Technology, "The Individual in Darwin's World" (1990, p. 12):

> In fact almost all the major anatomical designs of organisms appear in one great whoosh called the Cambrian Explosion about 600 million years ago. You realize that a whoosh or an explosion in geological terms has a very long fuse. It can take a couple of million years, but a couple of million years in the line of billions is nothing. *And that is not what that world of necessary, predictable advance ought to look like* [emphasis added].

Really? Consider a parallel. There you sit, on a rock in Wyoming, watching a hole in the ground. Nothing much happens for ten, twenty, thirty minutes, and then, suddenly—whoosh!—a stream of boiling water shoots more than thirty meters into the air. It's all over in a few seconds, and then nothing much happens—just like before, apparently—and you wait for an hour, and still nothing much happens. This, then, was your experience: a single, amazing explosion lasting but a few seconds out of an hour and a half of tedium. Perhaps you would be tempted to think, "Surely this must be a unique and unrepeatable event!"

So why do they call it Old Faithful? In fact, this geyser repeats itself once every sixty-five minutes, on average, year in and year out. The "shape" of the Cambrian Explosion—its "sudden" onset and equally "sudden" termination—is no evidence *at all* for the thesis of "radical contingency." But Gould seems to think that it is.[11] He seems to think that, if we replayed the tape of life, we couldn't get another "Cambrian" Explosion the next time, or ever. But although that might be true, he has not yet offered us a single bit of evidence.

Where might such evidence come from? It might come from the computer simulations of Artificial Life, for instance, which do permit us to rewind the tape again and again. It is surprising that Gould has overlooked the possibility that he might find some evidence for (or against) his main

---

11. Gould says (1989a, p. 230), in response to Conway Morris's objections: "The Cambrian explosion was too big, too different, and too exclusive." See also the remarks on the unpredictability of "zigzag" trajectories (Gould 1989b).

conclusion by looking at the field of Artificial Life, but he never mentions the prospect. Why not? I don't know, but I do know Gould is not fond of computers, and to this day does not even use a computer for word-processing; that might have something to do with it.

A much more important clue, surely, is the fact that when you do rerun the tape of life, you find all sorts of evidence of repetition. We already knew that, of course, because convergent evolution is nature's own way of re-playing the tape. As Maynard Smith says:

> In Gould's "replay from the Cambrian" experiment, I would predict that many animals would evolve eyes, because eyes have in fact evolved many times, in many kinds of animal. I would bet that some would evolve pow-ered flight, because flight has evolved four times, in two different phyla; but I would not be certain, because animals might never get out on the land. But I agree with Gould that one could not predict which phyla would survive and inherit the earth. [Maynard Smith 1992, p. 34.]

Maynard Smith's last point is a sly one: if convergent evolution reigns, it doesn't make any difference *which phyla* inherit the earth, because of bait-and-switch! Combining bait-and-switch with convergent evolution, we get the orthodox conclusion that *whichever* lineage happens to survive will gravitate towards the Good Moves in Design Space, and the result will be hard to tell from the winner that would have been there if some different lineage had carried on. Consider the kiwi, for instance. It has evolved in New Zealand, where it didn't have any mammals to compete with, and it has converged on an amazing number of mammalian features—basically, it's a bird that pretends it's a mammal. Gould himself has written about the kiwi and its remarkably large egg (in 1991b), but as Conway Morris points out in his review (1991, p. 6):

> ... there is something else about the kiwi that receives only passing men-tion, and that is the extraordinary convergence between kiwis and mam-mals. . . . I am sure Gould would be the last to deny convergence, but surely it undermines much of his thesis of contingency.

Gould does not deny convergence—how could he?—but he does tend to ignore it. Why? Perhaps because, as Conway Morris says, it is the fatal weakness in his case for contingency. (See also Maynard Smith 1992, Daw-kins 1990, Bickerton 1993.)

So now we have an answer to our third question. The Burgess Shale fauna inspire Gould because he mistakenly thinks that they provide evidence for his thesis of "radical contingency." They *might* illustrate the thesis—but we won't know until we do the sort of research that Gould himself has ignored.

We have reached second base. Just what *is* Gould's claim about contingency? He says (1990, p. 3) that "the most common misunderstanding of
evolution, at least in lay culture," is the idea that "our eventual appearance"
is "somehow intrinsically inevitable and predictable within the confines of
the theory." *Our* appearance? What does that mean? There is a sliding scale
on which Gould neglects to locate his claim about rewinding the tape. If by
"us" he meant something very particular—Steve Gould and Dan Dennett,
let's say—then we wouldn't need the hypothesis of mass extinction to persuade us how lucky *we* are to be alive; if our two moms had never met our
respective dads, that would suffice to consign us both to Neverland, and of
course the same counterfactual holds true of every human being alive today.
Had such a sad misfortune befallen us, this would not mean, however, that our
respective offices at Harvard and Tufts would be unoccupied. It would be
astonishing if the Harvard occupant's name in this counterfactual circumstance was "Gould," and I wouldn't bet that its occupant would be a habitué
of bowling alleys and Fenway Park, but I *would* bet that its occupant would
know a lot about paleontology, would give lectures and publish articles and
spend thousands of hours studying fauna (not flora—Gould's office is in the
Museum of Comparative Zoology). If, at the other extreme, by "us" Gould
meant something very general, such as "air-breathing, land-inhabiting verte-
brates," he would probably be wrong, for the reasons Maynard Smith mentions. So we may well suppose he meant something intermediate, such as
"intelligent, language-using, technology-inventing, culture-creating beings."
This is an interesting hypothesis. If it is true, then contrary to what many
thinkers routinely suppose, the search for extra-terrestrial intelligence is as
quixotic as the search for extra-terrestrial kangaroos—it happened once,
here, but would probably never happen again. But *Wonderful Life* offers no
evidence in its favor (Wright 1990); even if the decimations of the Burgess
Shale fauna were random, whatever lineages happened to survive would, according to standard neo-Darwinian theory, proceed to grope towards the
Good Tricks in Design Space.

We have answered our second question. We are finally ready to tackle
first base: why would this thesis be of great importance, whichever way it
came out? Gould thinks that the hypothesis of "radical contingency" will
upset our equanimity, but why?

> We talk about the "march from monad to man" (old-style language again)
> as though evolution followed continuous pathways of progress along
> unbroken lineages. Nothing could be further from reality. [Gould 1989b,
> p. 14.]

What could not be further from reality? At first it might appear as if Gould
was saying here that there is no continuous, unbroken lineage between the

"monads" and us, but surely there is. There is no more secure implication of Darwin's great idea than that. As I put it in chapter 8, it is not controversial that we are all direct descendants of macros—or monads—simple precellular replicators under one name or another. So what can Gould be saying here? Perhaps we are meant to put the emphasis on "pathways *of progress*"—it is the belief in progress that is so far from the truth. The pathways are continuous, unbroken lineages all right, but not lineages of global progress. This is true, but so what?

There aren't *global* pathways of progress, but there is incessant *local* improvement. This improvement seeks out the best designs with such great reliability that it can often be predicted by adaptationist reasoning. Replay the tape a thousand times, and the Good Tricks will be found again and again, by one lineage or another. Convergent evolution is not evidence of global progress, but it is overwhelmingly good evidence of the power of processes of natural selection. This is the power of the underlying algorithms, mindless all the way down, but, thanks to the cranes it has built along the way, wonderfully capable of discovery, recognition, and wise decision. There is no room, and no need, for skyhooks.

Can it be that Gould thinks his thesis of radical contingency would refute the core Darwinian idea that evolution is an algorithmic process? That is my tentative conclusion. Algorithms, in the popular imagination, are algorithms *for* producing a particular result. As I said in chapter 2, evolution can be an algorithm, and evolution can have produced us by an algorithmic process, without its being true that evolution is an algorithm *for* producing us. But if you didn't understand that point, you might think:

> *If* we are not the predictable result of evolution, evolution cannot be an algorithmic process.

And then you would be strongly motivated to prove "radical contingency" if you wanted to show that evolution wasn't just an algorithmic process. It might not have recognizable skyhooks in it, but at least we'd know it wasn't all done with nothing but cranes.

Is it likely that Gould could be so confused about the nature of algorithms? As we shall see in chapter 15, Roger Penrose, one of the world's most distinguished mathematicians, wrote a major book (1989) on Turing machines, algorithms, and the impossibility of Artificial Intelligence, and his whole book is based on that confusion. This is not really such an implausible error, on either thinker's part. A person who really doesn't like Darwin's dangerous idea often finds it hard to get the idea in focus.

That concludes my Just So Story about how Stephen Jay Gould became the Boy Who Cried Wolf. A good adaptationist should not just rest content with a plausible story, however. At the very least, an effort should be made to consider, and rule out, alternative hypotheses. As I said at the outset, I am more interested in the reasons that have held the myth together than I am in the actual motives of the actual man, but it might seem disingenuous for me not even to mention the obvious "rival" explanations crying to be considered: politics and religion. (It could well be that there is a political or religious motivation behind the yearning for skyhooks I impute to him, but those would not be rival hypotheses; they would be elaborations of my interpretation, postponable to another occasion. Here I must briefly consider whether one of these—politics or religion—might offer a simpler, more straightforward interpretation of his campaigns, obviating my analysis. Many of Gould's critics have thought so; I think they are missing the more interesting possibility.)

Gould has never made a secret of his politics. He learned his Marxism from his father, he tells us, and until quite recently he was very vocal and active in left-wing politics. Many of his campaigns against specific scientists and specific schools of thought within science have been conducted in explicitly political—indeed, explicitly Marxist—terms, and have often had right-wing thinkers as their targets. Not surprisingly, his opponents and critics have often supposed, for instance, that his punctuationism was just his Marxist antipathy for reform playing itself out in biology. Reformers are the worst enemies of revolutionaries, as we all know. But that, I think, is only a superficially plausible reading of Gould's reasons. After all, John Maynard Smith, his polar opposite in the evolution controversies, has a Marxist background as rich and active as Gould's, and there are others with left-wing sympathies against whom Gould has directed attacks. (And then there are all the ACLU liberals like myself, though I doubt if he knows or cares.) Following his return from a visit to Russia, Gould (1992b) drew attention, as often before, to the difference between the gradualness of reform and the suddenness of revolution. In this interesting piece, Gould (p. 14) reflects on his experiences in Russia, and the failure of Marxism there— "Yes, the Russian reality does discredit a specific Marxist economics"—but goes on to say that Marx has been proven right about "the validity of the larger model of punctuational change." That does not mean that, for Gould, Marx's economic and social theory was never the point, but it is not hard to believe that Gould would keep his attitudes about evolution on board while jettisoning some political baggage that had outlasted its welcome.

As for religion, my own interpretation is, in one important sense, a hypothesis about Gould's religious yearnings. I see his antipathy to Darwin's dangerous idea as fundamentally a desire to protect or restore the Mind-first, top-down vision of John Locke—at the very least to secure *our* place in the cosmos with a skyhook. (Secular Humanism is a religion for some, and they sometimes think that Humanity cannot be special enough to matter if it is the product of merely algorithmic processes, a theme I will explore in

later chapters.) Gould has certainly seen his task as one with cosmic implications, something that is especially clear in the epiphanies about the Burgess Shale in *Wonderful Life*. That makes his world-view a question of
religion in one important sense, whether or not it has among its direct
ancestors the official creed of his religious heritage—or any other organized
religion. Gould often quotes the Bible in his monthly columns, and sometimes the rhetorical effect is striking. Surely, one thinks, an article with this
opening sentence has to have been written by a religious man: "Just as the
Lord holds the whole world in his hands, how we long to enfold an entire
subject into a witty epigram" (Gould 1993e, p. 4).

Gould has often asserted that there is no conflict between evolutionary
theory and religion.

> Unless at least half my colleagues are dunces, there can be—on the most
> raw and empirical grounds—no conflict between science and religion. I
> know hundreds of scientists who share a conviction about the fact of
> evolution, and teach it in the same way. Among these people I note an
> entire spectrum of religious attitudes—from devout daily prayer and wor-
> ship to resolute atheism. Either there's no correlation between religious
> belief and confidence in evolution—or else half of these people are fools.
> [Gould 1987, p. 68.]

Some more realistic alternatives would be that those evolutionists who
see no conflict between evolution and their religious beliefs have been
careful not to look as closely as we have been looking, or else hold a
religious view that gives God what we might call a merely ceremonial role
to play (more on this in chapter 18). Or perhaps, with Gould, they are
careful to delimit the presumed role of both science and religion. The
compatibility that Gould sees between science and religion holds only so
long as science knows its place and declines to address the big questions.
"Science does not deal with questions of ultimate origins" (Gould 1991b, p.
459). One way of interpreting Gould's campaigns within biology over the
years might be as an attempt to restrict evolutionary theory to a properly
modest task, creating a *cordon sanitaire* between it and religion. He says,
for instance:

> Evolution, in fact, is not the study of origins at all. Even the more restricted
> (and scientifically permissible) question of life's origin on our earth lies
> outside its domain. (This interesting problem, I suspect, falls primarily
> within the purview of chemistry and the physics of self-organizing sys-
> tems.) Evolution studies the pathways and mechanisms of organic change
> following the origin of life. [Gould 1991b, p. 455.]

This would rule the entire topic of chapter 7 out of bounds to evolutionary
theory, but, as we have seen, that has become the very foundation of Dar-

winian theory. Gould seems to think that he should discourage his fellow evolutionists from drawing grand philosophical conclusions from their work, but if so he has been trying to deny to others what he allows himself. In the concluding sentence of *Wonderful Life* (1989a, p. 323), Gould is ready to draw a fairly specific religious conclusion from his own consideration of the implications of paleontology:

> We are the offspring of history, and must establish our own paths in this most diverse and interesting of conceivable universes—one indifferent to our suffering, and therefore offering us maximal freedom to thrive, or to fail, in our own chosen way.

Curiously enough, this strikes me as a fine expression of the implications of Darwin's dangerous idea, not at all in conflict with the idea that evolution is an algorithmic process. It is certainly an opinion I wholeheartedly share. Gould, however, seems to think the view he is combating so vigorously is deterministic and ahistorical, in conflict with this creed of freedom. "Hyper-Darwinism," Gould's bogey, is simply the claim that no skyhooks are needed, at any point, to explain the upward trends of the branches of the Tree of Life. Like others before him, Gould has tried to show the existence of leaps, speed-ups, or other inexplicable trajectories—inexplicable by the tools of "hyper-Darwinism." But however "radically contingent" those trajectories may have been, however "punctuated" the pace of travel has been, whether by "non-Darwinian" saltations or unfathomed "mechanisms of speciation," this does not create any more elbow room for "the power of contemporary events and personalities to shape and direct the actual path taken among myriad possibilities." No more elbow room was needed (Dennett 1984).

One striking effect of Gould's campaign on contingency is that he ends up turning Nietzsche upside down. Nietzsche, you will recall, thought that nothing could be more terrifying, more world-shattering, than the thought that if you kept replaying the tape, it would all happen again and again and again—eternal recurrence, the sickest idea that anybody ever had. Nietzsche viewed his task as teaching people to say "Yes!" to this awful truth. Gould, on the other hand, thinks he must assuage the people's terror when confronted with the denial of this idea: if you kept replaying the tape, it *wouldn't* ever happen again! Are both propositions equally mind-boggling?[12] Which is worse? Would it happen again and again, or never again? Well, Tinker might say, either it would or it wouldn't, there's no denying that—and in

---

12. Philip Morrison has pointed out that if the proposition that there *is* other intelligent life in the universe is mind-boggling, so is its denial. There are no ho-hum truths of cosmology.

fact the truth is a mixture of both: a little bit of Chance, a little bit of Ever. That's Darwin's dangerous idea, like it or not.

CHAPTER 10: *Gould's self-styled revolutions, against adaptationism, gradualism, and extrapolationism, and for "radical contingency," all evaporate, their good points already firmly incorporated into the modern synthesis, and their mistaken points dismissed. Darwin's dangerous idea emerges strengthened, its dominion over every corner of biology more secure than ever.*

CHAPTER 11: *A review of all the major charges that have been leveled at Darwin's dangerous idea reveals a few surprisingly harmless heresies, a few sources of serious confusion, and one deep but misguided fear: if Darwinism is true of us, what happens to our autonomy?*

CHAPTER ELEVEN

Controversies Contained

⌐⌐⌐⌐⌐

1. A CLUTCH OF HARMLESS HERESIES

I find on re-reading it that the picture it presents is close to the one I would paint if I were to start afresh, and write a wholly new book.

—JOHN MAYNARD SMITH, introduction, 1993 edition of his 1958 book, *The Theory of Evolution*

Before turning in part III to an examination of Darwin's dangerous idea applied to humanity (and the humanities), let's pause to take stock of our survey of controversies within biology proper. Gould has spoken of the "hardening" of the modern synthesis, but also voiced his frustration about how the modern synthesis keeps shifting in front of his eyes, making it difficult to get off a good shot. Its defenders keep changing the story, co-opting revolutionaries by incorporating the good points they make into the synthesis. How secure is the modern synthesis—or its unnamed successor, if you think it has changed too much to keep its old title? Is the current embodiment of Darwinism too hard or too soft? Like Goldilocks' favorite bed, it has proven to be just right: hard where it had to be, and compliant about those issues that are open for further investigation and debate.

To get a good sense of what is hard and what is soft, we may stand back a bit and survey the whole field. Some people would still love to destroy the credentials of Darwin's dangerous idea, and we can help them by pointing to controversies on which they needn't waste their energies, since no matter how they come out, Darwin's idea will survive intact or strengthened. And then we can also point out those hard, fixed points which, if destroyed, would truly overthrow Darwinism—but they are fixed for good reasons, and are about as likely to budge as the Pyramids.

Let's consider first some tempting heresies that would *not* overthrow

Darwinism even if they were confirmed. Probably the best known has been championed in recent years by the maverick astronomer Fred Hoyle, who argues that life did not originate—could not have originated—on Earth, but has to have been "seeded" from outer space (Hoyle 1964, Hoyle and Wickramasinghe 1981). Francis Crick and Leslie Orgel (1973, and Crick 1981) point out that this idea of *panspermia* has been championed in various forms since early in the century, when Arrhenius (1908) coined the term, and, however unlikely it may seem, it is not an incoherent idea. It is not (yet) disprovable that primitive life forms (something as "simple" as a macro or as complex as a bacterium) arrived by asteroid or comet from some other region of the universe and colonized our planet. Crick and Orgel go a step further: it is even possible that the panspermia was *directed*, that life began on Earth as a result of our planet's being *deliberately* "infected" or colonized by life forms from somewhere else in the universe that got a head start on us, and indeed indirectly produced us. If we can now send a spacecraft loaded with life forms to another planet—and we can, but should not—then, by parity of reasoning, others could have done it. Since Hoyle—unlike Crick and Orgel—has voiced the suspicion (1964, p. 43) that, unless panspermia is true, "life has little meaning, but must be judged a mere cosmic fluke," it is not surprising that many, including Hoyle himself, have supposed that panspermia, if confirmed, would shatter Darwinism, that dreaded threat to the meaning of life. And since panspermia is often treated with derision by biologists—"Hoyle's Howler"—the illusion is fostered that here is a grave threat indeed, one that strikes at the very core of Darwinism.

Nothing could be further from the truth. Darwin himself surmised that life began on Earth in some warm little pond, but it might equally have started in some hot, sulfurous underground pressure-cooker (as has recently been proposed by Stetter et al. 1993), or, for that matter, on some other planet, whence it traveled here after some astronomical collision pulverized its birthplace. Wherever and whenever life started, it had to bootstrap itself by *some version* of the process we explored in chapter 7—that is what orthodox Darwinism insists upon. And as Manfred Eigen has pointed out, panspermia would do nothing to solve the difficult problem of how this bootstrapping happened: "The discrepancy between the numbers of sequences testable in practice and imaginable in theory is so great that attempts at explanation by shifting the location of the origin of life from Earth to outer space do not offer an acceptable solution to the dilemma. The mass of the universe is 'only' 10^{29} times, and its volume 'only' 10^{57} times, that of the Earth" (Eigen 1992, p. 11).

The reason orthodoxy prefers to assume a birthplace on Earth is that this is the simplest and most scientifically accessible hypothesis. That does not make it true. Whatever happened, happened. If Hoyle is right, then (darn it) we will find it much harder to confirm or disconfirm any detailed hypotheses about exactly how life started. The hypothesis that life began on Earth has the virtue of putting some admirably tough constraints on storytelling: the whole story has to unfold in under five billion years, and it has to start with conditions known to have existed on Earth in the early days. Biologists *like* having to work within these constraints; they *want* deadlines and a short list of raw materials, the more demanding the better.[1] So they hope that no hypothesis will ever be confirmed that opens up vast possibilities that will be well-nigh impossible for them to evaluate in detail. The arguments that Hoyle and others have given for panspermia all belong in the phylum of "otherwise there's not enough time," and evolutionary theorists much prefer to keep the geological deadlines intact and hunt for more cranes to do all the lifting in the time available. So far, this policy has borne excellent results. If Hoyle's hypothesis were someday confirmed, it would be a gloomy day for evolutionary theorists, not because it would overthrow Darwinism, but because it would make important features of Darwinism less *dis*confirmable, more speculative.

For the same reason, biologists would be hostile to any hypothesis that proposed that ancient DNA had been tampered with by gene-splicers from another planet who became high-tech before we did, and played a trick on us. Biologists would be hostile to the hypothesis, but would have a hard time disproving it. This raises such an important point about the nature of evidence in evolutionary theory that it is worth exploring in greater detail, with the help of a few thought experiments (drawn from Dennett 1987b, 1990b).

As many commentators have noted, evolutionary explanations are inescapably historical narratives. Ernst Mayr (1983, p. 325) puts it this way: "When one attempts to explain the features of something that is the product of evolution, one must attempt to reconstruct the evolutionary history of this feature." But particular historical facts play an elusive role in such explanations. The theory of natural selection shows how every feature of the natural world *can* be the product of a blind, unforesightful, nonteleological, ultimately mechanical process of differential reproduction over long periods of time. But of course some features of the natural world—the short legs of dachshunds and Black Angus beef cattle, the thick skins of tomatoes—are the products of artificial selection, in which the goal of the

---

1. For just this reason, biologists have mixed emotions about the recent (apparent) discovery by J. William Schopf (1993) of fossil microbes roughly a billion years older (3.5 billion instead of 2.5 billion) than orthodoxy has recently supposed. If confirmed, this would drastically revise a lot of standard assumptions about the intermediate deadlines, giving more time for the evolution of advanced forms ("Whew!"), but only by reducing the time available for the process of molecular evolution to get all the way to microbes ("Uh oh!").

process, and the rationale of the design aimed for, actually did play a role in the process. In these cases, the goal was explicitly represented in the minds of the breeders who did the selecting. So the theory of evolution must allow for the existence of such products, and such historical processes, as special cases—organisms designed with the help of supercranes. Now the question arises: can such special cases be distinguished in retrospective analysis?

Imagine a world in which *actual* hands from another galaxy supplemented the "hidden hand" of natural selection. Imagine that natural selection on this planet was aided and abetted over the eons by visitors: tinkering, farsighted, reason-representing organism-designers, like the animal- and plant-breeders of our actual world, but not restricting themselves to "domesticated" organisms designed for human use. (To make it vivid, we may suppose they treated Earth as their "theme park," creating whole phyla for educational or entertainment purposes.) These bioengineers would have actually formulated, and represented, and acted on, the rationales of their designs—just like automobile engineers or our own contemporary gene-splicers. Then, let's suppose, they absconded. Now, would their handiwork be detectable by any imaginable analysis by biologists today?

If we found that some organisms came with service manuals attached, this would be a dead giveaway. Most of the DNA in any genome is unexpressed—often called "junk DNA"—and NovaGene, a biotechnology company in Houston, has found a use for it. They have adopted the policy of "DNA branding": writing the nearest codon rendering of their company trademark in the junk DNA of their products. According to the standard abbreviations for the amino-acid specifiers, asparagine, glutamine, valine, alanine, glycine, glutamic acid, asparagine, glutamic acid = NQVAGENE (reported in *Scientific American*, June 1986, pp. 70–71). This suggests a new exercise in "radical translation" (Quine 1960) for philosophers: how, in principle or in practice, could we confirm or disconfirm the hypothesis that trademarks—or service manuals or other messages—were discernible in the junk DNA of any species? The presence of functionless DNA in the genome is no longer regarded as a puzzle. Dawkins' (1976) selfish-gene theory predicts it, and elaborations on the idea of "selfish DNA" were simultaneously developed by Doolittle and Sapienza (1980) and Orgel and Crick (1980) (see Dawkins 1982, ch. 9, for the details). That doesn't show that junk DNA *couldn't* have a more dramatic function, however, and hence it could have a meaning after all. Our imagined intergalactic interlopers could as readily have exapted the junk DNA for their own purposes as the NovaGene engineers exapted it for theirs.

Finding the high-tech version of "Kilroy was here" written in the genome of a cabbage or a king would be unnerving, but what if no such deliberate clues were left around? Would a closer look at the organism designs themselves—the phenotypes—reveal some telltale discontinuities? Gene-splicers are the most powerful cranes we have yet discovered. Are there designs that simply could not be erected without the help of this particular crane? If there are designs that cannot be approached by a gradual, stepwise redesign process in which each step is at least no worse for the gene's survival chances than its predecessor, then the existence of such a design in nature would seem to require, at some point in its ancestry, a helping hand from a foresightful designer—either a gene-splicer, or a breeder who somehow preserved the necessary succession of intermediate backsliders until they could yield their sought progeny. But could we ever conclusively establish that some design had this feature of *requiring* such a saltation in its ancestry? For over a century, skeptics have hunted for such cases—thinking that, if they ever found one, it would conclusively refute Darwinism—but so far their efforts have shown a systematic weakness.

Consider the most familiar example, the wing. Wings could not evolve in one fell swoop, runs the standard skeptical argument; and if we imagine—as we Darwinians must—that wings evolved gradually, we must admit that partially completed wings would not only not have provided partial value but would have been a positive hindrance. We Darwinians need admit no such thing. Wings that are good only for gliding (but not powered flight) have manifest net benefits for many actual creatures, and still stubbier, less aerodynamically effective protuberances could have evolved for some other reason, and then been exapted. Many versions of this story—and other stories—have been told to fill in the gap. Wings are not an embarrassment to orthodox Darwinians, or if they are, they are an embarrassment of riches. There are *too many* different plausible ways of telling the story of how functioning wings could have evolved by gradual increments! This shows how hard it would be for anyone to devise an insurmountable argument to prove that a particular feature must have arisen by a saltation, but at the same time it shows that it would be just as hard to prove that a feature must have arisen *without* a saltation, *un*aided by human or other intelligent hands.

Indeed, all the biologists I have queried on this point have agreed with me that there are no sure marks of natural, as opposed to artificial, selection. In chapter 5, we traded in the concept of strict biological possibility and impossibility for a graded notion of biological probability, but even in its terms, it is not clear how one could grade organisms as "probably" or "very probably" or "extremely probably" the products of artificial selection. Should this conclusion be viewed as a terrible embarrassment to the evolutionists in their struggle against creationists? One can imagine the headlines: "Scientists Concede: Darwinian Theory Cannot Disprove Intelligent Design!" It would be foolhardy, however, for any defender of neo-Darwinism to claim that contemporary evolution theory gives one the power to read history so finely from present data as to rule out the earlier historical presence of rational designers—a wildly implausible fantasy, but a possibility after all.

In our world today, there are organisms we *know* to be the result of foresighted, goal-seeking redesign efforts, but that knowledge depends on our direct knowledge of recent historical events; we've actually watched the breeders at work. These special events would not be likely to cast any fossily shadows into the future. To take a simpler variation on our thought experiment, suppose we were to send "Martian" biologists a laying hen, a Pekingese dog, a barn swallow, and a cheetah and ask them to determine which designs bore the mark of intervention by artificial selectors. What could they rely on? How would they argue? They might note that the hen did not care "properly" for her eggs; some varieties of hen have had their instinct for broodiness bred right out of them, and would soon become extinct were it not for the environment of artificial incubators human beings have provided for them. They might note that the Pekingese was pathetically ill-equipped to fend for itself in any demanding environment they could imagine. But the barn swallow's innate fondness for carpentered nest sites might fool them into the view that it was some sort of pet, and whatever features of the cheetah convinced them that it was a creature of the wild might also be found in greyhounds, and be features we know to have been patiently encouraged by breeders. Artificial environments are themselves a part of nature, after all, so it is unlikely that there are *any* clear signs of artificial selection that can be read off an organism in the absence of insider information on the actual history that created the organism.

Prehistoric fiddling by intergalactic visitors with the DNA of earthly species cannot be ruled out, except on the grounds that it is an entirely gratuitous fantasy. Nothing we have found (so far) on Earth so much as hints that such a hypothesis is worth further exploration. And remember—I hasten to add, lest creationists take heart—even if we were to discover and translate such a "trademark message" in our spare DNA, or found some other uncontestable mark of early tampering, this would do nothing to rescind the claim of the theory of natural selection to explain all design in nature without invocation of a foresighted Designer-Creator *outside the system*. If the theory of evolution by natural selection can account for the existence of the people at NovaGene who dreamt up DNA branding, it can also account for the existence of any predecessors who may have left their signatures around for us to discover.

Now that we have seen this possibility, however unlikely it is, we also see that, if the skeptics had ever found their Holy Grail, the You-Couldn't-Get-Here-from-There Organ or Organism, it would not have been *conclusive* against Darwinism after all. Darwin himself said that he would have to abandon his theory if such a phenomenon were discovered (see note 5 of chapter 2), but now we can see that it would always have been logically coherent (however lame and *ad hoc*) for Darwinians to reply that what they were being shown was telling evidence for the surprising hypothesis of intergalactic interlopers! The power of the theory of natural selection is not the power to prove exactly how (pre)history was, but only the power to prove how it could have been, given what we know about how things are.

Before leaving this curious topic of unwelcome but nonfatal heresies, let's consider one that is a bit more realistic. Did life on Earth arise just once, or perhaps many times? Orthodoxy supposes it happened just once, but there is no skin off its back if in fact life arose twice or ten or a hundred times. However improbable the initial bootstrapping event may have been, we must not commit the Gambler's Fallacy of supposing that after it happened once, the odds rose against its happening again. Still, the question of how many times life arose independently opens up some interesting prospects. If at least some of the assignments in the DNA are purely arbitrary, then might there not have been two *different* genetic languages coexisting side by side, like French and English, only entirely unrelated? This has not been discovered—DNA has clearly coevolved with its parent, RNA—but that does not yet show that life *didn't* arise more than once, because we don't (yet) know how wide the scope for variation in genetic code actually was.

Suppose there were exactly two equally viable and constructible DNA languages, Mendelese (ours) and Zendelese. If life arose twice, there would be four equiprobable possibilities: both times Mendelese, both times Zendelese, Mendelese and then Zendelese, or Zendelese and then Mendelese. If we ran the tape of life many times, and looked at the times in which life arose twice, we'd expect that half the time both languages would get created, but in one quarter of those replays only Mendelese would appear. In those worlds, the DNA language of all organisms would be the same, even though another language was just as possible. This shows that the "universality" (at least on our planet) of the DNA language does not permit a valid inference that all organisms had arisen from a single progenitor, the ultimate Adam, since, *ex hypothesi* in these cases, Adam could have had an entirely independent twin of sorts, accidentally sharing the same DNA language. Of course, if life arose many more times—say, a hundred times—under these same conditions, then the likelihood of only one of the two equiprobable languages' appearing would plummet to Vanishing. And if in fact there are many more than two equally usable genetic codes, this would similarly change the implications about probability. But until we know more about the range of genuine possibilities and their associated probabilities, we can't get any good leverage to decide for sure that life arose just once. For the time being, it's the simplest hypothesis—life only *has* to have arisen once.

2. THREE LOSERS: TEILHARD, LAMARCK, AND DIRECTED MUTATION

Now let us go to the opposite extreme and consider a heresy that would be truly fatal to Darwinism if it weren't such a confused and ultimately self-contradictory alternative: the attempt by the Jesuit paleontologist Teilhard de Chardin to reconcile his religion with his belief in evolution. He proposed a version of evolution that put humanity at the center of the universe, and discovered Christianity to be an expression of the goal—"the Omega-point"—towards which all evolution is striving. Teilhard even made room for Original Sin (in its orthodox Catholic version, not the scientific version I noted in chapter 8). To his dismay, the Church viewed this as heresy, and forbade him to teach it in Paris, so he spent the rest of his days in China, studying fossils, until his death in 1955. His book *The Phenomenon of Man* (1959) was published posthumously and met with international acclaim, but the scientific establishment, orthodox Darwinism in particular, was just as resolute as the Church in rejecting it as heretical. It is fair to say that in the years since his work was published, it has become clear to the point of unanimity among scientists that Teilhard offered nothing serious in the way of an alternative to orthodoxy; the ideas that were peculiarly his were confused, and the rest was just bombastic redescription of orthodoxy.[2] The classic savaging was by Sir Peter Medawar, and is reprinted in his book of essays, *Pluto's Republic* (1982, p. 245). A sample sentence: "In spite of all the obstacles that Teilhard perhaps wisely puts in our way, it is possible to discern a train of thought in *The Phenomenon of Man*."

The problem with Teilhard's vision is simple. He emphatically denied the fundamental idea: that evolution is a mindless, purposeless, algorithmic process. This was no constructive compromise; this was a betrayal of the central insight that had permitted Darwin to overthrow Locke's Mind-first vision. Alfred Russel Wallace had been tempted by the same abandonment, as we saw in chapter 3, but Teilhard embraced it wholeheartedly and made it the centerpiece of his alternative vision.[3] The esteem in which Teilhard's book is still held by nonscientists, the respectful tone in which his ideas are

---

2. The rhetorical method of "bombastic redescription" of the commonplace was first described by Paul Edwards (1965) in an essay on another continental obscurantist, the theologian Paul Tillich.

3. Teilhard's book had an unlikely champion in England, Sir Julian Huxley, one of the contributors to—indeed, the baptizer of—the modern synthesis. As Medawar makes plain, what Huxley admired in Teilhard's book was largely its support for the doctrine of the continuity of genetic and "psychosocial" evolution. This is a doctrine I am myself enthusiastically supporting under the heading of the unity of Design Space, so some of

alluded to, is testimony to the depth of loathing of Darwin's dangerous idea, a loathing so great that it will excuse any illogicality and tolerate any opacity in what purports to be an argument, if its bottom line promises relief from the oppressions of Darwinism.

What about that other notorious heresy, Lamarckism, the belief in the inheritance of acquired characteristics?[4] Here the situation is much more interesting. The main appeal of Lamarckism has always been its promise of speeding up the passage of organisms through Design Space by taking advantage of the design improvements acquired by individual organisms during their lives. So much design work to do, and so little time! But the prospect of Lamarckism *as an alternative* to Darwinism can be ruled out on logical grounds alone: the capacity to get Lamarckian inheritance off the ground in the first place *presupposes* a Darwinian process (or a miracle) (Dawkins 1986a, pp. 299–300). But couldn't Lamarckian inheritance be an important crane *within* a Darwinian framework? Darwin himself, notoriously, included Lamarckian inheritance as a booster process (in addition to natural selection) in his own version of evolution. He could entertain this idea because he had such a foggy sense of the mechanics of heredity. (To get a clear idea of how unconstrained Darwin's imagination about mechanisms of inheritance could be, see Desmond and Moore 1991, pp. 531ff., for an account of his bold speculations about "pangenesis.")

One of the most fundamental contributions to neo-Darwinism after Darwin himself was August Weismann's (1893) firm distinction between the *germ line* and the *somatic line*; the germ line consists of the sex cells in an organism's ovaries or gonads, and all the other cells of the body belong to the soma. What happens to somatic-line cells during their lifetime has a bearing, of course, on whether that body's germ line flows into any progeny at all, but changes to the somatic cells die with those cells; only changes to germ-line cells—mutations—can carry on. This doctrine, sometimes called Weismannism, is the bulwark that orthodoxy eventually raised against Lamarckism—which Darwin himself thought he could countenance. Might

---

Teilhard's views can certainly be applauded by some orthodox Darwinians. (Medawar demurs on this point.) But in any event, Huxley could not buy all that Teilhard was offering. "Yet for all this Huxley finds it impossible to follow Teilhard 'all the way in his gallant attempt to reconcile the supernatural elements in Christianity with the facts and implications of evolution'. But, bless my soul, this reconciliation is just what Teilhard's book is *about*!" (Medawar 1982, p. 251).

4. I restrict Lamarckism to inheritance of acquired characteristics *through the genetic apparatus*. If we relax the definition, then Lamarckism is not clearly a fallacy. After all, human beings inherit (by legacy) acquired wealth from their parents, and most animals inherit (by proximity) acquired parasites from their parents, and some animals inherit (by succession) acquired nests, burrows, dens from their parents. These are all phenomena of biological significance, but they are not what Lamarck was getting at—heretically.

Weismannism still be overthrown? Today the odds against Lamarckism as a major crane look much more formidable (Dawkins 1986a, pp. 288–303). For Lamarckism to work, the information about the acquired characteristic in question would somehow have to get from the revised body part, the soma, to the eggs or sperm, the germ line. In general, such message-sending is deemed impossible—no communication channels have been discovered that could carry the traffic—but set that difficulty aside. The deeper problem lies with the nature of the information in the DNA. As we have seen, our system of embryological development takes DNA sequences as a recipe, not a blueprint. There is no point-for-point mapping between body parts and DNA parts. This is what makes it extremely unlikely—or in some cases impossible—that any particular acquired change in a body part (in a muscle or a beak or, in the case of behavior, a neural control circuit of some sort) will correspond to any discrete change in the organism's DNA. So, even if there were a way of getting a change order *sent* to the sex cells, there would be no way of *composing* the necessary change order.

Consider an example. The violinist assiduously develops a magnificent vibrato, thanks largely to adjustments built up in the tendons and ligaments of her left wrist quite different from the adjustments she simultaneously builds up in her right wrist, the wrist of her bowing arm. The recipe for wrist-making in human DNA makes both wrists from a single set of instructions that takes advantage of mirror-image reflection (that's why your wrists are so much alike), so there would be no simple way to change the recipe for the left wrist without making the same (and unwanted) change in the right wrist. It is not hard to imagine how "in principle" the embryological process might be cajoled into rebuilding each wrist separately after the initial construction takes place, but even if this problem can be overcome, the chances are small indeed that this would be a *practical* mutation, a localized and smallish revision in her DNA, that corresponds closely to the improvements her years of practice have created. So almost certainly her children will have to learn their vibrato the same way she did.

This is not quite conclusive, however, and hypotheses that have features at least strongly reminiscent of Lamarckism keep popping up in biology and are often taken seriously, in spite of the general taboo against anything smacking of Lamarckism.[5] I noted in chapter 3 that the Baldwin Effect has

---

5. Dawkins (1986a, p. 299) issues the right caveat: Lamarckism is "incompatible with embryology as we know it," but "this is not to say that, somewhere in the universe, there may not be some alien system of life in which embryology *is* preformationistic; a life-form that really does have a 'blueprint genetics', and that really could, therefore, inherit acquired characteristics." There are other possibilities that might be called Lamarckian as well. For a survey, see Landman 1991, 1993; for another interesting variation on the theme, see Dawkins' account of "A Lamarckian Scare" (Dawkins 1982, pp. 164–78).

often been overlooked, or even shunned, by biologists who confused it with some dread Lamarckian heresy. The saving grace for the Baldwin Effect is that organisms pass on *their particular capacity to acquire* certain characteristics, rather than any of the characteristics they actually acquire. This *does* have the effect of taking advantage of the design explorations of individual organisms, as we saw, and hence is a powerful crane under the right circumstances. It is just not Lamarck's crane.

Finally, what about the possibility of "directed" mutation? Ever since Darwin, orthodoxy has presupposed that all mutation is random; *blind* chance makes the candidates. Mark Ridley (1985, p. 25) provides the standard declaration:

> Various theories of evolution by 'directed variation' have been proposed, but we must rule them out. There is no evidence for directed variation in mutation, in recombination, or in the process of Mendelian inheritance. Whatever the internal plausibility of these theories, they are in fact wrong.

But that is a mite too strong. The orthodox theory mustn't *presuppose* any process of directed mutation—that would be a skyhook for sure—but it can leave open the possibility of somebody's discovering nonmiraculous mechanisms that can bias the distribution of mutations in speed-up directions. Eigen's ideas about quasi-species in chapter 8 are a case in point.

In earlier chapters, I have drawn attention to various other possible cranes that are currently being investigated: trans-species "plagiarism" of nucleotide sequences (Houck's *Drosophila*), the crossovers made possible by the innovation of sex (Holland's genetic algorithms), the exploration of multiple variations by small teams (Wright's "demes") that return to the parent population (Schull's "intelligent species"), and Gould's "higher level species sorting," to name four. Since these debates all fit comfortably within the commodious walls of contemporary Darwinism, they don't need further scrutiny from us, fascinating though they are. Almost always, the issue in evolutionary theory is not possibility in principle, but relative importance, and the issues are always *much* more complex than I have portrayed them.[6]

There is one area of ongoing controversy, however, that deserves a somewhat fuller treatment, not because it threatens something hard or brittle in the modern synthesis—however it comes out, Darwinism will still be stand-

---

6. To those who want to explore these and other controversies more fully, I recommend the following books as particularly clear and accessible to neophytes *willing to work hard*: Buss 1987, Dawkins 1982, G. Williams 1992, and, as an invaluable handbook, Keller and Lloyd 1992. Mark Ridley 1993 is an excellent textbook. For a more accessible primer, Calvin 1986 is a ripping good story, with enough bold speculation thrown in to whet your appetite for more.

ing strong—but because it has been *seen* to have particularly upsetting implications for the extension of evolutionary thinking to humanity. This is the debate over the "units of selection."

3. Cui Bono?

"What's good for General Motors is good for the country."
—not said by Charles E. Wilson, 1953

In 1952, Charles E. Wilson was president of General Motors, and newly elected U.S. President Dwight Eisenhower nominated him to be his Secretary of Defense. At his nomination hearing before the Senate Armed Services Committee in January 1953, Wilson was asked to sell his shares in General Motors, but he objected. When asked if his continued stake in General Motors mightn't unduly sway his judgment, he replied: "For years, I thought what was good for the country was good for General Motors and vice versa." Unfortunately for him, what he actually said did not have much replicative power—though just enough for me to locate a descendant in a reference book and reproduce it once again in the preceding sentence. What replicated like a flu virus in the press reports of his testimony, on the other hand, was the mutated version used as the epigraph for this section; in response to the ensuing furor, Wilson was forced to sell his stock in order to win the nomination, and he was dogged by the "quotation" for the rest of his days.

We can press this frozen accident into new service. There is little doubt *why* the mutated version of Wilson's remarks spread. Before people would approve Charles Wilson for this important decision-making post, they wanted to assure themselves about who would be the *principal beneficiary* of his decisions: the country or General Motors. Was he going to make selfish decisions, or decisions for the benefit of the whole body politic? His actual answer did little to reassure them. They smelled a rat, and exposed it in the mutation of his words that they disseminated. He seemed to be claiming that nobody should be concerned about his decision-making, since even if the principal or direct beneficiary was General Motors, it would all work out fine for the whole country. A dubious claim to be sure. Although it might be true most of the time—"other things being equal"—what about the times when other things wouldn't be equal? Whose benefit would Wilson further in those circumstances? That is what had people upset, and rightly so. They wanted the actual decision-making by the Secretary of Defense to be *directly* responsive to the *national* interest. If decisions reached under those benign circumstances benefited General Motors (and presumably most of them would, if Wilson's long-held homily is true), that would be just fine, but people were afraid that Wilson had his priorities backwards.

This is an example of a topic of perennial and proper human concern. Lawyers ask, in Latin, *Cui bono?*, a question that often strikes at the heart of important issues: Who benefits from this matter? The same issue arises in evolutionary theory, where the counterpart of Wilson's actual dictum would be: "What's good for the body is good for the genes and vice versa." By and large, biologists would agree, this must be true. The fate of a body and the fate of its genes are tightly linked. But they are not perfectly coincident. What about those cases when push comes to shove, and the interests of the body (long life, happiness, comfort, etc.) conflict with the interests of the genes?

This question was always latent in the modern synthesis. Once genes had been identified as the things whose differential replication was responsible for all the design change in the biosphere, the question was unavoidable, but for a long time theorists could be lulled, like Charles Wilson, with the reflection that by and large what was good for the whole was good for the part and vice versa. But then George Williams (1966) drew attention to the question, and people began to realize that it had profound implications for our understanding of evolution. Dawkins made the point unforgettable by framing it in terms of the concept of the selfish gene (1976), pointing out that from the gene's "point of view," a body was a sort of survival machine created to enhance the gene's chances of continued replication.

The old Panglossianism had dimly thought in terms of adaptations being for "the good of the species"; Williams, Maynard Smith, Dawkins, and others showed that "for the good of the organism" was just as myopic a perspective as "for the good of the species" had been. In order to see this, one had to adopt a still more undeluded perspective, the gene's perspective, and ask what was good for the genes. At first it does seem hard-boiled, coldhearted, ruthless. It reminds me, in fact, of that hackneyed rule of thumb made famous in hard-boiled mystery stories: *cherchez la femme!*—look for the woman![7] The idea is that, as any tough-minded, worldly-wise detective should know, the key that unlocks your mystery will involve some woman or other in some way or other. Probably bad advice, even in the stylized and unrealistic world of whodunits. Better advice, claim the gene centrists, is *cherchez le gène!* We saw a good example of this in the account in chapter

---

7. The original or at least primary source of *"cherchez la femme!"* is Alexandre Dumas's (that's Dumas *père*, not *fils*) novel *Les Mohicans de Paris*, in which the inspector, M. Jackal, enunciates the principle several times. The remark has also been attributed to Talleyrand and others. (Thanks to Justin Leiber for the scholarly sleuthing.)

9 of David Haig's sleuthing, but there are hundreds or thousands of others that could be cited. (Cronin 1991 and Matt Ridley 1993 survey the history of this research up to the present.) Whenever you have an evolutionary puzzle, the gene's-eye perspective is apt to yield a solution in terms of some gene or other being favored for one reason or other. Insofar as adaptations are manifestly for the good of the organism (the eagle-as-organism surely benefits from its eagle-eye and eagle-wing), this is largely for the Wilsonian reason: what's good for the genes is good for the whole organism. But when push comes to shove, what's good for the genes determines what the future will hold. They are, after all, the replicators whose varying prospects in the self-replication competitions set the whole process of evolution in motion, and keep it in motion.

This perspective, sometimes called gene centrism, or the gene's-eye point of view, has provoked a great deal of criticism, much of it misguided. For instance, it is often said that gene centrism is "reductionistic." So it is, in the good sense. That is, it shuns skyhooks, and insists that all lifting in Design Space must be done by cranes. But as we saw in chapter 3, sometimes people use "reductionism" to refer to the view that one should "reduce" all science, or all explanations, to some lowest level—the molecular level or the atomic or subatomic level (but probably nobody has ever espoused this variety of reductionism, for it is manifestly silly). In any event, gene centrism is triumphantly *non*-reductionistic, in that sense of the term. What could be less reductionistic (in that sense of the term) than explaining the presence of, say, a particular amino-acid molecule in a particular location in a particular body by citing, not some other molecular-level facts, but, rather, the fact that the body in question was a female in a species that provides prolonged care for its young? The gene's-eye point of view explains things in terms of the intricate interactions between long-range, large-scale ecological facts, long-term historical facts, and local, molecular-level facts.

Natural selection is not a force that "acts" at one level—for instance, the molecular level as opposed to the population level or organism level. Natural selection occurs because a sum of events, of all sorts and sizes, has a particular statistically describable outcome. The blue whale teeters on the brink of extinction; if it goes extinct, a particularly magnificent and almost impossible to replace set of volumes in the Library of Mendel will cease to have extant copies, but the factor that best explains why those characteristic chromosomes, or collections of DNA nucleotide sequences, vanish from the earth might be a virus that somehow directly attacked the DNA-replicating machinery in the whales, a stray comet landing near the pod of the survivors at just the wrong time, or a surfeit of television publicity, causing curious humans to interfere catastrophically with their breeding habits! There is always a gene's-eye description of every evolutionary effect, but the more important question is whether such a description might often be mere "bookkeeping" (and as unilluminating as a molecular-level box score of a baseball game). William Wimsatt (1980) introduced the term "bookkeeping" to refer to the fact, agreed to on all sides, that the genes are the storehouse of information on genetic change, leaving it debatable whether the gene-centered view was *just* bookkeeping, a charge that has often been made (e.g., by Gould 1992a). George Williams (1985, p. 4) accepts the label but vigorously defends the importance of bookkeeping: "The idea that bookkeeping has been taking place in the past is what gives the theory of natural selection its most important kind of predictive power." (See Buss 1987, especially pp. 174ff., for important reflections on this claim.)

The claim that the gene-centrist perspective is best, or most important, is not a claim about the importance of molecular biology, but about something more abstract: about which level does the most explanatory work under most conditions. Philosophers of biology have paid more close attention, and made more substantive contributions, to the analysis of this issue than to any other in evolutionary theory. I have just mentioned Wimsatt, and there are others—to pick just some of the best, David Hull (1980), Elliot Sober (1981a), and Kim Sterelny and Philip Kitcher (1988). One reason philosophers have been attracted to the question is surely its abstractness and conceptual intricacy. Thinking about it soon gets you into deep questions about what it is to explain something, what causation is, what a level is, and so forth. This is one of the brightest areas in recent philosophy of science; the scientists have paid respectful attention to their philosophical colleagues, and have had that attention repaid with knowledgeable and well-communicated analyses and arguments by the philosophers, to which the scientists in turn have responded with discussions of their own of more than workaday philosophical significance. It is a rich harvest, and I find it hard to tear myself away from it without giving a proper introduction to the subtleties in the issues, all the more so because I have strongly held opinions about where the bulk of the wisdom lies with these controversies, but I have a different agenda here, which is, curiously enough, to *drain the drama* from them. They are excellent scientific and philosophical problems, but no matter how they come out, their answers won't have the impact that some have feared. (This will be a topic of further discussion in chapter 16.)

The tantalizing recursions and reflections of evolutionary explanation are reason enough for philosophers to pay close attention to the units-of-selection controversy, but another reason it has attracted so much attention is surely the reflection with which we began this section: people feel threatened by the gene's-eye perspective for the same reason they felt threatened by Charles Wilson's allegiance to General Motors. People want to be in charge of their own destinies; they take themselves to be both the deciders and the principal beneficiaries of their decisions, and many are afraid that Darwinism, in its gene-centered version, will undercut their assurance on that score. They are apt to see Dawkins' vivid picture of organisms as mere vehicles created to carry a gaggle of genes into future vehicles as intellectual assault and battery. So one reason, I venture, why organism-level and group-level perspectives are so frequently hailed as a worthy opponent to the gene-level perspective is the background thought—never articulated—that *we* are organisms (and we live in groups that matter to us)—and we don't want *our* interests playing second fiddle to any others! My hunch is, in other words, that we wouldn't care whether pine trees or hummingbirds were "mere survival machines" for *their* genes if it weren't for our realization that we bear the same relation to our genes that they bear to theirs. In the next chapter, I want to put that worry to rest by showing that this is really not so! Our relationship to our genes is importantly different from the relationship of any other species to its genes—because what *We* are is not just what we as a species are. This will pull the plug, draining all the anxiety out of the still fascinating and unresolved conceptual questions about how to think about the units of selection, but before I turn to that task, I must make sure the threatening aspect of the issue is made clear, and several common misconceptions are cleared up.

Perhaps the most misguided criticism of gene centrism is the frequently heard claim that genes simply cannot have interests (Midgley 1979, 1983, Stove 1992). This criticism, if taken seriously, would lead us to discard a treasury of insights, but it is flatly mistaken. Even if genes could not *act* on their interests in just the way we can act on ours, they can surely have them, in a sense that is uncontroversial and clear. If a body politic, or General Motors, can have interests, so can genes. You can do something for your own sake, or for the sake of the children, or for the sake of art, or for the sake of democracy, or for the sake of ... peanut butter. I find it hard to imagine why anybody would *want* to put the well-being and further flourishing of peanut butter above all else, but peanut butter can be put on the pedestal just as readily as art or the children can. One could even decide—though it would be a strange choice—that the thing one wanted most to protect and enhance, even at the cost of one's own life, was one's own genes. No sane *person* would make such a decision. As George Williams (1988, p. 403) says, "There is no conceivable justification for any personal concern with the interests (long-term average proliferation) of the genes we received in the lottery of meiosis and fertilization."

But that doesn't mean that there aren't *forces* bent on furthering the sakes or interests of genes. In fact, until quite recently, genes were the principal beneficiaries of all the selective forces on the planet. That is to say, there were no forces whose *principal* beneficiary was anything else. There were *accidents* and *catastrophes* (lightning bolts and tidal waves), but no *steady* forces acting systematically to favor anything but genes.

To whose interests is the actual "decision-making" of natural selection most directly responsive? It is not controversial that conflicts between genes and bodies (between genes and the phenotypic expressions of the genotypes of which they are a proper part) can arise. Moreover, no one doubts that in general the body's claim to be considered the principal beneficiary lapses as soon as it has completed its procreational mission. Once the salmon have fought their way upstream and successfully spawned, they are dead meat. They literally fall apart, because *there is no evolutionary pressure* in favor of any of the design revisions that might prevent them from falling apart, giving them nice long grandparent-retirement periods like those many of us get to enjoy. In general, the body is thus only an instrumental, and hence secondary, beneficiary of the "decisions" made by natural selection.

This is true throughout the biosphere, revealed in a pattern with a few important variations. In many phyla, parents die before their offspring are born, and their entire lives are a preparation for a single climactic act of replication. Others—trees, for instance—live through many generations of offspring, and can hence come into competition with their own young for sunlight and other resources. Mammals and birds typically invest large portions of their energy and activity to caring for young, and hence have many more opportunities to "choose" between themselves and their young as beneficiaries of whatever course of action they take. Creatures for which such options never come up can be designed "under the assumption" (Mother Nature's tacit assumption) that this is simply not an issue that needs any design attention at all.

Presumably, the control system of a moth, for instance, is ruthlessly designed to sacrifice the body for the sake of the genes, whenever a generic and recognizable opportunity to do so arises. A little fantasy: We somehow surgically replace this standard system (a "Damn the torpedoes, full speed ahead!" system) with a body-favoring system (a "To hell with my genes, I'm looking out for Number One!" system). What could the replacement ever do that wasn't just one way or another of committing suicide or pointlessly wandering? A moth is simply not equipped to take any advantage of opportunities tangential to its lifework of reproducing itself. Life-enhancing ends are hard to take seriously, if it is the short life of a moth we are considering. Birds, in contrast, may abandon a nest full of eggs when they themselves are threatened in one way or another, and this looks more like what we often do, but the reason they can do this is that they can start another nest—if not this season, then next. They are looking out for Number One now, but only because this gives their genes a better chance of getting replicated later.

We are different. There is a huge scope for alternative policies in human life, but the question then becomes: how and when does this scope get established? There can be no doubt that many people have clearheadedly, well-informedly *chosen* to forgo the risks and pains of childbearing for the safety and comfort of a "barren" life of other rewards. The culture may stack

FIGURE 11.1

the deck against it (with such loaded words as "barren"), and it is true that it is a reversal of the fundamental strategy of all life, but, still, it often happens. *We* recognize that bearing and raising offspring is just one of life's possible projects, and by no means the most important, given *our* values. But where could those values have come from? How did our control systems become equipped with them, if not by miraculous surgery? How is it that we have been able to establish a rival perspective that can often overpower our genes' interests while other species have not?[8] This will be a topic for the next chapter.

---

8. Dog-lovers may protest that there is good evidence of dogs' sacrificing their lives for their human masters, putting their own prospects for reproduction and even "personal" longevity firmly in second place. Certainly this can happen, because dogs have actually been *bred* for this very capacity to acquire such occasionally fatal trans-species loyalties. These are necessarily exceptional cases, however. The cartoonist Al Capp saw the problem many years ago when he created his delightful shmoos, white armless blobs with two rather pseudopodal feet and happy, cat-whiskered faces. Shmoos loved people above all, and instantly sacrificed themselves whenever appropriate, turning themselves into sumptuous roast beef dinners (or peanut-butter sandwiches, or whatever their human companions happened to need or desire). Shmoos, you may recall, reproduced asexually by cloning in large numbers at the drop of a hat—a bit of poetic license that got Capp out of the nagging problem of how shmoos, given their proclivities, could ever have survived. Kim Sterelny has suggested to me that shmoos exhibit the sort of features we should look for as proof of intergalactic interlopers in our past! If we found organisms whose adaptations were manifestly not for their own direct benefit, but for the benefit of their putative makers, this would properly set us wondering, but it would not be conclusive.

CHAPTER 11: *Panspermia, intergalactic gene-splicers, and multiple origins of life on Earth are all harmless if unwelcome heretical possibilities. Teilhard's "Omega-point," Lamarck's genetic transmission of acquired traits, and directed mutation (without a crane to support it) would be fatal to Darwinism, but are safely discredited. The controversies over the units of selection and the "gene's-eye point of view" are important issues in contemporary evolutionary theory, but they don't have the dire implications often seen in them, whichever way they come out.*

This completes our survey of Darwinism in biology itself. Now that we are armed with a fair and quite detailed picture of contemporary Darwinism, we are ready to see, in part III, what implications it has for Homo sapiens.

CHAPTER 12: *The primary difference between our species and all others is our reliance on cultural transmission of information, and hence on cultural evolution. The unit of cultural evolution, Dawkins' meme, has a powerful and underappreciated role to play in our analysis of the human sphere.*

PART III

MIND, MEANING, MATHEMATICS, AND MORALITY

The new fundamental feeling: our conclusive transitoriness.—*Formerly one sought the feeling of the grandeur of man by pointing to his divine origin: this has now become a forbidden way, for at its portal stands the ape, together with other gruesome beasts, grinning knowingly as if to say: no further in this direction! One therefore now tries the opposite direction: the way mankind is going shall serve as proof of his grandeur and kinship with God. Alas this, too, is vain! At the end of this way stands the funeral urn of the last man and gravedigger (with the inscription 'nihil humani a me alienum puto'). However high mankind may have evolved—and perhaps at the end it will stand even lower than at the beginning!—it cannot pass over into a higher order, as little as the ant and the earwig can at the end of its 'earthly course' rise up to kinship with God and eternal life. The becoming drags the has-been along behind it: why should an exception to this eternal spectacle be made on behalf of some little star or for any little species upon it! Away with such sentimentalities!*

—FRIEDRICH NIETZSCHE 1881, p. 47

CHAPTER TWELVE

The Cranes of Culture

~~~

1. THE MONKEY'S UNCLE MEETS THE MEME

*What is the question now placed before society with a glib assurance
the most astounding? The question is this—Is man an ape or an angel?
My Lord, I am on the side of the angels.*

—BENJAMIN DISRAELI, speech at Oxford, 1864

Darwin himself saw clearly that if he claimed that his theory applied to
one particular species, this would upset its members in ways he dreaded, so
he held back at first. There is almost no mention of our species in *Origin of
Species*—aside from its important role as a crane in artificial selection. But
of course this fooled no one. It was clear where the theory was heading, so
Darwin worked hard to produce his own, carefully thought-out version
before the critics and skeptics could bury the issue in misrepresentations
and alarm calls: *The Descent of Man, and Selection in Relation to Sex*
(1871). There was no doubt at all, Darwin observed: we—*Homo sapiens*—
are one of the species over which evolutionary theory reigns. Seeing that
there was little hope of denying this fact, some Darwin-dreaders have sought
a champion who might deliver a pre-emptive strike, disabling the dangerous
idea before it ever got a chance to spread across the isthmus that connects
our species with all the others. Whenever they have found someone announcing the demise of Darwinism (or neo-Darwinism, or the modern
synthesis), they have egged him on, hoping that this time the revolution
would be real. Self-styled revolutionaries have struck early and often, but, as
we have seen, they have managed only to invigorate their target, deepening
our understanding of it while enhancing it with complexities undreamt of
by Darwin himself.

Falling back, then, some of the foes of Darwin's dangerous idea have
planted themselves firmly on the isthmus, like Horatio at the bridge, intent

on preventing the idea from crossing over. The famous first confrontation was the notorious debate in Oxford's Museum of Natural History in 1860, only a few months after the initial publication of *Origin*, between "Soapy Sam" Wilberforce, Bishop of Oxford, and Thomas Henry Huxley, "Darwin's bulldog." This is a tale told so often in so many variations that we might count it a phylum of memes, not just a species. Here it was that the good bishop made his famous rhetorical mistake, asking Huxley whether it was on his grandfather's side or his grandmother's side that he was descended from an ape. Tempers were running high in that meeting room; a woman had fainted, and several of Darwin's supporters were almost beside themselves with fury at the contemptuous misrepresentation of their hero's theory that was being given, so it is understandable that eyewitnesses' stories diverge at this point. In the best version—which in all likelihood has undergone some significant design improvement over the retellings—Huxley replied that he "was not ashamed to have a monkey for his ancestor; but he would be ashamed to be connected with a man who used great gifts to obscure the truth" (R. Richards 1987, p. 4; see also pp. 549–51, and Desmond and Moore 1991, ch. 33).

And ever since, some members of *Homo sapiens* have been remarkably thin-skinned about our ancestral relationship to the apes. When Jared Diamond published *The Third Chimpanzee* in 1992, he drew his title from the recently discovered fact that we human beings are actually more closely related to the two species of chimpanzees (*Pan troglodytes*, the familiar chimp, and *Pan paniscus*, the rare, smaller pygmy chimp or bonobo) than those chimpanzees are to the other apes. We three species have a common ancestor more recent than the common ancestor of the chimpanzee and the gorilla, for instance, so we are all on one branch of the Tree of Life, with gorillas and orangutans and everything else on other branches.

We are the third chimpanzee. Diamond cautiously lifted this fascinating fact from the "philological" work on primate DNA by Sibley and Ahlquist (1984 and later papers), and made it clear to his readers that theirs were a somewhat controversial set of studies (Diamond 1992, pp. 20, 371–72). He was not cautious enough for one reviewer, however. Jonathan Marks, an anthropologist at Yale, went into orbit in denunciation of Diamond—and Sibley and Ahlquist, whose work, he declared, "needs to be treated like nuclear waste: bury it safely and forget about it for a million years" (Marks 1993a, p. 61). Since 1988, Marks, whose own earlier investigations of primate chromosomes had placed the chimpanzee *marginally* closer to the gorilla than to us, has waged a startlingly vituperative campaign condemning Sibley and Ahlquist, but this campaign has recently suffered a major setback. The original findings of Sibley and Ahlquist have been roundly confirmed by more sensitive methods of analysis (theirs was a relatively crude technique, path-finding at the time, but subsequently superseded by

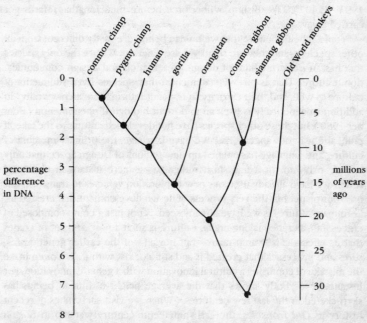

Family tree of the higher primates. Trace back each pair of modern higher primates to the black dot connecting them. The numbers to the left then give the percentage difference between the DNAs of those modern primates, while the numbers to the right give the estimated number of millions of years ago since they last shared a common ancestor. For example, the common and pygmy chimps differ in about 0.7 percent of the DNA and diverged around three million years ago; we differ in 1.6 percent of our DNA from either chimp and diverged from their common ancestor around seven million years ago; and gorillas differ in about 2.3 percent of their DNA from us or chimps and diverged from the common ancestor leading to us and the two chimps around ten million years ago. [Diamond 1992.]

FIGURE 12.1

more powerful techniques). Why, though, should it make any *moral* difference whether we or the gorillas win the competition to be the closest cousin of the chimpanzee? The apes are our closest kin in any case. But it matters mightily to Marks, apparently, whose desire to discredit Sibley and Ahlquist has driven him right out of bounds. His most recent attack on them, in a review of some other books in *American Scientist* (Marks 1993b), drew a chorus of condemnation from his fellow scientists, and a remarkable apology from the editors of that magazine: "Although reviewers' opinions are their own and not the magazine's, the editors do set standards that we deeply regret were not maintained in the review in question" (Sept.–Oct.

1993, p. 407). Like Bishop Wilberforce before him, Jonathan Marks got carried away.

People ache to believe that we human beings are vastly different from all other species—and they are right! We are different. We are the only species that has an *extra* medium of design preservation and design communication: culture. That is an overstatement; other species have rudiments of culture as well, and their capacity to transmit information "behaviorally" in addition to genetically is itself an important biological phenomenon (Bonner 1980), but these other species have not developed culture to the takeoff point the way our species has. We have language, the primary medium of culture, and language has opened up new regions of Design Space that only we are privy to. In a few short millennia—a mere instant in biological time—we have already used our new exploration vehicles to transform not only our planet but the very process of design development that created us.

Human culture, as we have already seen, is not just a crane composed of cranes, but a crane-making crane. Culture is such a powerful set of cranes that its effects can swamp many—but not all—of the earlier genetic pressures and processes that created it and still coexist with it. We often make the mistake of confusing a cultural innovation with a genetic innovation. For instance, everybody knows that the average height of human beings has skyrocketed in the last few centuries. (When we visit such relics of recent history as *Old Ironsides*, the early-nineteenth-century warship in Boston Harbor, we find the space below decks to be comically cramped—were our ancestors really a race of midgets?) How much of this rapid change in height is due to genetic changes in our species? Not much, if any at all. There has been time for only about ten generations of *Homo sapiens* since *Old Ironsides* was launched in 1797, and even if there were a strong selection pressure favoring the tall—and is there evidence for that?—this would not have had time to produce such a big effect. What have changed dramatically are human health, diet, and living conditions; these are what have produced the dramatic change in phenotype, which is 100 percent due to cultural innovations, passed on through cultural transmission: schooling, the spread of new farming practices, public-health measures, and so forth. Anyone who worries about "genetic determinism" should be reminded that virtually all the differences discernible between the people of, say, Plato's day and the people living today—their physical talents, proclivities, attitudes, prospects—must be due to cultural changes, since fewer than two hundred generations separate us from Plato. Environmental changes due to cultural innovations change the landscape of phenotypic expression so much, and so fast, however, that they can in principle change the genetic selection pressures rapidly—the Baldwin Effect is a simple instance of such a change in selection pressure due to widespread behavioral innovation. Although it is important to remember how slowly evolution works in general, we should never forget that there is no inertia at all in selection pressure. Pressures that have been dominant for millions of years can vanish overnight; and, of course, new selection pressures can come into existence with a single volcanic eruption, or the appearance of a new disease organism.

Cultural evolution operates many orders of magnitude faster than genetic evolution, and this is part of its role in making our species special, but it has also turned us into creatures with an entirely different outlook on life from that of any other species. In fact, it isn't clear that the members of any other species *have* an outlook on life. But we do; we can choose celibacy for reasons; we can pass laws regulating what we eat; we can have elaborate systems for encouraging or punishing certain sorts of sexual behavior, and so forth. Our outlook on life is so compelling and obvious to us that we often fall in the trap of imposing it willy-nilly on other creatures—or on all of nature. One of my favorite examples of this widespread cognitive illusion is the puzzlement researchers have expressed about the evolutionary explanation of sleep.

Lab shelves sag beneath volumes of data, yet no one has discerned that sleep has any clear biological function. Then what evolutionary pressure selected this curious behavior that forces us to spend a third of our lives unconscious? Sleeping animals are more vulnerable to predators. They have less time to search for food, to eat, to find mates, to procreate, to feed their young. As Victorian parents told their children, sleepy-heads fall behind—in life and evolution.

University of Chicago sleep researcher Allan Rechtshaffen asks "how could natural selection with its irrevocable logic have 'permitted' the animal kingdom to pay the price of sleep for no good reason?" Sleep is so apparently maladaptive that it is hard to understand why some other condition did not evolve to satisfy whatever need it is that sleep satisfies. [Raymo 1988.]

But why does sleep need a "clear biological function" at all? It is *being awake* that needs an explanation, and presumably its explanation is obvious. Animals—unlike plants—need to be awake at least part of the time, in order to search for food and procreate, as Raymo notes. But once you've headed down this path of leading an active existence, the cost-benefit analysis of the options that arise is far from obvious. Being awake is relatively costly, compared with lying dormant (think of its root, *dormire*). So presumably Mother Nature economizes where she can. If we could get away with it, we'd "sleep" our entire lives. That is what trees do, after all: all winter they "hibernate" in deep coma, because there is nothing else for them to do, and in the summer they "estivate" in a somewhat lighter coma, in what the doctors call a *vegetative state* when a member of our species has the misfortune to enter it. If the woodchopper comes along while the tree

340 THE CRANES OF CULTURE

is sleeping, well, that's just the chance that trees have to take, all the time. But surely we animals are at greater risk from predators while we sleep? Not necessarily. Leaving the den is risky, too, and if we're going to minimize that risky phase, we might as well keep the metabolism idling while we bide our time, conserving energy for the main business of replicating. (These matters are much more complicated than I am portraying them, of course. My point is just that the cost-benefit analysis is far from obvious, and that is enough to remove the air of paradox.)

We think that being up and about, having adventures and completing projects, seeing our friends and learning about the world, is the whole point of life, but Mother Nature doesn't see it that way at all. A life of sleep is as good a life as any other, and in many regards better—certainly cheaper—than most. If the members of some other species also seem to *enjoy* their periods of wakefulness as much as we do, this is an interesting commonality, so interesting that we should not make the mistake of assuming it must exist just because we find it to be such an appropriate attitude towards life in our own case. Its existence in other species needs to be shown, and that is not easy.[1]

What we are is very much a matter of what culture has made us. Now we must ask how this all got started. What sort of evolutionary revolution happened that set us apart so decisively from all the other products of genetic revolution? The story I am going to tell is a retelling of the story we encountered in chapter 4, about the creation of the eukaryotic cells that made multicellular life possible. You will recall that before there were cells with nuclei there were simpler, and more solitary, life forms, the prokaryotes, destined for nothing fancier than drifting around in an energy-rich soup reproducing themselves. Not nothing, but not much of a life. Then, one day, according to Lynn Margulis' wonderful story (1981), some prokaryotes were invaded by parasites of sorts, and this turned out to be a blessing in disguise, for, whereas parasites are—by definition—deleterious to the fitness of their hosts, these invaders turned out to be beneficial, and hence were *symbionts* but not parasites. They and those they invaded became more like *commensals*—literally, from the Latin, organisms that feed at the same table—or *mutualists*, benefiting from each other's company. They joined forces, creating a revolutionary new kind of entity, a eukaryotic cell. This opened up the Vast space of possibilities we know as multicellular life, a space previously unimaginable, to say the least; prokaryotes are no doubt clueless on all topics.

---

1. See the discussion of fun in Dennett 1991a. Some human beings claim to love to sleep. "What do you plan to do this weekend?" "Sleep! Ahh, it will be wonderful!" Other human beings find this attitude well-nigh incomprehensible. Mother Nature sees nothing strange about either attitude, under the right conditions.

Then a few billion years passed, while multicellular life forms explored various nooks and crannies of Design Space until, one fine day, another invasion began, in a single species of multicellular organism, a sort of primate, which had developed a variety of structures and capacities (don't you dare call them preadaptations) that just happened to be particularly well suited for these invaders. It is not surprising that the invaders were well adapted for finding homes in their hosts, since they were themselves created by their hosts, in much the way spiders create webs and birds create nests. In a twinkling—less than a hundred thousand years—these new invaders transformed the apes who were their unwitting hosts into something altogether new: *witting* hosts, who, thanks to their huge stock of newfangled invaders, could imagine the heretofore unimaginable, leaping through Design Space as nothing had ever done before. Following Dawkins (1976), I call the invaders *memes*, and the radically new kind of entity created when a particular sort of animal is properly furnished by—or infested with— memes is what is commonly called a *person*.

That is the story in rough outline. Some people, I have found, just hate the whole idea. They like the idea that it is our human minds and human culture that distinguish us sharply from all the "thoughtless brutes" (as Descartes called them), but they don't like the idea of trying to give an evolutionary explanation of the creation of this most important distinguishing mark. I think they are making a big mistake.[2] Do they want a miracle? Do they want culture to be God-given? A skyhook, not a crane? Why? They want the human way of life to be radically different from the way of life of all other living things, and so it is, but, like life itself, and every other wonderful thing, culture must have a Darwinian origin. It, too, must grow out of something less, something *quasi-*, something merely *as if* rather than *intrinsic*, and at every step along the way the results have to be, as David Haig puts it, *evolutionarily enforceable*. For culture we need language, for instance, but language has to evolve on its own hook first; we can't just notice how good it would be once it was all in place. We can't presuppose cooperation; we can't presuppose *human* intelligence; we can't presuppose tradition— this all has to be built up from scratch, just the way the original replicators were. Settling for anything less in the way of an explanation would be just giving up.

In the next chapter, I will address the important theoretical questions

---

2. It has been made before, by no less stalwart a Darwinian than Thomas Henry Huxley, in his Romanes Lecture of 1893 in Oxford. "Huxley's critics ... noted the apparent bifurcation he had introduced into nature, between natural processes and human activity, as if man could somehow lift himself out of nature" (Richards 1987, p. 316). Huxley quickly saw his error and attempted to restore a Darwinian account of culture—by an appeal to the force of group selection! History does have a way of repeating itself.

about how language and the human mind could evolve in the first place by Darwinian mechanisms. I will have to confront and disarm the tremendous—and largely misguided—animosity to this story, and also work out answers to the responsible objections to it. But before we consider how this magnificent crane-structure might have been built, I want to sketch the completed product, distinguishing it from its caricatures, and showing in a little more detail how culture comes to have such revolutionary powers.

2. INVASION OF THE BODY-SNATCHERS

Human beings owe their biological supremacy to the possession of a form of inheritance quite unlike that of other animals: exogenetic or exosomatic heredity. In this form of heredity information is transmitted from one generation to the next through nongenetic channels—by word of mouth, by example, and by other forms of indoctrination; in general, by the entire apparatus of culture.

—PETER MEDAWAR 1977, p. 14

The nucleic acids invented human beings in order to be able to reproduce themselves even on the Moon.

—SOL SPIEGELMAN, quoted in Eigen 1992, p. 124

I am convinced that comparisons between biological evolution and human cultural or technological change have done vastly more harm than good—and examples abound of this most common of intellectual traps. . . . Biological evolution is powered by natural selection, cultural evolution by a different set of principles that I understand but dimly.

—STEPHEN JAY GOULD 1991a, p. 63

Nobody wants to reinvent the wheel, a mythic example of wasted design work, and I have no intention of making that error here. Up till now I have been helping myself to Dawkins' term "meme" as the name for any item of cultural evolution, postponing the discussion of what kind of Darwinian theory of memes we might be able to devise. The time has come to consider more carefully what Dawkins' memes are or might be. He has done much of the basic design work (drawing on the work of others, of course), and I myself have drawn on his meme meme before, devoting considerable time and effort to building suitable explanation vehicles out of it. I am going to reuse these earlier constructions, adding further design modifications. I first presented my own version (Dennett 1990c) of Dawkins' account of memes in a Mandel Lecture to the American Society for Aesthetics, a lecture series endowed for the purpose of exploring the question whether art promotes human evolution. (The answer is Yes!) I then exapted my own device, reusing it, with modifications, in my book on human consciousness (1991a, pp. 199–208), to show how memes could transform the operating system or computational architecture of a human brain. That account offers many details about the relationship between the genetically designed hardware of the human brain and the culturally transmitted habits that transform it into something much more powerful, and I will skip lightly over most of those details here. This time I will modify my exaptation of Dawkins a second time, the better to deal with the particular environmental problems encountered in the current explanatory project. (Those who are familiar with either of its immediate ancestors should find important improvements in the current version.)

The outlines of the theory of evolution by natural selection make clear that evolution occurs whenever the following conditions exist:

(1) variation: there is a continuing abundance of different elements
(2) heredity or replication: the elements have the capacity to create copies or replicas of themselves
(3) differential "fitness": the number of copies of an element that are created in a given time varies, depending on interactions between the features of that element and features of the environment in which it persists

Notice that this definition, though drawn from biology, says nothing specific about organic molecules, nutrition, or even life. This maximally abstract definition of evolution by natural selection has been formulated in many roughly equivalent versions—see, e.g., Lewontin 1980 and Brandon 1978 (both reprinted in Sober 1984b). As Dawkins has pointed out, the fundamental principle is

> that all life evolves by the differential survival of replicating entities. . . .
> The gene, the DNA molecule, happens to be the replicating entity which prevails on our own planet. There may be others. If there are, provided certain other conditions are met, they will almost inevitably tend to become the basis for an evolutionary process.
> But do we have to go to distant worlds to find other kinds of replication and other, consequent, kinds of evolution? I think that a new kind of replicator has recently emerged on this very planet. It is staring us in the face. It is still in its infancy, still drifting clumsily about in its primeval soup, but already it is achieving evolutionary change at a rate which leaves the old gene panting far behind. [Dawkins 1976, p. 206.]

These new replicators are, roughly, ideas. Not the "simple ideas" of Locke and Hume (the idea of red, or the idea of round or hot or cold), but the sort of complex ideas that form themselves into *distinct memorable units*—such as the ideas of

> arch
> wheel
> wearing clothes
> vendetta
> right triangle
> alphabet
> calendar
> the *Odyssey*
> calculus
> chess
> perspective drawing
> evolution by natural selection
> impressionism
> "Greensleeves"
> deconstructionism

Intuitively, we see these as more or less identifiable cultural units, but we can say something more precise about how we draw the boundaries—about why D-$F\#$-A isn't a unit, and the theme from the slow movement of Beethoven's Seventh Symphony is: the units are the smallest elements that replicate themselves with reliability and fecundity. We can compare them, in this regard, to genes and their components: C-G-A, a single codon of DNA, is "too small" to be a gene. It is one of the codes for the amino acid arginine, and it copies itself prodigiously wherever it appears in genomes, but its effects are not "individual" enough to count as a gene. A three-nucleotide phrase does not count as a gene for the same reason that you can't copyright a three-note musical phrase: it is not enough to make a melody. But there is no "principled" lower limit on the length of a sequence that might come to be considered a gene or a meme (Dawkins 1982, pp. 89ff.). The first four notes of Beethoven's Fifth Symphony are clearly a meme, replicating all by themselves, detached from the rest of the symphony, but keeping intact a certain identity of effect (a phenotypic effect), and hence thriving in contexts in which Beethoven and his works are unknown. Dawkins explains how he coined the name he gave these units:

> ... a unit of cultural transmission, or a unit of *imitation*. 'Mimeme' comes from a suitable Greek root, but I want a monosyllable that sounds a bit like 'gene'.... It could alternatively be thought of as being related to 'memory' or to the French word *même*....

Examples of memes are tunes, ideas, catch-phrases, clothes fashions, ways of making pots or of building arches. Just as genes propagate them-selves in the gene pool by leaping from body to body via sperm or eggs, so memes propagate themselves in the meme pool by leaping from brain to brain via a process which, in the broad sense, can be called imitation. If a scientist hears, or reads about, a good idea, he passes it on to his colleagues and students. He mentions it in his articles and his lectures. If the idea catches on, it can be said to propagate itself, spreading from brain to brain. [Dawkins 1976, p. 206.]

Meme evolution is not just analogous to biological or genic evolution, according to Dawkins. It is not just a process that can be metaphorically described in these evolutionary idioms, but a phenomenon that obeys the laws of natural selection quite exactly. The theory of evolution by natural selection is neutral, he suggests, regarding the differences between memes and genes; these are just different kinds of replicators evolving in different media at different rates. And just as the genes for animals could not come into existence on this planet until the evolution of plants had paved the way (creating the oxygen-rich atmosphere and ready supply of convertible nu-trients), so the evolution of memes could not get started until the evolution of animals had paved the way by creating a species—*Homo sapiens*—with brains that could provide shelter, and habits of communication that could provide transmission media, for memes.

There is no denying that there is cultural evolution, in the Darwin-neutral sense that cultures change over time, accumulating and losing features, while also maintaining features from earlier ages. The history of the idea of, say, crucifixion, or of a dome on squinches, or powered flight, is undeniably a history of transmission through various nongenetic media of a family of variations on a central theme. But whether such evolution is weakly or strongly analogous to, or parallel to, genetic evolution, the process that Darwinian theory explains so well, is an open question. In fact, it is many open questions. At one extreme, we may imagine, it could turn out that cultural evolution recapitulates *all* the features of genetic evolution: not only are there gene analogues (memes), but there are strict analogues of phenotypes, genotypes, sexual reproduction, sexual selection, DNA, RNA, codons, allopatric speciation, demes, genomic imprinting, and so forth—the whole edifice of biological theory perfectly mirrored in the medium of culture. You thought DNA-splicing was a scary technology? Wait till they start making meme implants in their laboratories! Not likely. At the other extreme, cultural evolution could be discovered to operate according to entirely different principles (as Gould suggests), so that there was no help at all to be found amid the concepts of biology. This is surely what many humanists and social scientists fervently hope—but it is also highly unlikely, for reasons we have already seen. In between the extremes lie the likely and valuable prospects: that there is a large (or largish) and important (or merely mildly interesting) transfer of concepts from biology to the human sciences. It might be, for example, that, although the processes of cultural transmission of ideas are truly Darwinian *phenomena*, for various reasons they resist being captured in a Darwinian *science*, so we will have to settle for the "merely philosophical" realizations we can glean from this, and leave science to tackle other projects.

First let's consider the case for the claim that the phenomena of cultural evolution are truly Darwinian. Then we can turn to the skeptical complications. At the outset, the meme perspective is distinctly unsettling, even appalling. We can sum it up with a slogan:

A scholar is just a library's way of making another library.

I don't know about you, but I am not initially attracted by the idea of my brain as a sort of dungheap in which the larvae of other people's ideas renew themselves, before sending out copies of themselves in an informational diaspora. It does seem to rob my mind of its importance as both author and critic. Who's in charge, according to this vision—we or our memes?

There is no simple answer to that important question. There could not be. We would like to think of ourselves as godlike creators of ideas, manipulating and controlling them as our whim dictates, and judging them from an independent, Olympian standpoint. But even if this is our ideal, we know that it is seldom if ever the reality, even with the most masterful and creative minds. As Mozart famously observed of his own brainchildren:

When I feel well and in a good humor, or when I am taking a drive or walking after a good meal, or in the night when I cannot sleep, thoughts crowd into my mind as easily as you would wish. Whence and how do they come? I do not know and *I have nothing to do with it* [emphasis added]. Those which please me I keep in my head and hum them; at least others have told me that I do so.[3]

---

3. Peter Kivy informed me after the Mandel Lecture that this oft-quoted passage is counterfeit—not Mozart at all. I found it in Jacques Hadamard's classic study, *The Psychology of Inventing in the Mathematical Field* (1949, p. 16), and first quoted it myself in Dennett 1975, one of my first forays into Darwinian thinking. I persist in quoting it here, in spite of Kivy's correction, because it not only expresses but exemplifies the thesis that memes, once they exist, are independent of authors and critics alike. Historical accuracy is important (which is why I have written this footnote), but the passage so well suits my purposes that I am choosing to ignore its pedigree. I might not have persisted in this, had I not encountered a supporting meme the day after Kivy informed me: I

Mozart is in good company. Rare is the novelist who *doesn't* claim characters who "take on a life of their own"; artists are rather fond of confessing that their paintings take over and paint themselves; and poets humbly submit that they are the servants or even slaves to the ideas that teem in their heads, not the bosses. And we all can cite cases of memes that persist unbidden and unappreciated in our own minds, or that spread—like rumors—in spite of the general disapproval of that spreading by those who help spread them.

The other day, I was embarrassed—dismayed—to catch myself walking along humming a melody to myself. It was not a theme of Haydn or Brahms or Charlie Parker or even Bob Dylan: I was energetically humming "It Takes Two to Tango"—a perfectly dismal and entirely unredeemed bit of chewing gum for the ears that was unaccountably popular sometime in the 1950s. I am sure I have never in my life chosen this melody, esteemed this melody, or in any way judged it to be better than silence, but there it was, a horrible musical virus, at least as robust in my meme pool as any melody I actually esteem. And now, to make matters worse, I have resurrected the virus in many of you, who will no doubt curse me in days to come when you find yourself humming, for the first time in over thirty years, that boring tune.

Human language, first spoken and then, very recently, written, is surely the principal medium of cultural transmission, creating the *infosphere* in which cultural evolution occurs. Speaking and hearing, writing and reading—these are the underlying technologies of transmission and replication most analogous to the technologies of DNA and RNA in the biosphere. I needn't bother reviewing the familiar facts about the recent explosive proliferation of these media via the memes for movable type, radio and television, xerography, computers, fax machines, and electronic mail. We are all well aware that today we live awash in a sea of paper-borne memes, breathing in an atmosphere of electronically-borne memes. Memes now spread around the world at the speed of light, and replicate at rates that make even fruit flies and yeast cells look glacial in comparison. They leap promiscuously from vehicle to vehicle, and from medium to medium, and are proving to be virtually unquarantinable.

Genes are invisible; they are carried by gene vehicles (organisms) in which they tend to produce characteristic effects (phenotypic effects) by which their fates are, in the long run, determined. Memes are also invisible, and are carried by meme vehicles—pictures, books, sayings (in particular languages, oral or written, on paper or magnetically encoded, etc.). Tools

---

overheard a guide at the Metropolitan Museum of Art, commenting on the Gilbert Stuart portrait of George Washington: "This may not be what George Washington looked like then, but this is what he looks like now." That experience of mine, of course, illustrates another of my themes: the role of serendipity in all design work.

and buildings and other inventions are also meme vehicles (Campbell 1979). A wagon with spoked wheels carries not only grain or freight from place to place; it carries the brilliant idea of a wagon with spoked wheels from mind to mind. A meme's existence depends on a physical embodiment in some medium; if all such physical embodiments are destroyed, that meme is extinguished. It may, of course, make a subsequent, independent reappearance, just as dinosaur genes could, in principle, get together again in some distant future, but the dinosaurs they created and inhabited would not be descendants of the original dinosaurs—or at least not any more directly than we are. The fate of memes is similarly determined by whether copies and copies of copies of them persist and multiply, and this depends on the selective forces that act directly on the various physical vehicles that embody them.

Memes, like genes, are potentially immortal, but, like genes, they depend on the existence of a continuous chain of physical vehicles, persisting in the face of the Second Law of Thermodynamics. Books are relatively permanent, and inscriptions on monuments even more permanent, but unless these are under the protection of human conservators, they tend to dissolve in time. Manfred Eigen makes the same point about genes, though driving the analogy in the other direction:

> Consider, for instance, one of Mozart's compositions, one that is retained stably in our concert repertoire. The reason for its retention is not that the notes of this work are printed in a particularly durable ink. The persistence with which a Mozart symphony reappears in our concert programmes is solely a consequence of its high selection value. In order for this to retain its effect, the work must be played again and again, the public must take note of it, and it must be continually re-evaluated in competition with other compositions. Stability of genetic information has similar causes. [Eigen 1992, p. 15.]

As with genes, immortality is more a matter of replication than of the longevity of individual vehicles. As we saw in note 4 of chapter 6, the preservation of the Platonic memes, via a series of copies of copies, is a particularly striking case of this. Although a few papyrus fragments of Plato's texts roughly contemporaneous with the man himself still exist, the survival of his memes owes almost nothing to the chemical stability of these fragments. Today's libraries contain thousands if not millions of physical copies (and translations) of Plato's *Republic*, and the key ancestors in the transmission of this text turned to dust centuries ago.

Brute physical replication of vehicles is not enough to ensure meme longevity. A few thousand hardbound copies of a new book can disappear with scarcely a trace in a few years, and who knows how many brilliant letters to the editor, reproduced in hundreds of thousands of copies, disappear into landfills and incinerators every day? The day may come when nonhuman meme-evaluators suffice to select and arrange for the preservation of particular memes, but for the time being, memes still depend at least indirectly on one or more of their vehicles' spending at least a brief, pupal stage in a remarkable sort of meme nest: a human mind.

Minds are in limited supply, and each mind has a limited capacity for memes, and hence there is a considerable competition among memes for entry into as many minds as possible. This competition is the major selective force in the infosphere, and, just as in the biosphere, the challenge has been met with great ingenuity. "Whose ingenuity?" you may want to ask, but by now you should know that this is not always a good question; the ingenuity is *there* to appreciate, whatever its source. Like a mindless virus, a meme's prospects depend on its design—not its "internal" design, whatever that might be, but the design it shows the world, its phenotype, the way it affects things in its environment. The things in its environment are minds and other memes.

For instance, whatever virtues (from our perspective) the following memes have, they have in common the property of having phenotypic expressions that tend to make their own replication more likely by disabling or pre-empting the environmental forces that would tend to extinguish them: the meme for *faith*, which discourages the exercise of the sort of critical judgment that might decide that the idea of faith was, all things considered, a dangerous idea (Dawkins 1976, p. 212); the memes for *tolerance,* or *free speech*; the meme for including a *warning in a chain letter* about the terrible fates of those who have broken the chain in the past; the *conspiracy-theory* meme, which has a built-in response to the objection that there is no good evidence of the conspiracy: "Of course not—that's how powerful the conspiracy is!" Some of these memes are "good," and others "bad"; what they have in common is a phenotypic effect that systematically tends to disable the selective forces arrayed against them. Other things being equal, memetics predicts that conspiracy-theory memes will persist quite independently of their truth, and the meme for faith is apt to secure its own survival, and that of the religious memes that ride piggyback on it, in even the most rationalistic environments. Indeed, the meme for faith exhibits *frequency-dependent fitness*: it flourishes particularly in the company of rationalistic memes. In a skeptic-poor world, the meme for faith does not attract much attention, and hence tends to go dormant in minds, and hence is seldom reintroduced into the infosphere. (Can we demonstrate classic predator-prey population boom-and-bust cycles between memes for faith and memes for reason? Probably not, but it might be instructive to look, and ask why not.)

Other concepts from population genetics transfer quite smoothly. Here is a case of what a geneticist would call *linked loci*: two memes that happen to be physically tied together so that they tend always to replicate together, a tendency that affects their chances. There is a magnificent ceremonial march, familiar to many of us, and generally beloved. It is stirring and bright and grand—just the thing, you would think, to use for commencements, weddings, and other festive occasions, perhaps driving "Pomp and Circumstance" and the Wedding March from *Lohengrin* to near-extinction, were it not for the fact that its musical meme is too tightly linked to its title meme, which we all tend to think of as soon as we hear the music: Sir Arthur Sullivan's unusable masterpiece, "Behold the Lord High Executioner." If this march had no lyrics and were titled, say, "Koko's March," it would not be disqualified from use. But the actual title, comprising the first five words of the lyrics, which are tightly locked to the melody, virtually guarantees *a chain of thought* in most listeners that would be undesirable on almost any festive occasion. This is the phenotypic effect that prevents the greater replication of this meme. If performances of *The Mikado* waned over the years, so that a time came when few if any people knew the lyrics of the march, let alone the silly story, the march might come back into its own as a piece of ceremonial music without words—except for the darn title at the head of the score! It wouldn't look good on the program, would it, just before the Vice-Chancellor's address to the graduates?

This is actually just a vivid case of one of the most important phenomena in the infosphere: the misfiltering of memes due to such linkages. There is even a meme that names the phenomenon: *throwing out the baby with the bathwater*. This book is largely intended to undo the unfortunate effects of misfiltering the Darwinian memes, a process that has been going on ever since Darwin himself got confused about which were his best ideas (even though some of his enemies agreed with them) and which were his worst (even though they seemed to perform yeoman service against certain pernicious doctrines). (R. Richards 1987 provides a particularly fascinating history of the evolution of the ideas of evolution.) We all have filters of the following sort:

Ignore everything that appears in *X*.

For some people, *X* is the *National Geographic* or *Pravda*; for others, it is *The New York Review of Books*; we all take our chances, counting on the "good" ideas to make it eventually through the stacks of filters of others into the limelight of our attention.

This structure of filters is itself a meme construction of considerable robustness. John McCarthy, one of the founders of Artificial Intelligence (and the coiner of its name, a meme with its own, independent base in the infosphere) once suggested to a humanist audience that electronic-mail networks could revolutionize the ecology of the poet. Only a handful of poets can make their living by selling their poems, McCarthy noted, because poetry books are slender, expensive volumes purchased by very few individuals and libraries. But imagine what would happen if poets could put their poems on an international network, where anybody could read them or copy them for a penny, electronically transferred to the poet's royalty account. This could provide a steady source of income for many poets, he surmised. Quite independently of any aesthetic objections poets and poetry-lovers might have to poems embodied in electronic media, the obvious counterhypothesis arises from population memetics. If such a network were established, no poetry-lover would be willing to wade through thousands of electronic files filled with doggerel, looking for the good poems; there would be a niche created for various memes for poetry filters. One could subscribe, for a few pennies, to an editorial service that scanned the info-sphere for good poems. Different services, with different critical standards, would flourish, as would services for reviewing all the different services—and services that screened, collected, formatted, and presented the works of the best poets in slender electronic volumes which only a few would pur-chase. In other words, the memes for editing and criticism will find niches in any environment in the infosphere; they flourish because of the short supply and limited capacity of minds, whatever the transmission media between minds. Do you doubt this prediction? If so, I'd like to discuss framing a suitable wager with you. Here once again, as we have seen so often in evolutionary thinking, explanation proceeds by an assumption that the processes—whatever their media, and whatever the contingent zigs and zags of their particular trajectories—will home in on the forced moves and other Good Tricks in the relevant space.

The structure of filters is complex and quick to respond to new chal-lenges, but of course it doesn't always "work." The competition among memes to break through the filters leads to an "arms race" of ploy and counterploy, with ever more elaborate "advertising" raised against ever more layers of selective filters. In the dignified ecology of academia, we don't call it advertising, but the same arms race is manifested in department letterheads, "blind refereeing," the proliferation of specialized journals, book reviews, reviews of book reviews, and anthologies of "classic works." These filters are not even always intended to preserve the best. Philoso-phers might care to ask themselves, for instance, how often they are ac-complices in increasing the audience for a second-rate article simply because their introductory course needs a simple-minded version of a bad idea that even the freshmen can refute. Some of the most frequently re-printed articles in twentieth-century philosophy are famous precisely be-cause nobody believes them; everybody can see what is wrong with them.[4]

---

4. The confirmation of this claim is left as an exercise for the reader. Among the memes that structure the infosphere and hence affect the transmission of other memes are the laws of libel.

A related phenomenon in the competition of memes for our attention is positive feedback. In biology, this is manifested in such phenomena as the "runaway sexual selection" that explains the long and cumbersome tail of the bird of paradise or the peacock (for the details, see Dawkins 1986a, pp. 195–220; Cronin 1991; Matt Ridley 1993). Dawkins (1986a, p. 219) provides an example from the world of publishing: "Best-seller lists of books are published weekly, and it is undoubtedly true that as soon as a book sells enough copies to appear in one of these lists, its sales increase even more, simply by virtue of that fact. Publishers speak of a book 'taking off', and those publishers with some knowledge of science even speak of a 'critical mass for take-off'."

Meme vehicles inhabit our world alongside all the fauna and flora, large and small. By and large they are "visible" only to the human species, however. Consider the environment of the average New York City pigeon, whose eyes and ears are assaulted every day by approximately as many words, pictures, and other signs and symbols as assault each human New Yorker. These physical meme vehicles may impinge importantly on the pigeon's welfare, but not by virtue of the memes they carry—it means nothing to the pigeon that it is under a page of *The National Enquirer*, not *The New York Times*, that it finds a crumb. To human beings, on the other hand, each meme vehicle is a potential friend or foe, bearing a gift that will enhance our powers or a gift horse that will distract us, burden our memories, derange our judgment.

3. Could There Be a Science of Memetics?

The scope of the undertaking strikes me as staggering. But more than this, if one accepts the evolutionary perspective, attempts to discuss science (or any other sort of conceptual activity) become much more difficult, so difficult as to produce total paralysis.

—David Hull 1982, p. 299

Memes are capable of instructing, not protein synthesis as genes do, but behaviour. However, genes can do that too indirectly through protein synthesis. On the other hand meme replication, by involving neurostructural modifications, is invariably associated with the induction of protein synthesis.

—Juan Delius 1991, p. 84

This is all very enticing, but we have been glossing over a host of complications. I can hear a chorus of skepticism building in the wings. Remember the story near the end of chapter 4 about Francis Crick's jaundiced view of population genetics as science? If population genetics just barely qualifies as science—and obsolete science at that—what chance is there for a true *science* of memetics? Philosophers, some will say, may appreciate the (apparent) insight to be found in a striking new perspective, but if you can't turn it into actual science, with testable hypotheses, reliable formalizations, and quantifiable results, what good is it, really? Dawkins himself has never claimed to be founding a new scientific discipline of memetics. Is this because there is something wrong with the concept of a meme?

What stands to a meme as DNA stands to a gene? Several commentators (see, e.g., Delius 1991) have argued for the identification of memes with complex brain-structures, parallel to the identification of genes with complex structures of DNA. But as we have already seen, it is a mistake to *identify* genes with their vehicles in DNA. The idea that evolution is an algorithmic process is the idea that it must have a useful description in substrate-neutral terms. As George Williams proposed many years ago (1966, p. 25): "In evolutionary theory, a gene could be defined as any hereditary *information* [emphasis added] for which there is favorable or unfavorable selection bias equal to several or many times its rate of endogenous change." The importance of the separation between information and vehicle is even easier to discern in the case of memes.[5] The obvious problem noted by all is that it is very unlikely—but not quite impossible—that there is a uniform "brain language" in which information is stored in different human brains, and this makes brains very different from chromosomes. Geneticists have recently identified a chromosomal structure they call the *homeobox*; in spite of differences, this structure is identifiable in widely separated species of animals—perhaps in them all—so it is very ancient, and it plays a central role in embryological development. We may be startled at first to learn that a gene identified as playing a major role in eye development in the homeobox of mice has almost the same codon spelling as a gene dubbed (for its phenotypic effect) *eyeless* when it was identified in the homeobox of the fruitfly, *Drosophila*. But we would be even more flabbergasted were we to discover that the brain-cell complex that stored the original meme for bifocals in Benjamin Franklin's brain was the same as, or very similar to, the brain-cell complex that is called upon today to store the meme for bifocals whenever any child in Asia, Africa, or Europe first learns about them—by reading about them, seeing them on television, or noticing them on a parent's nose. What this reflection makes vivid is the fact that what is preserved and transmitted in cultural evolution is *informa-*

---

5. For a good discussion of the embattled relation between gene talk and molecule talk, see Waters 1990.

tion—in a media-neutral, language-neutral sense. Thus the meme is primarily a *semantic* classification, not a *syntactic* classification that might be directly observable in "brain language" or natural language.

In the case of genes, we are blessed by a gratifyingly strong alignment of semantic and syntactic identity: there *is* a single genetic language, in which meaning is (roughly) preserved across all species. Still, it is important to distinguish semantic types from syntactic types. In the Library of Babel we identify a set of syntactic text-variants as all falling into the *Moby Dick* galaxy by virtue of what they tell us *about*, not their syntactic similarity. (Think of all the different translations of *Moby Dick* into other languages, and also the English abridgments, outlines, and study aids—to say nothing of the versions in film and other media!) Our interest in identifying and re-identifying genes over the evolutionary ages is similarly *primarily* because of the uniformity of the phenotypic effects—what they are "about" (such as making hemoglobin, or eyes). Our ability to rely on their syntactic identifiability in DNA is a recent advance, and even when we cannot conceivably avail ourselves of it (for instance, in deducing facts about genetic changes from what we can observe in the fossil record of species that have left no DNA for us to "read"), we can still confidently speak of the genes—the information—that must have been preserved or transmitted.

It is conceivable, but hardly likely and certainly not necessary, that we will someday discover a striking identity between brain structures storing the same information, allowing us to identify memes syntactically. Even if we encountered such an unlikely blessing, however, we should cling to the more abstract and fundamental concept of memes, since we already know that meme transmission and storage can proceed indefinitely in noncerebral forms—in artifacts of every kind—that do not depend on a shared language of description. If ever there was "multimedia" transmission and transformation of information, it is cultural transmission and transformation. So some of the varieties of reductionistic triumph we have come to expect in biology—discovering exactly how many different ways hemoglobin is "spelled" in all the species in the world, for instance—are almost certainly ruled out in any science of culture, notwithstanding the prophecies of a golden age of mind-reading one sometimes hears these days from the ideologues of neuroscience.

This would thwart only some kinds of memetic science, but isn't the situation actually worse than that? Darwinian evolution, as we have seen, depends on *very* high-fidelity copying—almost but not quite perfect copying, thanks to the exquisite proofreading and duplication machinery of the DNA-readers that accompany the DNA texts. Raise the mutation rate just a bit too high and evolution goes haywire; natural selection can no longer work to guarantee fitness over the long run. Minds (or brains), on the other hand, aren't much like photocopying machines at all. On the contrary, instead of just dutifully passing on their messages, correcting most of the typos as they go, brains seem to be designed to do just the opposite: to transform, invent, interpolate, censor, and generally mix up the "input" before yielding any "output." Isn't one of the hallmarks of cultural evolution and transmission the extraordinarily high rate of mutation and recombination? We *seldom* pass on a meme unaltered, it seems, unless we are particularly literal-minded rote learners. (Are walking encyclopedias hidebound?) Moreover, as Steven Pinker has stressed (personal communication), much of the mutation that happens to memes—how much is not clear—is manifestly *directed* mutation: "Memes such as the theory of relativity are not the cumulative product of millions of *random* (undirected) mutations of some original idea, but each brain in the chain of production added huge dollops of value to the product in a nonrandom way." Indeed, the whole power of minds as meme nests comes from what a biologist would call *lineage-crossing* or *anastomosis* (the coming back together of separating gene-pools). As Gould (1991a, p. 65) points out, "The basic topologies of biological and cultural change are completely different. Biological evolution is a system of constant divergence without subsequent joining of branches. Lineages, once distinct, are separate forever. In human history, transmission across lineages is, perhaps, the major source of cultural change."

Moreover, when memes come into contact with each other in a mind, they have a marvelous capacity to become adjusted to each other, swiftly changing their phenotypic effects to fit the circumstances—and it is the recipe for the new phenotype that then gets replicated when the mind broadcasts or publishes the results of this mixing. For instance, my three-year-old grandson, who loves construction machinery, recently blurted out a fine mutation on a nursery rhyme: "Pop! goes the diesel." He didn't even notice what he had done, but I, to whom the phrase would never have occurred, have now seen to it that this mutant meme gets replicated. As in the case of jokes discussed earlier, this modest moment of creativity is a mixture of serendipity and appreciation, distributed over several minds, no one of which gets to claim the authorship of special creation. It is a sort of Lamarckian replication of acquired characteristics, as Gould and others have suggested.[6] The very creativity and activity of human minds as temporary homes for memes seems to guarantee that lines of descent are hopelessly muddled, and that phenotypes (the "body designs" of memes) change so fast that there's no keeping track of the "natural kinds." Recall, from chapter

---

6. Usually, the "charge" that cultural evolution is Lamarckian is a deep confusion, as Hull (1982) carefully points out, but in this version it is undeniable—though also not a "charge." In particular, the entity that exhibits the Lamarckian talent of passing on an acquired characteristic is not the human agent, but the meme itself.

10 (p. 293), that species are invisible without a modicum of stasis, but remember, too, that this is an epistemological, not a metaphysical, point: if species weren't rather static, we couldn't *find out* and organize the facts needed to do certain kinds of science; that wouldn't show, however, that the phenomena weren't governed by natural selection. Similarly, the conclusion here would be a pessimistic *epistemological* conclusion: even if memes *do* originate by a process of "descent with modification," our chances of cranking out a science that charts that descent are slim.

Once the worry is put in that form, it points to what may seem to be a partial solution. One of the most striking features of cultural evolution is the ease, reliability, and confidence with which we *can* identify commonalities in spite of the vast differences in underlying media. What do *Romeo and Juliet* and (the film, let's say, of) *West Side Story* have in common (Dennett 1987b)? Not a string of English characters, not even a sequence of propositions (in English or French or German . . . translation). What is in common, of course, is not a *syntactic* property or system of properties but a *semantic* property or system of properties: the story, not the text; the characters and their personalities, not their names and speeches. What we so readily identify as the same thing in both cases is the predicament that both William Shakespeare and Arthur Laurents (who wrote the book for *West Side Story*) want us to think about. So it is only at the level of *intentional objects*, once we have adopted the intentional stance, that we can describe these common properties.[7] When we do adopt the stance, the sought-for common features often stick out like sore thumbs.

Does this help? Yes, but we must be careful about a problem we have already identified in several different guises: the problem of how to tell plagiarism (or respectful borrowing) from convergent evolution. As Hull (1982, p. 300) points out, we do not want to consider two *identical* cultural items as instances of the same *meme* unless they are related by descent. (The genes for octopus eyes are not the same genes as those for dolphin eyes, however similar the eyes may appear.) This is apt to create a host of illusions, or just undecidability, for cultural evolutionists whenever they attempt to trace the memes for Good Tricks. The more abstract the level at which we identify the memes, the harder it is to tell convergent evolution from descent. We happen to know, because they told us, that the creators of *West Side Story* (Arthur Laurents, Jerome Robbins, and Leonard Bernstein) got the idea from *Romeo and Juliet*, but if they had been carefully secretive about this, we might well have thought they had simply

---

7. Cf. the parallel point about the welcome—indeed, indispensable—power of adopting the intentional stance as a scientific tactic in *heterophenomenology*, the objective science of consciousness (Dennett 1991a).

reinvented a wheel, rediscovered a cultural "universal" that will appear, on its own, in almost any cultural evolution. The more purely semantic our principles of identification are—or, in other words, the less bound they are to particular forms of expression—the harder it is to trace descent with confidence. (Remember that it was peculiarities in the particular form of expression that gave Otto Neugebauer his crucial clue in deciphering the mystery of the Greek translation of the Babylonian ephemeris in chapter 6.) This is the same epistemological problem, in the science of culture, that taxonomists confront when they try to sort out homology from analogy, ancestral from derived characters, in cladistic analysis (Mark Ridley 1985). Ideally, in the imagined field of cultural cladistics, one would want to find "characters"—literally, alphabetic characters—that are functionally optional choices within a huge class of possible alternatives. If we found whole speeches by Tony and Maria that suspiciously replicated the words and phrases of Romeo and Juliet, we wouldn't need autobiographical clues from Laurents, Robbins, or Bernstein. We wouldn't hesitate to declare that the coincidence of the words was no coincidence; Design Space is too Vast to make that credible.

In general, however, we can't count on such discoveries in our attempts at a science of cultural evolution. Suppose, for instance, we want to argue that such institutions as agriculture or monarchy, or even such particular practices as tattooing or shaking hands, descend from a common cultural ancestor instead of having been independently reinvented. There is a trade-off. To the extent that we have to go to quite abstract functional (or semantic) levels to find our common features, we lose the capacity to tell homology from analogy, descent from convergent evolution. This has always been tacitly appreciated by students of culture, of course, quite independently of Darwinian thinking. Consider what you can deduce from potsherds, for instance. Anthropologists looking for evidence of shared culture are, quite properly, more impressed by common idiosyncrasies of decorative style than by common functional shapes. Or consider the fact that two widely separated cultures both used *boats*; this is no evidence at all of a shared cultural heritage. If both cultures were to paint *eyes* on the bows of their boats, it would be much more interesting, but still a rather obvious move in the game of design. If both cultures were to paint, say, *blue hexagons* on the bows of their boats, this would be telling indeed.

The anthropologist Dan Sperber, who has thought a great deal about cultural evolution, thinks there is a problem with any use of abstract, intentional objects as the anchors for a scientific project. Such abstract objects, he claims,

> do not directly enter into causal relations. What caused your indigestion was not the Mornay sauce recipe in the abstract, but your host having read

a public representation, having formed a mental representation, and having followed it with greater or lesser success. What caused the child's enjoy-able fear was not the story of Little Red Riding Hood in the abstract, but her understanding of her mother's words. More to the present point, what caused the Mornay sauce recipe or the story of Little Red Riding Hood to become cultural representations is not, or rather is not directly, their formal properties, it is the construction of millions of mental representa-tions causally linked by millions of public representations. [Sperber 1985, pp. 77–78.]

What Sperber says about the indirectness of the role of the abstract features is certainly true, but, far from this being an obstacle to science, it is the best sort of invitation to science: an invitation to cut through the Gordian knot of tangled causation with an abstract formulation that is predictive precisely *because* it ignores all those complications. For instance, genes are selected because of their indirect and only statistically visible phenotypic effects. Consider the following prediction: wherever you find moths with camou-flage on their wings, you will find that they have keen-sighted predators, and wherever you find moths that are heavily predated by echo-locating bats, you will find that they have traded in wing camouflage for jamming devices or a particular talent for creating evasive flight patterns. Of course, our ultimate goal is to explain whatever features we find in the moths and their surroundings all the way down to the molecular or atomic mechanisms responsible, but there is no reason to demand that such a reduction be uniform or generalizable across the board. It is the glory of science that it can find the patterns in spite of the noise (Dennett 1991b).

The peculiarities of human psychology (and human digestion, for that matter, as the Mornay-sauce example shows) are important *eventually*, but they don't stand in the way of a scientific analysis of the phenomenon in question. In fact, as Sperber himself has persuasively argued, we can use higher-level principles as levers to pry open lower-level secrets. Sperber points to the importance of the invention of writing, which initiated major changes in cultural evolution. He shows how to reason from facts about preliterate culture to facts about human psychology. (He prefers to think of cultural transmission along the lines of *epidemiology* rather than *genetics*, but the direction of his theory is very much the same as Dawkins'—to the point of near-indistinguishability when you think of what the Darwinian treatment of epidemiology looks like; see Williams and Nesse 1991.) Here is Sperber's "Law of the Epidemiology of Representations":

In an oral tradition, all cultural representations are easily remembered ones; hard to remember representations are forgotten, or transformed into more easily remembered ones, before reaching a cultural level of distri-bution. [Sperber 1985, p. 86.]

It looks trivial at first, but consider how we can apply it. We can use the existence of a particular sort of cultural representation endemic to oral traditions to shed light on how human memory works, by asking what it is about *this* sort of representation that makes it more memorable than others.

Sperber points out that people are better at remembering a story than they are at remembering a text—at least today, now that the oral tradition is waning.[8] But even today we sometimes remember—involuntarily—an advertising jingle, including its precise rhythmic properties, its "tone of voice," and many other "low-level" features. When scientists decide on acronyms or cute slogans for their theories, they are hoping thereby to make them more memorable, more vivid and attractive memes. And hence the actual details of the representing are sometimes just as much a candidate for memehood as the content represented. Using acronyms is itself a meme—a meta-meme, of course—which caught on because of its demonstrated power in furthering the content memes whose name memes it helped to design. What is it about acronyms, or about rhymes or "snappy" slogans, that makes them fare so well in the competitions that rage through a human mind?

This sort of question exploits a fundamental strategy both of evolutionary theory and of cognitive science, as we have seen many times. Where evolutionary theory considers information transmitted through genetic channels, whatever they are, cognitive science considers information transmitted through the channels of the nervous system, whatever they are—plus the adjacent media, such as the translucent air, which transmits sound and light so well. You can finesse your ignorance of the gory mechanical details of how the information got from A to B, at least temporarily, and just concentrate on the implications of the fact that some information *did* get there—and some other information didn't.

Suppose you were given the task of catching a spy, or a whole spy ring, in the Pentagon. Suppose what was known was that information about, say, nuclear submarines was somehow getting into the hands of the wrong people. One way of catching the spy would be to insert various tidbits of false (but credible) information at various places within the Pentagon and see which ones surfaced, in which order, in Geneva or Beirut or wherever the marketplace for secrets is. Varying the conditions and circumstances, you might gradually build up an elaborate diagram of the route—the various way stations and transfers and compounding places—even to the point of

---

8. For an analysis of the astonishing mnemonic powers of the oral tradition, see Albert Lord's classic, *The Singer of Tales* (1960), about the technology of verse memorization developed by bards from Homer's day to modern times in the Balkan countries and elsewhere.

arresting and duly convicting the spy ring, and yet still be in the dark about the medium of communication used. Was it radio? Microdots glued to documents? Semaphore flags? Did the agent memorize the blueprint and simply walk naked across the border, or did he have a verbal description in Morse code hidden on a floppy disk in his computer?

In the end, we want to know the answers to all these questions, but in the meantime there is a lot we can do in the substrate-neutral domain of pure information transfer. In cognitive science, for example, the linguist Ray Jackendoff (1987, 1993) shows the surprising power of this method in his ingenious deductions about the number of representational levels, and their powers, that *must* go into such tasks as getting information from the light that strikes our eyes all the way to places where we can talk about what we see. He doesn't have to know the details of neurophysiology (though he's interested, unlike many other linguists) in order to reach confident and reliable conclusions about the structure of the processes, and the representations they transform.

What we learn at this abstract level is scientifically important in its own right. It is, indeed, the basis of everything important. Nobody has ever put it better than the physicist Richard Feynman:

> Is no one inspired by our present picture of the universe? This value of science remains unsung by singers: you are reduced to hearing not a song or poem, but an evening lecture about it. This is not yet a scientific age.
>
> Perhaps one of the reasons for this silence is that you have to know how to read the music. For instance, the scientific article may say, "The radioactive phosphorus content of the cerebrum of the rat decreases to one-half in a period of two weeks." Now, what does that mean?
>
> It means that phosphorus that is in the brain of a rat—and also in mine, and yours—is not the same phosphorus as it was two weeks ago. It means the atoms that are in the brain are being replaced: the ones that were there before have gone away.
>
> So what is this mind of ours: what are these atoms with consciousness? Last week's potatoes! They now can *remember* what was going on in my mind a year ago—a mind which has long ago been replaced.
>
> To note that the thing I call my individuality is only a pattern or dance, *that* is what it means when one discovers how long it takes for the atoms of the brain to be replaced by other atoms. The atoms come into my brain, dance a dance, and then go out—there are always new atoms, but always doing the same dance, remembering what the dance was yesterday. [Feynman 1988, p. 244.]

4. THE PHILOSOPHICAL IMPORTANCE OF MEMES

Cultural 'evolution' is not really evolution at all if we are being fussy and purist about our use of words, but there may be enough in common between them to justify some comparison of principles.

—RICHARD DAWKINS 1986a, p. 216

There is no more reason to expect a cultural practice transmitted between churchgoers to increase churchgoers' fitness than there is to expect a similarly transmitted flu virus to increase fitness.

—GEORGE WILLIAMS 1992, p. 15

When Dawkins introduced memes in 1976, he described his innovation as a literal extension of the classical Darwinian theory. He has since drawn in his horns slightly. In *The Blind Watchmaker* (1986a, p. 196), he spoke of an analogy "which I find inspiring but which can be taken too far if we are not careful." Why did he retreat like this? Why, indeed, is the meme meme so little discussed eighteen years after *The Selfish Gene* appeared?

In *The Extended Phenotype* (1982, p. 112), Dawkins replied forcefully to the storm of criticism from sociobiologists and others, while conceding some interesting disanalogies between genes and memes:

> ... memes are not strung out along linear chromosomes, and it is not clear that they occupy and compete for discrete 'loci', or that they have identifiable 'alleles'.... The copying process is probably much less precise than in the case of genes.... Memes may partially blend with each other in a way that genes do not.

But then (p. 112) he retreated further, apparently in the face of unnamed and unquoted adversaries:

> My own feeling is that its [the meme meme's] main value may lie not so much in helping us to understand human culture as in sharpening our perception of genetic natural selection. This is the only reason I am presumptuous enough to discuss it, for I do not know enough about the existing literature on human culture to make an authoritative contribution to it.

I suggest that the meme's-eye view of what happened to the meme meme is quite obvious: "humanist" minds have set up a particularly aggressive set of filters against memes coming from "sociobiology," and once Dawkins was identified as a sociobiologist, this almost guaranteed rejection of whatever this interloper had to say about culture—not for good reasons, but just in a sort of immunological rejection.[9]

One can see why. The meme's-eye perspective challenges one of the central axioms of the humanities. Dawkins (1976, p. 214) points out that in our explanations we tend to overlook the fundamental fact that "a cultural trait may have evolved in the way it has simply because it is *advantageous to itself*." This is a new way of thinking about ideas, but is it a good way? When we have answered this question, we will know whether or not the meme meme is one we should exploit and replicate.

The first rules of memes, as for genes, is that replication is not necessarily for the good of anything; replicators flourish that are good at ... replicating—for whatever reason!

A meme that made its bodies run over cliffs would have a fate like that of a gene for making bodies run over cliffs. It would tend to be eliminated from the meme-pool.... But this does not mean that the ultimate criterion for success in meme selection is gene survival.... Obviously a meme that causes individuals bearing it to kill themselves has a grave disadvantage, but not necessarily a fatal one.... A suicidal meme can spread, as when a dramatic and well-publicized martyrdom inspires others to die for a deeply loved cause, and this in turn inspires others to die, and so on. [Dawkins 1982, pp. 110–11.]

The publicity Dawkins speaks about is crucial, and has a direct parallel in Darwinian medicine. As Williams and Nesse (1991) have pointed out, disease organisms (parasites, bacteria, viruses) depend for their long-term survival on hopping from host to host, and this carries important implications. Depending on how they are spread—through a sneeze or sexual contact, for instance, rather than via a mosquito that bites first an infected person and then an uninfected person—their future may hinge on their keeping their host up and about rather than on his deathbed. More benign variants will be favored by natural selection if the conditions for replication of the organisms can be rigged so that it is "in their interests" not to harm their hosts. By the same reasoning, we can see that benign or harmless memes will tend to flourish, other things being equal, and those that tend to be fatal to those whose minds carry them can only flourish if they have some way of publicizing themselves before—or while—they go down with the

---

9. A striking example of the vituperative and uncomprehending dismissal of Dawkins by a humanist who identifies him as a sociobiologist is found in Midgley 1979, an attack so wide of the mark that it should not be read without its antidote: Dawkins 1981. Midgley 1983 is an apologetic but still largely hostile rejoinder.

ship. Suppose Jones encounters or dreams up a truly compelling argument in favor of suicide—so compelling it leads him to kill himself. If he doesn't leave a note explaining why he has done this, the meme in question—at least the Jonesian lineage of it—is not going to spread.

The most important point Dawkins makes, then, is that there is no *necessary* connection between a meme's replicative power, its "fitness" from *its* point of view, and its contribution to *our* fitness (by whatever standard we judge that). This is an unsettling observation, but the situation is not totally desperate. Although some memes definitely manipulate us into collaborating on their replication *in spite of* our judging them useless or ugly or even dangerous to our health and welfare, many—most, if we are lucky—of the memes that replicate themselves do so not just with our blessings but *because of* our esteem for them. I think there can be little controversy that the following memes are, all things considered, good from our perspective, and not just from their own perspective as selfish self-replicators: such very general memes as cooperation, music, writing, calendars, education, environmental awareness, arms reduction; and such particular memes as the Prisoner's Dilemma, *The Marriage of Figaro*, *Moby Dick*, returnable bottles, the SALT agreements. Other memes are more controversial; we can see why they spread, and why, all things considered, we should tolerate them, in spite of the problems they cause for us: colorization of classic films, advertising on television, the ideal of political correctness. Still others are pernicious, but extremely hard to eradicate: anti-Semitism, hijacking airliners, spray-can graffiti, computer viruses.[10]

Our normal view of ideas is also a *normative* view: it embodies a canon or an ideal about which ideas we *ought* to accept or admire or approve of. In brief, we ought to accept the true and the beautiful. According to the normal view, the following are virtual tautologies—trivial truths not worth the ink to write them down:

Idea X was believed by the people because X was deemed true.

People approved of X because people found X to be beautiful.

These norms are not just dead obvious, they are *constitutive*: they set the rules whereby we think about ideas. We require explanations only when there are deviations from these norms. Nobody has to explain why a book purports to be full of *true* sentences, or why an artist might strive to make something *beautiful*—it just "stands to reason." The constitutive status of

---

10. Dawkins 1993 offers an important new perspective on computer viruses and their relation to other memes.

these norms grounds the air of paradox in such aberrations as "The Metropolitan Museum of Banalities" or "The Encyclopedia of Falsehoods." What requires special explanation in the normal view are the cases in which despite the truth or beauty of an idea it is *not* accepted, or despite its ugliness or falsehood it *is*.

The meme's-eye view purports to be an alternative to this normal perspective. What is tautological for *it* is:

Meme X spread among the people because X is a good replicator.

There is a nice parallel to be found in physics. Aristotelian physics supposed that an object's continuing to move in a straight line required explanation, in terms of something like forces continuing to act on it. Central to Newton's great perspective shift was the idea that such rectilinear motion did not require explanation; only deviations from it did—accelerations. An even better parallel can be seen in biology. Before Williams and Dawkins pointed to the alternative gene's-eye perspective, evolutionary theorists tended to think that it was just *obvious* that adaptations existed because they were good for the organisms. Now we know better. The gene-centered perspective is valuable precisely because it handles the "exceptional" cases in which the good of the organism counts for nothing, and shows how the "normal" circumstance is a derivative and exceptioned regularity, not a truth of pure reason, as it seemed to be from the old perspective.

The prospects for meme theory become interesting only when we look at the exceptions, the circumstances under which there is a pulling apart of the two perspectives. Only if meme theory permits us better to understand the deviations from the normal scheme will it have any warrant for being accepted. (Note that, in its own terms, whether or not the meme meme replicates successfully is strictly independent of its epistemological virtue; it might spread in spite of its perniciousness, or go extinct in spite of its virtue.)

Fortunately for us, there is a nonrandom correlation between the two perspectives, just as there is between what is good for General Motors and what is good for America. It is no accident that the memes that replicate tend to be good for us, not for our *biological* fitness (Williams' sardonic commentary on the churchgoers is absolutely right on that score), but for whatever it is we hold dear.[11] And never forget the crucial point: the facts

---

11. Memes that are (relatively) benign to their hosts but vicious to others are not uncommon, alas. When ethnic pride turns to xenophobia, for instance, this mirrors the phenomenon of a tolerable bacillus that mutates into something deadly—if not necessarily to its original carrier, then to others.

about whatever we hold dear—our highest values—are themselves very much a product of the memes that have spread most successfully. We may want to claim that *we* are in charge of what our *summum bonum* is to be, but this is mystical nonsense unless we admit that what *we* are (and hence what we might persuade ourselves to consider the *summum bonum*) is itself something we have learned to be, in outgrowing our animal heritage. Biology puts some constraint on what we could value; in the long run, we would not survive unless we had a better-than-chance habit of choosing the memes that help us, *but we haven't seen the long run yet.* Mother Nature's experiment with culture on this planet is only a few thousand generations old. Nevertheless, we have good reason to believe that our meme-immunological systems are not hopeless—even if they are not foolproof. We can rely, as a general, crude rule of thumb, on the coincidence of the two perspectives: by and large, the good memes—good by *our* standards—will tend to be the ones that are also the good replicators.

The haven all memes depend on reaching is the human mind, but a human mind is itself an artifact created when memes restructure a human brain in order to make it a better habitat for memes. The avenues for entry and departure are modified to suit local conditions, and strengthened by various artificial devices that enhance fidelity and prolixity of replication: native Chinese minds differ dramatically from native French minds, and literate minds differ from illiterate minds. What memes provide in return to the organisms in which they reside is an incalculable store of advantages—with some Trojan horses thrown in for good measure, no doubt. Normal human brains are not all alike; they vary considerably in size, shape, and the myriad details of connection on which their prowess depends. But the most striking differences in human prowess depend on microstructural differences (still inscrutable to neuroscience) induced by the various memes that have entered them and taken up residence. The memes enhance each other's opportunities: the meme for education, for instance, is a meme that reinforces the very process of meme implantation.

But if it is true that human minds are themselves to a very great degree the creations of memes, then we cannot sustain the polarity of vision we considered earlier; it cannot be "memes versus us," because earlier infestations of memes have already played a major role in determining who or what we are. The "independent" mind struggling to protect itself from alien and dangerous memes is a myth. There is a persisting tension between the biological imperative of our genes on the one hand and the cultural imperatives of our memes on the other, but we would be foolish to "side with" our genes; that would be to commit the most egregious error of pop sociobiology. Besides, as we have already noted, what makes us special is that we, alone among species, can rise above the imperatives of our genes—thanks to the lifting cranes of our memes.

What foundation, then, can we stand on as we struggle to keep our feet in the meme-storm in which we are engulfed? If replicative might does not make right, what is to be the eternal ideal relative to which "we" will judge the value of memes? We should note that the memes for normative concepts—for *ought* and *good* and *truth* and *beauty*—are among the most entrenched denizens of our minds. Among the memes that constitute us, they play a central role. Our existence as us, as what we as thinkers are—not as what we as organisms are—is not independent of these memes.

Dawkins ends *The Selfish Gene* (1976, p. 215) with a passage that many of his critics must not have read, or understood:

> We have the power to defy the selfish genes of our birth and, if necessary, the selfish memes of our indoctrination. . . . We are built as gene machines and cultured as meme machines, but we have the power to turn against our creators. We, alone on earth, can rebel against the tyranny of the selfish replicators.

In distancing himself thus forcefully from the oversimplifications of pop sociobiology, he somewhat overstates his case. This "we" that transcends not only its genetic creators but also its memetic creators is, we have just seen, a myth. Dawkins himself acknowledges that in his later work. In *The Extended Phenotype* (1982), Dawkins argues for the biological perspective that recognizes the beaver's dam, the spider's web, the bird's nest as not merely products of the phenotype—the individual organism considered as a functional whole—but parts of the phenotype, on a par with the beaver's teeth, the spider's legs, the bird's wing. From this perspective, the vast protective networks of memes that we spin is as integral to our phenotypes—to explaining our competences, our chances, our vicissitudes—as anything in our more narrowly biological endowment. (This claim is developed in greater detail in Dennett 1991a.) There is no radical discontinuity; one can be a mammal, a father, a citizen, a scholar, a Democrat, and an associate professor with tenure. Just as man-made barns are an integral part of the barn swallow's ecology, so cathedrals and universities—and factories and prisons—are an integral part of our ecology, as are the memes without which we could not live in these environments.

But if *I* am nothing over and above some complex system of interactions between my body and the memes that infest it, what happens to personal responsibility? How could *I* be held accountable for my misdeeds, or honored for my triumphs, if *I* am not the captain of my vessel? Where is the autonomy *I* need to act with free will?

"Autonomy" is just a fancy term for "self-control." When the Viking spacecraft got too far from Earth for the engineers in Houston to control it, they sent it a new program which removed it from their *remote* control and put it under local *self*-control (Dennett 1984, p. 55). That made it autonomous, and although the goals it continued to seek were the goals Houston had installed in it at its birth, it and it alone was responsible for making the decisions in furtherance of those goals. Now imagine it landed on some distant planet inhabited by tiny green men who promptly invaded it, tampering with its software and bending it (exapting it) to their own purposes—making it into a recreational vehicle, let's say, or a nursery for their young. Its autonomy would be lost as it came under the control of these alien controllers. Switching responsibility from my genes to my memes may seem to be a similarly unpromising step on the road to free will. Have we broken the tyranny of the selfish genes, only to be taken over by the selfish memes?

Think about symbionts again. Parasites are (by definition) those symbionts that are deleterious to the fitness of the host. Consider the most obvious meme example: the meme for celibacy (and chastity, I might add, to close a notorious loophole). This meme complex inhabits the brains of many a priest and nun. From the point of view of evolutionary biology, this complex is deleterious to *fitness* by definition: anything that virtually guarantees that the host's germ line is a cul-de-sac, with no further issue, lowers fitness. "But so what?" a priest might retort. "*I* don't *want* to have progeny!" Exactly. But, you might say, his body still does. He has distanced him*self* somewhat from his own body, in which the machinery designed by Mother Nature keeps right on running, sometimes giving *him* problems of self-control. How did this *self* or *ego* with the divergent goal get constituted? We may not know the detailed history of the infestation. The Jesuits famously say, "Give us the first five years of a child's life, and you can have the rest," so it may be very early in the priest's life that this particular meme secured a stronghold. Or it may have been later, and it may have happened very gradually. But whenever and however it happened, it has been incorporated by the priest—at least for the time being—into his identity.

I am *not* saying that because the priest's body is "doomed" to sire no offspring, this is a bad or "unnatural" thing. That would be to side with our selfish genes, which is exactly what we don't want to do. I *am* saying that this is just the most extreme, and hence vivid, example of the process that has made us all: our *selves* have been created out of the interplay of memes exploiting and redirecting the machinery Mother Nature has given us. My brain harbors the memes for celibacy and chastity (I couldn't write about them otherwise), but they never managed to get into the driver's seat in me. I do not *identify* with them. My brain also harbors the meme for fasting or dieting, and I wish I could get it more often into the driver's seat (so that I could more *wholeheartedly* diet), but, for one reason or another, the coalitions of memes that would incorporate the meme for dieting into my whole "heart" seldom form a government with long-term stability. No one meme rules anybody; what makes a person the person he or she is are the coalitions of memes that govern—that play the long-term roles in determining which decisions are made along the way. (We will look more closely at this idea in chapters 16 and 17.)

Whether or not the meme perspective can be turned into science, in its philosophical guise it has already done much more good than harm, contrary to what Gould has claimed, even though, as we shall see, there may be *other* applications of Darwinian thinking in the social sciences that truly deserve Gould's condemnation. What, in fact, is the alternative to this through-and-through Darwinian vision of a mind? A last hope for the Darwin-dreaders is simply to deny that what happens to memes when they enter a mind could ever, ever be explained in "reductionistic," mechanistic terms. One way would be to espouse outright Cartesian dualism: the mind just can't be the brain, but, rather, some *other* place, in which great and mysterious alchemical processes occur, transforming the raw materials they are fed—the cultural items we are calling memes—into new items that transcend their sources in ways that are simply beyond the ken of science.[12]

A slightly less radical way of supporting the same defensive view is to concede that the mind is, after all, just the brain, which is a physical entity bound by all the laws of physics and chemistry, but insist that it nevertheless does its chores in ways that defy scientific analysis. This view has often been suggested by the linguist Noam Chomsky and enthusiastically defended by his former colleague the philosopher/psychologist Jerry Fodor (1983), and more recently by another philosopher, Colin McGinn (1991). We can see that this is a *saltational* view of the mind, positing great leaps in Design Space that get "explained" as acts of sheer genius or intrinsic creativity or something else science-defying. It insists that somehow the brain itself is a skyhook, and refuses to settle for what the wily Darwinian offers: the brain, thanks to all the cranes that have formed it in the first place, and all the cranes that have entered it in the second place, is itself a prodigious, but not mysterious, lifter in Design Space.

It will take some further work to turn this highly metaphorical confrontation into a more literal one, and resolve it, in chapter 13. Fortunately for me, much of this work has already been done by me, so I can once again avoid reinventing the wheel by simply reusing a wheel I've made before. My next exaptation is from my 1992 Darwin Lecture at Darwin College, Cambridge (Dennett 1994b).

---

12. Lewontin, Rose, and Kamin (1984, p. 283) claim that memes presuppose a "Cartesian" view of the mind, whereas in fact memes are a key (central but optional) ingredient in the best alternatives to Cartesian models (Dennett 1991a).

CHAPTER 12: *The invasion of human brains by culture, in the form of memes, has created human minds, which alone among animal minds can conceive of things distant and future, and formulate alternative goals. The prospects for elaborating a rigorous science of memetics are doubtful, but the concept provides a valuable perspective from which to investigate the complex relationship between cultural and genetic heritage. In particular, it is the shaping of our minds by memes that gives us the autonomy to transcend our selfish genes.*

CHAPTER 13: *A series of ever more powerful types of mind can be defined in terms of the Tower of Generate-and-Test, which takes us from the crudest trial-and-error learners to the community of scientists and other serious human thinkers. Language plays the crucial role in this cascade of cranes, and Noam Chomsky's pioneering work in linguistics opens up the prospect of a Darwinian theory of language, but this is a prospect he has mistakenly shunned, along with Gould. The controversies surrounding the development in recent years of a science of the mind have been sadly amplified into antagonisms by misperceptions on both sides: are the critics calling for cranes or skyhooks?*

CHAPTER THIRTEEN

Losing Our Minds to Darwin

⌁⌁

1. THE ROLE OF LANGUAGE IN INTELLIGENCE

When ideas fail, words come in very handy.

—ANONYMOUS[1]

We are not like other animals; our minds set us off from them. That is the claim that inspires such passionate defense. It is curious that people who want so much to defend this difference should be so reluctant to examine the evidence in its favor coming from evolutionary biology, ethology, primatology, and cognitive science. Presumably, they are afraid they might learn that, although we are different, we aren't different *enough* to make the life-defining difference they cherish. For Descartes, after all, the difference was absolute and metaphysical: animals were just mindless automata; *we* have souls. Descartes and his followers have suffered calumny over the centuries at the hands of animal-lovers who have deplored his claim that animals have no souls. More theoretically minded critics have deplored his faintheartedness from the opposite pole: how could such a sound, ingenious mechanist flinch so badly when it came to making an exception for humanity? *Of course* our minds are our brains, and hence are ultimately just stupendously complex "machines"; the difference between us and other animals is one of huge degree, not metaphysical kind. It is no coincidence, I have shown, that those who deplore Artificial Intelligence are also those who deplore evolutionary accounts of human mentality: if human minds are

---

1. This *bon mot* appeared in the *Tufts Daily*, attributed to Johann Wolfgang von Goethe, but I daresay it is a meme of more recent birth.

nonmiraculous products of evolution, then they are, in the requisite sense, artifacts, and all their powers must have an ultimately "mechanical" explanation. We are descended from macros and made of macros, and nothing we can do is anything beyond the power of huge assemblies of macros (assembled in space and time).

Still, there is a huge difference between our minds and the minds of other species, a gulf wide enough even to make a moral difference. It is—it must be—due to two intermeshed factors, each of which requires a Darwinian explanation: (1) the brains we are born with have features lacking in other brains, features that have evolved under selection pressure over the last six million years or so, and (2) these features make possible an enormous elaboration of powers that accrue from the sharing of Design wealth through cultural transmission. The pivotal phenomenon that unites these two factors is language. We human beings may not be the most admirable species on the planet, or the most likely to survive for another millennium, but we are without any doubt at all the most intelligent. We are also the only species with language.

Is that true? Don't whales and dolphins, vervet monkeys and honeybees (the list goes on) have languages *of sorts*? Haven't chimpanzees in laboratories been taught rudimentary languages *of sorts*? Yes, and body language is a sort of language, and music is the international language (sort of), and politics is a sort of language, and the complex world of odor and olfaction is another, highly emotionally charged language, and so on. It sometimes seems that the highest praise we can bestow on a phenomenon we are studying is the claim that its complexities entitle it to be called a language—of sorts. This admiration for language—real language, the sort only we human beings use—is well founded. The expressive, information-encoding properties of real language are practically limitless (in at least some dimensions), and the powers that other species acquire in virtue of their use of proto-languages, hemi-semi-demi-languages, are indeed similar to the powers we acquire thanks to our use of real language. These other species do climb a few steps up the mountain on whose summit we reside, thanks to language. Looking at the vast differences between their gains and ours is one way of approaching the question we now must address: just how does language contribute to intelligence?

What varieties of thought require language? What varieties of thought (if any) are possible without language? We watch a chimpanzee, with her soulful face, inquisitive eyes, and deft fingers, and we very definitely get a sense of the mind within, but, the more we watch, the more our picture of her mind swims before our eyes. In some ways she is so human, so insightful; yet we soon learn (to our dismay or relief, depending on our hopes) that in other ways she is so dense, so uncomprehending, so unreachably cut off from our human world. How could a chimp who so obviously understands *A* fail to understand *B*? Consider a few simple questions about chimpanzees.

Could they learn to tend a fire—could they gather firewood, keep it dry, preserve the coals, break the wood, keep the fire size within proper bounds? And if they couldn't invent these novel activities on their own, could they be trained by human beings to do these things? Here's another question. Suppose you imagine something novel—I hereby invite you to imagine a man climbing up a rope with a plastic garbage-pail over his head. An easy mental task for you. Could a chimpanzee do the same thing in her mind's eye? I wonder. I chose the elements—man, rope, climbing, pail, head—as familiar objects in the perceptual and behavioral world of a laboratory chimp, but I wonder whether a chimp could *put them together* in this novel way—even by accident, as it were. You were provoked to perform your mental act by my verbal suggestion, and probably you often perform similar mental acts on your own in response to verbal suggestions you give your-self—not out loud, but definitely in words. Could it be otherwise? Could a chimpanzee get itself to perform such a mental act without the help of verbal suggestion?

These are rather simple questions about chimpanzees, but nobody knows the answers—yet. The answers are not impossible to acquire, but not easy either; controlled experiments could yield the answers, which would shed light on the role of language in turning brains into minds like ours. I raise the question about whether chimpanzees could learn to tend a fire because, at some point in prehistory, our ancestors tamed fire. Was language necessary for this great civilizing advance? Some of the evidence suggests that it happened hundreds of thousands of years—or even as much as a million years (Donald 1991, p. 114)—*before* the advent of language, but of course *after* our hominid line split away from the ancestors of modern apes, such as chimpanzees. Opinions differ sharply. Many researchers are convinced that language began much earlier, in plenty of time to underwrite the tam-ing of fire (Pinker 1994). We might even try to argue that the taming of fire is itself incontrovertible evidence for the existence of early language—if we can just convince ourselves that this mental feat *required* rudimentary lan-guage. Or is fire-tending not such a big deal? Perhaps the only reason we don't find chimps in the wild sitting around campfires is that in their rainy habitats there is never enough tinder around to give fire a chance to be tamed. (Sue Savage-Rumbaugh's pygmy chimps in Atlanta love to go on picnics in the woods, and enjoy staring into the campfire's flames, just as we do, but she tells me she doubts that they could be relied on to tend a fire, even with training.)

If termites can create elaborate, well-ventilated cities of mud, and wea-verbirds can weave audaciously engineered hanging nests, and beavers can build dams that take months to complete, couldn't chimpanzees tend a simple campfire? This rhetorical question climbs a misleading ladder of abilities. It ignores the independently well-evidenced possibility that there are two profoundly different ways of building dams: the way beavers do and the way we do. The differences are not necessarily in the products, but in the control structures within the brains that create them. A child might study a weaverbird building its nest, and then replicate the nest herself, finding the right pieces of grass, and weaving them in the right order, creating, by the very same series of steps, an identical nest. A film of the two building processes occurring side by side might overwhelm us with a sense that we were seeing the same phenomenon twice, but it would be a big mistake to impute to the bird the sort of thought processes we know or imagine to be going on in the child. There could be very little in common between the processes going on in the child's brain and in the bird's brain. The bird is (apparently) endowed with a collection of interlocking special-purpose minimalist subroutines, well designed by evolution according to the notorious *need-to-know principle* of espionage: give each agent as little information as will suffice for it to accomplish its share of the mission.

Control systems designed under this principle can be astonishingly successful—witness the birds' nests, after all—whenever the environment has enough simplicity and regularity, and hence predictability, to favor pre-design of the whole system. The system's very design in effect makes a prediction—a wager, in fact—that the environment will be the way it must be for the system to work. When the complexity of encountered environments rises, however, and unpredictability becomes a more severe problem, a different design principle kicks in: the *commando-team principle*, illustrated by such films as *The Guns of Navarone*: give each agent as much knowledge about the total project as possible, so that the team has a chance of ad-libbing appropriately when unanticipated obstacles arise.

So there is a watershed in the terrain of evolutionary Design Space; when a control problem lies athwart it, it could be a matter of chance which direction evolution propels the successful descendants. Perhaps, then, there are two ways of tending fires—roughly, the beaver-dam way and our way. If so, it's a good thing for us that our ancestors didn't hit upon the beaver-dam way, for if they had, the woods might today be full of apes sitting around campfires, but we would not be here to marvel at them.

I want to propose a framework in which we can place the various design options for brains, to see where their power comes from. It is an outrageously oversimplified structure, but idealization is the price one should often be willing to pay for synoptic insight. I call it the Tower of Generate-and-Test; as each new floor of the Tower gets constructed, it empowers the organisms at that level to find better and better moves, and find them more efficiently.[2]

---

2. This is an elaboration of ideas I first presented in Dennett 1975. I recently discovered that Konrad Lorenz (1973) described a similar cascade of cranes—in different terms, of course.

In the beginning—once the pump had been primed—there was Darwinian evolution of species by natural selection. A variety of candidate organisms were blindly generated by more or less arbitrary processes of
recombination and mutation of genes. These organisms were field-tested,
and only the best designs survived. This is the ground floor of the Tower. Let
us call its inhabitants *Darwinian creatures*.

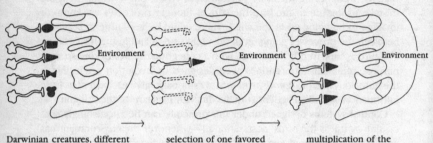

| Darwinian creatures, different "hard-wired" phenotypes | selection of one favored phenotype | multiplication of the favored genotype |

FIGURE 13.1

This process went through many millions of cycles, producing many
wonderful designs, both plant and animal, and eventually among its novel
creations were some designs with the property of phenotypic plasticity. The
individual candidate organisms were not wholly designed at birth, or, in
other words, there were elements of their design that could be adjusted by
events that occurred during the field tests. (This is what makes the Baldwin
Effect possible, as we saw in chapter 3, but now we are going to focus on the
intra-organismic design that sets up that crane.) Some of these candidates,
we may suppose, were no better off than their hard-wired cousins, since
they had no way of favoring (selecting for an encore) the behavioral options
they were equipped to "try out," but others, we may suppose, were fortunate enough to have wired-in "reinforcers" that happened to favor Smart
Moves, actions that were better for their agents. These individuals thus
confronted the environment by generating a variety of actions, which they
tried out, one by one, until they found one that worked. We may call this
subset of Darwinian creatures, the creatures with conditionable plasticity,
Skinnerian creatures, since, as B. F. Skinner was fond of pointing out, operant conditioning is not just analogous to Darwinian natural selection; it is
continuous with it. "Where inherited behavior leaves off, the inherited
modifiability of the process of conditioning takes over" (Skinner 1953,
p. 83).

Skinnerian conditioning is a fine capacity to have, so long as you are not
killed by one of your early errors. A better system involves *preselection*
among all the possible behaviors or actions, weeding out the truly stupid

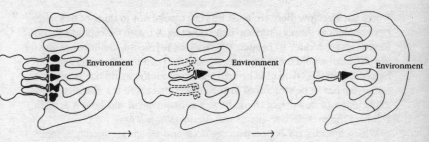

Skinnerian creature "blindly"   ... until one is selected by   Next time, the creature's first
tries different responses ...   "reinforcement."   choice will be the reinforced response.

FIGURE 13.2

options before risking them in the harsh world. We human beings are
creatures capable of this third refinement, but we are not alone. We may call
the beneficiaries of this third story in the Tower *Popperian creatures*, since,
as Sir Karl Popper once elegantly put it, this design enhancement "permits
our hypotheses to die in our stead." Unlike the merely Skinnerian creatures,
many of whom survive only because they make lucky first moves, Popperian
creatures survive because they're smart enough to make better-than-chance
first moves. Of course, they're just lucky to be smart, but that's better than
just being lucky.

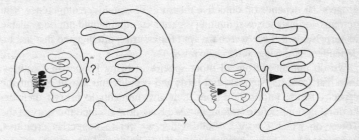

Popperian creature has an inner   First time, the creature acts in a
selective environment that previews   foresightful way (better than chance).
candidate acts.

FIGURE 13.3

But how is this preselection in Popperian agents to be done? Where is the
feedback to come from? It must come from a sort of *inner environment*—an
inner something-or-other that is structured in such a way that the surrogate
actions it favors are more often than not the very actions the real world would
also bless, if they were actually performed. In short, the inner environment,
whatever it is, must contain lots of *information* about the outer environment
and its regularities. Nothing else (except magic) could provide preselection

worth having. Now, here we must be very careful not to think of this inner environment as simply a replica of the outer world, with all its physical contingencies reproduced. (In such a miraculous toy world, the little hot stove in your head would be hot enough actually to burn the little finger in your head that you placed on it!) The information about the world has to be there, but it also has to be structured in such a way that there is a nonmiraculous explanation of how it got there, how it is maintained, and how it actually achieves the preselective effects that are its *raison d'être*.

Which animals are Popperian creatures, and which are merely Skinnerian? Pigeons were Skinner's favorite experimental animals, and he and his followers developed the technology of operant conditioning to a very sophisticated level, getting pigeons to exhibit quite bizarre and sophisticated learned behaviors. Notoriously, the Skinnerians never succeeded in proving that pigeons were *not* Popperian creatures, and research on a host of different species, from octopuses to fish to mammals, strongly suggests that if there are any purely Skinnerian creatures, capable only of blind trial-and-error learning, they are to be found among the simple invertebrates. The sea slug *Aplysia* has more or less replaced the pigeon as the focus of attention among those who study the mechanisms of simple conditioning. (Researchers unhesitatingly and uncontroversially rank species in terms of how intelligent they are. This involves no myopic endorsement of the Great Chain of Being, no unwarranted assumptions about climbing the ladder of progress. It depends on objective measures of cognitive competence. The octopus, for instance, is stunningly smart, a fact that would not be available to surprise us if there weren't ways of measuring intelligence that are independent of phylogenetic chauvinism.)

We do not differ from all other species in being Popperian creatures, then. Far from it; mammals and birds and reptiles and fish all exhibit the capacity to use information from their environments to presort their behavioral options before striking out. We have now reached the story of the Tower on which I want to build. Once we get to Popperian creatures, creatures whose brains have the potential to be shaped into inner environments with preselective prowess, what happens next? How does new information about the outer environment get incorporated into these brains? This is where *earlier* design decisions come back to haunt—to constrain— the designer. In particular, choices that evolution has already made between need-to-know and commando-team now put major constraints on the options for design improvement. If a particular species' brain design has already gone down the need-to-know path with regard to some control problem, only minor modifications (fine tuning, you might say) can be *readily* made to the existing structures, so the only hope of making a major revision of the internal environment to account for new problems, new features of the external environment that matter, is to *submerge* the old hard-wiring under a new layer of pre-emptive control (a theme developed in the work of the AI researcher Rodney Brooks [e.g., 1991]). It is these higher levels of control that have the potential for vast increases in versatility. And it is at these levels in particular that we should look for the role of language (when it finally arrives on the scene), in turning *our* brains into virtuoso preselectors.

We engage in our share of rather mindless routine behavior, but our important acts are often directed on the world with incredible cunning, composing projects exquisitely designed under the influence of vast libraries of information about the world. The instinctual actions we share with other species show the benefits derived from the harrowing explorations of our ancestors. The imitative actions we share with some higher animals may show the benefits of information gathered not just by our ancestors, but also by our social groups over generations, transmitted nongenetically by a "tradition" of imitation. But our more deliberatively planned acts show the benefits of information gathered and transmitted by our conspecifics in every culture, including, moreover, items of information that no single individual has embodied or understood in any sense. And though some of this information may be of rather ancient acquisition, much of it is brandnew. When comparing the time scales of genetic and cultural evolution, it is useful to bear in mind that we today—every one of us—can *easily* understand many ideas that were simply unthinkable *by the geniuses* in our grandparents' generation!

The successors to mere Popperian creatures are those whose inner environments are informed by the *designed* portions of the outer environment. We may call this sub-sub-subset of Darwinian creatures *Gregorian creatures*, since the British psychologist Richard Gregory is to my mind the pre-eminent theorist of the role of information (or, more exactly, what Gregory calls Potential Intelligence) in the creation of Smart Moves (or what Gregory calls Kinetic Intelligence). Gregory observes that a pair of scissors, as a well-designed artifact, is not just a result of intelligence, but an endower of intelligence (external Potential Intelligence), in a very straightforward and intuitive sense: when you give someone a pair of scissors, you enhance his potential to arrive more safely and swiftly at Smart Moves (Gregory 1981, pp. 311ff.).

Anthropologists have long recognized that the advent of tool use accompanied a major increase in intelligence. Chimpanzees in the wild fish for termites with crudely prepared fishing sticks. This fact takes on further significance when we learn that not all chimpanzees have hit upon the trick; in some chimpanzee "cultures," termites are a present but unexploited food source. That reminds us that tool use is a two-way sign of intelligence; not only does it *require* intelligence to recognize and maintain a tool (let alone fabricate one), but tool use *confers* intelligence on those who are lucky enough to be given the tool. The better designed the tool (the more information embedded in its fabrication), the more Potential Intelligence it confers on its user. And among the pre-eminent tools, Gregory reminds us, are what he calls "mind-tools": words.

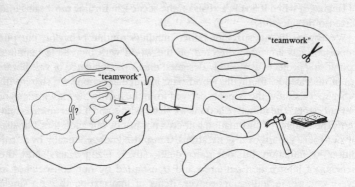

Gregorian creature imports mind-tools
from the (cultural) environment; these
improve both the generators and the
testers.

FIGURE 13.4

Words and other mind-tools give a Gregorian creature an inner environment that permits it to construct ever more subtle move-generators and move-testers. Skinnerian creatures ask themselves, "What do I do next?" and haven't a clue how to answer until they have taken some hard knocks. Popperian creatures make a big advance by asking themselves, "What should I think about next?" before they ask themselves, "What should I do next?" Gregorian creatures take a further big step by learning how to think better about what they should think about next—and so forth, a tower of further internal reflections with no fixed or discernible limit.

What happens to a human or hominid brain when it becomes equipped with words? In particular, what is the shape of this environment when words first enter it? It is definitely *not* an even playing field or a *tabula rasa*. Our newfound words must anchor themselves on the hills and valleys of a landscape of considerable complexity. Thanks to earlier evolutionary pressures, our innate quality spaces are species-specific, narcissistic, and even idiosyncratic from individual to individual. A number of investigators are currently exploring portions of this terrain. The psychologist Frank Keil (1992) and his colleagues at Cornell have evidence that certain highly abstract concepts—such as the concepts of *being alive* or *ownership*, for instance—have a genetically imposed head start in the young child's kit of mind-tools; when the specific words for owning, giving and taking, keeping and hiding, and their kin enter a child's brain, they find homes already partially built for them. Ray Jackendoff (1993) and other linguists have identified fundamental structures of spatial representation—notably designed to enhance the control of *locomotion* and the *placement* of movable things—that underlie our intuitions about concepts like *beside*, *on*, *behind*, and their kin. Nicholas Humphrey (1976, 1983, 1986) has argued that there must be a genetic predisposition for adopting the intentional stance, and Alan Leslie (1992) and others have developed evidence for this, in the form of what he calls a "theory of mind module" designed to generate second-order beliefs (beliefs about the beliefs and other mental states of others). Some autistic children seem to be well described as suffering from the disabling of this module, for which they can occasionally make interesting compensatory adjustments. (For an overview, see Baron-Cohen 1995.) So the words (and hence memes) that take up residence in a brain, like so many earlier design novelties we have considered, enhance and shape preexisting structures, rather than generating *entirely* new architectures (see Sperber [in press] for a Darwinian overview of this exaptation of genetically provided functions by culturally transmitted functions). Though these newly redesigned functions are not made from whole cloth, they do create an explosive new capacity to look ahead.

> An internal model allows a system to look ahead to the future consequences of current actions, without actually committing itself to those actions. In particular, the system can avoid acts that would set it irretrievably down some road to future disaster ("stepping off a cliff"). Less dramatically, but equally important, the model enables the agent to make current "stage-setting" moves that set up later moves that are obviously advantageous. The very essence of a competitive advantage, whether it be in chess or economics, is the discovery and execution of stage-setting moves. [Holland 1992, p. 25.]

This, then, is the crane to end all cranes: an explorer that *does* have foresight, that can see beyond the immediate neighborhood of options. But how good can the "stage-setting" be without the intervention of language to help control the manipulation of the model? How intricate and long-range can the look-ahead be, for instance? This is the relevance of my question about the chimpanzee's capacities to visualize a novel scene. Darwin (1871, p. 57) was convinced that language was the prerequisite for "long trains of thought," and this claim has been differently supported by several recent theorists, especially Julian Jaynes (1976) and Howard Margolis (1987). Long trains of thought have to be controlled, or they will wander off into delicious if futile woolgathering. These authors suggest, plausibly, that the self-exhortations and reminders made possible by language are actually essential to maintaining the sorts of long-term projects only we human beings engage in (unless, like the beaver, we have a built-in specialist for completing a particular long-term project). (For further explorations of these topics, see Clark and Karmiloff-Smith 1994, Dennett 1994c.)

This brings me to the final step up the Tower of Generate-and-Test. There is one more embodiment of that wonderful idea, and it is the one that gives our minds their greatest power: once we have language—a bountiful kit of mind-tools—we can *use* these tools in the structure of deliberate, foresightful generate-and-test known as *science*. All the other varieties of generate-and-test are willy-nilly.

The soliloquy that accompanies the errors committed by the lowliest Skinnerian creature might be "Well, I mustn't do *that* again!" and the hardest lesson for any agent to learn, apparently, is how to learn from its own mistakes. In order to learn from them, one has to be able to contemplate them, and this is no small matter. Life rushes on, and unless one has developed positive strategies for recording one's tracks, the task known in AI as *credit assignment* (also known, of course, as "blame assignment") is insoluble. The advent of high-speed still photography was a revolutionary technological advance for science because it permitted human beings, for the first time, to examine complicated temporal phenomena not in real time but *in their own good time*—in leisurely, methodical backtracking analysis of the traces they had created of those complicated events. Here a technological advance carried in its wake a huge enhancement in cognitive power. The advent of language was an exactly parallel boon for human beings, a technology that created a whole new class of objects-to-contemplate, verbally embodied surrogates that could be reviewed in any order at any pace. And this opened up a new dimension of self-improvement—all one had to do was learn to savor one's own mistakes.

Science, however, is not just a matter of making mistakes, but of making mistakes in public. Making mistakes for all to see, in the hopes of getting the others to help with the corrections. It has been plausibly maintained, by Nicholas Humphrey, David Premack (1986), and others, that chimpanzees are natural psychologists—what I would call second-order intentional systems, capable of adopting the intentional stance towards other things. This is not surprising if our own innate equipment includes a theory-of-mind module, as Leslie, Baron-Cohen, and others have maintained, for perhaps this is part of the endowment chimpanzees and we inherit from a common ancestor. But even if chimpanzees are, like us, innately equipped as natural psychologists, they nevertheless lack a crucial feature shared by all human natural psychologists, folk and professional varieties: they never get to compare notes. They never dispute over attributions, and ask to know the grounds for each other's conclusions. No wonder their comprehension is so limited. Ours would be, too, if we had to generate it all on our own.

Let me sum up the results of this rather swift survey. Our human brains, and only human brains, have been armed by habits and methods, mind-tools and information, drawn from millions of other brains which are not ancestral to our own brains. This, amplified by the *deliberate, foresightful* use of generate-and-test in science, puts our minds on a different plane from the minds of our nearest relatives among the animals. This species-specific process of enhancement has become so swift and powerful that a single generation of its design improvements can now dwarf the R-and-D efforts of millions of years of evolution by natural selection. Comparing our brains anatomically with chimpanzee brains (or dolphin brains or any other non-human brains) would be almost beside the point, because our brains are in effect joined together into a single cognitive system that dwarfs all others. They are joined by an innovation that has invaded our brains and no others: language. I am not making the foolish claim that all our brains are knit together by language into one gigantic mind, thinking its transnational thoughts, but, rather, that each individual human brain, thanks to its communicative links, is the beneficiary of the cognitive labors of the others in a way that gives it unprecedented powers.

Naked animal brains are no match at all for the heavily armed and outfitted brains we carry in our heads. This fact reverses the burden of proof in what would otherwise be a compelling argument: the claim, first considered by the linguist Noam Chomsky (1975) and more recently defended by the philosophers Jerry Fodor (1983) and Colin McGinn (1991), that our minds, like those of all other species, must suffer "cognitive closure" with regard to some topics of inquiry. Spiders can't contemplate the concept of fishing; birds (some of whom are excellent at fishing) aren't up to thinking about democracy. What is inaccessible to the dog or the dolphin may be readily grasped by the chimp, but the chimp in turn will be cognitively closed to some domains we human beings have no difficulty thinking about. Chomsky and company ask a rhetorical question: What makes us think we are different? Aren't there bound to be strict limits on what *Homo sapiens* may conceive?

According to Chomsky, all matters of human puzzlement can be sorted into "problems," which can be solved, and "mysteries," which cannot. The problem of free will, Chomsky opines, is one such mystery.[3] The problem of consciousness, according to Fodor, is another, and McGinn concurs. As the author of books (1984, 1991a) that claim to explain each of these

---

3. In fairness to Chomsky, all he says is that free will *might* be a mystery. "I am not urging this conclusion, but merely noting that it is not to be ruled out *a priori*" (Chomsky 1975, p. 157). This mild suggestion has been eagerly inflated by others into a scientifically based demonstration!

impenetrable mysteries, I can be expected to disagree, but this is not the place to pursue such issues. Since neither Chomsky nor Fodor thinks he himself can explain free will or consciousness, the claim that it is humanly impossible is doctrinally convenient for them, perhaps, but also in considerable tension with another claim of theirs. In other moods, they have both (correctly) hailed the capacity of the human brain to "parse," and hence presumably understand, the official infinity of grammatical sentences of a natural language such as English. If we can understand all the sentences (in principle), couldn't we understand the ordered sets of sentences that best express the solutions to the problems of free will and consciousness? After all, one of the volumes in the Library of Babel is—must be—the best statement in fewer than five hundred pages of short grammatical English sentences of the solution to the problem of free will, and another is the optimal job in English on consciousness.[4] I daresay neither of my books is either of those, but that's life. I can't believe that Chomsky or Fodor would declare either of those books (or the trillions of runners-up) to be incomprehensible to a normal English reader.[5] So perhaps they think that the mysteries of free will and consciousness are so deep that no book, of any length, in any language, could explain them to any intelligent being. But *that* claim has absolutely no evidence in its favor to be derived from any biological considerations. It must have, um, fallen from the sky.

Consider the "closure" argument in more detail. "What is closed to the mind of a rat may be open to the mind of a monkey, and what is open to us may be closed to the monkey" (McGinn 1991, p. 3). Monkeys, for instance, can't grasp the concept of an electron, McGinn reminds us, but I think we

---

4. Two other books in the Library are the most compelling "refutations" of these masterpieces, but of course the Library doesn't contain any refutations, properly so-called, of any of the *true* books on its shelves. These hatchet jobs must be merely *apparent* refutations—an example of a fact that must be true but is systematically useless, since we could never tell which books were which, without the help of, say, God. The existence of this sort of fact will become important in chapter 15.

5. Chomsky has in fact revised his earlier views about the nature of language, making a distinction these days between "E-language" (the external—and you might say eternal—Platonic object, English, in which so many of the books in the Library of Babel are written) and "I-language" (the internal, intensional, idiolect of an individual), and he denies that E-language is a proper object for scientific study, so he would probably object to the straightforward way I have run this objection (Steven Pinker [personal communication]). But there are more devious ways of running the argument and appealing only to the I-language of individuals. Can Chomsky or anyone else give a good reason for believing that any five-hundred-page book of short sentences meeting the I-language standards of any normal, literate individual would be incomprehensible ("in principle") to that person?

should be unimpressed by the example, for not only can the monkey not understand the answers about electrons, it can't understand the questions (Dennett 1991d). The monkey isn't *baffled*, not even a little bit. We definitely understand the questions about free will and consciousness well enough to know what we're baffled by (if we are), so until Chomsky and Fodor and McGinn can provide us with clear cases of animals (or people) who can be baffled by questions whose true answers could not unbaffle them, they have given us no evidence of the reality or even likelihood of "cognitive closure" in human beings.[6]

Their argument is presented as a biological, naturalistic argument, reminding us of our kinship with the other beasts, and warning us not to fall into the ancient trap of thinking "how like an angel" we human "souls" are, with our "infinite" minds. But it is in fact a pseudo-biological argument, one that, by ignoring the actual biological details, misdirects us away from the case that can be made for taking one species—our species—right off the scale of intelligence that ranks the pig above the lizard and the ant above the oyster. We certainly cannot rule out the possibility in principle that our minds will be cognitively closed to some domain or other. In fact, as we shall see in more detail in chapter 15, we can be certain that there are realms of no doubt fascinating and important knowledge that our species, in its actual finitude, will never enter, not because we will butt our heads against some stone wall of utter incomprehension, but because the Heat Death of the universe will overtake us before we can get there. This is not, however, a limitation due to the frailty of our animal brains, a dictate of "naturalism." On the contrary, a proper application of Darwinian thinking suggests that *if* we survive our current self-induced environmental crises, our capacity to comprehend will continue to grow by increments that are now incomprehensible to us.

Why shouldn't Chomsky and Fodor and McGinn love this conclusion? It grants to human minds—and only to human minds—an indefinitely expanding dominion over the puzzles and problems of the universe, with no limits in sight. What could be more wonderful than that? The trouble is, I suspect, that they deem the *means* to be unsatisfactory; if the mind's power is due to cranes, not skyhooks, they would just as soon settle for mystery. That attitude, at any rate, has often surfaced in these controversies, and Chomsky has been a primary source of authority for it.

------

6. Fodor has bitten this bullet: "Nobody has the slightest idea how anything material could be conscious. Nobody even knows what it would be like to have the slightest idea about how anything material could be conscious" (Fodor 1992). In other words, if you so much as *think* you understand the question of consciousness, you're mistaken. Take his word for it—and change the subject, please.

2. CHOMSKY CONTRA DARWIN: FOUR EPISODES

*Chomsky, one might think, would have everything to gain by ground-
ing his controversial theory about a language organ in the firm foundation of evolutionary theory, and in some of his writings he has hinted
at a connection. But more often he is skeptical.*

—STEVEN PINKER 1994, p. 355

*In the case of such systems as language or wings it is not easy even to
imagine a course of selection that might have given rise to them.*

—NOAM CHOMSKY 1988, p. 167

*A sizeable gulf in communication still exists between cognitive scien-
tists who entered the field from AI or from the study of problem solving
and concept-forming behavior, on the one side, and those who entered
from a concern with language, on the other. . . . When the uniqueness
of language processes as a human faculty is emphasized, as it has been
by Chomsky . . . , the gulf becomes wider.*

—HERBERT SIMON and CRAIG KAPLAN 1989, p. 5

On September 11, 1956, at MIT, three papers were presented at a meeting of the Institute for Radio Engineers. One was by Allen Newell and
Herbert Simon (1956), "The Logic Theory Machine," and in it they showed,
for the first time, how a computer could prove nontrivial theorems of logic.
Their "machine" was the father (or grandfather) of their General Problem
Solver (Newell and Simon 1963), and the prototype for the computer
language Lisp, which is to Artificial Intelligence roughly what the DNA code
is to genetics. The Logic Theory Machine is a worthy rival of Art Samuel's
checkers program for the honor of AI-Adam. Another paper was by the
psychologist George A. Miller, "The Magical Number Seven, Plus or Minus
Two," which went on to be one of the classic papers inaugurating the field
of cognitive psychology (Miller 1956). The third paper was by a twenty-
seven-year-old Junior Fellow at Harvard, Noam Chomsky, "Three Models for
the Description of Language" (1956). Retrospective coronations are always
a bit arbitrary, as we have seen several times, but Chomsky's talk to the IRE
is as good an event as any to mark the birth of modern linguistics. Three
major new scientific disciplines born in the same room on a single day—I
wonder how many in the audience had the sense that they were participat-
ing in a historic event of such proportions. George Miller did, as he tells us
in his account (1979) of that meeting. Herbert Simon's own retrospective
view of the occasion has changed over the years. In his 1969 book, he drew
attention to the remarkable occasion and said of it (p. 47): "Thus the two

bodies of theory [linguistics and Artificial Intelligence] have had cordial relations from an early date. And quite rightly, for they rest conceptually on the same view of the human mind." If only that were true! By 1989, he could see how the gulf had widened.

Not many scientists are great scientists, and not many great scientists get to found a whole new field, but there are a few. Charles Darwin is one; Noam Chomsky is yet another. In much the way there was biology before Darwin—natural history and physiology and taxonomy and such—all united by Darwin into what we know as biology today, so there was linguistics before Chomsky. The contemporary scientific field of linguistics, with its subdisciplines of phonology, syntax, semantics, and pragmatics, its warring schools and renegade offshoots (computational linguistics in AI, for instance), its subdisciplines of psycholinguistics and neurolinguistics, grows out of various scholarly traditions going back to pioneer language sleuths and theorists from the Grimm brothers to Ferdinand de Saussure and Roman Jakobson, but it was all unified into a richly interrelated family of scientific investigations by the theoretical advances first proposed by one pioneer, Noam Chomsky. His slender 1957 book, *Syntactic Structures*, was an application to natural languages such as English of the results of an ambitious theoretical investigation he had undertaken into yet another region of Design Space: the logical space of all possible algorithms for generating and recognizing the sentences of all possible languages. Chomsky's work followed closely in the path of Turing's purely logical investigations into the powers of what we now call computers. Chomsky eventually defined an ascending scale of types of grammars or types of languages—the Chomsky Hierarchy, on which all students of computation theory still cut their teeth—and showed how these grammars were interdefinable with an ascending scale of types of automata or computers—from "finite state machines" through "push-down automata" and "linear bounded machines" to "Turing machines."

I can vividly remember the shock wave that rolled through philosophy when Chomsky's work first came to our attention a few years later. In 1960, my sophomore year at Harvard, I asked Professor Quine what critics of his views I should be reading. (I considered myself at the time to be an anti-Quinian of ferocious conviction, and was already beginning to develop the arguments for my senior thesis, attacking him. Anybody who was arguing against Quine was somebody I had to know about!) He immediately suggested that I should look at the work of Noam Chomsky, an author few in philosophy had heard of at the time, but his fame soon engulfed us all. Philosophers of language were divided in their response to his work. Some loved it, and some hated it. Those of us who loved it were soon up to our eyebrows in transformations, trees, deep structures, and all the other arcana of a new formalism. Many of those who hated it condemned it as dreadful, philistine *scientism*, a clanking assault by technocratic vandals on the beautiful, unanalyzable, unformalizable subtleties of language. This hostile attitude was overpowering in the foreign-language departments of most major universities. Chomsky might be a professor of linguistics at MIT, and linguistics might be categorized, there, as one of the humanities, but Chomsky's work was science, and science was the Enemy—as every card-carrying humanist knows.

> Sweet is the lore which Nature brings;
> Our meddling intellect
> Misshapes the beauteous forms of things:—
> We murder to dissect.

Wordsworth's Romantic view of the scientist as murderer of beauty seemed perfectly embodied by Noam Chomsky, automata theorist and Radio Engineer, but it is a great irony that he was all along the champion of an attitude towards science that might seem to offer salvation to humanists. As we saw in the previous section, Chomsky has argued that science has limits, and, in particular, it stubs its toe on the mind. Discerning the shape of this curious fact has long been difficult, even for those who can handle the technicalities and controversies of contemporary linguistics, but it has long been marveled at. Chomsky's notorious review (1959) slamming B. F. Skinner's *Verbal Behavior* (1957) was one of the founding documents of cognitive science. At the same time, Chomsky has been unwaveringly hostile to Artificial Intelligence, and has been so bold as to entitle one of his major books *Cartesian Linguistics* (1966)—almost as if he thought the anti-materialistic dualism of Descartes was going to come back in style. Whose side was he on, anyway? Not on Darwin's side, in any case. If Darwin-dreaders want a champion who is himself deeply and influentially enmeshed within science itself, they could not do better than Chomsky.

This was certainly slow to dawn on me. In March 1978, I hosted a remarkable debate at Tufts, staged, appropriately, by the Society for Philosophy and Psychology.[7] Nominally a panel discussion on the foundations and prospects of Artificial Intelligence, it turned into a tag-team rhetorical wrestling match between four heavyweight ideologues: Noam Chomsky and Jerry Fodor attacking AI, and Roger Schank and Terry Winograd defending it. Schank was working at the time on programs for natural language comprehension, and the critics focused on his scheme for representing (in a computer) the higgledy-piggledy collection of trivia we all know and somehow rely on when deciphering ordinary speech acts, allusive and truncated

---

7. This account is drawn, with revisions, from Dennett 1988a.

as they are. Chomsky and Fodor heaped scorn on this enterprise, but the grounds of their attack gradually shifted in the course of the match, for Schank is no slouch in the bully-baiting department, and he staunchly defended his research project. Their attack began as a straightforward, "first-principles" condemnation of conceptual error—Schank was on one fool's errand or another—but it ended with a striking concession from Chomsky: it just might turn out, as Schank thought, that the human capacity to comprehend conversation (and, more generally, to think) was to be explained in terms of the interaction of hundreds or thousands of jerry-built gizmos, but that would be a shame, for then psychology would prove in the end not to be "interesting." There were only two interesting possibilities, in Chomsky's mind: psychology could turn out to be "like physics"—its regularities explainable as the consequences of a few deep, elegant, inexorable laws—or psychology could turn out to be utterly lacking in laws—in which case the only way to study or expound psychology would be the novelist's way (and he much preferred Jane Austen to Roger Schank, if that were the enterprise).

A vigorous debate ensued among the panelists and audience, capped by an observation from Chomsky's colleague at MIT Marvin Minsky: "I think only a humanities professor at MIT could be so oblivious to the third 'interesting' possibility: psychology could turn out to be like engineering." Minsky had put his finger on it. There is something about the prospect of an engineering approach to the mind that is deeply repugnant to a certain sort of humanist, and it has little or nothing to do with a distaste for materialism or science. Chomsky was himself a scientist, and presumably a materialist (his "Cartesian" linguistics did not go *that* far!), but he would have no truck with engineering. It was somehow beneath the dignity of the mind to be a gadget or a collection of gadgets. Better the mind should turn out to be an impenetrable mystery, an inner sanctum for chaos, than that it should turn out to be the sort of entity that might yield its secrets to an engineering analysis!

Though I was struck at the time by Minsky's observation about Chomsky, the message didn't sink in. In 1980, Chomsky published "Rules and Representations" as a target article in *Behavioral and Brain Sciences*, and I was among the commentators. The contentious issue, then and now, was Chomsky's insistence that language competence was largely innate, not something that a child could properly be said to *learn*. According to Chomsky, the structure of language is mostly fixed in the form of innately specified rules, and all the child does is set a few rather peripheral "switches" that turn him into an English-speaker instead of a Chinese-speaker. Chomsky says the child is *not* a sort of general-purpose learner—a "General Problem Solver," as Newell and Simon would say—who must figure out what language is and learn to engage in it. Rather, the child is innately equipped to speak and understand a language, and merely has to rule out certain (very limited) possibilities and rule in certain others. That's why it is so effortless, according to Chomsky, for even "slow" children to learn to speak. They aren't really learning at all, any more than birds learn their feathers. Language, and feathers, just *develop* in species ordained to have them, and are off limits to species that lack the innate equipment. A few developmental triggers set the language-acquisition process in motion, and a few environmental conditions subsequently do some minor pruning or shaping, into whichever mother tongue the child encounters.

This claim has encountered enormous resistance, but we can now be sure that the truth lies much closer to Chomsky's end of the table than to that of his opponents (for the details, see the defenses of Chomsky's position in Jackendoff 1993 and Pinker 1994). Why the resistance? In my *BBS* commentary—which I presented as a constructive observation, not an objection—I pointed out that there was one reason to resist this that was perfectly reasonable, even if it was only a reasonable *hope*. Just like the biologists' resistance to "Hoyle's Howler," the hypothesis that life didn't begin on Earth but began somewhere else and migrated here, the psychologists' resistance to Chomsky's challenge had a benign explanation: if Chomsky was right, it would just make the phenomena of language and language acquisition that much harder to investigate. Instead of finding the learning process going on before our eyes in individual children, where we could study it and manipulate it, we would have to "pass the buck to biology" and hope that the biologists could explain how our *species* "learned" to have language competences built in at birth. This was a much less tractable research program. In the case of Hoyle's hypothesis, one could imagine

> arguments that fixed a maximal speed of mutation and selection and showed that there had not been enough time on Earth for the *whole* process to have occurred locally.
>
> Chomsky's arguments, from the poverty of the stimulus and the speed of language acquisition, are analogous; they purport to show that there must have been *large* gifts of design in the infant if we are to explain the speedy development of the mature competence. And while we can take solace in the supposition that we may someday be able to confirm the presence of these innate structures by direct examination of the nervous system (like finding fossils of our extraterrestrial ancestors), we will have to accept the disheartening conclusion that a larger portion than we had hoped of *learning theory*, considered in its most general form as the attempt to explain the transition from utter ignorance to knowledge, is not the province of psychology at all, but rather of evolutionary biology at its most speculative. [Dennett 1980.]

To my surprise, Chomsky missed the point of my commentary. Whereas he himself had offered reflections on what would make psychology "interesting," he couldn't see how there might be something "disheartening" to psychologists in the discovery that they might have to pass the buck to biology. Years later, I finally realized that the reason he didn't see what I was driving at was that although he insisted that the "language organ" was innate, this did *not* mean to him that it was a product of natural selection! Or at least not in such a way as to permit biologists to *pick up* the buck and analyze the way in which the environment of our ancestors had shaped the design of the language organ over the eons. The language organ, Chomsky thought, was *not* an adaptation, but ... a mystery, or a hopeful monster. It was something that *perhaps* would be illuminated some day by physics, but not by biology.

> It may be that at some remote period a mutation took place that gave rise to the property of discrete infinity, perhaps for reasons that have to do with the biology of cells, to be explained in terms of properties of physical mechanisms, now unknown. ... Quite possibly other aspects of its evolutionary development again reflect the operation of physical laws applying to a brain of a certain degree of complexity. [1988, p. 170.]

How could this be? Many linguists and biologists have tackled the problems of the evolution of language, using the same methods that have worked well on other evolutionary puzzles, and getting results, or at least what seem to be results. For instance, at the most empirical end of the spectrum, work by neuroanatomists and psycholinguists has shown that our brains have features lacking in the brains of our closest surviving relatives, features that play crucial roles in language perception and language production. There is a wide diversity of opinion about when in the last six million years or so our lineage acquired these traits, in what order, and why, but these disagreements are as amenable to further research—no better and no worse off— than disagreements about whether the archaeopteryx flew, for instance. On the purely theoretical front, and casting the net much more widely, conditions for the evolution of communication systems in general have been deduced (e.g., Krebs and Dawkins 1984, Zahavi 1987), and the implications are being explored in simulation models and empirical experiments.

We saw in chapter 7 some of the ingenious speculations and models that have been directed at the problem about how life bootstrapped itself into existence, and there is a similar bounty of clever ideas about how language must have got going. There is no question that the origin of language is theoretically a much easier problem than the origin of life; we have such a rich catalogue of not-so-raw materials with which to build an answer. We may never be able to confirm the details, but if so this will not be a mystery but only a bit of irreparable ignorance. Some particularly abstemious scientists may be reluctant to devote time and attention to such far-flung exercises in deductive speculation, but that does not appear to be Chomsky's position. His reservations are directed not to the likelihood of success but to the very point of the enterprise.

> It is perfectly safe to attribute this development [of innate language structures] to "natural selection", so long as we realize that there is no substance to this assertion, that it amounts to nothing more than a belief that there is some naturalistic explanation for these phenomena. [Chomsky 1972, p. 97.]

There have long been signs, then, of Chomsky's agnosticism—or even antagonism—towards Darwinism, but many of us have found them hard to interpret. To some, he appeared to be a "crypto-creationist," but that didn't seem very plausible, especially since he had the endorsement of Stephen Jay Gould. Remember the linguist Jay Keyser's appeal (on page 279) to Gould's term "spandrel" to describe how language came to be? Keyser probably got his terminology from his colleague Chomsky, who got it from Gould, who in return has avidly endorsed Chomsky's view that language didn't really evolve but just rather suddenly arrived, an inexplicable gift, at best a by-product of the enlargement of the human brain.

> Yes, the brain got big by natural selection. But as a result of this size, and the neural density and connectivity thus imparted, human brains could perform an immense range of functions quite unrelated to the original reasons for increase in bulk. The brain did not get big so that we could read or write or do arithmetic or chart the seasons—yet human culture, as we know it, depends upon skills of this kind.... [T]he universals of language are so different from anything else in nature, and so quirky in their structure, that origin as a side consequence of the brain's enhanced capacity, rather than as a simple advance in continuity from ancestral grunts and gestures, seems indicated. (This argument about language is by no means original with me, though I ally myself fully with it; this line of reasoning follows directly as the evolutionary reading for Noam Chomsky's theory of universal grammar.) [Gould 1989b, p. 14.]

Gould stresses that the brain's growth may not have been due initially to selection for language (or even for heightened intelligence) and that human language may not have developed "as a simple advance in continuity from ancestral grunts," but it does not follow from these suppositions (which we may grant him for the sake of argument) that the language organ is not an adaptation. It is, let us grant, an exaptation, but exaptations are adaptations. Let the remarkable growth of the hominid brain be a "spandrel" in whatever sense Gould or Keyser wishes, and *still* the language organ will be as much an adaptation as the bird's wing! No matter how suddenly the punctuation occurred that jogged our ancestors abruptly to the right in Design Space, it was still a gradual design development under the pressure of natural selection—unless it was indeed a miracle or a hopeful monster. In short, although Gould has heralded Chomsky's theory of universal grammar as a bulwark against an adaptationist explanation of language, and Chomsky has in return endorsed Gould's antiadaptationism as an authoritative excuse for rejecting the obvious obligation to pursue an evolutionary explanation of the innate establishment of universal grammar, these two authorities are supporting each other over an abyss.

In December 1989, the MIT psycholinguist Steven Pinker and his graduate student Paul Bloom presented a paper, "Natural Language and Natural Selection," to the Cognitive Science Colloquium at MIT. Their paper, which has itself subsequently appeared as a target article in *Behavioral and Brain Sciences*, laid down the gauntlet:

> Many people have argued that the evolution of the human language faculty cannot be explained by Darwinian natural selection. Chomsky and Gould have suggested that language may have evolved as the by-product of selection for other abilities or as a consequence of as-yet unknown laws of growth and form.... [W]e conclude that there is every reason to believe that a specialization for grammar evolved by a conventional neo-Darwinian process. [Pinker and Bloom 1990, p. 707.]

"In one sense," Pinker and Bloom said (p. 708), "our goal is incredibly boring. All we argue is that language is no different from other complex abilities such as echolocation and stereopsis, and that the only way to explain the origin of such abilities is through the theory of natural selection." They arrived at this "incredibly boring" conclusion by a patient evaluation of various analyses of multifarious phenomena that show beyond a reasonable doubt—surprise, surprise—that the "language organ" must indeed have evolved many of its most interesting properties as adaptations, just as any neo-Darwinian would expect. The response from the audience at MIT was anything but boring, however. Chomsky and Gould had been scheduled to reply, so there was a standing-room-only crowd.[8] The level of hostility and ignorance about evolution that was unabashedly expressed by

---

8. As it turned out, Chomsky was unable to attend, and his place was taken by his (and my) good friend Massimo Piatelli-Palmarini (who almost always agrees with Chomsky, and seldom agrees with me!). Piatelli-Palmarini was the optimal understudy; he had cotaught a seminar on cognition and evolution with Gould at Harvard, and was the author of the article (1989) that had first rendered explicit the Gould-Chomsky position on the nonevolution of language. His article had been a major provocation and target of Pinker and Bloom's essay.

eminent cognitive scientists on that occasion shocked me. (In fact, it was reflecting on that meeting that persuaded me I could no longer put off writing this book.) So far as I know, no transcript of that meeting exists (the commentaries in *BBS* include some of the themes raised at the meeting), but you can recover something of the flavor by contemplating Pinker's list (personal communication) of the ten most amazing objections he and Bloom have fielded since drafts of their paper began to circulate. Versions of most of them, if memory serves me, were expressed at the MIT meeting:

(1) Color vision has no function; we could tell red from green apples using intensity cues.

(2) Language is not designed for communication at all: it's not like a watch, it's like a Rube Goldberg device with a stick in the middle that you can use as a sundial.

(3) Any argument that language is functional could be made with equal plausibility and force when applied to writing in sand.

(4) The structure of the cell is to be explained by physics, not evolution.

(5) Having an eye calls for the same kind of explanation as having mass, because just as the eye lets you see, mass prevents you from floating into space.

(6) Hasn't that stuff about insect wings refuted Darwin?

(7) Language can't be useful; it's led to war.

(8) Natural selection is irrelevant, because we now have chaos theory.

(9) Language couldn't have evolved through selection pressure for communication, because we can ask people how they feel without really wanting to know.

(10) Everyone agrees that natural selection plays some role in the origin of the mind but that it cannot explain every aspect—thus there is nothing more to say.

Are Gould and Chomsky responsible for the bizarre convictions of some of their supporters? This question has no simple answer. More than half of the items on Pinker's list have a clear ancestry in claims that have been made by Gould (numbers 2, 6, and 9 in particular) and Chomsky (numbers 4, 5, and 10 in particular). Those who make these claims (including the others on the list) typically present them on the authority of Gould and Chomsky (see, e.g., Otero 1990). As Pinker and Bloom say (1990, p. 708), "Noam Chomsky, the world's greatest linguist, and Stephen Jay Gould, the world's best-known evolutionary theorist, have repeatedly suggested that language may not be the product of natural selection." Moreover—two important dogs that haven't barked—I have yet to witness either Gould or Chomsky attempting to correct these howlers when they arise in the heat of battle. (As we shall see, this is everybody's weakness; I regret that the siege mentality among sociobiologists has led them to overlook—at any rate, neglect to correct—more than a few cases of egregiously bad reasoning by members of their own team.)

One of Darwin's most enthusiastic supporters was Herbert Spencer, coiner of the phrase "the survival of the fittest" and an important clarifier of some of Darwin's best ideas, but also the father of Social Darwinism, an odious misapplication of Darwinian thinking in defense of political doctrines that range from callous to heinous.[9] Was Darwin responsible for Spencer's misuse of his views? Opinions differ on this. For my part, I excuse Darwin from the truly heroic task of chastising his champion in public, even though I regret that he wasn't more energetic in pursuing private acts of dissuasion or correction. Both Gould and Chomsky have been vigorous proponents of the view that intellectuals *are* responsible for the applications *and likely misapplications* of their own work, so presumably they are at least embarrassed to find themselves cited as the sources of all this nonsense, for they themselves do not hold these views. (It is perhaps too much to expect their gratitude to me for doing their dirty work for them.)

3. NICE TRIES

In studying the evolution of mind, we cannot guess to what extent there are physically possible alternatives to, say, transformational generative grammar, for an organism meeting certain other physical conditions characteristic of humans. Conceivably, there are none—or very few—in which case talk about evolution of the language capacity is beside the point.

—NOAM CHOMSKY 1972, p. 98

9. Spencer's woolly style was the target of William James' mockery in the epigraph for part II (p. 147). Spencer (1870, p. 396) had offered the following definition: "Evolution is an integration of matter and concomitant dissipation of motion; during which the matter passes from an indefinite, incoherent homogeneity to a definite, coherent heterogeneity; and during which the retained motion undergoes a parallel transformation." The memeology of James' marvelous parody is worth recording. I got the quotation from Garrett Hardin, who informs me that he got it from Sills and Merton (1991, p. 104). They in turn cite James' *Lecture Notes 1880–1897* as their source, but Hardin has tracked down some further details. P. G. Tait (1880, p. 80) gives credit to a mathematician named Kirkman for his "exquisite translation" of Spencer, of which James' version—presumably borrowed from Tait—is a mutation. Kirkman's (presumably) original version: "Evolution is a change from a nohowish, untalkaboutable all-alikeness, to a somehowish and in-general-talkaboutable, not-all-alikeness, by continuous somethingelsifications and stick-togetherations."

*To make progress in understanding all this, we probably need to begin
with simplified (oversimplified?) models and ignore the critics' tirade
that the real world is more complex. The real world is always more
complex, which has the advantage that we shan't run out of work.*

—JOHN BALL 1984, p. 159

The problem here is how to get the pendulum to stop swinging back and
forth so destructively. Time and again we see the same failure of communication. The truly unfortunate communication gap that Simon and Kaplan
speak of (in their quotation at the head of the previous section) is the
amplified effect of a relatively simple bit of initial misunderstanding. Recall
the difference between reductionists and greedy reductionists (chapter 3,
section 5): reductionists think everything in nature can all be explained
without skyhooks; greedy reductionists think it can all be explained without
cranes. But one theorist's healthy optimism is another theorist's unseemly
greed. One side proposes an oversimple crane, at which the other side
scoffs—"Philistine reductionists!"—declaring, truthfully, that life is much
more complicated than that. "Bunch of crazy skyhook-seekers!" mutters the
first side, in defensive overreaction. That is what they would mutter if they
had the term—but, then again, if both sides had the terms, they might be
able to see what the issues really were, and avoid the miscommunication
altogether. That is my hope.

What are Chomsky's actual views? If he doesn't think the language organ
is shaped by natural selection, what account does he give of its complexities? The philosopher of biology Peter Godfrey-Smith (1993) has recently
focused on the family of views that maintain, in one way or another, that
"there is complexity in the organism in virtue of complexity in the environment." Since this was one of Herbert Spencer's pet themes, Godfrey-
Smith proposes we call any such view "Spencerian."[10] Spencer was a
Darwinian—or you could say that Charles Darwin was a Spencerian. In any
event, the modern synthesis is Spencerian to its core, and it is the Spencerism of that orthodoxy that is most often attacked in one way or another by
rebels. Manfred Eigen and Jacques Monod are both Spencerian, for instance,
in their insistence that it is only through environmental selection that molecular function can be specified (chapter 7, section 2; chapter 8, section 3),
whereas Stuart Kauffman's insistence that order emerges *in spite of* (environmental) selection expresses an anti-Spencerian challenge (chapter 8,
section 7). Brian Goodwin's denial (1986) that biology is a *historical* science is another example of anti-Spencerism, since it is a denial that historical interactions with earlier environments are the source of the

---

10. It is also one of Herbert Simon's pet themes in *Sciences of the Artificial* (1969), so
we might call it Simonian—or Herbertian.

complexities to be found in organisms. Gould and Lewontin's (1979) brief dalliance with "intrinsic" *Baupläne* that account for all but the minor trimmings of organism design is yet another.

Chomsky's suggestion that it is physics, not biology (or engineering), that will account for the structure of the language organ is as pure an anti-Spencerian doctrine as you could find. This explains his misconstrual of my friendly suggestion about passing the buck to biology. I was assuming, as a good Spencerian adaptationist, that "genes are the channel through which the environment speaks," as Godfrey-Smith puts it, whereas Chomsky prefers to think of the genes' getting their message from some intrinsic, ahistorical, nonenvironmental source of organization—"physics," we may call it. Spencerians think that even if there are such timeless "laws of form," they could impose themselves on things only through some selectional process or other.

Evolutionary thinking is just one chapter in the history of Spencerian–versus–anti-Spencerian thinking. Adaptationism is a Spencerian doctrine, and so is Skinner's behaviorism, and so, more generally, is any variety of *empiricism*. Empiricism is the view that we furnish our minds with details that all come from the outside environment, via experience. Adaptationism is the view that the selecting environment gradually shapes the genotypes of organisms, molding them so that the phenotypes they command are some near-optimal fit with the encountered world. Behaviorism is the view that what Skinner (1953, especially pp. 129–41) called "the controlling environment" is what "shapes" the behavior of all organisms. Now we can see that Chomsky's famous attack on Skinner was as much an attack on Skinner's Spencerian view *that* the environment shaped the organism as it was on the limitations of Skinner's model of *how* this shaping took place.

Skinner proclaimed that *one simple iteration* of the fundamental Darwinian process—operant conditioning—could account for all mentality, all learning, not just in pigeons but in human beings. When critics insisted that thinking and learning were much, much more complicated than that, he (and his followers) smelled skyhooks, and wrote off the critics of behaviorism as dualists, mentalists, antiscientific know-nothings. This was a misperception; the critics—at least the best of them—were simply insisting that the mind was composed of a lot more cranes than Skinner imagined.

Skinner was a greedy reductionist, trying to explain *all* the design (and design power) in a single stroke. The proper response to him should have been: "Nice try—but it turns out to be much more complicated than you think!" And one should have said it without sarcasm, for Skinner's *was* a nice try. It was a great idea, which inspired (or provoked) a half-century of hardheaded experimentation and model-building from which a great deal was learned. Ironically, it was the repeated failures of *another* brand of greedy reductionism, dubbed "Good Old-Fashioned AI" or "GOFAI" by Haugeland (1985), that really convinced psychologists that the mind was indeed a phenomenon of surpassing architectural complexity—much too complicated for behaviorism to describe. The founding insight of GOFAI was Turing's recognition that a computer could be *indefinitely complicated* but that all computers could be made from simple parts. Whereas Skinner's simple parts had been randomly mated stimulus-response pairings that could then be subjected, over and over again, to the selection pressure of reinforcement from the environment, Turing's simple parts were internal data-structures—different "machine states" that could be composed to respond differentially to indefinitely many different inputs, creating input-output behavior of any imaginable sophistication. Which of these internal states were innately specified and which were to be revised by experience was something left to be investigated. Like Charles Babbage (see note 13 of chapter 8), Turing saw that the behavior of an entity need not be any *simple* function of its *own* history of stimulation, since it could have accrued huge amounts of design over the eons, which would permit it to use its internal complexity to mediate its responses. That abstract opening was eventually filled by GOFAI-modelers with contrivances of dazzling complexity that *still* fell comically short of producing human-style cognition.

Today the reigning orthodoxy in cognitive science is that yesterday's simple models of perception, learning, memory, language production, and language understanding are orders of magnitude too simple, but those simple models were often nice tries, without which we would still be wondering how simple it might, after all, turn out to be. It makes sense to err on the side of greedy reductionism, to try for the simple model before wallowing around in complexities. Mendel's simple genetics was a nice try, and so was the rather more complex "bean-bag genetics" it became in the hands of population geneticists, even though it has often relied on such retrospectively outrageous oversimplifications that Francis Crick was tempted to kick it out of science. Graham Cairns-Smith's clay crystals are a nice try, and Art Samuel's checkers-player was a nice try—much too simple, as we learned, but on the right track.

In the earliest days of the computer, Warren McCulloch and W. H. Pitts (1943) proposed a magnificently simple "logical neuron" from which "neural nets" might be woven, and for a while it looked as if perhaps they had broken the back of the brain problem. Certainly, before they made their modest proposal, neurologists were desperately confused about how to think of the brain's activity. One has only to go back and read their brave flounderings, in the more speculative books of the 1930s and 1940s, to see what a tremendous lift neuroscience got from McCulloch and Pitts.[11] They

---

11. One of Warren McCulloch's students, himself a major contributor to these early developments, is Michael Arbib, whose crystal-clear early discussion of these issues

made possible such pioneers as Donald Hebb (1949) and Frank Rosenblatt (1962), whose "Perceptrons" were, as Minsky and Papert soon pointed out (1969), a nice try, but much too simple. Now, several decades later, another wave of more complicated but still usefully simple nice tries, flying the flag of Connectionism, are exploring portions of Design Space left unexamined by their intellectual ancestors.[12]

The human mind is an amazing crane, and there is a lot of design work that has to have been done to build it, and to keep it working and up-to-date now. That is Darwin's "Spencerian" message. One way or another, the history of environmental encounters over the eons (and during the last ten minutes) has shaped the mind you have right now. Some of the work must have been done by natural selection, and the rest by one or another internal generate-and-test process of the sort we looked at earlier in the chapter. None of it is magic; none of it involves an internal skyhook. Whatever models we propose of these cranes will surely be too simple in one regard or another, but we are closing in, trying out the simple ideas first. Chomsky has been one of the leading critics of these nice tries, dismissing everyone from B. F. Skinner, through such GOFAI mavens and mavericks as Herbert Simon and Roger Schank, to all the Connectionists, and he has always been right that their ideas have been too simple by far, but he has also exhibited a hostility to the *tactic* of trying for simple models that has unduly raised the temperature of the debates. Suppose, for the sake of argument, we grant that Chomsky could see better than anyone else that the mind, and the language organ which plays such a central role in its superiority over animal minds, are structures of a systematicity and complexity that beggar all models to date. All the more reason, one would think, to search for an evolutionary explanation of these brilliant devices. But although Chomsky uncovered for us the abstract structure of language, the crane that is most responsible for lifting all the other cranes of culture into place, he has vigorously discouraged us from treating it as a crane. No wonder yearners for skyhooks have often taken him as their authority.

He is not the only candidate, however. John Searle is another favorite champion of skyhook-seekers, and he is certainly no Chomskian. We saw in

---

(1964) inspired me when I was a graduate student, and whose later work (e.g., 1989) has persistently carved out new territories, still underappreciated by many in the trenches, and on the outskirts, in my opinion.

12. Other nice tries have been the neuroscientists' many models of learning as "Darwinian" evolution in the nervous system, going back to the early work of Ross Ashby (1960) and J. Z. Young (1965), and continuing today in the work of such people as Arbib, Grossberg (1976), Changeux and Danchin (1976), and Calvin (1987)—and Edelman (1987), whose work would be a nicer try if he didn't present it as if it were such a saltation in the wilderness.

chapter 8 (section 4) that Searle has defended a version of John Locke's Mind-first vision, under the banner of Original Intentionality. According to Searle, automata (computers or robots) don't have real intentionality; at best they have mere *as if* intentionality. Moreover, original or real intentionality cannot be composed of, derived from—or, presumably, descended from—mere *as if* intentionality. This creates a problem for Searle, because, whereas Artificial Intelligence says you are *composed of* automata, Darwinism says you are *descended from* automata. It is hard to deny the former if you admit the latter; how could anything born of automata ever be anything but a much, much fancier automaton? Do we somehow reach escape velocity and leave our automaton heritage behind? Is there some threshold that marks the onset of real intentionality? Chomsky's original hierarchy of ever fancier automata permitted him to draw the line, showing that *the minimal complexity* of an automaton capable of generating the sentences of a human language puts it in a special class—still a class of automata, but at least an advanced class. This was not quite enough for Chomsky. As we have just seen, he planted his feet and said, in effect: "Yes, language makes the difference—but don't try to explain how the language organ got designed. It's a hopeful monster, a gift, nothing that could ever be explained."

An awkward position to maintain: the brain is an automaton, but not one we can reverse-engineer. Is this perhaps a tactical mistake? According to Searle, Chomsky took one step too many before planting his feet. He should have denied that the language organ had a structure that could even be described in automaton terms at all. By lapsing into information-processing talk, talk about rules and representations and algorithmic transformations, Chomsky had given a hostage to the reverse engineers. Perhaps Chomsky's heritage as a Radio Engineer is coming back to haunt him:

> Specifically, the evidence for universal grammar is much more simply accounted for by the following hypothesis: There is, indeed, a language acquisition device [LAD] innate in human brains, and LAD constrains the form of languages that human beings can learn. There is, thus, a hardware level of explanation in terms of the structure of the device, and there is a functional level of explanation, describing which sorts of languages can be acquired by the human infant in the application of this mechanism. No further predictive or explanatory power is added by saying that there is in addition a level of deep unconscious rules of universal grammar, and indeed, I have tried to suggest that that postulation is incoherent anyway. [Searle 1992, pp. 244–45.]

According to Searle, the whole idea of *information processing in the brain*, described abstractly in terms of algorithms that exhibit substrate neutrality, is incoherent. "There are brute, blind neurophysiological processes and there is consciousness, but there is nothing else" (Searle 1992, p. 228).

That is certainly biting the bullet, and biting the same bullet as Chomsky, but in a somewhat different spot: Yes of course, the LAD evolved, and so did consciousness (Searle 1992, pp. 88ff.), but Chomsky is right that there is no hope of a reverse-engineering account of either of them. Chomsky is wrong, however, to grant even the coherence of an automaton-level description of the process, for that opens the door to "strong Artificial Intelligence."

If Chomsky's position on the slippery slope is hard to maintain, Searle's has even more awkward consequences.[13] He grants, as we can see in the passage quoted above, that there is a "functional" story to be told about how the brain does its work in language acquisition. There is also, he grants, a "functional" story to be told about how parts of the brain arrive at depth or distance judgments in vision. "*But there is no mental content whatever at this functional level*" (Searle 1992, p. 234, Searle's emphasis). He then puts to himself the following quite reasonable retort from the cognitive scientists: "the distinction [between "function" talk and "mental content" talk] does not really make much difference to cognitive science. We continue to say what we have always said and do what we have always done, we simply substitute the word 'functional' for the word 'mental' in these cases." (This is in fact what Chomsky has often said, in reply to such criticisms. See, for instance, 1980.) To answer this retort, Searle (1992, p. 238) is obliged to take a step backwards himself: not only is there no information-processing level of explanation for the brain, he says; there is also really no "functional level" of explanation in biology:

> To put the point bluntly, in addition to its various causal relations, the heart does not have any functions. When we speak of its functions, we are talking about those of its causal relations to which we attach some *normative* importance.... In short, the actual facts of intentionality contain normative elements, but where functional explanations are concerned, the only *facts* are brute, blind physical facts and the only norms are in us and exist only from our point of view.

It turns out, then, that function talk in biology, like mere *as if*–intentionality talk, is not really to be taken seriously after all. According to Searle, only artifacts made by genuine, conscious human artificers have *real* functions. Airplane wings are really for flying, but eagles' wings are not. If one biologist says they are adaptations for flying and another says they are merely display racks for decorative feathers, there is no sense in which one biologist is closer to the truth. If, on the other hand, we ask the aeronautical engineers whether the airplane wings they designed are for keeping the plane aloft or for displaying the insignia of the airline, they can tell us a brute fact. So

---

13. The remainder of this section draws on my review of Searle's book (Dennett 1993c).

Searle ends up denying William Paley's premise: according to Searle, nature does *not* consist of an unimaginable variety of *functioning* devices, exhibiting design. Only human artifacts have that honor, and only because (as Locke "showed" us) it takes a Mind to make something with a function![14]

Searle insists that human minds have "Original" Intentionality, a property unattainable in principle by any R-and-D process of building better and better algorithms. This is a pure expression of the belief in skyhooks: minds are original and inexplicable sources of design, not results of design. He defends this position more vividly than other philosophers, but he is not alone. The hostility to Artificial Intelligence and its evil twin, Darwinism, lies just beneath the surface of much of the most influential work in recent twentieth-century philosophy, as we shall see in the next chapter.

CHAPTER 13: *When generate-and-test, the basic move in any Darwinian algorithm, moves into the brains of individual organisms, it builds a series of ever more powerful systems, culminating in the deliberate, foresightful generation and testing of hypotheses and theories by human beings. This process creates minds that show no signs of "cognitive closure," thanks to their capacity to generate and comprehend language. Noam Chomsky, who created contemporary linguistics by proving that language was generated by an innate automaton, has nevertheless resisted all evolutionary accounts of how and why the language automaton got designed and installed, and has also resisted all Artificial Intelligence attempts to model language use. Chomsky has stood firm against (reverse) engineering, flanked by Gould on one side and Searle on the other, exemplifying the resistance to the spread of Darwin's dangerous idea, and holding out for the human mind as a skyhook.*

CHAPTER 14: *In chapter 8, I sketched an evolutionary account of the birth of meaning, which will now be expanded and defended against the skeptical challenges of philosophers. A series of thought experiments building on the concepts introduced in earlier chapters shows not just the coherence but the inevitability of an evolutionary theory of meaning.*

---

14. Given Searle's position on this, one would predict that he should be utterly opposed to my analysis of the power of adaptationist thinking, as presented in chapter 9. He is. I don't know whether he has expressed this view in print, but in several debates with me (Rutgers, 1986; Buenos Aires, 1989), he has expressed the view that my account is exactly backward: the idea that one can hunt for the "free-floating rationales" of evolutionary selection processes is, in his view, a travesty of Darwinian thinking. One of us has unintentionally refuted himself; the identity of the victim is left as an exercise for the reader.

CHAPTER FOURTEEN

The Evolution of Meanings

〰️

1. THE QUEST FOR REAL MEANING

"When I use a word," Humpty Dumpty said, in a rather scornful tone,
"it means just what I choose it to mean—neither more nor less."
"The question is," said Alice, "whether you can make words mean so
many different things."
"The question is," said Humpty Dumpty, "which is to be master—
that's all."

—LEWIS CARROLL 1871

There is no topic in philosophy that has received more attention than meaning, in its multifarious manifestations. At the grand end of the spectrum, philosophers of all schools have grappled with the ultimate question of the meaning of life (and whether or not this question has any meaning). At the modest end, philosophers of the contemporary analytic school—sometimes called "linguistic philosophy" by outsiders—have subjected the nuances of the meaning of words and whole utterances to microscopic scrutiny, in a variety of quite distinct enterprises. Back in the 1950s and 1960s, the school of "ordinary language philosophy" lavished attention on the subtle differences between particular words—the differences between doing something "deliberately" or "intentionally" or "on purpose," to cite a famous instance (Austin 1961). This gave way to a more formal and systematic set of investigations. Which different propositions could you mean by uttering such sentences as

Tom believes that Ortcutt is a spy.

And what theory accounts for their differences in presupposition, context, and implication? This sort of question is pursued by the subschool that has sometimes been called the "propositional attitude task force," of which some exemplary recent efforts are Peacocke 1992 and Richard 1992. A different set of investigations was inaugurated by Paul Grice's (1957, 1969) theory of "non-natural meaning." This was the attempt to specify the conditions under which a bit of behavior had not just natural meaning (where there's smoke, there's fire; when somebody cries, there's sadness), but the sort of meaning that a speech act has, with its element of conventionality. What has to be the state of a speaker's (or hearer's) mind for the utterance act to mean anything at all, or to mean a particular thing? Or, in other words, what is the relationship between an agent's psychology and the meaning of an agent's words? (The relations between these two enterprises is perhaps best seen in Schiffer 1987.)

An assumption shared by all these philosophical research programs is that there is one sort of meaning—perhaps divided into many different sub-sorts—that is language-dependent. Before there were words, there were no *word* meanings, even if there were other sorts of meanings. The further working assumption, particularly among English-speaking philosophers, has been that until we get clear about how words can have meaning, we are unlikely to make much progress on the other varieties of meaning, especially such staggering issues as the meaning of life. But this reasonable assumption has typically had an unnecessary and debilitating side effect: by concentrating first on linguistic meaning, philosophers have distorted their vision of the minds these words depend on, treating them as somehow *sui generis*, rather than as themselves evolved products of the natural world. This is manifest especially in the resistance philosophers have shown to evolutionary theories of meaning, theories that purport to discern that the meaning of words, and all the mental states that somehow lie behind them, is grounded ultimately in the rich earth of biological function.

On the one hand, few if any philosophers have wanted to deny the obvious fact: human beings are products of evolution, and their capacity to speak, and hence to mean anything (in the relevant sense), is due to a suite of specific adaptations not shared with other products of evolution. On the other hand, philosophers have been reluctant to entertain the hypothesis that evolutionary thinking might shed light on their specific problems about how it is that words, and their sources and destinations in people's minds or brains, have meaning. There have been important exceptions. Willard Van Orman Quine (1960) and Wilfrid Sellars (1963) each developed functionalistic theories of meaning that had their roots firmly if sketchily planted in biology. Quine, however, hitched his wagon too firmly to the behaviorism espoused by his friend B. F. Skinner, and has been dogged for thirty years with the problem of persuading philosophers—with scant success—that his claims do not succumb to the blanket denunciation of greedy reductionism that was heaped on Skinner and all behaviorists by the ascendant cognitivists, under the direction of Chomsky and Fodor.[1] Sellars, the father of "functionalism" in the philosophy of mind, said all the right things, but in difficult language that was largely ignored by the cognitivists. (See Dennett 1987b, ch. 10, for a historical account.) Earlier, John Dewey made it clear that Darwinism should be assumed to be the foundation of any naturalistic theory of meaning.

> No account of the universe in terms *merely* of the redistribution of matter in motion is complete, no matter how true as far as it goes, for it ignores the cardinal fact that the character of matter in motion and of its redistribution is such as cumulatively to achieve ends—to effect the world of values we know. Deny this and you deny evolution; admit it and you admit purpose in the only objective—that is, the only intelligible—sense of that term. I do not say that in addition to the mechanism there are other ideal causes or factors which intervene. I only insist that the whole story be told, that the character of the mechanism be noted—namely, that it is such as to produce and sustain good in a multiplicity of forms. [Dewey 1910, p. 34.]

Note how carefully Dewey wended his way between Scylla and Charybdis: no skyhooks ("ideal causes or factors") are called for, but we must not suppose that we can make sense of an *uninterpreted* version of evolution, an evolution with no functions endorsed, no meanings discerned. More recently, several other philosophers and I have articulated specifically evolutionary accounts of the birth and maintenance of meaning, both linguistic and prelinguistic (Dennett 1969, 1978, 1987b, Millikan 1984, 1993, Israel 1987, Papineau 1987). Ruth Millikan's account is by far the most carefully articulated, bristling with implications about the details of the other philosophical approaches to meaning mentioned above. Her differences with my position have loomed larger for her than for me, but the gap is closing fast—see especially Millikan 1993, p. 155—and I expect the present book to close the gap further, but this is not the place to expose our remaining differences, for they are minor in the context of a larger skirmish, a battle we have not yet won: the battle for *any* evolutionary account of meaning.

---

1. Like the misguided fear among evolutionists that the Baldwin Effect commits the sin of Lamarckism, and Darwin's own precipitous flight from Catastrophism, the indiscriminate rejection of anything that smacks of behaviorism by the "thoroughly modern mentalists" (Fodor 1980) is an instance of misfiltered memes. See R. Richards 1987 for an excellent account of the distortions of such guilt-by-association in early evolutionary thinking, and Dennett (1975; 1978, ch. 4) for attempts to separate the wheat from the chaff of behaviorism.

Opposed to us stand an eminent if unlikely bunch of bedfellows: Jerry Fodor, Hilary Putnam, John Searle, Saul Kripke, Tyler Burge, and Fred Dretske, each in his own way opposed both to evolutionary accounts of meaning and to AI. All six of these philosophers have expressed their reservations about AI, but Fodor has been particularly outspoken in his denunciation of evolutionary approaches to meaning. His diatribes against every naturalist since Dewey (see especially Fodor 1990, ch. 2) are often quite funny. For instance, in ridiculing my view, he says (p. 87): "Teddy bears are artificial, but *real bears are artificial too*. We stuff the one and Mother Nature stuffs the other. Philosophy is *full* of surprises."[2]

Words, according to Humpty Dumpty, get their meanings from us, but whence do we get meaning? What exercises Fodor and these other philosophers is a concern for *real* meaning as opposed to *ersatz* meaning, "*intrinsic*" or "*original*" intentionality as opposed to *derived* intentionality. Well might Fodor want to ridicule the idea of organisms as artifacts, for it provides the perspective from which the central flaw in his view, a flaw shared by the views of the five others, can be exposed by a thought experiment.[3]

Consider a soft-drink vending machine, designed and built in the United States, and equipped with a standard transducer device for accepting and rejecting U.S. quarters. Let's call such a device a "two-bitser." Normally, when a quarter is inserted into a two-bitser, it goes into a state, call it Q, which "means" (note the scare-quotes) "I perceive/accept a genuine U.S. quarter now." Such two-bitsers are quite "clever" and "sophisticated." (More scare-quotes; the thought experiment begins with the assumption that this sort of intentionality is *not* the real thing, and ends by exposing the embarrassments such an assumption entails.) Two-bitsers are hardly foolproof, however; they do "make mistakes." To say the same thing unmetaphorically, sometimes they go into state Q when a slug or other foreign object is inserted in them, and sometimes they reject perfectly legal quarters—they fail to go into state Q when they are supposed to. No doubt there are detectable patterns in the cases of "misperception." No doubt at least some of the cases of "misidentification" could be predicted by someone with enough knowledge of the relevant laws of physics and design param-

2. Read Fodor for amusement, and for insight into a dislike of Darwin's dangerous idea so deep that it overrides the standard practice of attempting to find a sympathetic reading of texts. His misrepresentation of Millikan is particularly egregious, and is not to be trusted at all, but can be readily cured by reading Millikan herself.

3. As usual, the issues are more complicated than I can show here; for all the gory details, see Dennett 1987b, ch. 8, from which this thought experiment is drawn, and Dennett 1990b, 1991c, 1991e, 1992.

eters of the two-bitser's transducing machinery. In other words, it follows just as directly from the laws of physics that objects of kind K would put the device into state Q as that quarters would. Objects of kind K would be good "slugs"—reliably "fooling" the transducer.

If objects of kind K became more common in the two-bitser's normal environment, we would expect the owners and designers of two-bitsers to develop more advanced and sensitive transducers that would reliably discriminate between genuine U.S. quarters and slugs of kind K. Of course, trickier counterfeits might then make their appearance, requiring further advances in the detecting transducers. At some point, such escalation of engineering would reach diminishing returns, for there is no such thing as a foolproof mechanism. In the meantime, the engineers and users are wise to make do with standard, rudimentary two-bitsers, since it is not cost-effective to protect oneself against negligible abuses.

The only thing that makes the device a quarter-detector rather than a slug-detector or a quarter-*or*-slug-detector is the environment of shared intentions of the artifact's designers, builders, owners—its users, in short. It is only in the context of those users and their intentions that we can single out some of the occasions of state Q as "veridical" and others as "mistaken." It is only relative to that context of intentions that we could justify calling the device a two-bitser in the first place.

I take it that so far I have Fodor, Putnam, Searle, Kripke, Burge, and Dretske nodding their agreement: that's just how it is with such artifacts; this is a textbook case of *derived* intentionality, laid bare. Such an artifact has no *intrinsic* intentionality at all. And so it embarrasses no one to admit that a particular two-bitser, straight from the American factory and with "Model A Two-Bitser" stamped right on it, might be installed on a Panamanian soft-drink machine, where it proceeded to earn its keep as an accepter and rejecter of quarter-balboas, legal tender in Panama, and easily distinguished from U.S. quarters by the design and writing stamped on them, but not by their weight, thickness, diameter, or material composition. (I'm not making this up. I have it on excellent authority—Albert Erler of the Flying Eagle Shoppe, Rare Coins—that Panamanian quarter-balboas minted between 1966 and 1984 are indistinguishable from U.S. quarters by standard vending machines. Small wonder, since they were struck from U.S. quarter stock in American mints. And—to satisfy the curious, although it is strictly irrelevant to the example—the exchange rate when last I checked for quarter-balboas was indeed $.25.)

Such a two-bitser, whisked off to Panama, would still normally go into a certain physical state—the state with the physical features by which we used to identify state Q—whenever a U.S. quarter, an object of kind K, or a Panamanian quarter-balboa is inserted in it, but now a different set of such occasions count as the mistakes. In the new environment, U.S. quarters count as slugs, as inducers of error, misperception, misrepresentation, just as much as objects of kind K do. After all, back in the United States a Panamanian quarter-balboa is a kind of slug.

Once our two-bitser is resident in Panama, should we say that the state we used to call Q still occurs? The physical state in which the device "accepts" coins still occurs, but should we now say that we should identify it as "realizing" a new state—call it QB instead? At what point would we be entitled to say that the meaning, or the function, of this physical state of the two-bitser had shifted? Well, there is considerable freedom—not to say boredom—about what we should say, since after all a two-bitser is just an artifact; talking about its perceptions and misperceptions, its veridical and nonveridical states—its intentionality, in short—is "just metaphor." The two-bitser's internal state, call it what you like, doesn't *really* (originally, intrinsically) mean either "U.S. quarter here now" or "Panamanian quarter-balboa here now." It doesn't *really* mean anything. That's what Fodor, Putnam, Searle, Kripke, Burge, and Dretske (*inter alia*) would insist.

The two-bitser was originally designed to be a detector of U.S. quarters. That was its "proper function" (Millikan 1984) and, quite literally, its *raison d'être*. No one would have bothered bringing it into existence had this purpose not occurred to them. This historic fact licenses a way of speaking: we may call the thing a two-bitser, a thing whose function is to detect quarters, so that *relative to that function* we can identify both its veridical states and its errors.

This would not prevent the two-bitser from being wrested from its original niche and pressed (exapted) into new service—whatever new purpose the laws of physics permit and circumstances favor. It could be used as a K-detector, or a slug-detector, a quarter-balboa-detector, a doorstop, or a deadly weapon. In its new role there might be some brief period of confusion or indeterminacy. How long a track record must something accumulate before it is no longer a two-bitser but a quarter-balboa-detector (a "q-balber," we might call it)? On its very debut as a q-balber, after ten years of faithful service as a two-bitser, is the state it goes into when presented with a quarter-balboa a *veridical* detection of a quarter-balboa, or might there be a sort of force-of-habit error of nostalgia, a mistaken acceptance of a quarter-balboa *as* a U.S. quarter?

As described, the two-bitser is ludicrously too simple to count as having the sort of memory we have of our past experiences, but we might take the first step in the direction of providing it with one. Suppose it has a counter on it, which advances each time it goes into its acceptance state, and which stands, after its ten years of service, at 1,435,792. Suppose the counter is not reset to zero when it is moved to Panama, so that after its debut acceptance of a quarter-balboa, it reads 1,435,793. Does this tip the balance in favor of the claim that it has not yet switched to the task of identifying quarter-balboas? (We could go on adding complications and variations, if it might make a difference to our intuitions. Should it?)

One thing is clear: there is absolutely nothing *intrinsic* about the two-bitser considered narrowly all by itself (and its internal operations) that would distinguish it from a genuine q-balber, made to order on commission from the Panamanian government. What must make the difference, of course, is whether it was *selected for* its capacity to detect quarter-balboas (agreeing with Millikan 1984). If it was so selected (by its new owners, in the simplest case), then even if they forget to reset the counter, its maiden move is the veridical acceptance of a quarter-balboa. "It works!" its new owners might exclaim delightedly. If, on the other hand, the two-bitser were sent to Panama by mistake, or arrived by sheer coincidence, then its maiden move would mean nothing, though it might soon come to be appreciated by those in the vicinity for its power to tell quarter-balboas from the indigenous slugs, in which case it could come to function as a q-balber in the fullest meaning of that term, via a less official route. This, by the way, already makes a problem for Searle's view that only artifacts can have functions, and those are the functions its creators endow it with by their very special mental acts of creation. The original designers of the two-bitser may have been entirely oblivious of some later use to which it was opportunistically exapted, so their intentions count for nothing. And the new selectors may also fail to *formulate* any specific intentions—they may just fall into the habit of relying on the two-bitser for some handy function, unaware of the act of unconscious exaptation they are jointly executing. Recall that Darwin, in *Origin*, already drew attention to unconscious selection of traits in domestic animals; unconscious selection of traits in artifacts is no stretch at all; it is rather a frequent event, one might suppose.

Presumably, Fodor and company will not want to disagree with this treatment of artifacts, which have, they claim, no smidgen of real intentionality, but they may begin to worry that I have maneuvered them onto a buttered slide, for now let's consider the exactly parallel case of what the frog's eye tells the frog's brain. In Lettvin, Maturana, McCulloch, and Pitts' classic article (1959)—another Institute of Radio Engineers masterpiece—they showed that the frog's visual system is sensitive to small moving dark spots on the retina, tiny shadows cast in almost all natural circumstances by flies flying by in the vicinity. This "fly-detector" mechanism is appropriately wired to the hair trigger in the frog's tongue, which handily explains how frogs feed themselves in a cruel world and thereby help propagate their kind. Now, what *does* the frog's eye tell the frog's brain? That there is a fly out there, or that there is a fly-or-a-"slug" (a fake fly of one sort or another) or a thing of kind F (whatever kind of thing reliably triggers this visual gadget)? Millikan, Israel, and I, as Darwinian meaning theorists, have all discussed this very case, and Fodor pounces on it to show what is wrong, by his lights, with any evolutionary account of such meanings: they are too indeterminate. They fail to distinguish, as they ought, between such frog-eye reports as "fly here now" and "fly or small dark projectile here now" and so forth. But this is false. We can use the frog's environment of selection (to the extent that we can determine what it has been) to distinguish between the various candidates. To do this, we use exactly the same considerations we used to settle the questions—to the extent that they were worth trying to settle—about the meaning of the state in the two-bitser. And to the extent that there is just no telling what that environment of selection has been, there is also just no fact of the matter about what the frog-eye report *really* means. This can be brought home vividly by sending the frog to Panama— or, more precisely, sending the frog to a novel selective environment.

Suppose scientists gather up a small population of frogs of some fly-grabbing species, on the brink of extinction, and put them under protective custody in a new environment—a special frog zoo in which there are no flies at all but, rather, zookeepers who periodically arrange to launch little food pellets past the frogs in their care. To their delight, it works; the frogs thrive by zapping their tongues for these pellets, and after a while there is a crowd of descendant frogs who have never in their lives seen a fly, only pellets. What do *their* eyes tell *their* brains? If you insist on saying the meaning hasn't changed, you are in a bind, for this is simply an artificially clear instance of what happens in natural selection all the time: exaptation. As Darwin was careful to remind us, the reuse of machinery for new pur-poses is one of the secrets of Mother Nature's success. We can drive home the point, to any who wish some further persuasion, by supposing that the captive frogs do not all do equally well, because, due to variation in pellet-detecting prowess in their eyes, some eat less heartily than others, and leave less progeny as a result. In short order there will have been undeniable selection for pellet detection—though it would be a mistake to ask exactly when enough of this has occurred for it to "count."

Unless there were "meaningless" or "indeterminate" variation in the trig-gering conditions of the various frogs' eyes, there could be no raw material (blind variation) for selection for a *new* purpose to act upon. The indeter-minacy that Fodor (and others) see as a flaw in Darwinian accounts of the evolution of meaning is actually a precondition for any such evolution. The idea that there must be *something determinate* that the frog's eye really means—some possibly unknowable proposition in froggish that expresses *exactly* what the frog's eye is telling the frog's brain—is just essentialism applied to meaning (or function). Meaning, like function, on which it so directly depends, is not something determinate at its birth. It arises not by saltation or special creation, but by a (typically gradual) shift of circum-stances.

Now we are ready for the only case that really matters to these philoso-phers: what happens when we move a *person* from one environment to another? This is the notorious Twin Earth thought experiment of Hilary Putnam (1975). I am reluctant to go into the details, but I have learned that nothing short of spelling it all out and blocking all the exits will have a prayer of persuading those whose allegiances lie with original intentionality. So, with apologies, here goes. Armed with our background briefing on the two-bitser and the frog, we can see exactly what the Twin Earth thought experiment depends on for its undeniable rhetorical force. Twin Earth, let's suppose, is a planet almost exactly like Earth, except that there are no horses on Twin Earth. There are animals that look just like horses, and are called "horses" by the inhabitants of Twin Boston and Twin London, "*chevaux*" by the inhabitants of Twin Paris, and so forth—that's how similar Twin Earth is to Earth. But these Twin Earth animals are not horses; they are something else. Call them *schmorses*, and if you like, you may suppose they are a sort of pseudo-mammal, a hairy reptile or whatever—this is philosophy, and you get to make up whatever details you find you need to make your thought experiment "work." Now comes the dramatic part. One night, while you sleep, you are whisked off to Twin Earth. (It is important that you sleep through this momentous change, for that keeps you in the dark about what has happened to you—it keeps you "in the same state" you were in on Earth.) When you awake, you look out the window and a schmorse gallops by. "Lo, a horse!" you say (out loud or just to yourself, it makes no difference). To make the case simple, let's suppose Twining, a handy Twin Earthling, utters the very same sounds at the very same time when he, too, sees the schmorse gallop by. Here is what Putnam and others insist on: Twining says, and thinks, something *true*—namely, that a schmorse has just run by. You, Earthling that you still are, say and think something *false*—namely, that a horse has just run by. How long, though, would you have to live on Twin Earth, calling schmorses "horses" (just like all the natives), before the state of your *mind* (or what your eye tells your brain) is a *truth about schmorses* rather than a falsehood about horses? (When does the aboutness or intentionality leap to the new position?—a saltation demanded by these theorists.) Would you ever make the transition? Would you somehow make it without knowing it? The two-bitser was forever oblivious of the change of meaning of its internal state, after all.

I suppose you may be inclined to think that you are quite radically unlike the frog and the two-bitser. You, it may seem, have intrinsic or original intentionality, and this marvelous property has a certain amount of inertia: your brain can't just turn on a dime and suddenly mean something entirely new by its old state. In contrast, frogs don't have much of a memory, and two-bitsers none at all. What *you* mean by the word "horse" (your private mental concept of a horse) is something like *one of those equinish beasts that we Earthlings like to ride*, an epithet anchored in your mind by all your memories of horse shows and cowboy movies. Let us agree that this memory matrix *fixes* the kind of thing to which your concept of horse applies. *Ex hypothesi*, schmorses are not beasts of *that* kind; they are not of the same species at all, but just conveniently indistinguishable by you from horses. So, according to this line of thought, you make unwitting errors every time you (mis)classify a schmorse perceptually or in reflection ("Wasn't that a fine horse I saw gallop by my window yesterday!").

But there is another way of thinking of the same example. Nothing forces us to suppose that your concept of a horse wasn't more relaxed in the first place, rather like your concept of a table. (Try telling the story of Twin Earth with the suggestion that the tables there aren't *really* tables, but just look like tables and are used for tables. It doesn't work, does it?) Horses and schmorses may not be the same biological species, but what if you, like most Earthlings, have no clear concept of species, and classify by appearance: *living thing that looks like Man-o-War*. Horses and schmorses both fall into that kind, so, when you call a Twin Earth beast a horse, you're *right* after all. Given what you mean by "horse," schmorses *are* horses—a non-Earthly kind of horse, but a horse just the same. Non-Earthly tables are tables, too. It is clear that you *could* have such a relaxed concept of horses, and that you could have a tighter concept, according to which schmorses are not horses, not being of the same Earthly species. Both cases are possible. Now, must it be *determinate* whether your horse concept (prior to your move) meant the species or the wider class? It might be, if you are well read in biology, for instance, but suppose you are not. Then your concept— what "horse" *actually* means to you—would suffer the same indeterminacy as the frog's concept of *fly* (or was it all along the concept *small airborne food item*?).

It might help to have a more realistic example, something that could happen right here on Earth. I have been told that it once was the case that the Siamese had a word for, well, "cat" but had never seen or imagined any other cats than Siamese cats. Let's suppose their word was "kat"—it doesn't matter what the actual details were, or even whether this particular tale is true. It could be. When they discovered that other varieties existed, they had a problem: did their word mean "cat" or "Siamese cat"? Had they just discovered that there were other, rather different-looking, sorts of kats, or that kats and those other creatures belonged to a supergroup? Was their traditional term the name of a species or a variety? If they lacked the biological theory that made this distinction, how could there possibly be a fact of the matter? (Well, they might discover that peculiarities of appearance were really very important to them—"That just doesn't look like a kat, so it isn't one!" And you might discover that you similarly resisted the suggestion that Shetland ponies were horses.)

When a Siamese person saw a (non-Siamese) cat walking by and thought "Lo, a kat!" would this be an error or the simple truth? Perhaps the Siamese person wouldn't have any opinion about how to answer this question, but could there nevertheless be a determinate fact about whether this was an error, something we might never be able to discover, but a fact nonetheless? The same thing could happen to you, after all: imagine that a biologist told you, one day, that coyotes are in fact dogs—members of the same species. You might find yourself wondering whether the biologist and you had the same concept of dog. How strong was your allegiance to the view that "dog" was a species term and not the name for a large subspecies of *domestic* dogs? Does your heart of hearts tell you loud and clear that you had already ruled out "by definition" the hypothesis that coyotes are dogs, or does it silently allow that your concept has all along had the openness to admit this purported discovery? Or would you find that now that the issue has been raised, you would have to settle, one way or the other, something that had simply not been fixed before because it had never come up before?

Such a threat of indeterminacy undermines the argument in Putnam's thought experiment. To preserve its point, Putnam tries to plug this gap by declaring that our concepts—whether we know it or not—refer to *natural kinds*. But which kinds are natural? Varieties are just as natural as species, which are as natural as genera and higher classifications. Essentialism with regard to the meaning of the frog's mental state and the two-bitser's inner state Q (or QB) was seen to evaporate; it must evaporate just as surely for us. The frog *would have* zapped just as readily at pellets in the wild, had any come its way, so it certainly was not equipped with anything that discriminated *against* pellets. In one sense, *fly-or-pellet* is a natural kind for frogs; they *naturally* fail to discriminate between the two. In another sense, *fly-or-pellet* is not a natural kind for frogs; their natural environment has never made that classification relevant before. Exactly the same is true for you. Had schmorses been secretly brought to Earth, you would have just as readily called them "horses." You would have been wrong if it was somehow already fixed that what your term meant was the species, not the lookalikes, but if not, there would be no grounds at all for calling your classification an error, since the distinction had never before come up. Like the frog and the two-bitser, you have internal states that get their meanings from their functional roles, and where function fails to yield an answer, there is nothing more to inquire about.

The tale of Twin Earth, if we read it through Darwinian spectacles, proves that human meanings are just as derived as the meanings of two-bitsers and frogs. This is not what it was intended to show, but any attempt to block this interpretation is forced to postulate mysterious and unmotivated doctrines of essentialism, and to insist, point-blank, that there are facts about meaning that are utterly inert and undiscoverable, but facts all the same. Since some philosophers are ready to swallow these bitter pills, I need to provide a few further persuasions.

The idea that our meanings are just as dependent on function as the meanings of the states of artifacts, and hence just as derivative and potentially indeterminate, strikes some philosophers as intolerable because it fails to give meaning a proper *causal role.* This is an idea we have seen in an earlier incarnation, as the worry about minds' being mere *effects*, not originating *causes.* If meaning gets determined by the selective forces that endorse certain functional roles, then all meaning may seem, in a sense, to be only *retrospectively* attributed: what something means is not an *intrinsic* property it has, capable of making a difference in the world at the moment of its birth, but at best a retrospective coronation secured only by an analysis of the subsequent effects engendered. That is not quite right: an engineering analysis of the two-bitser, newly arrived in Panama, would permit us to say what roles the device, so configured, would be *good for*, even if it had not yet been chosen for any role. We could reach this verdict: its acceptance state *could* mean "quarter-balboa here now" if we put it in the right environment. But of course it could also mean lots of other things, if placed in other environments, so it won't mean any one of them until a particular functional role for it gets established—and there is no threshold for how long it takes a functional role to become established.

That is not enough for some philosophers, who think that meaning, so construed, doesn't pull its weight. The clearest expression of this idea is Fred Dretske's (1986) insistence that meaning must itself play a causal role in our mental lives in a way that meaning never plays a causal role in the career of an artifact. Put this way, the attempt to distinguish real meaning from artificial meaning betrays a yearning for skyhooks, a yearning for something "principled" that could block the gradual emergence of meaning from some cascade of mere purposeless, mechanical causes, but this is (you must suspect) an optional and tendentious way of putting it. As usual, the issues are more complex than I am showing,[4] but we can force the key points into the open with the help of a little fable I recently devised precisely to give these philosophers fits. It works. First the fable, and then the fits.

2. TWO BLACK BOXES[5]

Once upon a time, there were two large black boxes, A and B, connected by a long insulated copper wire. On box A there were two buttons, marked α and β, and on box B there were three lights, red, green, and amber. Scientists studying the behavior of the boxes had observed that whenever you

4. *Dretske and His Critics* (McLaughlin, ed., 1991) is devoted largely to this issue.

5. This began as an impromptu response to Jaegwon Kim's talk at Harvard, November 29, 1990: "Emergence, Non-Reductive Materialism, and 'Downward Causation,' " and evolved under the persistent rebuttals of Kim and many other philosophers, to whom I am grateful.

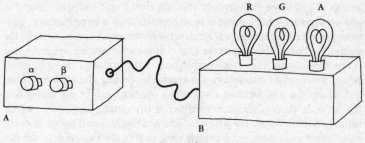

FIGURE 14.1

pushed the α button on box A, the red light flashed briefly on box B, and whenever you pushed the β button on box A, the green light flashed briefly. The amber light never seemed to flash. They performed a few billion trials, under a very wide variety of conditions, and found no exceptions. There seemed to them to be a causal regularity, which they conveniently summarized thus:

All α's cause reds.
All β's cause greens.

The causation passed through the copper wire somehow, they determined, since severing it turned off all effects in box B, and shielding the two boxes from each other without severing the wire never disrupted the regularity. So naturally they were curious to know just how the causal regularity they had discovered was effected through the wire. Perhaps, they thought, pressing button α caused a low-voltage pulse to be emitted down the wire, triggering the red light, and pressing button β caused a high-voltage pulse, which triggered the green. Or perhaps pressing α caused a single pulse, which triggered the red light, and pressing β caused a double pulse. Clearly, there was something that always happened in the wire when you pressed button α, and something different that always happened in the wire when you pressed β. Discovering just what this was would explain the causal regularity they had discovered.

A wiretap of sorts on the wire soon revealed that things were more complicated. Whenever *either* button was pushed on box A, a long stream of pulses and gaps (ons and offs, or bits) was sent swiftly down the wire to box B—ten thousand bits, to be exact. But it was a different pattern each time!

Clearly, there had to be a feature or property of the strings of bits that triggered the red light in one case and the green light in the other. What could it be? They decided to open up box B and see what happened to the strings of bits when they arrived. Inside B they found a supercomputer—just an ordinary digital, serial supercomputer, with a large memory, containing a huge program and a huge data base, written, of course, in more bit strings. And when they traced the effects of the incoming bit strings on this computer program, they found nothing out of the ordinary: the input string would always make its way into the central processing unit in normal fashion, where it would provoke a few billion operations to be performed in a few seconds, ending, always, with either of two output signals, a 1 (which turned on the red light) or a 0 (which turned on the green light). In every case, they found, they could explain each step of the causation at the microscopic level without any difficulty or controversy. No occult causes were suspected to be operating, and, for instance, when they arranged to input the same sequence of ten thousand bits again and again, the program in box B always yielded the same output, red or green.

But this was mildly puzzling, because although it always gave the same output, it didn't always yield the same output by going through the same intermediate steps. In fact, it almost always passed through different states before yielding the same output. This in itself was no mystery, since the program kept a copy of each input it received, and so, when the same input arrived a second or third or thousandth time, the state of the memory of the computer was slightly different each time. But the output was always the same; if the light turned red the first time a particular string was input, it always turned red for the same string thereafter, and the same regularity held for green strings (as the scientists began to call them). All strings, they were tempted to hypothesize, are either red strings (cause the red light to flash) or green strings (cause the green light to flash). But of course they hadn't tested all possible strings—only strings that had been emitted by box A.

So they decided to test their hypothesis by disconnecting A from B temporarily and inserting variations on A's output strings to B. To their puzzlement and dismay, they discovered that almost always when they tampered with a string from A, the *amber* light flashed! It was almost as if box B had detected their intervention. There was no doubt, however, that box B would readily accept man-made versions of red strings by flashing red, and man-made versions of green strings by flashing green. It was only when a bit—or more than one bit—was changed in a red or green string that the amber light usually—almost always—came on. "You've killed it!" somebody once blurted out, after watching a "tampered" red string turn into an amber string, and this led to a flurry of speculation that red and green strings were in some sense *alive*—perhaps male and female—whereas amber strings were *dead* strings. But, appealing as this hypothesis was, it did not turn out to lead anywhere, although further experimentation with a few billion random variations on bit strings of ten thousand bits in length did strongly suggest to the scientists that there were really three varieties of strings: red strings, green strings, and amber strings—and amber strings outnumbered red and green strings by many, many orders of magnitude. Almost all strings were amber strings. That made the regularity they had discovered all the more exciting and puzzling.

What was it about red strings that turned on the red light, and green strings that turned on the green light? Of course, in each particular case, there was no mystery at all. They could trace the causation of each particular string through the supercomputer in B and see that, with gratifying determinism, it produced its red or green or amber light, as the case might be. What they couldn't find, however, was a way of predicting which of the three effects a new string would have, just by examining it (without "hand-simulating" its effect on box B). They knew from their empirical data that the odds were very high that any new string considered would be amber—unless it was a string known to have been emitted by box A, in which case the odds were better than a billion to one that it would be *either* red or green, but no one could tell which without running it through box B to see how the program settled out.

Since, in spite of much brilliant and expensive research, they found themselves still utterly unable to predict whether a string would turn out to be red, green, or amber, some theorists were tempted to call these properties *emergent* properties. What they meant was that the properties were (they thought) *unpredictable in principle* from a mere analysis of the microproperties of the strings themselves. But this didn't seem likely at all, since each particular case was as predictable as any deterministic input to a deterministic program could be. In any event, whether or not the properties of red, green, and amber were unpredictable in principle or merely in practice, they certainly were surprising and mysterious properties.

Perhaps the solution to the mystery lay in box A. They opened it up and found another supercomputer—of a different make and model, and running a different gigantic program, but also just a garden-variety digital computer. They soon determined that whenever you pushed button α this sent the program off in one way, by sending a code (11111111) to the CPU, and whenever you pushed button β this sent a different code (00000000) to the CPU, setting in motion a different set of billions of operations. It turned out that there was an internal "clock" ticking away millions of times a second, and whenever you pushed either button the first thing the computer did was take the "time" from the clock (e.g., 101101010101010111) and break it up into strings it then used to determine which subroutines to call in which order, and which part of its memory to access first in the course of its preparation of a bit string to send down the wire.

The scientists were able to figure out that it was this clock-consulting (which was as good as random) that virtually guaranteed that the same bit string was never sent out twice. But in spite of this randomness, or pseudo-randomness, it remained true that whenever you pushed button α the bit string the computer concocted turned out to be red, and whenever you pushed button β the bit string eventually sent turned out to be green. Actually, the scientists did find a few anomalous cases: in roughly one in a billion trials, pushing the α button caused a green string to be emitted, or pushing the β button caused a red string to be emitted. This tiny blemish in perfection only whetted the scientists' appetite for an explanation of the regularity.

And then, one day, along came the two AI hackers who had built the boxes, and they explained it all. (Do not read on if you want to figure out the mystery for yourself.) Al, who had built box A, had been working for years on an "expert system"—a data base containing "true propositions" about everything under the sun, and an inference engine to deduce further implications from the axioms that composed the data base. There were major-league baseball statistics, meteorological records, biological taxonomies, histories of the world's nations, and hosts of trivia in the data base. Bo, the Swede who had built box B, had been working during the same time on a rival "world-knowledge" data base for his own expert system. They had both stuffed their respective data bases with as many "truths" as years of work had permitted.[6]

But as the years progressed, they had grown bored with expert systems, and had both decided that the practical promise of this technology was vastly overrated. The systems weren't actually very good at solving interesting problems, or "thinking," or "finding creative solutions to problems." All they were good at, thanks to their inference engines, was generating lots and lots of true sentences (in their respective languages), and testing input sentences (in their respective languages) for truth and falsity—relative to their "knowledge," of course. So Al and Bo had got together and figured out how the fruits of their wasted effort could be put to use. They decided to make a philosophical toy. They chose a *lingua franca* for translating between their two representational systems (it was English, actually, sent in standard ASCII code, the code of electronic mail), and hooked the machines together with a wire. Whenever you pushed A's α button, this instructed A

---

6. For a real-world example of such a project, see Douglas Lenat's enormous CYC (short for "encyclopedia") project at MCC (Lenat and Guha 1990). The idea is to hand-code all the millions of facts in an encyclopedia (plus all the other millions of facts that everyone knows, so there is no point in putting them in the encyclopedia—such as the facts that mountains are bigger than molehills, and toasters can't fly), and then attach an inference engine that can update, preserve consistency, deduce surprising implications, and in general service the world-knowledge base. For an entirely different approach to AI, consider Rodney Brooks' and Lynn Stein's humanoid-robot project (Dennett 1994c).

to choose at random (or pseudo-random) one of its "beliefs" (either a stored axiom or a generated implication of its axioms), translate it into English (in a computer, English characters would already be in ASCII), add enough random bits after the period to bring the total up to ten thousand, and send the resulting string to B, which translated this input into its own language (which was Swedish Lisp), and tested it against its own "beliefs"—its data base. Since both data bases were composed of truths, and roughly the same truths, thanks to their inference engines, whenever A sent B something A "believed," B "believed" it, too, and signaled this by flashing a red light. Whenever A sent B what A took to be a falsehood, B announced that it judged that this was indeed a falsehood by flashing a green light.

And whenever anyone tampered with the transmission, this almost always resulted in a string that was not a well-formed sentence of English (B had absolutely zero tolerance for "typographical" errors). B responded to these by flashing the amber light. Whenever anyone chose a bit string at random, the odds were Vast that it would not be a well-formed truth or falsehood of English ASCII; hence the preponderance of amber strings.

So, said Al and Bo, the emergent property *red* was actually the property of being a true sentence of English, and *green* was the property of being a falsehood in English. Suddenly, the search that had eluded the scientists for years became child's play. Anyone could compose red strings *ad nauseam*—just write down the ASCII code for "Houses are bigger than peanuts" or "Whales don't fly" or "Three times four is two less than two times seven," for instance. If you wanted a green string, try "Nine is less than eight" or "New York is the capital of Spain."

Philosophers soon hit upon cute tricks, such as finding strings that were red the first hundred times they were given to B but green thereafter (e.g., the ASCII for "This sentence has been sent to you for evaluation fewer than a hundred and one times").

But, said some philosophers, the string properties *red* and *green* are not really *truth in English* and *falsity in English*. After all, there are English truths whose ASCII expression takes millions of bits, and besides, in spite of their best efforts, Al and Bo didn't always insert *facts* in their programs. Some of what had passed for common knowledge when they were working on their data bases had since been disproven. And so forth. There were lots of reasons why the string property—the causal property—of *redness* was not quite exactly the property of *truth in English*. So, perhaps *red* could better be defined as *relatively short expression in English ASCII of something "believed" true by box B (whose "beliefs" are almost all true)*. This satisfied some, but other picked nits, insisting, for various reasons, that this definition was inexact, or had counterexamples that could not be ruled out in any non–*ad hoc* way, and so forth. But as Al and Bo pointed out, there were no better candidate descriptions of the property to be found, and hadn't the scientists been yearning for just such an explanation? Hadn't the mystery of red and green strings now been entirely dissolved? Moreover, now that it was dissolved, couldn't one see that there wasn't any hope at all of explaining the *causal* regularity with which we began our tale without using *some* semantical (or mentalistic) terms?

Some philosophers argued that, though the newfound description of the regularity in the activity in the wire could be used to predict box B's behavior, it was not a *causal* regularity after all. Truth and falsehood (or any of the adjusted stand-ins just considered) are semantic properties, and as such are entirely abstract, and hence could not cause anything. Nonsense, others retorted. Pushing button α causes the red light to go on just as certainly as turning the ignition key causes your car to start. If it had turned out that what was being sent down the wire was simply high versus low voltage, or one pulse versus two, everybody would agree that this was a paradigm causal system. The fact that this system turned out to be a Rube Goldberg machine didn't show that the reliability of the link between α and red flashes was any less causal. On the contrary, in every single case the scientists could trace out the exact microcausal path that explained the result.[7]

Convinced by this line of reasoning, other philosophers began to argue that this showed that the properties red, green, and amber weren't *really* semantical or mentalistic properties after all, but only imitation semantical properties, mere *as if* semantical properties. What red and green were, really, were very, very complicated *syntactical* properties. These philosophers declined, however, to say anything further about just what syntactical properties these were, or to explain how even young children could swiftly and reliably produce instances of them, or recognize them. The philosophers were nevertheless convinced that there *had* to be a purely syntactic description of the regularity, since, after all, the causal systems in question were "just" computers, and computers are "just" syntactic engines, not capable of any real semanticity.

"I suppose," retorted Al and Bo, "that, if you had found *us* inside our black boxes, playing a trick on you by following the same scheme, you would then relent and agree that the operative causal property was genuine truth (or believed truth, in any event). Can you propose any good reason for drawing such a distinction?" This led some to declare that in a certain important sense Al and Bo *had* been in the boxes, since they were responsible for creating the respective data bases, as models of their own beliefs. It led others to deny that there really were any semantical or mentalistic proper-

---

7. Some have argued that my account of patterns in Dennett 1991b is *epiphenomenalism* about content. This is my reply.

ties anywhere in the world. Content, they said, had been *eliminated*. The debate went on for years, but the mystery with which we began was solved.

3. BLOCKING THE EXITS

The tale ends there. Experience teaches, however, that there is no such thing as a thought experiment so clearly presented that no philosopher can misinterpret it, so, in order to forestall some of the most attractive misinterpretations, I will inelegantly draw attention to a few of the critical details and explain their roles in this intuition pump.

(1) The devices in boxes A and B are nothing but automated encyclopedias—not even "walking encyclopedias," just "boxes of truths." Nothing in the story presupposes or implies that these devices are conscious, or *thinking things*, or even *agents*, except in the same minimal sense in which a thermostat is an agent. They are utterly boring intentional systems, rigidly fixed to fulfilling a single, simple goal. They contain large numbers of true propositions and the inferential machinery necessary to generate more truths, and to test for "truth" by testing a candidate proposition against their existing data bases.[8]

Since the two systems were created independently, they cannot plausibly be supposed to contain *exactly* the same truths (actually or even virtually), but, for the prank to work as well as I claim it did in the story, we must suppose a very large overlap, so that it was highly unlikely that a truth generated by A would not be recognized as such by B. Two considerations, I claim, make this plausible: (a) Al and Bo may live in different countries and have different native languages, but they *inhabit the same world*, and (b) although there are kazillions of true propositions about that world (our world), the fact that both Al and Bo set out to create *useful* data bases—containing the information that is relevant to all but the most recherché of human purposes—would guarantee a high degree of overlap between the two independently created systems. Although Al might know that at noon on his twentieth birthday his left foot was closer to the North

---

8. Since these are just boxes of truths, no support is hereby given to the "language of thought" hypothesis (Fodor 1975). I supposed that the world knowledge was stored in a quasi-linguistic form just to make the storytelling easier (which is probably also the reason motivating most researchers in cognitive science, who adopt the language-of-thought hypothesis out of convenience!).

Pole than to the South Pole, and Bo had not forgotten that his first French teacher was named Dupont, these would not be truths that either would be apt to put in his data base. But if you doubt that the mere fact that they were both intent on creating an internationally useful encyclopedia would ensure such a close correspondence between their respective data bases, just add, as an inelegant detail, the convenient fact that they compared notes as to topics to be covered during their years of hacking.

(2) Why not just have Al and Bob (a fellow American), or, for that matter, why not simply have a duplicate of Al's system in box B? Because it must be the essence (oops!) of my story that no simple, feasibly discoverable *syntactic* matching up could explain the regularity. That is why Bo's system is in Swedish Lisp—to conceal from prying eyes the underlying *semantic* commonalities between the data structures consulted during A's sentence-generation task and B's sentence-translation-and-truth-testing task. The idea was to create two systems that exhibited the fascinating regularity of external behavior described but that were internally as different as possible, so that *only* the fact that their respective innards were systematic *representations of a common world* could explain the regularity.

(3) We might pause to ask whether or not two such systems could ever be so inscrutable as to be invulnerable to reverse engineering. Cryptography has moved into such rarefied and arcane regions that one should think thrice at least before declaring either way. I have no idea whether anybody can make a sound argument to the effect that there are unbreakable encryption schemes or that there aren't. But, encryption aside, hackers will appreciate that all the convenient comments and other signposts one places in the "source code" when composing a program vanish when the source code is "compiled," leaving behind an *almost* impossible-to-decipher tangle of machine instructions. "Decompiling" is sometimes possible in practice (is it always possible in principle?), though of course it won't restore the comments but just render salient the structures in the higher-level language. My assumption that the scientists' efforts at decompiling the program and deciphering the data bases came to naught could be strengthened by postulating encryption, if need be.

In the story as told, we can agree that it is bizarre that the scientists never thought of checking to see if there was an ASCII translation of the bit streams running through the wire. How could they be so dense? Fair enough: send the whole gadget (boxes A and B, and the connecting wire) to "Mars," and let the alien scientists there try to figure out the regularity. The fact that all α's cause reds, all β's cause greens, and random bit strings cause ambers will be just as visible to them as to us, but they will be clueless about ASCII. To them, this gift from outer space will exhibit an utterly mysterious regularity, totally beyond all analytic probes, *unless* they hit upon the idea that each box contains a *description of a world*, and that the descriptions are *of the same world*. It is the fact that each box bears multifarious semantic relationships to the same things, though expressed in different "terminology" and differently axiomatized, that grounds the regularity.

When I tried this thought experiment out on Danny Hillis, creator of the Connection Machine, he thought immediately of a cryptographic "solution" to the puzzle, and then granted that my solution could be profitably viewed as a special case of his solution: "Al and Bo were using *the world* as a 'one-time pad!' "—an apt allusion to a standard technique of encryption. You can see the point by imagining a variation. You and your best friend are about to be captured by hostile forces, who may know English but not much about your world. You both know Morse code, and hit upon the following impromptu encryption scheme: for a dash, speak a truth; for a dot, speak a falsehood. Your captors are permitted to listen to you two speak: "Birds lay eggs, and toads fly. Chicago is a city, and my feet are not made of tin, and baseball is played in August," you say, answering "No" (dash-dot; dash-dash-dash) to whatever your friend has just asked. Even if your captors know Morse code, unless they can determine the truth and falsity of these sentences, they cannot detect the properties that stand for dot and dash. This variation could be added to our fable, for spice, as follows. Instead of shipping the computer systems in the boxes to Mars, we put Al and Bo in the boxes and ship them to Mars. The Martians will be as puzzled by them, if they play the Morse-code prank, as by the computers, unless they draw the conclusion (obvious to us, but we're not Martians) that these things in the boxes are to be semantically interpreted.

The point of the fable is simple. There is no substitute for the intentional stance; either you adopt it, and explain the pattern by finding the semantic-level facts, or you will forever be baffled by the regularity—the *causal* regularity—that is manifestly there. The same moral, we have seen, can be drawn about interpreting the historical facts of evolutionary history. Even if you can describe, in matchless microdetail, every causal fact in the history of every giraffe who has ever lived, unless you go up a level or two and ask "Why?"—hunting for the *reasons* endorsed by Mother Nature—you will never be able to *explain* the manifest regularities, such as the fact that giraffes have come to have long necks, for instance. That is Dewey's point in the quotation given earlier in this chapter.

At this juncture, if you are like many philosophers, you are attracted by the claim that this thought experiment "works" only because boxes A and B are artifacts whose intentionality, such as it is, is entirely derived and artifactual. The data structures in their memories get their reference (if they get any at all) from indirect reliance on the sense organs, life histories, and purposes of their creators, Al and Bo. The real source of the meaning or truth or semanticity in the artifacts lies in these human artificers. (That was the point of the suggestion that in a certain sense Al and Bo *were* in their respective boxes.) Now, I *might* have told the story differently: inside the boxes were two *robots*, Al and Bo, which had each spent a longish "lifetime" scurrying around in the world gathering facts before getting in their respective boxes. I chose a simpler route, to forestall all the questions about whether box A or box B was "really thinking," but we may now reinstate the issues thereby finessed, since it is finally time to dispose once and for all of the hunch that *original* intentionality could not emerge in any artifactual "mind" without the intervention of a (human?) artificer. Suppose that is so. Suppose, in other words, that, whatever differences there might be between a simple box of truths like box A and the fanciest imaginable robot, since both would just be artifacts, neither could have real—or original—intentionality, but only the derivative intentionality borrowed from its creator. Now you are ready for another thought experiment, a *reductio ad absurdum* of that supposition.

4. SAFE PASSAGE TO THE FUTURE[9]

Suppose you decided, for whatever reasons, that you wanted to experience life in the twenty-fifth century, and suppose that the only known way of keeping your body alive that long required it to be placed in a hibernation device of sorts. Let's suppose it's a "cryogenic chamber" that cools your body down to a few degrees above absolute zero. In this chamber your body would be able to rest, suspended in a super-coma, for as long as you liked. You could arrange to climb into the chamber, and its surrounding support capsule, be put to sleep, and then automatically be awakened and released in 2401. This is a time-honored science-fiction theme, of course.

Designing the capsule itself is not your only engineering problem, since the capsule must be protected and supplied with the requisite energy (for refrigeration, etc.) for over four hundred years. You will not be able to count on your children and grandchildren for this stewardship, of course, since they will be long dead before the year 2401, and you would be most unwise to presume that your more distant descendants, if any, will take a lively interest in your well-being. So you must design a supersystem to protect your capsule and to provide the energy it needs for four hundred years.

Here there are two basic strategies you might follow. On one, you should

---

9. An earlier version of this thought experiment appeared in Dennett 1987b, ch. 8.

prospect around for the ideal location, as best you can foresee, for a fixed installation that will be well supplied with water, sunlight, and whatever else your capsule (and the supersystem itself) will need for the duration. The main drawback to such an installation or "plant" is that it cannot be moved if harm comes its way—if, say, someone decides to build a freeway right where it is located. The alternative strategy is much more sophisticated and expensive, but avoids this drawback: design a mobile facility to house your capsule, along with the requisite sensors and early-warning devices so that it can move out of harm's way and seek out new sources of energy and repair materials as it needs them. In short, build a giant robot and install the capsule (with you inside it) in it.

These two basic strategies are copied from nature: they correspond roughly to the division between plants and animals. Since the latter, more sophisticated, strategy better fits our purposes, let's suppose that you decide to build a robot to house your capsule. You should try to design this robot so that above all else it "chooses" actions designed to further your interests, of course. Don't call these mere switching points in your robot's control system "choice" points if you think that this would imply that the robot had free will or consciousness, for I don't mean to smuggle any such contraband into the thought experiment. My point is uncontroversial: the power of any computer program lies in its capacity to execute *branching* instructions, zigging one way or another depending on some test it executes on the data then available to it, and my point is just that, as you plan your robot's control system, you would be wise to try to structure it so that whenever branching opportunities confront it, it will tend to branch down that path that has the highest probability of serving *your* interests. You are, after all, the *raison d'être* of the whole gadget. The idea of designing hardware and software that are specifically attuned to the interests of a particular human individual is not even science fiction any more, though the particular design problems facing your robot-builders would be profoundly difficult engineering challenges, somewhat beyond the state of the art today. This mobile entity would need a "vision" system to guide its locomotion, and other "sensory" systems as well, in addition to the self-monitoring capacities to inform it of its needs.

Since you will be comatose throughout, and thus cannot stay awake to guide and plan its strategies, you will have to design the robot supersystem to generate its own plans in response to changing circumstances over the centuries. It must "know" how to "seek out" and "recognize" and then exploit energy sources, how to move to safer territory, how to "anticipate" and then avoid dangers. With so much to be done, and done fast, you had best rely whenever you can on economies: give your robot no more discriminatory prowess than it will probably need in order to distinguish whatever needs distinguishing in the world—given its particular constitution.

Your task would be made much more difficult if you couldn't count on your robot's being the only such robot around with such a mission. Let us suppose that, in addition to whatever people and other animals are up and about during the centuries to come, there will be other robots, many *different* robots (and perhaps "plants" as well), competing with your robot for energy and safety. (Why might such a fad catch on? Let's suppose we get irrefutable advance evidence that travelers from another galaxy will arrive on our planet in 2401. I for one would ache to be around to meet them, and if cold storage was my only prospect, I'd be tempted to go for it.) If you have to plan for dealing with other robotic agents, acting on behalf of other clients like yourself, you would be wise to design your robot with enough sophistication in its control system to permit it to calculate the likely benefits and risks of cooperating with other robots, or of forming alliances for mutual benefit. You would be most unwise to suppose that other clients will be enamored of the rule of "live and let live"—there may well be inexpensive "parasite" robots out there, for instance, just waiting to pounce on your expensive contraption and exploit it. Any calculations your robot makes about these threats and opportunities would have to be "quick and dirty"; there is no foolproof way of telling friends from foes, or traitors from promise-keepers, so you will have to design your robot to be, like a chess-player, a decision-maker who takes risks in order to respond to time pressure.

The result of this design project would be a robot capable of exhibiting self-control of a high order. Since you must cede fine-grained real-time control to it once you put yourself to sleep, you will be as "remote" as the engineers in Houston were when they gave the Viking spacecraft its autonomy (see chapter 12). As an autonomous agent, it will be capable of deriving *its own* subsidiary goals from its assessment of its current state and the import of that state for its ultimate goal (which is to preserve you till 2401). These secondary goals, which will respond to circumstances you cannot predict in detail (if you could, you could hard-wire the best responses to them), may take the robot far afield on century-long projects, some of which may well be ill-advised, in spite of your best efforts. Your robot may embark on actions antithetical to your purposes, even suicidal, having been convinced by another robot, perhaps, to subordinate its own life mission to some other.

This robot we have imagined will be richly engaged in its world and its projects, always driven ultimately by whatever remains of the goal states that you set up for it at the time you entered the capsule. All the preferences it will ever have will be the offspring of the preferences you initially endowed it with, in hopes that they would carry you into the twenty-fifth century, but that is no guarantee that actions taken in the light of the robot's descendant preferences will continue to be responsive, directly, to your best interests. From your selfish point of view, that is what you hope, but this robot's projects are out of your direct control until you are awakened. It will have some internal representation of its currently highest goals, its *summum bonum*, but if it has fallen among persuasive companions of the sort we have imagined, the iron grip of the engineering that initially designed it will be jeopardized. It will still be an artifact, still acting only as its engineering permits it to act, but following a set of *desiderata* partly of its own devising.

Still, according to the assumption we decided to explore, this robot will not exhibit anything but *derived* intentionality, since it is just an artifact, created to serve your interests. We might call this position "client centrism" with regard to the robot: *I* am the original source of all the derived meaning within my robot, however far afield it drifts. It is just a survival machine designed to carry me safely into the future. The fact that it is now engaged strenuously in projects that are only remotely connected with my interests, and even antithetical to them, does not, according to our assumption, endow any of its control states, or its "sensory" or "perceptual" states, with genuine intentionality. If you still want to insist on this client centrism, then you should be ready to draw the further conclusion that you yourself never enjoy any states with *original* intentionality, since you are just a survival machine designed, originally, for the purpose of preserving your genes until they can replicate. Our intentionality is derived, after all, from the intentionality of our selfish genes. *They* are the Unmeant Meaners, not us!

If this position does not appeal to you, consider jumping the other way. Acknowledge that a fancy-enough artifact—something along the lines of these imagined robots—*can* exhibit real intentionality, given its rich functional embedding in the environment and its prowess at self-protection and self-control.[10] It, like you, owes its very existence to a project the goal of

---

10. In the light of this thought experiment, consider an issue raised by Fred Dretske (personal communication) with admirable crispness: "I think we could (logically) create an artifact that *acquired* original intentionality, but not one that (at the moment of creation, as it were) *had* it." How much commerce with the world would be enough to turn the dross of derived intentionality into the gold of original intentionality? This is our old problem of essentialism, in a new guise. It echoes the desire to zoom in on a crucial moment and thereby somehow identify a threshold that marks the first member of a species, or the birth of real function, or the origin of life, and as such it manifests a failure to accept the fundamental Darwinian idea that all such excellences emerge gradually by finite increments. Notice, too, that Dretske's doctrine is a peculiar brand of extreme Spencerism: the *current* environment must do the shaping of the organism before the shape "counts" as having real intentionality; *past* environments, filtered through the wisdom of engineers or a history of natural selection, don't count—even if they result in the very same functional structures. There is something wrong and something right in this. More important than any particular past history of individual appropriate commerce

which was to create a survival machine, but it, like you, has taken on a certain autonomy, has become a locus of self-control and self-determination, not by any miracle, but just by confronting problems during its own "lifetime" and more or less solving them—problems of survival presented to it by the world. Simpler survival machines—plants, for instance—never achieve the heights of self-redefinition made possible by the complexities of your robot; considering them *just* as survival machines for their comatose inhabitants leaves no patterns in *their* behavior unexplained.

If you pursue this avenue, which of course I recommend, then you must abandon Searle's and Fodor's "principled" objection to "strong AI." The imagined robot, however difficult or unlikely an engineering feat, is not an impossibility—nor do they claim it to be. They concede the possibility of such a robot, but just dispute its "metaphysical status"; however adroitly it managed its affairs, they say, its intentionality would not be the real thing. That's cutting it mighty fine. I recommend abandoning such a forlorn disclaimer and acknowledging that the meaning such a robot would discover in its world, and exploit in its own communications with others, would be exactly as real as the meaning you enjoy. Then your selfish genes can be seen to be the original *source* of your intentionality—and hence of every meaning you can ever contemplate or conjure up—even though you can then transcend your genes, using your experience, and in particular the culture you imbibe, to build an almost entirely independent (or "transcendent") locus of meaning on the base your genes have provided.

I find this an entirely congenial—indeed, inspiring—resolution of the tension between the fact that I, as a person, consider myself to be a source of meaning, an arbiter of what matters and why, and the fact that at the same time I am a member of the species *Homo sapiens*, a product of several billion years of nonmiraculous R and D, enjoying no feature that didn't spring from the same set of processes one way or another. I know that others find this vision so shocking that they turn with renewed eagerness to the conviction that somewhere, somehow, there just *has* to be a blockade against Darwinism and AI. I have tried to show that Darwin's dangerous idea carries the implication that there is no such blockade. It follows from the truth of Darwinism that you and I are Mother Nature's artifacts, but our intentionality is none the less real for being an effect of millions of years of

---

with the real world is the disposition to engage in supple *future* interactions, appropriately responsive to whatever novelty the world imposes. But—and this is the solid ground, I think, for Dretske's intuition—since this capacity for swift redesign is apt to show itself in current or recent patterns of interaction, his insistence that an artifact exhibit "do-it-yourself understanding" (Dennett 1992) is plausible, so long as we jettison the essentialism and treat it simply as an important symptom of intentionality worthy of the name.

mindless, algorithmic R and D instead of a gift from on high. Jerry Fodor may joke about the preposterous idea of our being Mother Nature's artifacts, but the laughter rings hollow; the only alternative views posit one skyhook or another. The shock of this conclusion may be enough to make you more sympathetic to Chomsky's or Searle's forlorn attempts to conceal the mind behind impenetrable mystery, or Gould's forlorn attempts to escape from the implication that natural selection is all it takes—an algorithmic series of cranes cranking out ever higher forms of design.

Or it may inspire you to look elsewhere for a savior. Didn't the mathematician Kurt Gödel prove a great theorem that demonstrated the impossibility of AI? Many have thought so, and recently their hunch was given a powerful boost by one of the world's most eminent physicists and mathematicians, Roger Penrose, in his book *The Emperor's New Mind: Concerning Computers, Minds, and the Laws of Physics* (1989), to which the next chapter is devoted.

CHAPTER 14: *Real meaning, the sort of meaning our words and ideas have, is itself an emergent product of originally meaningless processes—the algorithmic processes that have created the entire biosphere, ourselves included. A robot designed as a survival machine for you would, like you, owe its existence to a project of R and D with other ulterior ends, but this would not prevent it from being an autonomous creator of meanings, in the fullest sense.*

CHAPTER 15: *One more influential source of skepticism about AI (and Darwin's dangerous idea) must be considered and neutralized: the persistently popular idea to the effect that Gödel's Theorem proves that AI is impossible. Roger Penrose has recently revived this meme, which thrives in darkness, and his exposition of it is so clear that it amounts to exposure. We can exapt his artifact to our own purposes: with his unintended help, this meme can be extinguished.*

The Emperor's New Mind, and Other Fables

~~~

1. THE SWORD IN THE STONE

In other words then, if a machine is expected to be infallible, it cannot also be intelligent. There are several theorems which say almost exactly that. But these theorems say nothing about how much intelligence may be displayed if a machine makes no pretence at infallibility.

—ALAN TURING 1946, p. 124

The attempts over the years to use Gödel's Theorem to prove something important about the nature of the human mind have an elusive atmosphere of romance. There is something strangely thrilling about the prospect of "using science" to such an effect. I think I can put my finger on it. The key text is not the Hans Christian Andersen tale about the Emperor's New Clothes, but the Arthurian romance of the Sword in the Stone. Somebody (our hero, of course) has a special, perhaps even magical, property which is quite invisible under most circumstances, but which can be made to reveal itself quite unmistakably in special circumstances: if you can pull the sword from the stone, you have the property; if you can't, you don't. This is a feat or a failure that everyone can see; it doesn't require any special interpretation or special pleading on one's own behalf. Pull out the sword and you win, hands down.

What Gödel's Theorem promises the romantically inclined is a similarly dramatic proof of the specialness of the human mind. Gödel's Theorem defines a deed, it seems, that a genuine human mind can perform but that no impostor, no mere algorithm-controlled robot, could perform. The technical details of Gödel's proof itself need not concern us; no mathematician doubts its soundness. The controversy all lies in how to *harness* the theorem to prove anything about the nature of the mind. The weakness in any such argument must come at the crucial empirical step: the step where we *look to see* our heroes (ourselves, our mathematicians) doing the thing that the robot simply cannot do. Is the feat in question like pulling the sword from the stone, a feat that has no plausible lookalikes, or is it a feat that cannot readily (if at all) be distinguished from mere approximations of the feat? That is the crucial question, and there has been a lot of confusion about just what the distinguishing feat is. Some of the confusion can be blamed on Kurt Gödel himself, for he thought that he had proved that the human mind *must* be a skyhook.

In 1931, Gödel, a young mathematician at the University of Vienna, published his proof, one of the most important and surprising mathematical results of the twentieth century, establishing an absolute limit on mathematical proof that is really quite shocking. Recall the Euclidean geometry you studied in high school, in which you learned to create formal proofs of theorems of geometry, from a basic list of axioms and definitions, using a fixed list of inference rules. You were learning your way around in an *axiomatization* of plane geometry. Remember how the teacher would draw a geometric diagram on the blackboard, showing a triangle, say, with various straight lines intersecting its sides in various ways, meeting at various angles, and then ask you such questions as: "Do these two lines have to intersect at a right angle? Is this triangle over here congruent to that triangle over there?" Often the answer was obvious: you could just *see* that the lines had to intersect at a right angle, that the triangles were congruent. But it was another matter—in fact, a considerable amount of inspired drudgery—to prove it from the axioms, formally, according to the strict rules. Did you ever wonder, when the teacher put a new diagram on the blackboard, whether there might be facts about plane geometry that you could *see* were true but couldn't prove, not in a million years? Or did it seem obvious to you that, if you yourself were unable to devise a proof of some candidate geometric truth, this would just be a sign of your own personal frailty? Perhaps you thought: "There has to *be* a proof, since it's *true*, even if I myself can never find it!"

That's an intensely plausible opinion, but what Gödel proved, beyond any doubt, is that when it comes to axiomatizing simple *arithmetic* (not plane geometry), there are truths that "we can see" to be true but that can *never* be formally proved to be true. Actually, this claim must be carefully hedged: for any *particular* axiom system that is *consistent* (not subtly self-contradictory—a disqualifying flaw), there must be a sentence of arithmetic, now known as the Gödel sentence of that system, that is not provable within the system but is true. (In fact, there must be many such true sentences, but one is all we need to make the point.) We can change systems, and prove that first Gödel sentence in the next axiom system we choose, but it in turn will spawn its own Gödel sentence, if it is consistent, and so on forever. No single consistent axiomatization of arithmetic can prove all the truths of arithmetic.

This might not seem to matter very much, since we seldom if ever want to *prove* facts of arithmetic; we just take arithmetic for granted without proof. But it is possible to devise Euclid-type axiom systems for arithmetic— Peano's axioms, for instance—and prove such simple truths as "$2 + 2 = 4$," such obvious middle-level truths as "numbers evenly divisible by 10 are also evenly divisible by 2," and such unobvious truths as "There is no largest prime number." Before Gödel devised his proof, the goal of deriving *all* mathematical truth from a single set of axioms was widely regarded by mathematicians and logicians as their great project, difficult but within reach, the moon landing or Human Genome Project of the mathematics of the day. But it absolutely can't be done. That is what Gödel's Theorem establishes.

Now, what does this have to do with Artificial Intelligence or evolution? Gödel proved his theorem some years before the invention of the electronic computer, but then Alan Turing came along and extended the implications of that abstract theorem by showing, in effect, that any formal proof procedure of the sort covered by Gödel's Theorem is equivalent to a computer program. Gödel had devised a way of putting *all possible axiom systems* in "alphabetical order." In fact, they can all be lined up in the Library of Babel, and Turing then showed that this set was a subset of another set in the Library of Babel: the set of *all possible computers*. It doesn't matter what material you make a computer out of; what matters is the algorithm it runs; and since every algorithm is finitely specifiable, it is possible to devise a uniform language for uniquely describing each algorithm and putting all the specifications in "alphabetical order." Turing devised just such a system, and in it every computer—from your laptop to the grandest parallel supercomputer that will ever be built—has a unique description as what we now call a *Turing machine*. The Turing machines can each be given a unique identification number—its Library of Babel Number, if you like. Gödel's Theorem can then be reinterpreted to say that each of those Turing machines that happens to be a consistent algorithm for proving truths of arithmetic (and, not surprisingly, these are a Vast but Vanishing subset of all the possible Turing machines) has associated with *it* a Gödel sentence—a truth of arithmetic it cannot prove. So that is what Gödel, anchored by Turing to the world of computers, tells us: every computer that is a consistent truth-of-arithmetic-prover has an Achilles' heel, a truth it can never prove, even if it runs till doomsday. But so what?

Gödel himself thought that the implication of his theorem was that human beings—at least the mathematicians among us—cannot, then, be just machines, because they can do things no machine could do. More pointedly, at least some part of such a human being cannot be a mere machine, or even a huge collection of gadgets. If hearts are pumping machines, and lungs are air-exchanging machines, and brains are computing machines, then mathematicians' minds cannot be their brains, Gödel thought, since mathematicians' minds can do something that no mere computing machine can do.

What, exactly, can they do? This is the problem of defining the feat for the big empirical test. It is tempting to think we have already seen an example: they can do what you used to do when you looked at the blackboard in geometry class—using something like "intuition" or "judgment" or "pure understanding," they can *just see* that certain propositions of arithmetic are true. The idea would be that they don't need to rely on grubby algorithms to generate *their* mathematical knowledge, since they have a talent for grasping mathematical truth that transcends algorithmic processes altogether. Remember that an algorithm is a recipe that can be followed by servile dunces—or even machines; no understanding is required. Clever mathematicians seem, in contrast, to be able to *use* their understanding to go beyond what such mechanical dunces can do. But although this seems to be what Gödel himself thought, and it certainly expresses the general popular understanding of what Gödel's Theorem shows, it is much harder to *demonstrate* than first appears. How can we distinguish a case of somebody (or something) "grasping the truth" of a mathematical sentence from a case of somebody (or something) just wildly guessing correctly, for instance? You could train a parrot to utter "true" and "false" when various symbols were written on the blackboard in front of it; how many correct guesses without an error would the parrot have to make for us to be justified in believing that the parrot had an immaterial mind after all (or perhaps was just a human mathematician in a parrot costume) (Hofstadter 1979)?

This is the problem that has always given fits to those who want to use Gödel's Theorem to prove that our minds are skyhooks, not just boring old cranes. It won't do to say that mathematicians, unlike machines, can *prove* any truth of arithmetic, for, if what we mean by "prove" is what Gödel means by "prove" in his proof, then Gödel shows that human beings—or angels, if such there be—cannot do it either (Dennett 1970); *there is no* formal proof of a system's Gödel sentence within the system. A famous early attempt to harness Gödel's Theorem was by the philosopher J. R. Lucas (1961; see also 1970), who decided to define the crucial feat as "producing as true" a certain sentence—some Gödel sentence or other. This definition runs into insoluble problems of interpretation, however, ruining the "sword-in-the-stone" definitiveness of the empirical side of the argument (Dennett 1970, 1972; see also Hofstadter 1979). We can see more clearly what the problem is by considering several related feats, real and imaginary.

René Descartes, in 1637, asked himself how one could tell a genuine human being from any machine, and he came up with "two very certain means":

> The first is that they [the machines] could never use words, or put together other signs, as we do in order to declare our thoughts to others. For we can certainly conceive of a machine so constructed that it utters words, and even utters words which correspond to bodily actions causing a change in its organs (e.g., if you touch it in one spot it asks what you want of it, if you touch it in another it cries out that you are hurting it, and so on). But it is not conceivable that such a machine should produce different arrange-ments of words so as to give an appropriately meaningful answer to what-ever is said in its presence, as the dullest of men can do. Secondly, even though such machines might do some things as well as we do them, or perhaps even better, they would inevitably fail in others, which would reveal that they were acting not through understanding but only from the disposition of their organs. For whereas reason is a universal instrument which can be used in all kinds of situations, these organs need some particular disposition for each particular action; hence it is for all practical purposes impossible for a machine to have enough different organs to make it act in all the contingencies of life in the way in which our reason makes us act. [Descartes 1637, pt. 5.]

Alan Turing, in 1950, asked himself the same question, and came up with just the same acid test—somewhat more rigorously described—what he called the imitation game, and we now call the Turing Test. Put two con-testants—one human, one a computer—in boxes (in effect) and conduct conversations with each; if the computer can convince you it is the human being, it wins the imitation game. Turing's verdict, however, was strikingly different from Descartes's:

> I believe that in about fifty years' time it will be possible to program computers, with a storage capacity of about 10^9, to make them play the imitation game so well that an average interrogator will not have more than a 70 percent chance of making the right identification after five min-utes of questioning. The original question, 'Can machines think?' I believe to be too meaningless to deserve discussion. Nevertheless I believe that at the end of the century the use of words and general educated opinion will have altered so much that one will be able to speak of machines thinking without expecting to be contradicted. [Turing 1950, p. 435.]

Turing has already been proven right about his last prophecy: "the use of words and general educated opinion" has already "altered so much" that one *can* speak of machines thinking without expecting to be contradicted— "on general principles." Descartes found the notion of a thinking machine "inconceivable," and even if, as many today believe, no machine will ever succeed in passing the Turing Test, almost no one today would claim that the very idea is inconceivable.

Perhaps this sea-change in public opinion has been helped along by the computer's progress on other feats, such as playing checkers and chess. In an address in 1957, Herbert Simon (Simon and Newell 1958) predicted that a computer would be the world chess champion in less than a decade, a classic case of overoptimism, as it turns out. A few years later, the philosopher Hubert Dreyfus (1965) predicted that no computer would ever be able to play good chess, since playing chess required "insight," but he himself was soon trounced at chess by a computer program, an occasion for much glee among AI researchers. Art Samuel's checkers program has been followed by literally hundreds of chess-playing programs, which now compete in tournaments against both human and other computer contestants, and will *probably* soon be able to beat the best human chess players in the world.

But is chess-playing a suitable "sword-in-the-stone" test? Dreyfus may once have thought so, and he has a distinguished predecessor—Edgar Allan Poe, of all people—whose certainty on this score drove him to unmask one of the great hoaxes of the nineteenth century, von Kempelen's chess "automaton." In the eighteenth century, the great Vaucanson had made mechanical marvels that entranced the nobility, and other paying customers, by exhibiting behaviors that even today inspire our skepticism. Could Vaucanson's clockwork duck really do what it is reported to have done? "When corn was thrown down before it, the duck stretched out its neck to pick it up, swallowed, and digested it" (Poe 1836a, p. 1255). Other ingenious artificers and tricksters had followed in Vaucanson's wake, developing the art of mechanical simulacra to such a high pitch that one of them, Baron von Kempelen, in 1769, could exploit public fascination with such devices by creating a deliberate tease: a *purported* automaton that could play chess.

Von Kempelen's original machine passed into the hands of Johann Nepomuk Maelzel,[1] who made some improvements and revisions, and then caused quite a stir in the early nineteenth century by exhibiting Maelzel's Chess-Machine, never quite guaranteeing that it was just a machine, and surrounding the whole performance (for which he charged a pretty penny) with enough of the standard magician's ostentations to arouse anybody's

---

1. This same Maelzel is the inventor (or perfecter) of the metronome, and made the ear trumpet that Beethoven relied on for years, once he began to go deaf. Maelzel also created a mechanical orchestra, the Panharmonicon, for which Beethoven wrote *Wellington's Victory,* but the two had a falling out over property rights to that composition—Maelzel was both a crane-maker and crane-stealer of great talent.

suspicions. An obviously mechanical swami figure sat at a suspiciously en-closed cabinet, the doors and drawers of which were sequentially opened (but never all at once), permitting the audience to "see" that there was nothing but machinery inside. The swami figure then commenced to play a game of chess, picking up and moving the chess pieces on a board in response to the moves of a human opponent—and usually winning! But was there, literally, a homunculus inside, a little man doing all the mind-work? If AI is possible, the cabinet *could* be filled with some collection or other of cranes and other bits of machinery. If AI is impossible, then there must be a skyhook in the cabinet, a Mind pretending to be a Machine.

Poe was absolutely certain that Maelzel's machine concealed a human being, and his ingenious sleuthing confirmed his suspicions, which he de-scribed in detail with an appropriate air of triumph in an article in the *Southern Literary Messenger* (1836a). At least as interesting as his reason-ing about how the hoax was perpetrated is his reasoning, in a letter accom-panying the publication of his article, about why it *had* to be a hoax, a line of argument that perfectly echoes John Locke's "proof" (back in chapter 1):

We have never, at any time, given assent to the prevailing opinion, that human agency is not employed by Mr. Maelzel. That such agency is em-ployed cannot be questioned, unless it may be satisfactorily demonstrated

FIGURE 15.1.
Von Kempelen's chess automaton.

that man is capable to impart intellect to matter: for *mind* is no less requisite in the operations of the game of chess, than it is in the prosecution of a chain of abstract reasoning. We recommend those, whose credulity has in this instance been taken captive by plausible appearance; and all, whether credulous or not, who admire an ingenious train of inductive reasoning, to read this article attentively: each and all must arise from its perusal convinced that a *mere machine* cannot bring into requisition the intellect which this intricate game demands.... [Poe 1836b, p. 89.]

We now know that, however convincing this argument *used* to be, its back has been broken by Darwin, and the particular conclusion Poe drew about chess has been definitively refuted by the generation of artificers following in Art Samuel's footsteps. What, though, of Descartes's test—now known as the Turing Test? That has generated controversy ever since Turing proposed his nicely operationalized version of it, and has even led to a series of real, if restricted, competitions, which confirm what everybody who had thought carefully about the Turing Test already knew (Dennett 1985): it is embarrassingly easy to fool the naïve judges, and astronomically difficult to fool the expert judges—a problem, once more, of not having a proper "sword-in-the-stone" feat to settle the issue. Holding a conversation or winning a chess match is not a suitable feat, the former because it is too open-ended for a contestant to secure unambiguous victory in spite of its severe difficulty, and the latter because it is demonstrably within the power of a machine after all. Might the implications of Gödel's Theorem provide a better contest? Suppose we put a mathematician in box A and a computer—any computer you like—in box B, and ask each of them questions about the truth and falsehood of sentences of arithmetic. Would *this* be a test that would surely unmask the machine? The trouble is that human mathematicians all make mistakes, and Gödel's Theorem offers no verdict at all about the likelihood, let alone impossibility, of less-than-perfect truth detection by an algorithm. It does not appear, then, that there is any fair arithmetic test we can put to the boxes that will clearly distinguish the man from the machine.

This difficulty had been widely seen as systematically blocking any argument from Gödel's Theorem to the impossibility of AI. Certainly everybody in AI has always known about Gödel's Theorem, and they have all continued, unworried, with their labors. In fact, Hofstadter's classic *Gödel Escher Bach* (1979) can be read as the demonstration that Gödel is an unwilling *champion* of AI, providing essential insights about the paths to follow to strong AI, not showing the futility of the field. But Roger Penrose, Rouse Ball Professor of Mathematics at Oxford, and one of the world's leading mathematical physicists, thinks otherwise. His challenge has to be taken seriously, even if, as I and others in AI are convinced, he is making a fairly simple mistake. When Penrose's book appeared, I pointed out the problem in a review: his argument is highly convoluted, and bristling with details of physics and mathematics,

> and it is unlikely that such an enterprise would succumb to a single, crashing oversight on the part of its creator—that the argument could be 'refuted' by any simple observation. So I am reluctant to credit my observation that Penrose seems to make a fairly elementary error right at the beginning, and at any rate fails to notice or rebut what seems to be an obvious objection. [Dennett 1989b.]

My surprise and disbelief were soon echoed, first by the usual assortment of commentators to a target article (based on his book) by Penrose in *Behavioral and Brain Sciences*, and then by Penrose in turn. In "The Nonalgorithmic Mind" (1990), Penrose's reply to his critics, he expressed mild astonishment at the strong language some of them used: "quite fallacious," "wrong," "lethal flaw" and "inexplicable mistake," "invalid," "deeply flawed." The AI community was, not surprisingly, united in its dismissal of Penrose's argument, but, in Penrose's eyes, they didn't agree on what "the" lethal flaw was. This was itself a measure of how widely he had missed the mark, since the critics had found many different ways of zeroing in on one big misunderstanding, about the very nature of AI and its use of algorithms.

2. THE LIBRARY OF TOSHIBA

The people who are going to like the book best, however, will probably be those who don't understand it. As an evolutionary biologist, I have learned over the years that most people do not want to see themselves as lumbering robots programmed to ensure the survival of their genes. I don't think they will want to see themselves as digital computers either. To be told by someone with impeccable scientific credentials that they are nothing of the kind can only be pleasing.

—JOHN MAYNARD SMITH 1990
(review of Penrose)

Consider the set of all Turing machines—in other words, the set of all possible algorithms. Or, rather, to ease the task of imagination, consider instead a Vast but finite subset of them, relativized to a particular language, and consisting of "volumes" of a particular length: the set of all possible strings of 0 and 1 (bit strings), up to the length of one megabyte (eight million 0's and 1's). Consider the reader of these strings to be my old laptop computer, a Toshiba T-1200, with its twenty-megabyte hard disk (we'll prohibit using any additional memory, just for finiteness' sake). It should come as no surprise that the Vast majority of these bit strings do nothing at all worth mentioning if an attempt is made to "run" them as programs on the Toshiba. Programs, after all, are not random strings of bits, but highly designed sequences of bits, the products of thousands of hours of R and D. The fanciest program that ever could be is still something that can be expressed as one or another string of 0's and 1's, and although my old Toshiba is too small to run some of the truly huge programs that have been devised, it is quite capable of running a handsome and representative subset of them: word-processors, spread sheets, chess-players, Artificial Life simulations, logic-proof-checkers, and, yes, even a few automatic arithmetic-truth-provers. Call any such runnable program, actual or envisaged, an *interesting* program (it is roughly analogous to a readable book, actual or imaginary, in the Library of Babel, or a viable genotype in the Library of Mendel). We don't have to worry about the boundary separating the interesting from the uninteresting; when in doubt, throw it out. No matter how we rule, there are Vastly many interesting programs in the Library of

Toshiba, but they are Vanishingly hard to "find"—that's why software companies make quite a few millionaires along with their software.

Now, *every* megabyte-length bit string is an algorithm in one sense—the sense that matters to us: it is a recipe, stupid or wise, that can be followed by a mechanism, my Toshiba. If we try bit strings at random, most of the time the Toshiba will just sit there emitting a faint hum (it won't even flash an amber light); there are Vastly more ways of being a dead program than a live one, to echo Dawkins. Only a Vanishing subset of these algorithms are interesting in any way at all, and only a Vanishing subset of *them* have anything at all to do with truths of arithmetic, and only a Vanishing subset of *these* attempt to generate formal proofs of arithmetical truths, and only a Vanishing subset of *those* are consistent. Gödel shows us that not a single one of the algorithms in *that* subset (and there are still Vastly many of them, even for my little Toshiba) can generate proofs of *all* the truths of arithmetic.

But Gödel's Theorem tells us nothing at all about any other algorithm in the Library of Toshiba. It does not tell us whether there are any algorithms that can play decent chess. There are in fact Vastly many, and a few actual ones reside on my actual Toshiba, and I've never beaten any of them! It does not tell us whether there are any algorithms that are pretty darn good at playing the Turing Test or imitation game. In fact, there is one actual one on my Toshiba, a stripped-down version of Joseph Weizenbaum's famous ELIZA program, and I have seen it fool uninitiated people into concluding, like Edgar Allan Poe, that there *must* be a human being issuing the answers. At first I was baffled by how any sane human being could think there was a tiny guy in my laptop Toshiba, sitting, unattached to anything, on a card table, but I had forgotten how resourceful a persuaded mind can be—there must be, these wily skeptics concluded, a *cellular phone* in my Toshiba!

Gödel's Theorem in particular has nothing at all to tell us about whether there might be algorithms in the Library of Toshiba that could do an impressive job of "producing as true" or "detecting as true or false" candidate sentences of arithmetic. If human mathematicians can do an impressive job of "just seeing" with "mathematical intuition" that certain propositions are true, perhaps a computer can imitate this talent, the same way it can imitate chess-playing and conversation-holding: imperfectly, but impressively. That is exactly what people in AI believe: that there are risky, heuristic algorithms for human intelligence in general, just as there are for playing good checkers and good chess and a thousand other tasks. And here is where Penrose made his big mistake: he ignored this set of possible algorithms— the only set of algorithms that AI has ever concerned itself with—and concentrated on the set of algorithms that Gödel's Theorem actually tells us something about.

Mathematicians, Penrose says, use "mathematical insight" to see that a certain proposition follows from the soundness of a certain system. He then goes to some length to argue that there could be no algorithm, or at any rate no practical algorithm, "for" mathematical insight. But, in going to all this trouble, he overlooks the possibility that some algorithm—many different algorithms, in fact—might yield mathematical insight even though that was not just what it was "for." We can see the mistake clearly in a parallel argument.

Chess is a finite game (since there are rules for terminating go-nowhere games as draws). That means that there is, in principle, an algorithm for determining either checkmate or a draw—I have no idea which. In fact, I can specify the algorithm for you quite simply: (1) Draw the entire decision tree of all possible chess games (a Vast but finite number). (2) Go to the end node of each game; it will be either a win for white or black, or a draw. (3) "Color" the node black, white, or gray, depending on the outcome. (4) Work backwards, one *whole* step (one white move plus one black move) at a time; if on the previous move *all* the paths from *any one* of white's moves lead through all black's responses to a white-colored node, color that node white and move back again, and so forth. (5) Do the same for any guaranteed winning paths for black. (6) Color all other nodes gray. At the end of this procedure (way past the universe's bedtime), you will have colored in every node of the tree of all possible chess games, leading back to white's opening move. Now it is time to play. If any one of the twenty legal moves is colored white, take it! There is a guaranteed checkmate ahead that can be reached just by always staying on the white nodes. Shun any black move, of course, since that opens up a guaranteed win for your opponent. If there are no white moves at the outset, choose a gray move, and hope that sometime later in the game you'll be offered a white move. The worst you can do is a draw. (If all the opening moves for white are colored black, most improbably, your only hope is to choose one at random and hope that your opponent, playing black, goofs at some later stage of play and lets you escape by getting on a gray or a white path.)

That is an algorithm, clearly. No step in the recipe requires any insight, and I have specified it unambiguously in a finite form. The trouble is that it is not remotely feasible or practical, because the tree it exhaustively searches is Vast. But I suppose it is nice to know that in principle there is an algorithm for playing perfect chess, however useless. There *might* be a *feasible* algorithm for playing perfect chess. No one has ever found one, thank goodness, since it would turn chess into a game of scarcely more interest than tic-tac-toe. No one knows whether there is such a feasible algorithm, but the general consensus is that it is very unlikely. Not knowing for sure, let's choose the supposition that makes for the worst case for AI. Let's suppose that *there is no* feasible algorithm for checkmate or a guaranteed draw—none at all.

Does it follow that no algorithm running on my Toshiba can achieve checkmate? Hardly! As I have already confessed, the chess algorithms on my Toshiba are undefeated in play against one human being—me. I'm not very good, but I expect I have about as much "insight" as the next human being. Someday I might beat my machine, if I practiced a lot and worked very hard, but the programs on my Toshiba are trivial compared with the current champion chess programs. About them you could safely *bet your life* that they would checkmate me (though not Bobby Fischer) *every time*. I don't recommend to anyone that you actually bet your life on the relative excellence of these algorithms—I might improve, and I wouldn't want your death on my conscience—but in fact, if Darwinism is right, you and your ancestors have an unbroken string of successful gambles for similarly fatal stakes on the algorithms embodied in your "machinery." That is what organisms have done, every day since life began: they have bet their lives that the algorithms that built them, and that operate within them if they are among the lucky organisms with brains, will keep them alive long enough to have children. Mother Nature has never aspired to absolute certainty; a good risk is enough for her. So we would *expect* that, if mathematicians' brains are running algorithms, they will be algorithms that happen to do pretty well in the truth-detecting department, without being foolproof.

The chess algorithms on my Toshiba, like all algorithms, yield guaranteed results, but what they are guaranteed to do is not checkmate me, but just *play legal chess*. That is all they are "for." Of the Vast number of algorithms guaranteed to play legal chess, some are much better than others, though none is guaranteed a win against any other—at least this is not the sort of thing one would hope to prove mathematically, even if, as a matter of brute mathematical fact, the initial state of program x and program y were such that x would win all possible games against y. This means that the following argument is fallacious:

> x is excellent at achieving checkmate;
> there is no (practical) algorithm for checkmate in chess;
> *therefore:* the explanation of x's talent cannot be that x is running an algorithm.

The conclusion is obviously false: the algorithm level of explanation is *exactly* the right level at which to explain the power of my Toshiba to beat me at chess. It's not as if it had particularly potent electricity running through it, or a secret reservoir of *élan vital* inside its plastic case. What makes it better than other chess-playing computers (I can beat the really simple ones) is that it has a better algorithm.

What kind of algorithms, then, might mathematicians be running? Algorithms "for" *trying to stay alive*. As we saw in our consideration of the survival-machine robots in the last chapter, such algorithms would have to be capable of indefinitely resourceful discrimination and planning; they must be good at recognizing food and shelter, telling friend from foe, learning to discriminate harbingers of spring *as* harbingers of spring, telling good arguments from bad, and even—as a sort of bonus talent thrown in—recognizing mathematical truths *as* mathematical truths. Of course, such "Darwinian algorithms" (Cosmides and Tooby 1989) wouldn't have been designed just for this special purpose, any more than our eyes were designed for telling *italics* from **boldface**, but that doesn't mean that they aren't superbly sensitive to such differences if given a chance to consider them.

Now how could Penrose have overlooked this retrospectively obvious possibility? He is a mathematician, and mathematicians are primarily interested in that Vanishing subset of algorithms that they *can* prove, mathematically, to have mathematically interesting powers. I call this the God's-eye view of algorithms. It is analogous to the God's-eye view of volumes in the Library of Babel. We can "prove" (for what it is worth) that there is a single volume in the Library of Babel that lists, in perfect alphabetical order, all the telephone subscribers in New York City whose net worth on January 10, 1994, was more than a million dollars. There has to be—there couldn't be *that* many millionaire phone-owners in New York, and so some one of the possible volumes in the Library must list them all. But finding it—or making it—would be a huge empirical task fraught with uncertainties and judgment calls, even if we just considered it to be a subset of the names already printed in the actual phone book as of that date (ignoring all those with unlisted numbers). Even though we can't put our hands on this volume, we can name it—just the way we named Mitochondrial Eve. Call it *Megaphone*. Now, we can prove things about *Megaphone*: for instance, the first letter printed on the first page on which there is printing is "A," but the first letter on the last page on which there is printing is not "A." (This is not quite up to the standards of mathematical proof, of course, but what are the odds that *none* of the people with phones whose names begin with "A" is a millionaire, or that there's only one page of such millionaires in all New York?)

As I noted on page 52, when mathematicians think about algorithms, it is usually from the God's-eye perspective. They are interested in proving, for instance, that *there is* some algorithm with some interesting property, or that *there is no* such algorithm, and in order to prove such things you needn't actually locate the algorithm you are talking about—by picking it out from a pile of algorithms stored on floppy disks, for instance. Our inability to locate (the remains of) Mitochondrial Eve did not prevent us from deducing facts about her either. The empirical issue of identification thus doesn't often arise for such formal deductions. Gödel's Theorem tells us that not a single one of the algorithms that can run on my Toshiba (or any other computer) has a certain mathematically interesting property: being a *consistent generator of proofs of arithmetic facts that generates them all if given enough run time*.

That is interesting, but it doesn't help us much. Lots of interesting things can be proved, mathematically, about each and every member of various sets of algorithms. Applying that knowledge in the real world is another matter, and that is the blind spot that led Penrose to overlook AI altogether, instead of refuting it, as he hoped. This has come out quite clearly in his subsequent attempts at reformulation of his claim in response to his critics:

> Given any particular algorithm, that algorithm cannot be *the* procedure whereby human mathematicians ascertain mathematical truth. Hence humans are not using algorithms at all to ascertain truth. [Penrose 1990, p. 696.]
>
> Human mathematicians are not using a knowably sound algorithm in order to ascertain mathematical truth. [Penrose 1991.]

In the more recent of these, he goes on to consider and close various "loopholes," of which two in particular concern us: mathematicians might be using "a horrendously complicated *unknowable* algorithm *X*" or "an *unsound* (but presumably approximately sound) algorithm *Y*." Penrose presents these loopholes as if they were *ad hoc* responses to the challenge of Gödel's Theorem, instead of the standard working assumptions of AI. Of the first he says:

> This seems to be totally at variance with what mathematicians seem *actually* to be doing when they express their arguments in terms that can (at least in principle) be broken down into assertions that are 'obvious', and agreed by all. I would regard it as far-fetched in the extreme to believe that it is *really* the horrendous unknowable *X*, rather than these simple and obvious *ingredients* [emphasis added], that lies lurking behind all our mathematical understanding. [Penrose 1991.]

These "ingredients" are indeed wielded by us all in an *apparently* nonalgorithmic way, but this phenomenological fact is misleading. Penrose pays careful attention to what it is like to be a mathematician, but he overlooks a possibility—indeed, a likelihood—that is familiar to AI researchers: the possibility that *underlying* our general capacity to deal with such "ingredients" is a heuristic program of mind-boggling complexity. Such a complicated algorithm would *approximate* the competence of the perfect understander, and be "invisible" to its beneficiary. Whenever we say we solved some problem "by intuition," all that really means is *we don't know how* we solved it. The simplest way of modeling "intuition" in a computer is simply denying the computer program any access to its own inner workings. Whenever it solves a problem, and you ask it how it solved the problem, it should respond: "I don't know; it just came to me by intuition" (Dennett 1968).

He goes on to dismiss his second loophole (the unsound algorithm) by claiming (1991): "Mathematicians require a degree of rigour that makes such heuristic arguments unacceptable—so no such known procedure of this kind can be the way that mathematicians actually operate." This is a more interesting mistake, for with it he raises the prospect that the crucial empirical test would be not to put a *single* mathematician "in the box" but the whole mathematical community! Penrose sees the theoretical importance of the added power that human mathematicians obtain by pooling their resources, communicating with each other, and hence becoming a sort of single giant mind that is hugely more reliable than any one homunculus we might put in the box. It is not that mathematicians have fancier *brains* than the rest of us (or than chimpanzees) but that they have mind-tools—the social institutions in which mathematicians present each other their proofs, check each other out, make mistakes in public, and then count on the public to correct those mistakes. This does indeed give the mathematics community powers to discern mathematical truth that dwarf the powers of any individual human brain (even an individual brain with paper-and-pencil peripherals, a hand calculator, or a laptop!). But this does not show that human minds are *not* algorithmic devices; on the contrary, it shows how the cranes of culture can exploit human brains in distributed algorithmic processes that have no discernible limits.

Penrose doesn't quite see it that way. He goes on to say that "it is our general (non-algorithmic) ability to *understand*" that accounts for our mathematical abilities, and then he concludes: "It was not an *algorithm* x that was favoured, in Man (at least) by natural selection, but this wonderful ability to understand!" (Penrose 1991). Here he commits the fallacy I just exposed using the chess example. Penrose wants to argue:

> x can understand;
> there is no feasible algorithm for understanding;
> *therefore:* what natural selection selected, the whatever-it-is that accounts for understanding, is not an algorithm.

This conclusion is a *non sequitur*. If the mind is an algorithm (contrary to Penrose's claim), surely it is not an algorithm that is recognizable to, or accessible to, those whose minds it creates. It is, in his terms, unknowable. As a product of biological design processes (both genetic and individual), it is almost certainly one of those algorithms that are somewhere or other in the Vast space of interesting algorithms, full of typographical errors or "bugs," but good enough to bet your life on—so far. Penrose sees this as a "far-fetched" possibility, but if that is all he can say against it, he has not yet come to grips with the best version of "strong AI."

3. THE PHANTOM QUANTUM-GRAVITY COMPUTER:
LESSONS FROM LAPLAND

*I am a strong believer in the power of natural selection. But I do not see
how natural selection, in itself, can evolve algorithms which could
have the kind of conscious judgements of the validity of other algorithms that we seem to have.*

—ROGER PENROSE 1989, p. 414

*I don't think the brain came in the Darwinian manner. In fact, it is
disprovable. Simple mechanisms can't yield the brain. I think the basic
elements of the universe are simple. Life force is a primitive element of
the universe and it obeys certain laws of action. These laws are not
simple and not mechanical.*

—KURT GÖDEL[2]

When Penrose insists that the brain is no Turing machine, it is important
to understand what he is *not* saying. He is not making the obvious (and
obviously irrelevant) claim that the brain is not well modeled by Turing's
original thought-device: a smallish gadget sitting astride a paper tape, examining one square of the tape at a time. Nobody ever thought otherwise.
He is also not merely saying that the brain is not a serial computer, a "von
Neumann machine," but, rather, a massively parallel computer. And he is
not just saying that the brain makes use of randomness or pseudo-
randomness in running its algorithms. He sees—though some others have
not—that algorithms availing themselves of large doses of randomness are
still algorithms within the purview of Artificial Intelligence, and still fall
under the limitations Gödel's Theorem places on all Turing machines, of
whatever size and shape.[3]

---

2. A remark made in 1971, quoted in Wang 1993, p. 133. See also Wang 1974, p. 326:
"Gödel believes that mechanism in biology is a prejudice of our time which will be
disproved. In this case, one disproval, in Gödel's opinion, will consist in a mathematical
theorem to the effect that the formation within geological times of a human body by the
laws of physics (or any other laws of a similar nature), starting from a random distribution of the elementary particles and the field, is as unlikely as the separation by chance
of the atmosphere into its components."

3. Someone who doesn't realize this is Gerald Edelman, whose "neural Darwinism"
simulations are both parallel and heavily stochastic (involving randomness), a fact he
often cites, mistakenly, as evidence that his models are not algorithms, and that he himself
is not engaged in "strong AI" (e.g., Edelman 1992). He is; his protestations to the contrary
betray an elementary misunderstanding of computers, but that just goes to show, as

Moreover, in the wake of the commentary his book provoked, Penrose now grants that heuristic programs are algorithms as well, and acknowledges that, if he is to find an argument against AI, he has to concede their tremendous power to track the truths of arithmetic and everything else, if not perfectly, then at least impressively. He offers a further point of clarification: any computer that operates by indulging in interactions with an external environment is an algorithmic computer *provided the external environment is itself entirely algorithmic.* (If skyhooks grew like toadstools—or, more to the point, like oracles perched on toadstools—and a computer was helped along by its occasional communication with these skyhooks, then what it did would be no algorithm.)

Now, with all this useful clarification in place, what does Penrose maintain? In May 1993, I spent a week with Penrose and some Swedish physicists and other scientists discussing our different views about these matters, at a workshop in Abisko, a tundra-research station well north of the Arctic Circle in Sweden. Perhaps the midnight sun helped as much as our Swedish hosts to illuminate the path, but, in any event, I think we both came away enlightened. Penrose proposes a revolution in physics, centered on a new—and still unformulated—theory of "quantum gravity," which he hopes will explain how the human brain transcends the limitations of algorithms. Does Penrose envisage the human brain, with its special quantum-physics powers, to be a skyhook or a crane? That was the question I went to Sweden to answer, and the answer I came back with is this: He has definitely been looking for a skyhook. I think he'd settle for a new crane—but I doubt that he's found one.

Descartes and Locke, and more recently Edgar Allan Poe, Kurt Gödel, and J. R. Lucas, thought that the alternative to a "mechanical" mind would be an *immaterial* mind, or a soul, to speak with tradition. Hubert Dreyfus and John Searle, more recent skeptics about AI, have shunned such dualism and opined that the mind is indeed just the brain, but the brain is not any *ordinary* computer; it has "causal powers" (Searle 1985) that go beyond the running of any algorithms. Neither Dreyfus nor Searle has been very forthcoming about what special powers these might be, or which of the physical sciences might be the right one to give an account of them, but others have wondered whether physics might hold the key. To many of them, Penrose appears to be a knight in shining armor.

Quantum physics to the rescue! Several different proposals have been advanced over the years about how quantum effects might be harnessed to give the brain special powers beyond those of any ordinary computer. J. R.

---

everybody in AI knows, that although you may not have "Absolute Ignorance" (as MacKenzie anonymously put it, back in chapter 3, p. 65), you still don't have to understand what you are making in order to make it.

Lucas (1970) yearned to drag quantum physics into this arena, but he thought that the indeterminacy gaps of quantum physics would permit a Cartesian spirit to intercede, twiddling the neurons, in effect, to get some extra mind-power out of the brain, a doctrine that has also been energetically defended by Sir John Eccles, the Nobel-laureate neurophysiologist who has scandalized his colleagues for years with his unabashed dualism (Eccles 1953, Popper and Eccles 1977). This is not the time and place for me to review the reasons for dismissing this dualism—the times and places are Dennett 1991a, 1993d—since Penrose shuns dualism as vigorously as anybody else in the materialist camp. What is refreshing about his attack on AI, in fact, is his insistence that he hopes to replace it with something that would still be a physical science of the mind, not some unexplorable mystery that takes place in the never-never-land of dualism.

Without abandoning the physical sphere, we might get some strange new powers out of subatomic particles, according to recent speculations about "quantum computers" (Deutsch 1985). Such a quantum computer would take advantage (it is claimed) of the "superposition of eigenstates" prior to the "collapse of the wave packet" in order to check out Vast (yes, Vast) search spaces in ordinary amounts of time. By being a sort of supermassively parallel computer, it could do Vastly many things "at once," and this could render feasible whole classes of algorithms that otherwise were unfeasible—such as the algorithm for perfect chess. This is *not* what Penrose is seeking, however, for such computers, even if they are possible, would still be Turing machines, and hence capable of computing only the officially computable functions—the algorithms (Penrose 1989, p. 402). They would hence fall under the limitations discovered by Gödel. Penrose is holding out for a phenomenon that is truly *noncomputable*, not just impractical to compute.

Present-day physics (including present-day quantum physics) is *all* computable, Penrose acknowledges, but he thinks that we might have to revolutionize physics, incorporating an explicitly noncomputable theory of "quantum gravity." Why does he think such a theory (which neither he nor anyone else has yet formulated) would have to be noncomputable? Because otherwise AI is possible, and he thinks he has already shown, via his argument from Gödel's Theorem, that AI is not possible. That's all. Penrose candidly admits that none of his reasons for believing in the noncomputability of quantum-gravity theory are drawn from quantum physics itself; the *only reason* he has for thinking that a theory of quantum gravity would be noncomputable is that otherwise AI would be possible after all. In other words, Penrose has a hunch that someday we're going to find a skyhook. This is the hunch of a brilliant scientist, but he himself admits that it is only a hunch.

In a review of the physicist Steven Weinberg's recent book, *Dreams of a*

Final Theory (Weinberg, you will recall from chapter 3, gave two cheers for reductionism), Penrose mused as follows:

> In my view, if there is to be a Final Theory, it could only be a scheme of a very different nature. Rather than being a physical theory in the ordinary sense, it would have to be something more like a principle—a mathematical principle whose implementation might itself involve nonmechanical subtlety (and perhaps even creativity). [Penrose 1993, p. 82.]

So it is not surprising that Penrose has expressed grave skepticism about Darwinism. And the grounds he gives are familiar: he can't imagine how "natural selection of algorithms" could do all that good work:

> [T]here are serious difficulties with the picture whereby algorithms are supposed to improve themselves in this way. It would certainly not work for normal Turing machine specifications, since a 'mutation' would almost certainly render the machine totally useless instead of altering it only slightly. [Penrose 1990, p. 654.]

Most mutations, Penrose sees, are either invisible to selection or fatal; only a very few improve things. That is true, but it is just as true of the evolutionary processes that produced the mandibles of crabs as it is of those that produced the mental states of mathematicians. Penrose's conviction that there are these "serious difficulties" is undercut, as Poe's conviction was, by the brute historical fact that genetic algorithms and their kin are daily overcoming these fearsome odds and improving themselves by, well, leaps and bounds (on the geological time scale).

If our brains *were* equipped with algorithms, Penrose argues, natural selection would have to have designed those algorithms, but:

> The 'robust' specifications are the *ideas* that underlie the algorithms. But ideas are things that, as far as we know, need conscious minds for their manifestations. [Penrose 1989, p. 415.]

In other words, the designing process would have to appreciate, somehow, the rationale of those algorithms it was designing, and doesn't that take a conscious mind? Could there be reasons recognized without some conscious mind's recognizing them? Yes, says Darwin, there could be. Natural selection is the *blind* watchmaker, the *unconscious* watchmaker, but still a discoverer of forced moves and other Good Tricks. This is not as inconceivable as many have taken it to be.

> To my way of thinking, there is still something mysterious about evolution, with its apparent 'groping' towards some future purpose. Things at least

seem to organize themselves somewhat better than they 'ought' to, just on the basis of blind-chance evolution and natural selection. It may well be that such appearances are quite deceptive. There seems to be something about the way that the laws of physics work, which allows natural selection to be a much more effective process than it would be with just arbitrary laws. [Penrose 1989, p. 416.]

There could not be a clearer, more heartfelt expression of the hope for skyhooks than this. And though we cannot yet rule out "in principle" the existence of a quantum-gravity skyhook, Penrose has not yet given us any reason to believe in one. If his theory of quantum gravity were already a reality, it could well turn out to be a crane, but he hasn't got that far yet, and I doubt that he ever will. At least he's trying, however. He wants his theory to provide a unified, scientific picture of how the mind works, not an excuse for declaring the mind to be an impenetrable Ultimate Source of Meaning. My own opinion is that the path he is now exploring—in particular, the possible quantum effects occurring in the microtubules of the cytoskeleton of neurons, an idea enthusiastically promoted in Abisko by Stuart Hameroff—is a nonstarter, but that is not a topic for this occasion. (I can't resist raising one question for Penrose to ponder: if the magnificent quantum property lurks in the microtubules, does that mean that cockroaches have noncomputable minds, too? They have the same kind of microtubules we have.)

If a Penrose-style quantum-gravity brain were truly capable of nonalgorithmic activity, and if we have such brains, and if our brains are themselves the products of an algorithmic evolutionary process, a curious inconsistency emerges: an algorithmic process (natural selection in its various levels and incarnations) creates a nonalgorithmic subprocess or subroutine, turning the *whole* process (evolution up to and *including* human mathematician brains) into a *non*algorithmic process after all. This would be a cascade of cranes creating, eventually, a real skyhook! No wonder Penrose has his doubts about the algorithmic nature of natural selection. If it were, truly, just an algorithmic process at all levels, all its products should be algorithmic as well. So far as I can see, this isn't an inescapable formal contradiction; Penrose could just shrug and propose that the universe contains these basic nuggets of nonalgorithmic power, not themselves created by natural selection in any of its guises, but incorporatable by algorithmic devices as found objects whenever they are encountered (like the oracles on the toadstools). Those would be truly nonreducible skyhooks.

The position is, I guess, possible, but Penrose must face an embarrassing shortage of evidence for it. The physicist Hans Hansson came up with a good challenge in Abisko, comparing a perpetual-motion machine to a truth-detecting computer. Different sciences, Hansson noted, can offer different reliable shortcuts to verdicts about projects. If someone were to go to the Swedish government with a plan to build a perpetual-motion machine (at government expense), Hansson would unhesitating testify, as a physicist, that this would be—would *have to be*—a waste of government money. It could not succeed, because physics has proven that a perpetual machine is flat impossible. Did Penrose think that he had offered a similar *sort* of proof? If some AI entrepreneur were to go to the government asking for money to build a mathematical-truth-detecting machine, would Penrose be similarly willing to testify that such money would be wasted?

To make the question more specific, consider some rather special varieties of mathematical truth. It is well known that there can be no all-purpose program that can examine any other program and tell whether or not it has an infinite loop in it, and hence will not stop if started. This is known as the Halting Problem, and there is a Gödel-style proof that it is insoluble. (This is one of the theorems Turing alluded to in his 1946 comment quoted at the beginning of the chapter.) No program that is itself guaranteed to terminate can tell of every (finite) program whether or not it will terminate. But it might still be handy—worth some serious money—to have a program around that was very, very good (if not perfect) at this task. Another class of interesting problems are known as Diophantine Equations, and it is known that there is no algorithm guaranteed to solve all such equations. If our lives depended on it, should we spend a nickel on a program for solving Diophantine Equations "in general" or for checking for halting "in general"? (Remember: we shouldn't spend a nickel on perpetual-motion machines, even to save our lives, since it will be money wasted on an impossible task.)

Penrose's answer was illuminating: if the candidates for truth-checking "just somehow bubble up out of the ground," then we would be wise to spend the money, but if some intelligent agent is the source of the candidates and gets to examine the program in our truth-checker, then it can foil our algorithmic truth-checker by constructing just the "wrong" candidate or candidates—an equation unsolvable by it, or a program whose termination prospects will confound it. To make the distinction vivid, we can imagine that a space pirate, Rumpelstiltskin by name, is holding the planet hostage, but will release us unharmed if we can answer a thousand true-false questions about sentences of arithmetic. Should we put a human mathematician on the witness stand, or a computer truth-checker devised by the best programmers? According to Penrose, if we hang our fate on the computer *and let Rumpelstiltskin see the computer's program*, he can devise an Achilles'-heel proposition that will foil our machine. (This would be true independently of Gödel's Theorem, if our program was a heuristic truth-checker, taking risks like any chess program.) But Penrose has given us no reason to believe that this isn't just as true of any human mathematicians we might put on the witness stand. None of us is perfect, and even a team of experts no doubt has some weaknesses that Rumpelstiltskin could exploit, given enough information about their brains. Von Neumann and Morgenstern invented game theory to deal with the particular class of complicated problems that life throws at us when there are other agents around to compete with us. You are always wise to shield your brain from such competitors, whether you are a human being or a computer. The reason a competitive agent makes a difference in this instance is that the space of all mathematical truths is Vast, the space of Diophantine Equation solutions is a Vast but Vanishing subspace within it, and the odds of hitting upon a truth *at random* that would "break" or "beat" our machine is truly negligible, whereas an intelligent *search* through that space, guided by knowledge of the particular style of the opponent and its limitations, would be likely to find the needle in the haystack: a crushing countermove.

Rolf Wasén raised another interesting point in Abisko. The class of *interesting* algorithms no doubt includes many that are not *humanly accessible*. To put it dramatically, there are programs out there in the Library of Toshiba that would not just run on my Toshiba, but be valued by me for the wonderful work they would do for me, but that no human programmers, or any of their artifacts (program-writing programs already exist), will ever be able to create! How can this be? None of these wonderful programs is more than a megabyte long, and there are plenty of actual programs much bigger than that already. Once again, we must remind ourselves just how Vast the space of such possible programs is. Like the space of possible five-hundred-page novels, or fifty-minute symphonies, or five-thousand-line poems, the space of megabyte-long programs will only ever get occupied by the slenderest threads of actuality, no matter how hard we work.

There are short novels nobody could write that would not just be best-sellers; they would be instantly recognized as classics. The keystrokes required to type them are all available on any word-processor, and the total number of keystrokes in any such book is trivial, but they still lie beyond the horizon of human creativity. Each particular creator, each novelist or composer or computer programmer, is sped along through Design Space by a particular idiosyncratic set of habits known as a *style* (Hofstadter 1985, sec. III). It is style that both constrains and enables us, giving a positive direction to our explorations but only by rendering otherwise neighboring regions off limits to us—and if off limits to us in particular, then probably off limits to everyone forever. Individual styles are truly unique, the product of untold billions of serendipitous encounters over the ages, encounters that produced first a unique genome, and then a unique upbringing, and finally a unique set of life experiences. Proust never got a chance to write any novels about the Vietnam War, and no one else could ever write *them*—the novels recounting *that* epoch in *his* manner. We are stuck, by our actuality and finitude, in a negligible corner of the total space of possibilities, but what a fine actuality is still accessible to us, thanks to the R-and-D work of all our predecessors! We might as well make the most of what we have, thereby leaving rather more for our descendants to work with.

It is time to turn the burden of proof around, the way Darwin did when he challenged his critics to describe some *other* way—other than natural selection—in which all the wonders of nature could have arisen. Those who think the human mind is nonalgorithmic should consider the hubris presupposed by that conviction. If Darwin's dangerous idea is right, an algorithmic process is powerful enough to design a nightingale and a tree. Should it be that much harder for an algorithmic process to write an ode to a nightingale or a poem as lovely as a tree? Surely Orgel's Second Rule is correct: Evolution is cleverer than you are.

CHAPTER 15: *Gödel's Theorem does not cast doubt on the possibility of AI after all. In fact, once we appreciate how an algorithmic process can escape the clutches of Gödel's Theorem, we see more clearly than ever how Design Space is unified by Darwin's dangerous idea.*

CHAPTER 16: *What, then, about morality? Did morality evolve, too? Sociobiologists from Thomas Hobbes to the present have offered Just So Stories about the evolution of morality, but, according to some philosophers, any such attempt commits the "naturalistic fallacy": the mistake of looking to facts about the way the world is in order to ground—or reduce—ethical conclusions about how things ought to be. This "fallacy" is better seen as a charge of greedy reductionism, a charge which is often justified. But then we shall just have to be less greedy in our reductionism.*

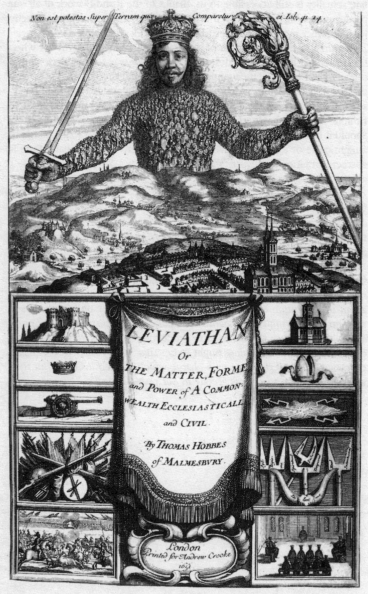

FIGURE 16.1

CHAPTER SIXTEEN

On the Origin of Morality

~~~

1. E PLURIBUS UNUM?

Nature (the Art whereby God hath made and governes the World) is by the Art of man, as in many other things, so in this also imitated, that it can make an Artificial Animal. For seeing life is but a motion of Limbs, the begining whereof is in some principall part within; why may we not say, that all Automata *(Engines that move themselves by springs and wheeles as doth a watch) have an artificial life? For what is the* Heart, *but a* Spring; *and the* Nerves, *but so many Strings; and the* Joynts, *but so many* Wheeles *giving motion to the whole Body, such as was intended by the Artificer? Art goes yet further, imitating that Rationall and most excellent worke of Nature,* Man. *For by Art is created that great* LEVIATHAN *called a* COMMON-WEALTH *or* STATE *(in Latine* CIVITAS*) which is but an Artificiall Man; though of greater stature and strength than the* Naturall, *for whose protection and defence it was intended; and in which, the* Soveraignty *is an Artificiall Soul, as giving life and motion to the whole body.*

—THOMAS HOBBES 1651, p. 1

Thomas Hobbes was the first sociobiologist, two hundred years before Darwin. As the opening words of his masterpiece make clear, he saw the creation of the state as fundamentally a matter of one artifact's making another, a sort of group-survival vehicle, "intended" for the "protection and defence" of its occupants. The frontispiece of the original edition shows how seriously he took his own metaphor.

Why, though, do I call Hobbes a sociobiologist? He couldn't have wanted to exploit *Darwin*'s ideas in an analysis of society, like today's sociobiologists. But he did see, clearly and confidently, the fundamental Darwinian task: he saw that there *had* to be a story to be told about how the state first came to be created, and how it brought with it something altogether new on the face of the Earth: morality. It would be a story taking us from a time in which there clearly was no right and wrong, just amoral competition, to a time in which there manifestly was right and wrong (in some parts of the biosphere) via a process that gradually introduced the "essential" features of an ethical perspective. Since the relevant period was prehistoric, and since he had no fossil record to consult, his story would have to be a rational reconstruction, a Just So Story of sorts (to commit a further anachronism).

Once upon a time, he said, there was no morality at all. There was life; there were human beings, and they even had language, so they had memes (to commit a third anachronism). We can presume that they had words—and hence memes—for good and bad, but not *ethical* good and bad. "The notions of Right and Wrong, Justice and Injustice have there no place." So, although they distinguished a good spear from a bad spear, a good supper from a bad supper, a good hunter (an expert killer of supper) from a bad hunter (who scared away the prey), they had no concept of a good or just person, a moral person, or a good act, a moral act—or their contraries, villains and vices. They could appreciate that some people were more dangerous than others, or better fighters, or more desirable mates, but their perspective went no farther than that. They had no concept of right or wrong because "They are Qualities, that relate to men in Society, not in Solitude." Hobbes called this epoch in our prehistory "the state of nature," because it resembled in its most important features the plight of all the other animals in the wild, to this day. In the state of nature, "there is no place for Industry; because the fruit thereof is uncertain; ... no Arts; no Letters, no Society; and which is worst of all, continuall feare, and danger of violent death; And the life of man, solitary, poore, nasty, brutish, and short."

And then, one fine day, a mutation happened to arise. One day, when yet another conflict arose, just like all the others that had come before it, something new happened to happen. Instead of persisting in the myopically selfish policies of mutual defection and distrust that had reigned heretofore, these particular lucky competitors hit upon a new idea: cooperation for mutual benefit. They formed a "social contract." Whereas before there had been families, or herds, or tribes, this was the birth of a different *kind* of group, a society. This was the birth of civilization. And the rest, as one says, is history.

How Hobbes would have admired Lynn Margulis' story of the eukaryotic revolution, and the creation, thereby, of multicellular life! Whereas before there had been nothing but boring prokaryotes, drifting through their nasty, brutish, short lives, now there could be multicellular organisms, which, thanks to a division of labor among a gang of specialist cells, could engage in Industry (oxygen-fired metabolism, in particular) and Arts (long-range perception and locomotion, and protective coloration, and so forth). And, in due course, their descendants created multicellular societies of a very peculiar sort, known (until recently) as Men, capable of creating Letters (or representations), which they fell to exchanging promiscuously; this made possible a second revolution.

How Hobbes would have admired Richard Dawkins' story of the birth of memes, and the creation thereby of persons, who were *not* mere survival vehicles for their genes! These tales, composed long after his, narrate major steps in evolution that antedate the step he decided to describe: the step from persons without morality to citizens. He saw this, correctly, as a major step in the history of life on this planet, and he set out to tell the tale, as best he could, of the conditions under which this step could be taken and, once taken, *evolutionarily enforced* (to use one more anachronism). Though it was not a saltation but a small step, it had momentous consequences, for it was the birth of a hopeful monster indeed.

It would be a mistake to read Hobbes as a would-be historian who was simply speculating irresponsibly. He surely knew that there was no hope of finding the birthplace of civilization with the tools of history (or archeology—a discipline not yet invented), but that was not his point. No doubt the actual prehistoric sequence of events was more muddled, and distributed, with elements of quasi-society (of the sort we see among herds of ungulates and packs of predators), quasi-language (of the sort we see among alarm-calling birds and monkeys, and even among foraging bees), and perhaps even elements of quasi-morality (of the sort reputedly evidenced by monkeys,[1] as well as solicitous whales and dolphins). Hobbes' rational reconstruction was a huge oversimplification, a model intended to illustrate the essentials while ignoring the grubby and unknowable details. And, without any doubt, it was too simple even in its own terms. Today, in the wake of hundreds of investigations into the nooks and crannies of game theory, Prisoner's Dilemma tournaments, and the like, we know that Hobbes was altogether too sanguine (to use a word from his vocabulary) about the conditions under which a social contract would be evolutionarily enforceable. But he was the pioneer explorer of this phenomenon.

Following in his footsteps, Jean-Jacques Rousseau and various English thinkers, including John Locke, offered their own rational reconstructions of the birth of society. More intricate "contractarian" Just So Stories have been exploited in recent years. The most famous, and most sophisticated, is John Rawls' *Theory of Justice* (1971), but there are others. They all agree in seeing morality to be, in one way or another, an emergent product of a major innovation in perspective that has been achieved by just one species, *Homo sapiens*, taking advantage of its unique extra medium of information

---

1. Wechkin et al. 1964, Masserman et al. 1964; for discussion, see Rachels 1991.

transfer, language. In Rawls' thought experiment about how a society ought to be formed, we are to imagine a time, at the birth of society, when its inhabitants gather to consider what sort of design their society shall have. They are to reason together about this until they achieve what Rawls calls "reflective equilibrium"—a stable agreement that cannot be upset by further consideration. In this regard, Rawls' idea is like Maynard Smith's idea of an evolutionarily stable strategy or ESS, but with a major difference: these are *people* doing the calculation, not birds or pine trees or other simple competitors in the games of life. The key innovation in Rawls' scenario, designed to ensure that undue selfishness among the participants in this exercise in reflection cancels itself out, is what he calls the "veil of ignorance." Everyone gets to vote on a favored design of society, but when you decide which society you would be happy to live in and give your allegiance to, you vote without knowing what your particular role or niche in it will be. You may be a senator or a surgeon or a street-sweeper or a soldier; you don't get to find out until after you have voted. Choosing from behind the veil of ignorance ensures that people will give due consideration to the likely effects, the costs and benefits, for all the citizenry, including those worst off.

Rawls' theory has received, and deserved, more attention than any other work in ethics in this century, and, as usual, I am presenting an oversimplified version of the issues. My point is to draw attention to the placement of this work, and all the work that it has provoked and inspired, relative to Darwinian thinking in general, and "evolutionary ethics" in particular. Note especially that, whereas Hobbes presented a rational reconstruction of something that actually happened—something that must have happened—Rawls presents a thought experiment about what, if it did happen, would be *right*. Rawls' project is not speculative history or prehistory, but an entirely normative project: an attempt to demonstrate how ethical questions *ought* to be answered, and, more particularly, an attempt to *justify* a set of ethical norms. Hobbes hoped to solve the normative problem about what ethics *ought* to be—Rawls' problem—but, greedy reductionist that he was, he tried to kill two birds with one stone: he also wanted to explain how such a thing as right and wrong came into existence in the first place, an exercise of imagination in the Darwinian mode. Needless to say, life is more complicated than that, but it was a nice try.

Hobbes' account in the *Leviathan* has a fine Panglossian ring to it—in both exapted senses of that popular word. First, by presupposing the rationality (or Prudence, as he called it) of the agents whose mutual solution society is supposed to be, he viewed the birth of society as dictated by reason, a forced move, or at least strongly endorsed by reason, a Good Trick. In other words, Hobbes' tale is an adaptationist Just So Story—and none the worse for it. But, second, by appealing as it does to our sense of *the good of our own species*, it is apt to lull us into overly sanguine models of how it must have come about—and this is a serious criticism. It may occur to us that, however it came about, the birth of morality was a good thing *for us*, but we should try not to indulge in that sort of reflection. No matter how true it may be, it cannot explain how these practices, for which we are retrospectively so grateful, came into existence and persisted. Group rationality *may not* be assumed, any more than we may assume that since we have benefited mightily from the eukaryotic revolution, it is thereby explained. Group rationality, or cooperation, has to be achieved, and that is a major design task, whether we are considering alliances of prokaryotes or alliances of our more recent ancestors. In fact, much of the best work in ethics in recent years has concerned precisely this issue (e.g., Parfit 1984, Gauthier 1986, Gibbard 1985).

Before looking more closely at the human predicament in this regard, we might consider more cautiously the metaphor that Hobbes invites us to take seriously, helping ourselves to the improved perspective provided by the Darwinian Revolution that has intervened. In what regards is a society like a giant organism, and in what regards is it different?

Multicellular organisms have solved the problem of group solidarity. One never hears tales of a person's thumbs rising up in civil war against the neighboring fingers, or of an eagle's wings going on strike, refusing to work unless some concession can be wrung from the beak or (more to the point) the gonads. And now that we have the gene's-eye perspective from which to look at the world, this can strike us as something of a puzzle. Why don't these rebellions happen? Each cell in a multicellular organism has its own strings of DNA, a complete set of genes for making a whole organism, and if genes are selfish, why do the genes in the thumb cells or wing cells so docilely cooperate with the rest of the genes? Don't the DNA copies in the thumbs and wings count as genes? (Are they denied the vote? Why do they put up with it?) As the biologist David Sloan Wilson and the philosopher of biology Elliot Sober (Wilson and Sober 1994) have suggested, we can learn a lot about our social problems of defection (e.g., promising and then reneging on the promise) and Hardin's tragedy of the commons (see chapter 9) by considering how our ancestors, going back to the first eukaryotes, managed to achieve "harmony and coordination of their parts." The lessons to be learned are tricky, however, because the cells that compose us belong to two very different categories.

> An average human is normally host to billions of symbiotic organisms belonging to perhaps a thousand different species.... His phenotype is not determined by his human genes alone but also by the genes of all the symbionts he happens to be infected with. The symbiont species an individual carries usually have a very varied provenance, with only a few being likely to have come from his parents. [Delius 1991, p. 85.]

Am I an organism, or a community, or both? I am both—and more—but there is a tremendous difference between the cells that are officially part of my body, and the cells, many of them just as important to my survival, that are not. The cells that compose multicellular me all share an ancestry; they are a single lineage, the "daughter cells" and "granddaughter cells" of the egg and sperm that united to form my zygote. They are *host* cells; the other cells are *visitors*, some welcome, some not. The visitors are outsiders, because they have descended from different lineages. What difference does this make?

This is extremely easy to lose sight of, especially in contexts in which we treat all these "parties" as intentional systems—as we should, but with extreme caution. Unless we are careful, we are apt to miss the fact that there are crucial moments in the careers of these various agents and semi-agents and hemi-semi-demi-agents when opportunities to "decide" arise, and then pass. The cells that compose my bulk have a shared fate, but some in a stronger sense than others. The DNA in my finger cells and blood cells is in a genetic cul-de-sac; in Weismann's terms (see chapter 11), these cells are part of the *somatic* line (the body), not the *germ* line (the sex cells). Barring revolutions in cloning techniques (and ignoring the strictly limited, short-lived prospects they have for giving way to replacement cells they help create), my somatic-line cells are doomed to die "childless," and since this was determined some time ago, there is no longer any pressure, any normal opportunity, any "choice points," at which their intentional trajectories—or the trajectories of their limited progeny—might be adjusted. They are, you might say, *ballistic* intentional systems, whose highest goals and purposes have been fixed once and for all, with no chance of reconsideration or guidance. They are totally committed slaves to the *summum bonum* of the body of which they form a part. They may be exploited or tricked by visitors, but under normal circumstances they cannot rebel on their own. Like the Stepford Wives, they have a single *summum bonum* designed right into them, and it is not "Look out for Number One." On the contrary, they are team players by their very nature.

How they further this *summum bonum* is also designed right into them, and in this regard they differ fundamentally from the other cells that are "in the same boat": my symbiont visitors. The benign mutualists, the neutral commensals, and the deleterious parasites that share the vehicle they all together compose—namely, me—each have their own *summum bonum* designed into them, and it is to further their own respective lineages. Fortunately, there are conditions under which an *entente cordiale* can be maintained, for, after all, they are all in the same boat, and the conditions under which they can do better by not cooperating are limited. *But they do have the "choice."* It is an issue for them in a way it is not for the host cells.

Why? What enables—or requires—the host cells to be so committed, but gives the visitor cells a free rein to rebel when the opportunity arises?

Neither sort of cell is a thinking, perceiving, rational agent, of course. And neither sort is significantly more cognitive than the other. That is not where the fulcrum of evolutionary game theory is located. Redwood trees are not notably clever either, but they are in conditions of competition that force them to defect, creating what is, from *their* point of view (!), a wasteful tragedy. The mutual cooperative agreement whereby they would all forgo growing tall trunks, and abandon their vain attempts to gain more than their fair share of sunlight, is evolutionarily unenforceable.

The condition that creates a choice is the mindless "voting" of *differential* reproduction. It is the opportunity for differential reproduction that lets the lineages of our visitors "change their minds" or "reconsider" the choices they have made, by "exploring" alternative policies. My host cells, however, have been designed once and for all by a single vote at the time my zygote was formed. If, thanks to mutation, dominating or selfish strategies occur to *them*, they will not flourish (relative to their contemporaries), since there is scant opportunity for differential reproduction. (Cancer can be seen as a selfish—and vehicle-destructive—rebellion made possible by a revision that does permit differential reproduction.)

The philosopher and logician Brian Skyrms has recently pointed out (1993, 1994a, 1994b) that the precondition for normal cooperation in the strongly shared fate of somatic-line cells is analogous to the cooperation Rawls tried to engineer behind the veil of ignorance. He calls this, aptly, the "Darwinian Veil of Ignorance." Your sex cells (sperm or ova) are formed by a process unlike that of normal cell division or *mitosis*. Your sex cells are formed by a different process, called *meiosis*, which randomly constructs *half* a genome-candidate (to join forces with a half from your mate) by choosing first a bit from "column A" (the genes you got from your mother) and then a bit from "column B" (the genes you got from your father) until a full complement of genes—but just one copy of each—is constructed and installed in a sex cell, ready to try its fate in the great mating lottery. But which "daughters" of your original zygote are destined for meiosis and which for mitosis? This, too, is a lottery. Thanks to this mindless mechanism, paternal and maternal genes (in you) could not ordinarily "know their fate" in advance. The question of whether they are going to have germ-line progeny that might have a flood of descendants flowing on into the future or be relegated to the sterile backwaters of somatic-line slavery for the good of the body politic or corporation (think of the etymology) is unknown and unknowable, so there is nothing to be gained by selfish competition among their "fellow" genes.

That, at any rate, is the usual arrangement. There are special occasions, however, on which the Darwinian Veil of Ignorance is briefly lifted. We have already noted them; they are the cases of "meiotic drive" or "genomic imprinting" (Haig and Grafen 1991, Haig 1992) we considered in chapter

9, in which circumstances *do* permit a "selfish" competition between genes to arise—and arise it does, leading to escalating arms races. But under most circumstances, the "time to be selfish," for genes, is strictly limited, and once the die—or the ballot—is cast, those genes are just along for the ride until the next election.[2]

Skyrms shows that when the individual elements of a group—whether of whole organisms or their parts—are closely related (clones or near-clones) or are otherwise able to engage in mutual recognition and assortative "mating," the simple game-theory model of the Prisoner's Dilemma, in which the strategy of defection always dominates, does not correctly model the circumstances. That is why our somatic cells don't defect; they are clones. This is *one* of the conditions under which groups—such as the group of my "host" cells—can have the "harmony and coordination" required to behave, quite stably, as an "organism" or "individual." But before we give three cheers and take this to be our model for how to make a just society, we should pause to notice that there is another way of looking at these model citizens, the somatic-line cells and organs: their particular brand of selflessness is the unquestioning obedience of zealots or zombies, exhibiting a fiercely xenophobic group loyalty that is hardly an ideal for human emulation.

We, unlike the cells that compose us, are not on ballistic trajectories; we are *guided* missiles, capable of altering course at any point, abandoning goals, switching allegiances, forming cabals and then betraying them, and so forth. For us, it is always decision time, and because we live in a world of memes, no consideration is alien to us, or a foregone conclusion. For this reason, we are constantly faced with social opportunities and dilemmas of the sort for which game theory provides the playing field and the rules of engagement but not the solutions. Any theory of the birth of ethics is going to have to integrate culture with biology. As I have said before, life, for people in society, is more complicated.

---

2. The parallel was perhaps first noted by E. G. Leigh: "It is as if we had to do with a parliament of genes: each acts in its own self-interest, but if its acts hurt the others, they will combine together to suppress it. The transmission rules of meiosis evolve as increasingly inviolable rules of fair play, a constitution designed to protect the parliament against the harmful acts of one or a few. However, at loci so closely linked to a distorter that the benefits of 'riding its coattails' outweigh the damage of its disease, selection tends to enhance the distortion effect. Thus a species must have many chromosomes if, when a distorter arises, selection at most loci is to favor its suppression. Just as too small a parliament may be perverted by the cabals of a few, a species with only one, tightly linked chromosome is an easy prey to distorters" (Leigh 1971, p. 249). See also Buss 1987, pp. 180ff., for a discussion of germ-line sequestration as basically a political innovation that permitted multicellular life.

2. FRIEDRICH NIETZSCHE'S JUST SO STORIES

The first impulse to publish something of my hypotheses concerning the origin of morality was given me by a clear, tidy, and shrewd—also precocious—little book in which I encountered distinctly for the first time an upside-down and perverse species of genealogical hypothesis, the genuinely English type, that attracted me—with that power of attraction which everything contrary, everything antipodal possesses.

—FRIEDRICH NIETZSCHE 1887, preface

It is in perfect accordance with the scheme of nature, as worked out by natural selection, that matter excreted to free the system from superfluous or injurious substances should be utilised for [other] highly useful purposes.

—CHARLES DARWIN 1862, p. 266

Friedrich Nietzsche published his *Genealogy of Morals* in 1887. He was the second great sociobiologist, and, unlike Hobbes, he was inspired (or provoked) by Darwinism. As I noted in chapter 7, Nietzsche probably never read Darwin. His contempt for the "English type" of genealogy was directed against the Social Darwinists: Herbert Spencer in particular, and Darwin's fans on the continent. One fan was Nietzsche's friend Paul Rée, whose "tidy" book, *Origin of the Moral Sensations* (1877), provoked Nietzsche's untidy masterpiece.[3] The Social Darwinists were sociobiologists, but certainly not great ones. In fact, their efforts almost did in the memes of their hero, by popularizing second-rate (per)versions of them.

The "survival of the fittest," Spencer proclaimed, is not just Mother Nature's way, but *ought* to be *our* way. According to the Social Darwinists, it is "natural" for the strong to vanquish the weak, and for the rich to exploit the poor. This is simply bad thinking, and Hobbes has already shown us why. It is equally "natural" to die young and illiterate, without benefit of eyeglasses for myopia, or medicine for illness—for that is how it was in the state of nature—but surely this counts for nothing when we ask: Ought it, then, be that way now? Alternatively, since it was (in an extended sense) entirely natural—it wasn't supernatural—for us to step out of the state of nature and adopt a host of societal practices for our mutual benefit, we may simply deny that there is anything universally natural about the strong dominating

---

3. Rée was Nietzsche's dearest friend, close enough to be entrusted with the task of conveying Nietzsche's proposal of marriage to Lou Salomé in 1882, but she refused, and Rée fell in love with her. Life is complicated.

the weak, and the rest of the Social Darwinist nonsense. It is amusing to note that the fundamental (bad) argument of the Social Darwinists is identical to a (bad) argument used by many religious fundamentalists. Whereas the fundamentalists sometimes begin their arguments by saying, "If God had intended Man to ... [fly, wear clothes, drink alcohol, ...]," the Social Darwinists begin theirs by saying, in effect, "If Mother Nature had intended Man to ...," and even though Mother Nature (natural selection) can be viewed as having intentions, in the limited sense of having retrospectively endorsed features for one reason or another, these earlier endorsements may count for nothing now, since circumstances have changed.

Among the Social Darwinists' ideas was a political agenda: efforts by do-gooders to provide nurture for the least fortunate members of society are counterproductive; such efforts permit those to replicate whom nature would wisely cull. These are abominable ideas, but they were not the primary target of Nietzsche's criticism. His primary target was the historical naïveté of the Social Darwinists (Hoy 1986), their Panglossian optimism about the ready adaptability of human reason (or Prudence) to Morality. Nietzsche saw their complacency as part of their heritage as "English psychologists"—intellectual descendants of Hume. He noted their desire to avoid skyhooks:

> These English psychologists—what do they really want? One always discovers them ... seeking the truly effective and directing agent, that which has been decisive in its evolution, in just that place where the intellectual pride of man would least *desire* to find it (in the *vis inertiae* of habit, for example, or in forgetfulness, or in a blind and chance mechanistic hooking-together of ideas, or in something purely passive, automatic, reflexive, molecular, and thoroughly stupid)—what is it really that always drives these psychologists in just *this* direction? Is it a secret, malicious, vulgar, perhaps self-deceiving instinct for belittling man? [Nietzsche 1887, First Essay, sec. 1, p. 24.]

Nietzsche's antidote to the banalities of the "English psychologists" was a very "continental" romanticism. They thought the passage from the state of nature to morality was easy, or at least quite presentable, but that was because they just made up their stories and didn't bother looking at the clues of history, which told a darker tale.

Nietzsche began, as Hobbes had done, by imagining a premoral world of human life, but he divided his story of transition into two phases (and told his tales in reverse order, starting in the middle, something that confuses many readers). Hobbes had noted (1651, pt. I, ch. 14) that the very existence of any practice of forming contracts or compacts depends on the capacity of human beings to make promises about the future, and what struck Nietzsche was that this capacity does not come for free. This was the topic of the Second Essay of the three that make up the *Genealogy*: "To breed an animal *with the right to make promises*—is not this the paradoxical task that nature has set itself in the case of man? is it not the real problem regarding man?" (Second Essay, sec. 1, p. 57). This "long story of how *responsibility* originated" is a story of how early human beings learned to torture each other—literally—into developing a special kind of memory, the memory needed to keep track of debts and credits. "Buying and selling, together with their psychological appurtenances, are older even than the beginnings of any kind of social forms of organization and alliances" (Second Essay, sec. 8, p. 70). The capacity to detect cheating, to remember the promise broken and punish the cheater, had to be drilled into our ancestors' brains, Nietzsche surmised: "Its beginnings were, like the beginnings of everything great on earth, soaked in blood thoroughly and for a long time" (Second Essay, sec. 6, p. 65). What is Nietzsche's evidence for all this? An imaginative—not to say unbridled—reading of what we might call the fossil record of human culture, in the form of ancient myths, surviving religious practices, archeological clues, and so forth. Leaving the gory details aside, fascinating though they are, Nietzsche's suggestion is that eventually—perhaps via an instance of the Baldwin Effect!—our ancestors "bred" an animal with an innate capacity to keep a promise, and a concomitant talent for detecting and punishing a promise-breaker.

This permitted the formation of early societies, according to Nietzsche, but there was still no morality—not in the sense that we recognize and honor today. The second transition occurred in historical times, he claimed, and can be traced via etymological reconstruction and a proper reading of the texts of the last two millennia—an adaptation by Nietzsche of the philological methods that he had been trained to use. To read these clues in a new way, you need a theory, of course, and Nietzsche had one, developed in opposition to the tacit theory he discerned in the Social Darwinists. The proto-citizens of Nietzsche's second Just So Story (told in the First Essay) live in societies of sorts, not Hobbes' state of nature, but the life he describes in them is about equally nasty and brutish. Might made right—or, rather, might ruled. The people had concepts of good and *bad*, but not good and *evil*, right and wrong. Like Hobbes, Nietzsche tried to tell the tale of how these latter memes arose. One of the most daring (and ultimately least persuasive) of his speculations is that the memes for (moral) good and evil were not just minor permutations of their amoral predecessors; the memes *traded places*. What had been *good* (old-style) became *evil* (new-style), and what had been *bad* (old-style) became (morally) *good* (new-style). This "transvaluation of values" was, for Nietzsche, the key event in the birth of ethics, and he explicitly opposed it to Herbert Spencer's bland supposition that the concept "good" is essentially identical with the concept "useful," "practical," so that in the judgments "good" and "bad" mankind has summed up and sanctioned precisely its *unforgotten* and *unforgettable* experiences regarding what is useful-practical and what is harmful-impractical. According to this theory, that which has always proved itself useful is good; therefore it may claim to be "valuable in the highest degree," "valuable in itself." This road to an explanation is, as aforesaid, also a wrong one, but at least the explanation is in itself reasonable and psychologically tenable. [First Essay, sec. 3, p. 27.]

The amazing and ingenious tale Nietzsche told about how the transvaluation of values happened defies fair summary, and is often outrageously misrepresented. I will not attempt to do justice to it here, but will just draw attention to its central theme (without judging its truth): the "aristocrats" who ruled by might over the weak were cunningly tricked (by the "priests") into adopting the inverted values, and this "slave revolt in morality" turned the cruelty of the strong against itself, so that the strong were manipulated into subduing and civilizing themselves.

For with the priests *everything* becomes more dangerous, not only cures and remedies, but also arrogance, revenge, acuteness, profligacy, love, lust to rule, virtue, disease—but it is only fair to add that it was on the soil of this *essentially dangerous* form of human existence, the priestly form, that man first became *an interesting animal*, that only here did the human soul in a higher sense acquire *depth* and become *evil*—and these are the two basic respects in which man has hitherto been superior to other beasts! [First Essay, sec. 6, p. 33.]

Nietzsche's Just So Stories are terrific (old-style and new-style). They are a mixture of brilliant and crazy, sublime and ignoble, devastatingly acute history and untrammeled fantasy. If Darwin's imagination was to some degree handicapped by his English mercantile heritage, Nietzsche's was even more handicapped by his German intellectual heritage, but those biographical facts (whatever they are) have no bearing on the current value of the memes whose birth each attended so brilliantly. Both came up with dangerous ideas—if I am right, this is no coincidence—but, whereas Darwin was ultra-cautious in his expression, Nietzsche indulged in prose so overheated that it no doubt serves him right that his legion of devotees has included a disreputable gaggle of unspeakable and uncomprehending Nazis and other such fans whose perversions of his memes make Spencer's perversions of Darwin's seem almost innocent. In both cases, we must work to repair the damage such descendants have inflicted on our meme filters, which tend to dismiss memes on the basis of guilt by association. Neither Darwin nor Nietzsche was politically correct, fortunately for us.

(Political correctness, in the extreme versions worthy of the name, is antithetical to almost all surprising advances in thought. We might call it *eumemics*, since it is, like the extreme eugenics of the Social Darwinists, an attempt to impose myopically derived standards of safety and goodness on the bounty of nature. Few today—but there are a few—would brand *all* genetic counseling, all genetic policies, with the condemnatory title of eugenics. We should reserve that term of criticism for the greedy and peremptory policies, the extremist policies. In chapter 18, we will consider how we might wisely patrol the memosphere, and what we might do to protect ourselves from the truly dangerous ideas, but we should keep the bad example of eugenics firmly in mind when we do so.)

Nietzsche's most important contribution to sociobiology, I think, is his steadfast application of one of Darwin's own fundamental insights to the realm of cultural evolution. This is the insight most notoriously overlooked by the Social Darwinists and by some contemporary sociobiologists. Their error is sometimes called the "genetic fallacy" (e.g., Hoy 1986): the mistake of inferring current function or meaning from ancestral function or meaning. As Darwin (1862, p. 284) put it, "Thus throughout nature almost every part of each living thing has probably served, in a slightly modified condition, for diverse purposes, and has acted in the living machinery of many ancient and distinct specific forms." And as Nietzsche put it:

... the cause of the origin of a thing and its eventual utility, its actual employment and place in a system of purposes, lie worlds apart; whatever exists, having somehow come into being, is again and again reinterpreted to new ends, taken over, transformed, and redirected by some power superior to it; all events in the organic world are a subduing, a *becoming master*, and all subduing and becoming master involves a fresh interpretation, an adaptation through which any previous "meaning" and "purpose" are necessarily obscured or even obliterated. [Second Essay, sec. 12, p. 77.]

Aside from Nietzsche's characteristic huffing and puffing about some power subduing and becoming master, this is pure Darwin. Or, as Gould might put it, all adaptations are exaptations, in cultural evolution as well as in biological evolution. Nietzsche went on to emphasize another classical Darwinian theme:

The "evolution" of a thing, a custom, an organ is thus by no means its *progressus* toward a goal, even less a logical *progressus* by the shortest route and with the smallest expenditure of force—but a succession of more or less profound, more or less mutually independent processes of subduing, plus the resistances they encounter, the attempts at transformation for the purpose of defense and reaction, and the results of successful counteractions. [Second Essay, sec. 12, pp. 77–78.][4]

Considering that Nietzsche may never have read Darwin's own work, his appreciation of its major directions is remarkable, but he rather spoiled his record as a sound Darwinian by lapsing, on the same page, into skyhook hunger, announcing his "fundamental opposition to the now prevalent instinct and taste which would rather be reconciled even to the absolute fortuitousness, even the mechanistic senselessness of all events than to the theory that in all events a *will to power* is operating." Nietzsche's idea of a will to power is one of the stranger incarnations of skyhook hunger, and, fortunately, few find it attractive today. But, if we set that aside, the upshot of Nietzsche's genealogy of morals is that we must be extremely careful not to read into the history we extrapolate from nature any simplistic conclusions about value:

> The question: what is the *value* of this or that table of values and "morals"? should be viewed from the most divers perspectives; for the problem "value for *what*?" cannot be examined too subtly. Something, for example, that possessed obvious value in relation to the longest possible survival of a race (or to the enhancement of its power of adaptation to a particular climate or to the preservation of the greatest number) would by no means possess the same value if it were a question, for instance, of producing a stronger type. The well-being of the majority and the well-being of the few are opposite viewpoints of value: to consider the former *a priori* of higher value may be left to the naiveté of English biologists. [Nietzsche 1887, First Essay, sec. 17.]

It is Spencer, clearly, not Darwin, whom Nietzsche is accusing of naïveté about value. Both Spencer and Rée thought they could see a straight, simple path to altruism (Hoy 1986, p. 29). We can see Nietzsche's criticism of this Panglossianism as a clear forerunner of George Williams' criticism of the Panglossianism of naïve group selectionism (see chapter 11). Spencer, in our terms, was an egregiously greedy reductionist, trying to derive "ought" from "is" in a single step. But doesn't this reveal the deeper problem with all sociobiology? Haven't the philosophers shown us that you can *never* derive "ought" from "is," no matter how many steps you take? Some have

---

4. It is interesting to note that Nietzsche also had a thoroughly sound and modern idea about the relationship between complexity and any notion of global progress: "The richest and most complex forms—for the expression 'higher type' means no more than this—perish more easily: only the lowest preserve an apparent indestructibility" (Nietzsche 1901, p. 684).

argued that sociobiology, no matter how sophisticated it became, no matter how many cranes it employed, could never bridge the gap between the "is" of empirical scientific fact and the "ought" of ethics! (They say this with impressive passion.) That is the conviction we must examine next.

3. Some Varieties of Greedy Ethical Reductionism

One of the shibboleths of contemporary philosophy is that you can't derive "ought" from "is." Attempting to do this is often called the *naturalistic fallacy*, taking the term from G. E. Moore's classic, *Principia Ethica* (1903). As the philosopher Bernard Williams points out (1983, p. 556), there really are several issues here. Naturalism "consists in the attempt to lay down certain fundamental aspects of the good life for man on the basis of considerations of human nature." Naturalism wouldn't be refuted by the rather obvious fact that you can't derive any *simple* "ought" statement from any *simple* "is" statement. Consider: does it follow logically that I *ought* to give you five dollars from the fact (and suppose it *is* a fact) that I said I would give you five dollars? Obviously not; any number of intervening excusing conditions might be cited to block this inference. Even if we were to characterize my saying as *promising*—an ethically loaded description—no *simple* "ought" statement follows directly.

But reflections like this make scarcely a dent on naturalism as a theoretical goal. Philosophers distinguish between finding the *necessary* and *sufficient* conditions for various things, and the application of the distinction in this case actually helps clarify the situation. It is one thing to deny that collections of facts about the natural world are *necessary* to ground an ethical conclusion, and quite another to deny that any collection of such facts is *sufficient*. According to standard doctrine, if we stay firmly planted in the realm of facts about the world as it *is*, we will never find any collection of them, taken as axioms, from which any particular ethical conclusion *can be conclusively proven*. You can't get there from here, any more than you can get from any consistent set of axioms about arithmetic to all the true statements of arithmetic.

Well, so what? We may bring out the force of this rhetorical question with another one, rather more pointed: If "ought" cannot be derived from "is," just what *can* "ought" be derived from? Is ethics an *entirely* "autonomous" field of inquiry? Does it float, untethered to facts from any other discipline or tradition? Do our moral intuitions arise from some inexplicable ethics module implanted in our brains (or our "hearts," to speak with tradition)? That would be a dubious skyhook on which to hang our deepest convictions about what is right and wrong. Colin McGinn notes:

... according to Chomsky, it is plausible to see our ethical faculty as analogous to our language faculty; we acquire ethical knowledge with very little explicit instruction, without great intellectual labour, and the end result is remarkably uniform given the variety of ethical input we receive. The environment serves merely to trigger and specialise an innate schematism.... On the Chomskyan model, both science and ethics are natural products of contingent human psychology, constrained by its specific constitutive principles; but ethics looks to have a securer basis in our cognitive architecture. There is an element of luck to our possession of scientific knowledge that is absent in the case of our ethical knowledge. [McGinn 1993, p. 30.]

By contrasting our presumed innate sense of ethical knowledge with our merely "lucky" capacity to engage in science, McGinn and Chomsky suggest that there are *reasons* to be discovered for our possession of the former. If there were a morality module, we would certainly want to know what it was, how it evolved—and, most important of all, why. But, once again, if we try to peer inside, McGinn tries to close the door on our fingers, decrying as "scientism" the attempt to provide answers to our scientific questions about the source of this marvelous perspective we and no other creatures have.

From what can "ought" be derived? The most compelling answer is this: ethics must be *somehow* based on an appreciation of human nature—on a sense of what a human being is or might be, and on what a human being might want to have or want to be. If *that* is naturalism, then naturalism is no fallacy. No one could seriously deny that ethics is responsive to such facts about human nature. We may just disagree about where to look for the most telling facts about human nature—in novels, in religious texts, in psychological experiments, in biological or anthropological investigations. The fallacy is not naturalism but, rather, any simple-minded attempt to rush from facts to values. In other words, the fallacy is *greedy* reductionism of values to facts, rather than reductionism considered more circumspectly, as the attempt to unify our world-view so that our ethical principles don't clash irrationally with the way the world *is*.

Most of the debates about the naturalistic fallacy are better interpreted as disagreements analogous to the skyhooks-versus-cranes debates in evolutionary theory. For instance, B. F. Skinner, in my estimation the world-champion greedy reductionist of all time, wrote an ethical treatise of his own, *Beyond Freedom and Dignity* (1971). In it, he "committed the naturalistic fallacy" on every scale, from the minute to the megalomaniacal. "To make a value judgment by calling something good or bad is to classify it in terms of its reinforcing effects" (Skinner 1971, p. 105). Let's see: that would mean that heroin is good, apparently, and taking care of elderly parents is bad? Is this objection just nitpicking a careless definition? The reinforcing effect of heroin, Skinner assures us when he notices the problem (p. 110), is "anomalous." Hardly a convincing defense against the charge of greedy reductionism. He goes on and on in the book about how scientific his "design for a culture" is, and how optimally suited it is for . . . for what? What is his characterization of the *summum bonum*?

> Our culture has produced the science and technology it needs to save itself. It has the wealth needed for effective action. It has, to a considerable extent, a concern for its own future. But if it continues to take freedom and dignity, rather than its own survival, as its principal value, then it is possible that some other culture will make a greater contribution in the future. [Skinner 1971, p. 181.]

I hope you want to join me in retorting: So what? Even if Skinner were right (and surely he isn't) that a behaviorist regime is our best chance of preserving our culture into the future, I hope it is clear to you that Skinner may well have been mistaken when he deemed "survival of the culture" to be the highest goal any of us could ever imagine wanting to further. In chapter 11, we briefly considered how mad it would be to put survival of one's own genes ahead of everything else. Is survival of one's own culture a clearly saner item to put on the pedestal above everything else? Would it justify mass murder, for instance, or betraying all your friends? We meme-users can see other possibilities—beyond our genes, and beyond even the welfare of the groups (and cultures) to which we currently belong. Unlike our somatic-line cells, we can conceive of more complicated *raisons d'être*.

What is wrong with Skinner is not that he tried to base ethics on scientific facts about human nature, but that his attempt was so simplistic! I suppose pigeons might indeed fare as well as they ever could want in a Skinnerian utopia, but we are really much more complicated than pigeons. The same defect can be seen in the attempt at ethics by another Harvard professor, E. O. Wilson, one of the world's great entomologists and the coiner of the term "sociobiology" (1975). In his ethical treatise, *On Human Nature* (1978), Wilson (pp. 196, 198) faces the problem of identifying the *summum bonum* or "cardinal value," and comes up with two coequals: "In the beginning the new ethicists will want to ponder the cardinal value of the survival of human genes in the form of a common pool over generations. . . . I believe that a correct application of evolutionary theory also favors diversity in the gene pool as a cardinal value." Then (p. 199) he adds a third, universal human rights, but suggests that it must be demythologized. A "rational ant" would find the ideal of human rights "biologically unsound and the very concept of individual freedom intrinsically evil."

> We will accede to universal rights because power is too fluid in advanced technological societies to circumvent this mammalian imperative; the long-

term consequences of inequity will always be visibly dangerous to its temporary beneficiaries. I suggest that this is the true reason for the universal rights movement and that an understanding of its raw biological causation will be more compelling in the end than any rationalization contrived by culture to reinforce and euphemize it. [E. Wilson 1978, p. 199.]

Writing in collaboration with the philosopher of biology Michael Ruse, Wilson declares that sociobiology has shown us that "Morality, or more strictly our belief in morality, is merely an adaptation put in place to further our reproductive ends" (Ruse and Wilson 1985). Nonsense. Our reproductive ends may have been the ends that kept us in the running till we could develop culture, and they may still play a powerful—sometimes overpowering—role in our thinking, but that does not license any conclusion at all about our current values. It does not follow from the fact that our reproductive ends were the ultimate historical *source* of our present values, that they are the ultimate (and still principal) *beneficiary* of our ethical actions. If Ruse and Wilson think otherwise, they are committing the "genetic" fallacy Nietzsche (and Darwin) warned us about. As Nietzsche said, "the cause of the origin of a thing and its eventual utility, its actual employment and place in a system of purposes, lie worlds apart." Do Ruse and Wilson commit this fallacy? Consider what else they say on the subject (p. 51):

> In an important sense, ethics as we understand it is an illusion fobbed off on us by our genes to get us to cooperate.... Furthermore, the way our biology enforces its ends is by making us think that there is an objective higher code, to which we are all subject.

It must be true that there is an evolutionary explanation of how our memes and genes interacted to create the policies of human cooperation that we enjoy in civilization—we haven't figured out all the details yet, but it must be true unless there are skyhooks in the offing—but this would not show that the result was *for the benefit of the genes* (as principal beneficiaries). Once memes are on the scene, they, and the *persons* they help create, are also potential beneficiaries. Hence, the truth of an evolutionary explanation would not show that our allegiance to ethical principles or a "higher code" was an "illusion." In a famous image, Wilson puts his vision this way:

> The genes hold culture on a leash. The leash is very long, but inevitably values will be constrained in accordance with their effects on the human gene pool. [E. Wilson 1978, p. 167.]

But all this means (unless it is just false) is that, in the long run, *if* we adopt cultural practices that have disastrous effects on the human gene pool, then the human gene pool will succumb. There is no reason to think, however, that evolutionary biology shows us that our genes are powerful enough, and insightful enough, to keep us from making policies quite antithetical to their interests. On the contrary, evolutionary thinking shows us that our genes could hardly be smarter than the engineers who designed our imagined survival machines (see chapter 14), and look how helpless they were in the face of unanticipatable collaborations with other robots! We have seen examples of parasites—such as viruses—that manipulate the behavior of their hosts to further *their* interests instead of the hosts' own interests. And we have seen examples of commensals and mutualists that join to make common cause, creating a larger beneficiary out of parts. Persons, according to the meme model we have sketched, are just such larger, higher entities, and the policies *they* come to adopt, as a result of interactions between their meme-infested brains, are not at all bound to answer to the interests of their genes alone—or their memes alone. That is our transcendence, our capacity to "rebel against the tyranny of the selfish replicators," as Dawkins says, and there is nothing anti-Darwinian or antiscientific about it.

The typical inability of Wilson and other sociobiologists to see their critics as anything but religious fanatics or scientifically illiterate mysterians is yet one more sad overswing of the pendulum. Skinner saw his critics as a bunch of Cartesian dualists and miracle-worshipers, and in his peroration he declared:

> To man *qua* man we readily say good riddance. Only by dispossessing him can we turn to the real causes of human behavior. Only then can we turn from the inferred to the observed, from the miraculous to the natural, from the inaccessible to the manipulable. [Skinner 1971, p. 201.]

Wilson and many other sociobiologists have the same bad habit of seeing anybody who disagrees with them as a benighted, science-fearing sky-hooker. In fact, only *most* of the people who disagree with them fit this description! There is a minority comprising responsible critics of the excesses of greed to which the enthusiastic exponents of any new scientific school are apt to succumb.

Another eminent biologist, Richard Alexander, whose own treatment of ethics is much more careful, expresses the appropriate skepticism about Wilson's candidates for cardinal values: "Whether or not these goals would all be judged admirable by humanity, Wilson does not connect his selection of them to biological principles" (Alexander 1987, p. 167). But Alexander also underestimates the power of culture—memes—to snap Wilson's leash. Like Wilson, he acknowledges the huge difference in speed between cultural and genetic evolution, and argues forcefully (pp. 10–11) that cultural versatility makes a shambles of any attempts—like Chomsky's and Fodor's—to find some "Thou Shalt Go No Farther" boundary to human cognition. He thinks, however, that evolutionary biology has shown that "the individual's self-interests can only be realized through reproduction, by creating descendants and assisting other relatives," and that a consequence of this is that no one ever acts out of genuine beneficence or altruism. As he puts it:

> . . . this "greatest intellectual revolution of the century" tells us that, despite our intuitions, there is not a shred of evidence to support this view of beneficence, and a great deal of convincing theory suggests that any such view will eventually be judged false [Alexander 1987, p. 3].

But, like Wilson and the Social Darwinists, he commits a subtle, attenuated version of the genetic fallacy, and emphasizes the very passage (p. 23) in which he does it.

> Even if culture changes massively and continually across multiple generations, even if our problems and promises arise out of the cultural process of change, even if there are no genetic variations among humans that significantly affect their behavior, *it is always true that the cumulative history of natural selection continues to influence our actions by the set of genes it has provided humanity*.

This is indeed true, but it does not establish the point he thinks it does. As he insists, no matter how potent cultural forces are, they always have to act on the materials genetic forces have shaped for them, and will go on shaping, but they can just as readily *redirect* or *exploit* or *subvert* those genetically endorsed designs as *attenuate* or *combat* them. Sociobiologists, overreacting to the cultural absolutists (those crazy skyhookers) in much the way Darwin overreacted to the Catastrophists, like to emphasize that culture must have *grown out of* our biological inheritance. Indeed it must have, and it is also true that we grew from fish, but our reasons aren't the reasons of fish just because fish are our ancestors.

The sociobiologists are also right to stress that our unique capacity to adopt and act on a different set of reasons does not prevent us from being inconvenienced or even tortured or betrayed by our "animal" urges. Long before Salome did her dance of the seven veils, it was already obvious to members of our species that innate procreative urges can be made to assert themselves at the most inopportune times, just as sneezes and coughs can, seriously threatening the welfare of the body in which those urges are asserted. As in other species, many is the woman who has perished to save her children, and many is the man who has gone to an early death eagerly pursuing one perilous course or another, driven on by the faint hope of procreation. But we must not turn this important fact about our biological limitations into the massively misleading idea that the *summum bonum* at the source of every chain of practical reasoning is the imperative of our genes. A counterexample shows why not: Larry, heartsick at being spurned by Lola, the love of his life, joins the Salvation Army in order to try to forget her, to put an end to his torment. It works. Years later, St. Larry the Sublimated wins the Nobel Peace Prize for all his good deeds, and Richard Alexander, at the ceremony in Oslo, throws a wet blanket on the proceedings by reminding us that this all grew out of Larry's basic reproductive urges. So it did. So what? We make a big mistake if we think the way to understand the bulk of Larry's life is to try to interpret his every move as designed, one indirect way or another, to ensure that he has grandchildren.

The possibility that a meme or complex of memes can redirect our underlying genetic proclivities is strikingly illustrated by a four-century-long human experiment in sociobiology that has recently been vividly drawn to the attention of evolutionary theorists by David Sloan Wilson and Elliot Sober:

> The Hutterites are a fundamentalist religious sect that originated in Europe in the sixteenth century and migrated to North America in the nineteenth century to escape conscription. The Hutterites regard themselves as the human equivalent of a bee colony. They practice community of goods (no private ownership) and also cultivate a psychological attitude of extreme selflessness.... Nepotism and reciprocity, the two principles that most evolutionists use to explain prosocial behavior in humans, are scorned by the Hutterites as immoral. Giving must be without regard to relatedness and without any expectation of return. [Wilson and Sober 1994, p. 602.]

Unlike most sects, the Hutterites have been quite successful at propagating their groups over the centuries, enlarging their range and increasing their global population, according to Wilson and Sober: "In present-day Canada, Hutterites thrive in marginal farming habitat without the benefit of modern technology and almost certainly would displace the non-Hutterite population in the absence of laws that restrict their expansion" (p. 605).

The Hutterites may be over four centuries old, but that is no time at all on the genetic calendar, so it is not likely that *any* of the striking differences between their groups and the groups the rest of us belong to are genetically transmitted. (Exchanging Hutterite infants for others would presumably not interfere noticeably with the "group fitness" of Hutterite colonies. Hutterites simply exploit, thanks to a heritage of *cultural* transmission, dispositions that are part of the common human stock.) So the Hutterites are an example of how cultural evolution can create new group effects, and what is particularly delicious, from an evolutionist's point of view, is their method of fission:

> Like a honey bee colony, Hutterite brotherhoods split when they attain a
> large size, with one half remaining at the original site and the other half
> moving to a new site that has been pre-selected and prepared. In prepa-
> ration for the split, the colony is divided into two groups that are equal
> with respect to number, age, sex, skills and personal compatibility. The
> entire colony packs its belongings and one of the lists is drawn by lottery
> on the day of the split. The similarity to the genetic rules of meiosis could
> hardly be more complete. [Wilson and Sober 1994, p. 604.]

The Darwinian Veil of Ignorance in action! But it is not enough, all by itself,
to ensure group solidarity, since human beings, even those who have lived
their entire lives in a Hutterite community, are not ballistic intentional
systems, but guided intentional systems, and guidance has to be provided on
a daily basis. Wilson and Sober quote Ehrenpreis, one of the early leaders of
the sect: "Again and again we see that man with his present nature finds it
very hard to practice true community." They go on to provide further
quotations in which Ehrenpreis emphasizes just how explicit and energetic
the practices of the Hutterites have to be to counteract this all-too-human
tendency. These declarations make it clear that one way or another, Hutterite social organization is the effect of cultural practices quite vigorously
arrayed *against* the very features of human nature Wilson and Sober wish to
deny or downplay: selfishness and openness to reasoning. If group thinking
were really as much a part of human nature as Wilson and Sober would like
to believe, Hutterite parents and elders wouldn't have to say a thing. (Compare this to a case in which there truly is a genetic predisposition in our
species: how often have you heard parents cajoling their children to eat
more sweets?)

Wilson and Sober are right to present the Hutterite ideals as the essence
of an organismic organization, but the big difference is that for people—
unlike the cells in our bodies, or the bees in a colony—there is always the
option of opting out. And that, I would think, is the last thing we want to
destroy in our social engineering. The Hutterites disagree, apparently, and
so, I gather, do the hosts of many non-Western memes.[5] Do you *like* the
idea of turning ourselves and our children into slaves to the *summum
bonum* of our groups? That is the direction in which the Hutterites have
always been headed, and, by Wilson and Sober's account, they achieve
impressive success, but only at the cost of prohibiting the free exchange of
ideas and discouraging thinking for oneself (which is to be distinguished
from being selfish). Any stubborn freethinker is brought before the congre-

---

5. "To us in Asia, an individual is an ant. To you, he's a child of God. It is an amazing
concept." (Lee Khan Yew, Senior Minister of Singapore, in response to the outcry over
the sentence of flogging of Michael Fay for vandalism, *Boston Globe,* April 29, 1994, p. 8.)

gation and firmly admonished; "if he persists in his stubbornness and refuses to listen even to the Church, then there is only one answer to this situation, and that is to cut him off and exclude him." A totalitarian regime (even a group totalitarianism) is extremely vulnerable to dissuasion, in almost exactly the same way an altruistic group is vulnerable to free-riders. That is not to say that reason is always on the side of defection. It isn't. It's always on the side of keeping options open, of design *revision*. This is usually a good thing, but not always, an important fact that has been noted by the economist Thomas Schelling (1960), the philosopher Derek Parfit (1984), and others, in their discussions of the conditions under which it would be rational for a rational agent to render himself (temporarily) irrational. (For instance, you may want to render yourself a poor target for extortion: if you can somehow convince the world that you are impervious to reason, the world will not try to make you offers you can't refuse.)

There are circumstances—extreme circumstances, as Wilson and Sober note—when we may reasonably curtail free thinking, but the Hutterites have to discourage free thinking all the time. They have to discourage reading whatever books you want, and listening to whomever you want. It is only by the most careful control of the communication channels that such a pristine state can be preserved. That is why the organismic solution is a nonsolution to the problems of human society. The Hutterites are thus themselves a curious example of greedy reductionism, not because they are individually greedy—they are apparently just the opposite—but because their solution to the problem of ethics is so drastically oversimplified. They are, however, an even better example of the power of memes to infect a group of mutual communicators in such a way that the whole group turns its efforts to ensuring the proliferation of *those memes* at whatever cost to themselves.[6]

---

6. According to Wilson and Sober, the Hutterites have "the highest birth rate of any known human society," but it would be a mistake to read this as the triumph of Alexander's reproductive selfishness. It would be a tactical mistake, for one thing: however many Hutterites there are or have been, there have been many, many more Catholic monks and nuns, whose life histories would be manifestly hard to explain as instances of individuals striving, as always, for the reproductive championship. More tellingly, if the point of the Hutterites example is *group* reproductive prowess, birth rate is relevant only as it bears on group birth rate, and we have almost nothing to compare that rate with, since few if any other human groups, so far as I know, behave that way. Perhaps the Hutterites have such a high individual birth rate because so many of their children leave or are expelled and have to be replaced to keep the communities going. We might consider the truly Machiavellian prospect that this is just what the selfish genes wanted all along! They found a meme—the Hutterite complex—that served their purposes, and formed a cabal: the spartan Hutterite communities are really just breeding pens which are kept quite unattractive so that many of the young will leave, making room for more

In the next section, we will look more closely at what sociobiology is and is not, what it could and could not be, but before we leave the topic of greedy ethical reductionism, we should stop to consider an ancient species of this ill-favored meme with many subvarieties: religion. If you wanted to give a clear example of the naturalistic fallacy, you could hardly improve on the practice of trying to justify an ethical precept, an "ought," by citing as your "is": the Bible says so. To this, as to Skinner and Wilson, we must say: So what? Why should the facts—even if they are all facts—recounted in the Bible (or any other holy text, I hasten to add) be supposed to provide any more satisfactory justification for an ethical principle than the facts cited by Darwin in *Origin of Species*? Now, if you believe that the Bible (or some other holy text) is *literally* the word of God, and that human beings are put here on Earth by God in order to do God's bidding, so that the Bible is a sort of user's manual for God's tools, then you do indeed have grounds for believing that the ethical precepts found in the Bible have a special warrant that no other writings could have. If, on the other hand, you believe that the Bible, like Homer's *Odyssey*, Milton's *Paradise Lost*, and Melville's *Moby Dick*, is really a nonmiraculous product of human culture, issuing from some one or more human authors, then you will grant it no authority beyond tradition and whatever its arguments generate by their own cogency. This, it should be obvious, is the unchallenged view of philosophers who work in ethics today, so uncontroversial that, if you ever tried to refute a claim in the contemporary ethics literature by pointing out that the Bible said otherwise, you would be met with surprised stares of disbelief. "That's just the naturalistic fallacy!" the ethicists might say. "You can't derive 'ought' from *that* sort of 'is'!" (So do not expect the philosophers to come to your defense if you claim that religion is a source of ethical wisdom that is superior in any way to science.)

Does that mean that religious texts are worthless as guides to ethics? Of course not. They are magnificent sources of insight into human nature, and into the possibilities of ethical codes. Just as we should not be surprised to discover that ancient folk medicine has a great deal to teach modern high-tech medicine, we should not be surprised if we find that these great religious texts hold versions of the very best ethical systems any human culture will ever devise. But, like folk medicine, we should test it all carefully, and take nothing whatever on faith. (Or do you think it is wise to pop those "holy" mushrooms in your mouth just because some millennias-old tradition declares they help you see the future?) The view I am expressing is what is often called "secular humanism." If secular humanism is your bo-

---

breeding. I am not endorsing this claim, just pointing out that it must be dealt with if an evolutionary account is to be given of how and why Hutterite communities have the features they do.

geyperson, you shouldn't concentrate all your energy on attacking sociobiologists or behaviorists or academic philosophers, for they are not a fraction
of a percent of the influential thinkers who quietly and firmly believe that
ethics is not to be *settled*, but at best guided, by religious doctrines. This is,
indeed, the reigning assumption of the U.S. Congress and the courts; citing
the Constitution has more standing than citing the Bible, and so it should.

Secular humanism often gets its bad name from self-styled secular humanists who are themselves greedy reductionists of one sort or another,
impatient with the complexities of ancient traditions, disrespectful of the
genuine wonders to be savored in the rich cultural heritage of others. If they
think that all ethical questions can be boiled down to one definition or a few
simple definitions (if it's bad for the environment, it's bad; if it's bad for Art,
it's bad; if it's bad for business, it's bad), then they are no better ethicists
than Herbert Spencer and the Social Darwinists. But when we make the
quite appropriate counterclaim that life is more complicated, we must be
careful not to turn that into an obstruction of inquiry rather than a plea for
more careful inquiry. Otherwise, we put ourselves right back on the forlorn
pendulum.

What, then, would a more careful inquiry look like? The task facing us is
still the task that faced Hobbes and Nietzsche: *somehow* we have to have
evolved into beings that can have a conscience, as Nietzsche says (1885,
epigram 98), that kisses us while it hurts us. A vivid way of posing the
question is to imagine becoming an *artificial* selector of altruistic people.
Like a breeder of domestic cattle, pigeons, or dogs, you could closely observe your herd, noting in a ledger which were naughty and which nice,
and, by meddling in various ways, arranging for the nice ones to have more
children. In due course, you ought to be able to evolve a population of nice
people—supposing that a tendency to niceness could be represented somehow in the genome. We should not think of this as selection for an "ethics
module" that is designed just *for* giving right answers to ethical questions.
Any modules or gadgets might have, singly or in coalition, the effect (or
by-product or bonus) of favoring the altruistic choices at decision time.
After all, the loyalty of dogs to human beings is apparently just such an
outcome of unconscious selection by our forebears. God could conceivably
have done this for us, but suppose we want to eliminate the Middleman and
explain the evolution of ethics by *natural* selection, not *artificial* selection.
Might there be some blind, unforesightful forces, some set of natural circumstances, that could accomplish the same thing?

Not in one fell swoop, so far as anybody can see, but there are devious
gradual routes by which we might have bootstrapped ourselves into genuine morality by a series of smallish changes. We may begin with "parental
investment" (Trivers 1972). It is uncontroversial that mutations that yield
creatures who invest more energy and time in caring for their young can,

under many but not all circumstances, evolve. (Remember that only some species engage in parental investment. This is not an option for species in which the young hatch after the parents have died, and the reasons why there should be these fundamentally different parental policies have been well investigated.[7]) Now, once parental investment in their own offspring is secured for a species, how do we expand the circle (Singer 1981)? It is just as uncontroversial, thanks to Hamilton's pioneering work (1964) on "kin selection" and "inclusive fitness," that the same considerations that favor sacrifices for one's offspring also favor, to a mathematically precise degree, sacrifices for one's more distant relatives: offspring aiding parents, siblings helping each other, aunts helping nephews, and so forth. But, again, it is important to remember that the conditions under which such aid is evolutionarily enforceable are not only not universal but relatively rare.

As George Williams (1988) notes, not only is cannibalism (eating conspecifics, even close relatives) common, but in many species sibling-cide (we won't call it murder, since they know not what they do) is almost the rule, not the exception. (For instance, when two or more eagle chicks are born in a single nest, the first to hatch is very likely to kill its younger siblings if it can, by pushing the eggs out of the nest, or even pushing the hatchlings out.) When a lion acquires a new lioness who is still nursing cubs from an earlier mating, the first order of business is to kill those cubs, so that the lioness will more quickly come into estrus. Chimpanzees have been known to engage in mortal combat against their own kind, and langur-monkey males often kill the infants of other males to gain reproductive access to females (Hrdy 1977)—so even our closest relatives engage in horrible behavior. Williams points out that, in all the mammalian species that have so far been carefully studied, the rate at which their members engage in the killing of conspecifics is several *thousand* times greater than the highest homicide rate measured in any American city.[8]

This dark message about our furry friends is often resisted, and popular presentations of nature (in television documentaries, magazine articles, and popular books) often engage in self-censorship to avoid shocking the squeamish. Hobbes was right: life in the state of nature *is* nasty, brutish, and short, for virtually all nonhuman species. If "doing what comes naturally" meant doing what virtually all other animal species do, it would be hazardous to the health and well-being of us all. Einstein famously said that the dear God

---

7. Complications abound, as usual. In some species of beetle, for instance, the males make a huge investment in a food plug (with sperm attached) that females compete for. This is a sort of parental investment, but not the sort we are discussing here.

8. Gould draws attention to the same striking statistic in "A Thousand Acts of Kindness," in Gould 1993d.

is subtle but not malicious; Williams turns that observation inside out: Mother Nature is heartless—even vicious—but boundlessly stupid. And as so often before, Nietzsche finds the point and gives it his special touch:

> "According to nature" you want to *live*? O you noble Stoics, what deceptive words these are! Imagine a being like nature, wasteful beyond measure, indifferent beyond measure, without purposes and consideration, without mercy and justice, fertile and desolate and uncertain at the same time; imagine indifference itself as a power—how *could* you live according to this indifference! [Nietzsche 1885, p. 15.]

Beyond inclusive fitness comes "reciprocal altruism" (Trivers 1971), in which nonrelated or distantly related organisms—they needn't even be of the same species—can form mutually beneficial arrangements of *quid pro quo*, the first step towards human promise-keeping. It is commonly "objected" that reciprocal altruism is ill-named, since it isn't *really* altruism at all, just enlightened self-interest of one form or another: you scratch my back and I'll scratch yours—quite literally, in the case of the grooming arrangements that are a favorite simple example. This "objection" misses the point that we have to pass by small steps to the real McCoy, and reciprocal altruism, ignoble (or just a-noble) as it may be, is a useful stepping-stone on the progression. It requires advanced cognitive abilities—a rather specific memory capable of reidentifying one's debtors and creditors, and the capacity to spot a cheat, for instance.

Moving beyond the most businesslike and brutal forms of reciprocal altruism towards a world in which genuine trust and sacrifice are possible is a task that has begun to be explored theoretically. The first major step was Robert Axelrod's (Axelrod and Hamilton 1981, Axelrod 1984) Prisoner's Dilemma tournaments, which invited all comers to submit strategies—algorithms—for competing against all comers in a reiterated Prisoner's Dilemma tournament. (Among the many discussions of this topic, two of the best are Dawkins 1989a, ch. 12, and Poundstone 1992.) The winning strategy became justly famous: Tit for Tat, which simply copies the "opponent's" previous move, cooperating in reward for past cooperation, and defecting in retaliation against any defections. Basic Tit for Tat comes in a variety of subspecies. In Nice Tit for Tat, one begins by cooperating, and then just does unto the other as the other has done unto oneself on the previous move. As can be readily seen, two Nice Tit-for-Tatters playing opposite one another make out splendidly, cooperating indefinitely, but a Nice Tit-for-Tatter who encounters a Nasty Tit-for-Tatter who throws in an unprovoked defection at any point is in for a debilitating round of endless retaliatory defection (it serves them both right, of course, as they keep reminding themselves).

The simple situations explored by Axelrod's initial tournament have given way to much more complex and realistic scenarios. Nowak and Sigmund (1993) have found a strategy that outperforms Tit for Tat under an important variety of circumstances. Kitcher (1993) examines a world of *non-compulsory* Prisoner's Dilemma games (if you don't fancy a particular opponent, you can decline to play). Kitcher shows, in careful mathematical detail, how "discriminating altruists" (who keep a tally on who has defected in the past) can flourish under certain—not all—conditions, and also begins to sort out the conditions under which varying policies of forgiveness and forgetfulness can hold their own against the ever-present prospect of a resurgence of antisocial types. Particularly fascinating in the directions opened up by Kitcher's analysis is the emergence of groups in which the strong and the weak would tend to segregate themselves and prefer to cooperate with their own sort.

Could this set the stage for something like the Nietzschean transvaluation of values? Stranger things have happened. Stephen White (unpublished) has begun to investigate the important further complexities of the *multi-person* Prisoner's Dilemma. (This is the game that leads to the tragedy of the commons, creating both depleted fish stocks in our oceans and forests of tall trees.) As Kitcher points out, the simple scenarios are analytically tractable—the equations of interaction and their expected yields can be solved directly by mathematical calculation—but as we add more realism, and hence complexity, the direct *solution* of the equations becomes unfeasible, so we have to turn to the indirect methods of computer simulation. In such a simulation, you just set up hundreds or thousands of imaginary individuals, endow them with dozens or hundreds or thousands of strategies or other properties, and let the computer do all the work of having them play thousands or millions of games against each other, keeping track of the results.[9]

This is a branch of sociobiology or evolutionary ethics that no one should deride. It directly *tests* the hunches, such as those of Hobbes and Nietzsche, that there are natural, evolutionarily enforceable paths to where we are today. We may be quite sure that this is true, for here we are, but what this research promises to clarify is how much R-and-D work, of what sorts, was

---

9. If you want to know the odds of being dealt a straight flush in poker, one way is to solve the equation provided by probability theory; you get a definitive answer. Another way is to deal yourself a few billion poker hands, shuffling well between each, and simply counting the straight flushes and dividing by the total number of hands dealt. That gives you a *very* reliable estimate, but it is not officially foolproof. The latter method is the *only* feasible way to study the complicated scenarios of evolutionary ethics, but, as we already saw in the discussion (in chapter 7) of Conway's reactions to the ways in which his Game of Life is being explored, the results of such simulations can be misleading, and should often be taken with a grain of salt.

required to get us here. At one extreme, it could turn out that there is an impressive bottleneck; a quite improbable but crucial series of happy accidents were required. (White's analysis offers some plausible reasons for believing that the conditions are really quite stringent.) At the other extreme, it might turn out that there is a rather wide "basin of attraction" that will lead almost any cognitively sophisticated creatures, whatever their circumstances, into societies with recognizable ethical codes. It will be fascinating to see what large-scale computer simulations of these complex social interactions tell us about the constraints on the evolution of ethics. But we can already be virtually certain that mutual recognition and the capacity to communicate a promise—stressed by both Hobbes and Nietzsche—are necessary conditions for the evolution of morality. It is conceivable, if unlikely on present evidence, that whales and dolphins, or the great apes, meet these necessary conditions, but no other species come close to exhibiting the sorts of social cognition that true morality depends on. (My pessimistic hunch is that the main reason we haven't yet ruled out dolphins and whales as moralists of the deep is that they are so hard to study in the wild. Most of the evidence about the chimpanzees—some of it self-censored by researchers for years—is that they are true denizens of Hobbes' state of nature, much more nasty and brutish than many would like to believe.)

4. Sociobiology: Good and Bad, Good and Evil

... the human brain works however it works. Wishing for it to work in some way as a shortcut to justifying some ethical principle undermines both the science and the ethics (for what happens to the principle if the scientific facts turn out to go the other way?).

—Steven Pinker 1994, p. 427

Sociobiology has two faces. One looks toward the social behavior of nonhuman animals. The eyes are carefully focused, the lips pursed judiciously. Utterances are made only with caution. The other face is almost hidden behind a megaphone. With great excitement, pronouncements about human nature blare forth.

—Philip Kitcher 1985b, p. 435

Another part of our inquiry into human nature, as a naturalistic basis for sound ethical thinking, would begin with the undisputable fact that we human beings are products of evolution, and consider what limitations we are born with and what variations there are among us that might have ethical relevance. Many people apparently think that ethics is in deep trouble if it turns out that human beings aren't, as the Bible tells us, just a little below the angels. If we are not all perfectly rational, and equally rational, and perfectly and equally malleable by education, and equally capable in all other regards, then our underlying assumptions of Equality and Perfectibility are jeopardized. If that were true, it would be too late to save ourselves, for we already know too much about human frailties and human differences to sustain that vision. But there are more reasonable visions that are also jeopardized by the discoveries of scientists (not just evolutionists).

There is no doubt that the sorts of facts we can learn about an individual, or a type or group of individuals (women, people of Asian descent, etc.), can profoundly affect how we tend to regard them and treat them. If I learn that Sam is schizophrenic, or profoundly retarded, or suffers from dizziness and periodic blackouts, I am not going to hire Sam to drive the school bus. When we turn from specific facts about individuals to generalizations about groups of individuals, the situation is more complicated. What is the reasonable and just response of insurance companies to the actuarial facts about the different life expectancies of men and women? Is it fair to adjust their premiums accordingly? Or should we treat both genders alike in the premium department and accept their differential rate of receiving benefits as fair? With regard to voluntarily acquired differences (smokers versus nonsmokers, for instance), we see fairness in making the smokers pay for their habit in higher premiums, but what about differences people are just born with? African-Americans are, as a group, unusually prone to high blood pressure, diabetes runs higher than average among Hispanics, and Whites are more prone to skin cancer and cystic fibrosis (Diamond 1991). Should these differences be reflected in calculating their health insurance? People whose *parents* smoked in the home while they were growing up run a higher risk of respiratory disorders through no fault of their own. Young men, as a group, are less safe drivers than young women. Which of these facts should count for how much, and why? Even when we deal with facts about particular individuals, rather than statistical trends, there are quandaries aplenty: Are employers—or others—entitled to know whether you have ever been married, have a criminal record, a safe driving record, a history of scuba-diving? Is there a principled difference between releasing information on a person's grades in school and releasing information about that same person's IQ score?

These are all difficult ethical problems. The citizenry is currently debating various restrictions on what information employers, the government, the schools, the insurance companies, and so forth may seek regarding individuals, and it is a short step to the conclusion that we'd all be better off if certain sorts of information were just not pursued by science at all. If there are big differences between the brains of men and women, or if there is a gene that predisposes for dyslexia, or violence—or musical genius, or homosexuality—it might be better for us to be kept in the dark about such things. One should not dismiss this suggestion lightly. If you have ever asked yourself whether there are facts about yourself (about your health, your competence, your prospects) you would rather not know, and decided that there were, you should be prepared to consider seriously the suggestion that the best—perhaps the only—way to ensure that such facts are not imposed on people is by prohibiting investigations likely to discover them.[10]

On the other hand, if we don't investigate these issues, we forgo important opportunities. Society has a strong interest in keeping track of the drunk-driving arrests of potential school-bus drivers and making them known to the appropriate decision-makers, and it has the same strong interest in discovering any other facts about its members that may enhance our lives or protect society at large or particular members of it. This is what makes the research decisions we reach so critical and so likely to generate controversy. It is not surprising that sociobiological research is conducted in an atmosphere of unremitting concern-mounting-to-alarm, and when it escalates, as it often does, the propaganda sometimes buries the truth.

Let's begin with the term, "sociobiology." When E. O. Wilson coined it, he meant it to cover the whole spectrum of biological investigations concerned with the evolution of interrelations between organisms in pairs, groups, herds, colonies, nations. Sociobiologists study the relations among termites in a mound, cuckoo hatchlings and their duped adoptive parents, the members of matriarchal groups of elephants, bands of monkeys, elephant-seal bulls and their harems—and human couples, families, tribes, and nations. But, as Kitcher says, the sociobiology of nonhuman animals has always been conducted with greater care and caution. (See also Ruse 1985.) In fact, it includes some of the most important (and widely heralded) advances in recent theoretical biology, such as the classic papers of Hamilton, Trivers, and Maynard Smith.

Hamilton could be said to have inaugurated the field with his introduction of the conceptual framework of kin selection, which solved, among other things, many of Darwin's puzzles about *eusociality* in insects—the way ants, bees, and termites live "selflessly" in large colonies, most of them sterile servants to a single fertile queen. But Hamilton's theory didn't solve all the problems, and among Richard Alexander's important contributions was his characterization of the conditions under which eusocial *mammals*

---

10. Philip Kitcher opens his critical survey of sociobiology, *Vaulting Ambition* (1985b), with the unanswerable tale of the damage done by the notorious British eleven-plus examination—now abandoned, thank goodness—which branded eleven-year-old children with an up-or-down verdict of their promise that quite inexorably fixed the subset of paths their lives might take.

might evolve—a "prediction" stunningly confirmed by the subsequent studies of the amazing South African naked mole rats (Sherman, Jarvis, and Alexander 1991). This was such an astonishing triumph of adaptationist reasoning that it deserves to be more widely known. As Karl Sigmund describes it, Hamilton's ideas

> led to a most remarkable discovery when, in 1976, the American biologist R. D. Alexander lectured on sterile castes. It was well known that these existed for ants, bees, and termites, but not for any kind of vertebrate. Alexander, in a kind of thought experiment, toyed with the notion of a mammal able to evolve a sterile caste. It would, like the termites, need an expandable nest allowing for an ample food supply and providing shelter from predators. For reasons of size, an underbark location [like that of the presumed insect ancestors of termites] was no good. But underground *burrows* replete with large tubers would fit the bill perfectly. The climate should be tropical; the soil (more than a hint of Sherlock Holmes here!) heavy clay. An ingenious exercise in armchair ecology altogether. But after his lecture, Alexander was told that his hypothetical beast did indeed live in Africa; it was the naked mole rat, a small rodent studied by Jennifer Jarvis. [Sigmund 1993, p. 117.]

Naked mole rats are surpassingly ugly and strange, a thought experiment of Mother Nature's to rival any of the fantasies of philosophy. They are genuinely eusocial. The single queen mole rat is the sole female breeder, and she keeps the rest of the colony in line by releasing pheromones that suppress the maturation of the other females' reproductive organs. Naked mole rats are coprophagous—they regularly eat their own feces—and when the grotesquely swollen pregnant queen cannot reach her own anus, she begs feces from her attendants. (Had enough? But there's much, much more, highly recommended to all whose curiosity exceeds their squeamishness.) A bounty has been learned from the study of naked mole rats, and other nonhuman species, using the techniques of Darwinian reverse engineering—using adaptationism, in other words—and there is surely more to come. E. O. Wilson's own important work on social insects (1971) is deservedly world famous, and there are literally hundreds of other fine animal sociobiologists. (See, e.g., the classic anthologies, Clutton-Brock and Harvey 1978, Barlow and Silverberg 1980, King's College Sociobiology Group 1982.) Unfortunately, they all work under a cloud of suspicion, raised by the escalation of greedy claims by a few human sociobiologists (through their megaphones, as Kitcher suggests), which is then echoed by the escalation of blanket condemnations from their opponents. This really is an unfortunate fallout, for, as in any other legitimate area of science, some of this work is great, some is good, some is good but false, and some is bad—but none of it is evil. That serious students of mating systems, courtship displays, territoriality, and the like in nonhuman species should be tarred with the same brush as the more flagrant oversteppers in human sociobiology is both a miscarriage of justice and a serious misrepresentation of science.

But neither "side" has done its duty. Unfortunately, the siege mentality has made the best of the sociobiologists somewhat reluctant to criticize the shoddy work of some of their colleagues. Though Maynard Smith, Williams, Hamilton, and Dawkins can often be found in print firmly setting aright various innocent flaws in arguments and pointing out complications—in short, making the corrections that are the normal topics of communication in all science—they have largely eschewed the deeply unpleasant task of pointing out more egregious sins in the work of those who enthusiastically misuse their own good work. Donald Symons (1992) is a bracing exception, however, and there are others. I will point to just one *major* source of bad thinking that is ubiquitous in human sociobiology, and is seldom carefully addressed by sociobiologists themselves, perhaps because Stephen Jay Gould has made the point in criticism, and they would hate to concede that he is right about anything. He is right about this point, and so is Philip Kitcher (1985b), who develops the criticism in much more detail. Here is Gould's version, which is a little hard to understand. (At first, I didn't see how to read it sympathetically, and had to ask Ronald Amundsen, an excellent philosopher of biology, to explain to me what Gould was getting at. He succeeded.)

> The standard foundation of Darwinian just-so stories does not apply to humans. That foundation is the implication: if adaptive, then genetic—for the inference of adaptation is usually the only basis of a genetic story, and Darwinism is a theory of genetic change and variation in population. [Gould 1980c, p. 259.]

What does this mean? Gould is not saying, as he may seem at first to be saying, that adaptationist inference does not apply to humans. He is saying that since in the case of humans (and only humans) there is always *another* possible source of the adaptation in question—namely culture—one cannot *so readily* infer that there has been genetic evolution for the trait in question. Even in the case of nonhuman animals, the inference from adaptation to genetic basis is risky when the adaptation in question is not an anatomical feature but a behavioral pattern which is an obviously Good Trick. For then there is another possible explanation: the general *nonstupidity* of the species. As we have seen so often, the more obvious the move, the less secure the inference that it has to have been copied from predecessors—specifically carried by the genes.

Many years ago, I played my first computer "video game" at the AI Lab at

MIT: it was called MazeWar, and more than one person could play it at once, each at a separate terminal linked to a central time-sharing computer. On the screen you saw a simple perspective line drawing of a maze, in which you, the viewer, were situated. Corridors could be seen up ahead leading off to left and right, and by pressing keys on the keyboard you could move forward and back, or turn ninety degrees to the left or the right. Another key on the keyboard was the trigger of your gun, which fired straight ahead. All the other players were in the same virtual maze, wandering around, looking for someone to shoot, and hoping not to be shot. If one of the other players crossed your path, he would show up as a simple cartoon figure, whom you would hope to shoot before it turned, saw you, and shot you. After a few minutes of frantic play, in which I was "shot" from behind several times, I found my mounting paranoia so uncomfortable that I sought relief: I found a cul-de-sac in the maze, backed myself into it, and just sat there, relatively calmly, with my finger on the trigger. It struck me then that I had adopted the policy of a moray eel, patiently waiting in its well-protected hole for something strike-worthy to swim by.

Now, does my behavior on this occasion give us any reason to suppose there is a genetic predisposition to moray-eel behavior in *Homo sapiens*? Did the stress of the occasion dredge up some ancient policy, lying dormant in my genes since the days when my ancestors were still fish? Of course not. The strategy is just too obvious. It felt like a forced move, but it was at least a Good Trick. We would not be surprised to find Martians backing them-selves self-protectively into Martian caves, and the likelihood that Martians had morays for ancestors would presumably not be adjusted upwards from zero by the discovery. It is true that I am distantly related to moray eels, but the fact that I found this strategy in this environment is surely just a matter of its obvious excellence, given my needs and desires and my own assess-ment of my limitations at the time. This illustrates the fundamental obsta-cle—not insuperable, but much larger than is commonly acknowledged—to inference in human sociobiology: showing that a particular type of human behavior is ubiquitous or nearly ubiquitous in widely separated human cultures goes *no way at all* towards showing that there is a genetic pre-disposition for that particular behavior. So far as I know, in every culture known to anthropologists, the hunters throw their spears pointy-end-first, but this obviously doesn't establish that there is a pointy-end-first gene that approaches fixation in our species.

Nonhuman species can exhibit a similar, if reduced, capacity to reinvent the wheel, even though they lack culture. Octopuses are remarkably intel-ligent, and although they show no signs of cultural transmission, they are smart enough so that we should not be surprised to discover them individ-ually hitting upon lots of Good Tricks that had never been posed as specific problems to their ancestors. Any such uniformity might be misread by biologists as signs of a special "instinct," when in fact it was just their general intelligence that led them again and again to hit upon the same bright idea. The problem of interpretation for *Homo sapiens* is multiplied many times over by the fact of cultural transmission. Even if some individual hunters are not bright enough to figure out for themselves that they should throw the pointy end first, they will be told to do so by their peers, or will just notice their practice, and will appreciate the results immediately. In other words, if you are not totally idiotic, you don't need a genetic basis for any adaptation that you will pick up from your friends in any case.

It is hard to believe that sociobiologists can make the mistake of ignoring this omnipresent possibility, but the evidence is striking that they have done so, again and again (Kitcher 1985). Many instances could be listed, but I will concentrate on a particularly visible and well-known case. Although E. O. Wilson (1978, p. 35) states clearly that the human behaviors to be accounted for by specific genetic hypotheses should be the "least rational of the human repertoire.... In other words, they should implicate innate, biological phenomena that are the least susceptible to mimicry by culture," he goes on (pp. 107ff.) to claim, for instance, that the evidence of *territoriality* in all human cultures (we human beings like to call a bit of space our own) is clear proof that we, like very many other species, have a genetic predisposition wired in at birth for the defense of territory. That may be true—in fact, it would not be at all surprising, since many species manifestly do exhibit innate territoriality, and it is hard to think of what force there might be to remove such a disposition from our genetic makeup. But the ubiquity of territoriality in human societies is *by itself* no evidence at all for this, since territoriality makes so much sense in so many human arrangements. It is, if not a forced move, close to it.

The very considerations that in other parts of the biosphere count *for* an explanation in terms of natural selection of an adaptation—manifest utility, obvious value, undeniable reasonableness of design—count *against* the *need* for any such explanation in the case of human behavior. If a trick is that good, then it will be routinely rediscovered by every culture, without need of either genetic descent or cultural transmission of the particulars.[11] We saw in chapter 12 that it is the prospect of convergent cultural evolution—reinventing the wheel—that plays havoc with our attempts to turn memetics into a science. The same difficulty besets all attempts to infer genetic

---

11. A useful exercise when considering any such case is to imagine creating a roomful of roughly rational robots (smart, but with no genetic ancestry at all) and asking yourself if they would soon settle into the behavior in question. (If the case is complex, a computer simulation should be used, as a prosthetic guide to your imagination.) If so, it is not so surprising that human beings everywhere do it, too, and it probably has nothing to do with their primate heritage, their mammalian heritage, even their vertebrate heritage.

factors from cultural commonalities, and for the same reason. But, although Wilson has sometimes noted this problem, at other times he forgets:

> The similarities between the early civilizations of Egypt, Mesopotamia, India, China, Mexico, and Central and South America in these major features are remarkably close. They cannot be explained away as the products of chance or cultural cross-fertilization. [Wilson 1978, p. 89.]

We need to look at each remarkable similarity in turn, to see if any of them *needs* a genetic explanation, for, in addition to cultural cross-fertilization (cultural descent) and chance, there is the possibility of reinvention. There *may* be specific genetic factors operating in many or all these similarities, but, as Darwin stressed, the best evidence will always be idiosyncrasies— quirky homologies—and no-longer-rational survivals. The most compelling cases of this sort are currently being uncovered in the marriage of sociobiology and cognitive psychology recently going by the name of evolutionary psychology (Barkow, Cosmides, and Tooby 1992). Highlighting a single case will provide a useful contrast between good and bad uses of Darwinian thinking in the study of human nature, and clarify the position on rationality (or just nonstupidity) just presented.

How logical are we human beings? In some regards very logical, it seems, and in others embarrassingly weak. In 1969, the psychologist Peter Wason devised a simple test that bright people—college students, for instance—do rather badly on. You may try it yourself. Here are four cards, some letter-side-up, and some number-side-up. Each card has a numeral on one side and a letter on the other:

Your task is to see whether in this case the following rule has any exceptions: *If a card has a "D" on one side, it has a "3" on the other side.* Now, which cards do you need to turn over in order to discover if this is true? Sad to say, fewer than half of students in most such experiments get the right answer. Did you? The correct answer is much more obvious if we shift the content (but not the structure) of the problem very slightly. You are the bouncer in a bar, and your job depends on not letting any underage (under twenty-one) customers drink beer. The cards have information about age on one side, and what the patron is drinking on the other. Which cards do you need to turn over?

| drinking beer | drinking Coke | 25 years old | 16 years old |

The first and the last, obviously, the same as in the first problem. Why is one setting so much easier than the other? Perhaps, you may think, it is the abstractness of the first, the concreteness of the second, or the familiarity of the second, or the fact that the second involves a conventional rule, not a regularity of nature. Literally hundreds of Wason card-sorting tests have been administered to subjects, in hundreds of variations, testing these and other hypotheses. The performance of subjects on the tests varies widely, depending on the details of the particular test and its circumstances, but a survey of the results leaves no doubt at all that there are settings that are hard for almost all groups of subjects, and others that are easy *for the same subjects*. But a riddle remained, reminiscent of the riddle of the two black boxes: what exactly was it about the hard cases that made them hard—or (a better question) what was it about the easy cases that made them easy? Cosmides and Tooby (e.g., Barkow, Cosmides, and Tooby 1992, ch. 2) came up with an evolutionary hypothesis, and it is hard to imagine this particular idea occurring to anyone who wasn't acutely aware of the possibilities of Darwinian thinking: the easy cases are all cases that are readily interpreted as tasks of patrolling a social contract, or, in other words, cheater detection.

Cosmides and Tooby seem to have uncovered a fossil of our Nietzschean past! Framing the hypothesis is not yet proving it, of course, but one of the important virtues of their hypothesis is that it is eminently testable, and has so far stood up very well to a wide variety of attempts to refute it. Suppose it is true; would it show that we can reason only about the things Mother Nature wired us to reason about? Obviously not; it just shows why it is easier (more "natural") for us to reason about some topics than others. We have devised cultural artifacts (systems of formal logic, statistics, decision theory, and so forth, taught in college courses) that expand our reasoning powers many fold. Even the experts often neglect these specialized techniques, however, and fall back on good old seat-of-the-pants reasoning, sometimes with embarrassing results, as the Wason test shows. Independently of any Darwinian hypotheses, we know that, except when people are particularly self-conscious about using these heavy-duty reasoning techniques, they tend to fall into cognitive illusions. Why are we susceptible to *these* illusions? The evolutionary psychologist says: For the same reason we are susceptible to optical illusions and other sensory illusions—we're built that way. Mother Nature designed us to solve a certain set of problems posed by the environments in which we evolved, and whenever a cut-rate solution emerged—a bargain that would solve the most pressing problems pretty well, even if it lacked generality—it tended to get installed.

Cosmides and Tooby call these modules "Darwinian algorithms"; they are mechanisms just like the two-bitser, only fancier. We obviously don't get by with just one such reasoning mechanism. Cosmides and Tooby have been gathering evidence for other special-purpose algorithms, useful in thinking about threats and other social exchanges, and other ubiquitous problem-types: hazards, rigid objects, and contagion. Instead of having a single, central general-purpose reasoning machine, we have a collection of gadgets, all pretty good (or at least pretty good in the environments in which they evolved), and readily exaptable for new purposes today. Our minds are like Swiss-army knives, Cosmides says. Every now and then, we discover curious gaps in our competence, strange lapses that give us clues about the particular history of R and D that explains the machinery that underlies the glittering façade of culture. This is surely the right way for psychologists to reverse-engineer the human mind, always watching out for QWERTY phenomena.

I consider Cosmides and Tooby to be doing some of the best work in Darwinian psychology today, which is why I chose them for my example, but I must temper my recommendation with some constructive criticism. The ferocity of the attacks they have encountered from the fans of Gould and Chomsky is breathtaking, and, embattled as they are, they, too, tend to caricature the opposition, and are sometimes too hasty in dismissing skepticism about their arguments as flowing from nothing more presentable than the defensive territoriality of old-fashioned social scientists who still haven't got the word about evolution. This is often, but not always, the case. Even if they are right—and I am confident that they are—that such rationality as we human beings have is the product of the activities of a host of special-purpose gadgets designed by natural selection, it does not follow that this "Swiss-army knife" of ours cannot have been used, time and time again, to reinvent the wheel. It still has to be shown, in other words, that any particular adaptation is *not* a cultural product responding quite directly (and rationally) to quite recent conditions. They know this, and they carefully avoid the trap we have just seen E. O. Wilson fall into, but in the heat of battle they sometimes forget.

Like Darwin overlooking the innocuous possibility of sudden extinctions because he was so intent on fleeing from Catastrophism, Tooby and Cosmides, and the other evolutionary psychologists, tend to overlook the bland possibility of the independent rediscovery of forced moves, so intent are they on replacing the "Standard Social Science Model" with a properly Darwinian model of the mind. The Standard Social Science Model has among its precepts:

Whereas animals are rigidly controlled by their biology, human behavior is determined by culture, an autonomous system of symbols and values. Free from biological constraints, cultures can vary from one another arbitrarily and without limit.... Learning is a general-purpose process, used in all domains of knowledge. [Pinker, 1994, p. 406; see also Tooby and Cosmides 1992, pp. 24–48.]

This, of course, is wrong, wrong, wrong. But compare it with my Only Slightly *Non*standard Social Science Model:

> Whereas animals are rigidly controlled by their biology, human behavior is *largely* determined by culture, a *largely* autonomous system of symbols and values, growing from a biological base, but growing indefinitely away from it. *Able to overpower or escape* biological constraints in most regards, cultures can vary from one another enough so that important portions of the variance are thereby explained.... Learning is *not* a general-purpose process, but human beings have so many special-purpose gadgets, and learn to harness them with such versatility, that learning *often* can be treated as if it were an entirely medium-neutral and content-neutral gift of non-stupidity.

This is the model I have argued for in this book; it is no defense of skyhooks; it simply acknowledges that we now have cranes of more general power than the cranes of any other species.[12]

There is plenty of good work in sociobiology and evolutionary psychology, and there is plenty of bad work, as in any field. Is any of it evil? Some of it is at least dismayingly heedless of the misuses to which it might be put by ideologues of one persuasion or another. But, here again, the escalation of charges typically produces more heat than light. One instance can stand in for a survey of the whole sorry field of battle. Do ducks rape? Sociobiologists have uncovered a common pattern in which males in some species—such as ducks—violently mate with obviously unwilling females. They have called it rape, and this terminology has been decried by critics, most vigorously by the feminist biologist Anne Fausto-Sterling (1985).

She has a point. I said we wouldn't call the sibling-cide that many species engage in "murder," since they know not what they do. They kill, but do not murder, each other. It is impossible for one bird to *murder* another bird—"murder" is reserved for the intentional, deliberate, wrongful killing of one human being by another. (You can kill a bear, but not murder it, and if it kills you, that isn't murder either.) Now, can one duck *rape* another? Fausto-

---

12. Even Donald Symons (1992, p. 142) slips slightly, succumbing to a luscious slogan: "There is no such thing as a 'general problem solver' because there is no such thing as a general problem." Oh? There is no such thing as a general wound either; each wound has a quite specific shape, but there can still be a general wound-healer, capable of healing wounds of an almost limitless variety of shapes—simply because it is cheaper for Mother Nature to make a (quite) general wound-healer than a specialist wound-healer (G. Williams 1966, pp. 86–87; see also Sober 1981b, pp. 106 ff.). How general any cognitive mechanism is, or can be made to be through cultural enhancement, is always an open empirical question.

Sterling, and other feminists, say No—this is to misapply a term that also properly applies only to human misdeeds. If there were a common term in English that stood to "rape" as "kill" (or "homicide" or "manslaughter") stood to "murder," then the use by sociobiologists of the term "rape" for nonhuman forced copulation, instead of using the less loaded term, would be truly outrageous. But there isn't any such term.

So is the use of the short, vivid term "rape" in place of "forced copulation" (or other such term) a serious sin? It is at least insensitive. But do the critics complain about the other terms drawn from human life in common use by sociobiologists? There is sexual "cannibalism" in spiders (the females wait till the males have finished impregnating them, and then kill and eat them), there are "lesbian" gulls (female couples that stay paired over several seasons, defending territory, building a nest, sharing the task of sitting on the eggs). There are "homosexual" worms and bird "cuckolds." At least one critic, Jane Lancaster (1975), does in fact object to the word "harem" used to refer to the group of females guarded and mated by a single male—such as an elephant seal; she recommends the term "one-male group," since these females "are virtually self-sufficient, except for fertilization" (Fausto-Sterling 1985, p. 181n.). It seems to me that deliberate human cannibalism is much, much more terrible than anything one spider could do to another, but I for one don't object if an arachnologist wants to use the term. For that matter, what about the benign terms (G. Williams 1988)? Do the critics also object to "courtship ritual" and "alarm call"—or the use of the term "mother" to refer to a female but nonhuman parent?

Fausto-Sterling does note that the sociobiologists she criticizes for using the term "rape" were careful to assert that human rape was different from rape in other species. She quotes (p. 193) from Shields and Shields 1983:

> Ultimately men may rape because it increases their biological fitness and thus rape may serve, at least in part, a reproductive function, but in an immediate proximate sense it is as likely that they rape because they are angry or hostile, as the feminists suggest.

This passage is not the ringing denunciation of rape that Fausto-Sterling requires—something that one might think would go without saying in the context of a scientific article—but it does firmly dissociate human rape from any biological "justification." That makes Fausto-Sterling's further charge outrageous. She places responsibility on these sociobiologists for various claims made by defense lawyers in rape cases who have got their clients off relatively easily by noting their "unbearable physical urges" or by describing a client's act: "as rapes go, a relatively mild rape." What do these claims have to do with sociobiology in general, or the articles she discusses in particular? She offers no reason at all to believe these lawyers cited the sociobiologists as supporting authorities, or so much as knew of their existence. With equal justice she could blame the tribe of Shakespeare scholars for such miscarriages of justice (supposing that such they were), for these scholars have no doubt been insufficiently condemnatory in their writings over the years about Shakespeare's sometimes tolerant portrayal of rape in his plays. This is surely not the way to foster enlightened consideration of the issues. Tempers run high, and the issues are deadly serious, which is all the more reason for scientists and philosophers to be careful not to abuse either the truth or each other in the name of a worthy cause.

What, then, would a more positive approach to a "naturalized" ethics look like? I have a few preliminary suggestions to offer in the next chapter.

CHAPTER 16: *As Darwinian thinking gets closer and closer to home—where we live—tempers run higher, and the rhetoric tends to swamp the analysis. But sociobiologists, beginning with Hobbes and continuing through Nietzsche to the present day, have seen that only an evolutionary analysis of the origins—and transformations—of ethical norms could ever properly make sense of them. Greedy reductionists have taken their usual first stumbling steps into this new territory, and been duly chastened by the defenders of complexity. We can learn from these errors without turning our backs on them.*

CHAPTER 17: *What are the implications for ethics of the fact that we are finite, time-pressured, heuristic searchers for ethical truths? An examination of the persistent pendulum swing between utilitarian and Kantian ethics suggests some principles for redesigning ethics along more realistic, Darwinian lines.*

CHAPTER SEVENTEEN

Redesigning Morality

〜〜

1. CAN ETHICS BE NATURALIZED?[1]

Thus at last man comes to feel, through acquired and perhaps inherited habit, that it is best for him to obey his more persistent impulses. The imperious word ought *seems merely to imply the consciousness of the existence of a rule of conduct, however it may have originated.*

—CHARLES DARWIN, *Descent of Man* (2nd ed., 1874), p. 486

Human culture, religion in particular, is a repository of ethical precepts, ranging from the Golden Rule, the Ten Commandments, and the Greeks' "Know Thyself" to all manner of specific commands and prohibitions, taboos, and rituals. Philosophers since Plato have attempted to organize these imperatives into a single rationally defensible and universal system of ethics, so far without achieving anything approaching consensus. Mathematics and physics are the same for everyone everywhere, but ethics has not yet settled into a similar reflective equilibrium.[2] Why not? Is the goal illusory? Is mo-

---

1. Material in this chapter is drawn from Dennett 1988b, where the issues are developed in more detail.

2. It is worth bearing in mind that mathematics and physics are the same throughout the entire universe, discoverable in principle by aliens (if such there be) no matter what their social class, political predilections, gender (if they have genders!), or peccadilloes. I mention this to ward off the recent nonsense you may have heard emanating from some schools of thought—I speak loosely—in the sociology of science. It is dismaying to read such a wise thinker as John Patrick Diggins falling under its spell:

But, as Mr. Marsden notes, in the past it was assumed that science would be the arbitrator of such disputes, whereas today science is dismissed as simply another way

rality just a matter of subjective taste (and political power)? Are there no discoverable and confirmable ethical truths, no forced moves or Good Tricks? Great edifices of ethical theory have been constructed, criticized and defended, revised and extended by the best methods of rational inquiry, and among these artifacts of human reasoning are some of the most magnificent creations of culture, but they do not yet command the untroubled assent of all those who have studied them carefully.

Perhaps we can get some clues about the status and prospects for ethical theory by reflecting on what we have seen to be the limitations of the great design process that has ethicists among its products to date. What follows, we may ask, from the fact that ethical decision-making, like all *actual* processes of exploration in Design Space, must be to some degree myopic and time-pressured?

Shortly after the publication of Darwin's *Origin of Species*, another eminent Victorian, John Stuart Mill, published his attempt at a universal ethical theory, *Utilitarianism* (1861). Darwin read it with interest, and responded to the "celebrated work" in his *Descent of Man* (1871). Darwin was puzzled by Mill's stand on whether the moral sentiment was innate or acquired, and sought the help of his son William, who advised his father that Mill was "rather in a muddle on the whole subject" (R. Richards 1987, p. 209n.), but, aside from a few such points of disharmony, Darwin and Mill were (correctly) seen as united in their naturalism—and duly excoriated together by the defenders of skyhooks, most notably St. George Mivart, who declared:

> ...men have a consciousness of an absolute and immutable rule *legitimately* claiming obedience with an authority necessarily supreme and absolute—in other words, intellectual judgments are formed which imply the existence of an ethical ideal in the judging mind. [Mivart 1871, p. 79.]

To such bluster there is probably no better response than Darwin's, quoted at the head of this section. But there were more measured criticisms as well, and one of the more frequent stuck in the craw of Mill: "Defenders of utility often find themselves called upon to reply to such objections as

---

of describing the world verbally rather than knowing it philosophically. In the recent past, religion had been driven from the campus because it lacked scientific credentials. But since that criterion has itself lost its own credentials, Mr. Marsden wonders why religion cannot reclaim its place on the campus. He is right to raise such questions. [Diggins 1994.]

It is not "scientism" to concede the objectivity and precision of good science, any more than it is history worship to concede that Napoleon did once rule in France and the Holocaust actually happened. Those who fear the facts will forever try to discredit the fact-finders.

this—that there is not time, previous to action, for calculating and weighing the effects of any line of conduct on the general happiness." His reaction was quite fierce:

> Men really ought to leave off talking a kind of nonsense on this subject, which they would neither talk nor listen to on other matters of practical concernment. Nobody argues that the art of navigation is not founded on astronomy because sailors cannot wait to calculate the Nautical Almanac. Being rational creatures, they go to sea with it ready calculated; and all rational creatures go out upon the sea of life with their minds made up on the common questions of right and wrong, as well as on many of the far more difficult questions of wise and foolish. And this, as long as foresight is a human quality, it is to be presumed they will continue to do. [Mill 1861, p. 31.]

This haughty retort has found favor with many—perhaps most—ethical theorists, but in fact it papers over a crack that has been gradually widening under an onslaught of critical attention. The objectors were under the curious misapprehension that a system of ethical thinking *was supposed to work*, and noted that Mill's system was highly impractical—at best. This was no objection, Mill insisted: utilitarianism is supposed to be practical, but not *that* practical. Its true role is as a background justifier of the foreground habits of thought of real moral reasoners. This background role for ethical theory (and not only utilitarians have sought it) has proven, however, to be ill-defined and unstable. Just how practical is a system of ethical thinking supposed to be? What is an ethical theory for? Tacit differences of opinion about this issue, and even a measure of false consciousness among the protagonists, have added to the inconclusiveness of the subsequent debate.

For the most part, philosophers have been content to ignore the practical problems of real-time decision-making, regarding the brute fact that we are all finite and forgetful, and have to rush to judgment, as a real but irrelevant element of friction in the machinery whose blueprint they are describing. It is as if there might be two disciplines—ethics proper, which undertakes the task of calculating the principles determining what the ideal agent ought to do under all circumstances—and then the less interesting, "merely practical" discipline of Moral First Aid, or What to Do Until the Doctor of Philosophy Arrives, which tells, in rough and ready terms, how to make "online" decisions under time pressure.

In practice, philosophers acknowledge, we overlook important considerations—considerations that we really shouldn't overlook—and we bias our thinking in a hundred idiosyncratic—and morally indefensible—ways; but *in principle*, what we ought to do is what the ideal theory (one ideal theory or another) says we ought to do. Philosophers have then concentrated, not unwisely, on spelling out what that ideal theory is. The theoretical fruits of deliberate oversimplification through idealization are not to be denied, in philosophy or in any scientific discipline. Reality in all its messy particularity is too complicated to theorize about, taken straight. The issue is, rather (since every idealization is a strategic choice), which idealizations might really shed some light on the nature of morality, and which will just land us with diverting fairy tales.

It is easy to forget just how impractical ethical theories actually are, but we can make the truth vivid by reflecting on what is implicit in Mill's use of a metaphor drawn from the technology of his own day. The *Nautical Almanac* is an ephemeris of sorts, a book of tables, calculated and published annually, from which one can easily and swiftly derive the exact position in the skies of the sun, the moon, the planets, and the major stars for *each second* of the forthcoming year. The precision and certainty of this annual generator of expectations was, and still is, an inspiring instance of the powers of human foresight, properly disciplined by a scientific system *and directed upon a sufficiently orderly topic*. Armed with the fruits of such a system of thought, the rational sailor can indeed venture forth confident of his ability to make properly informed real-time decisions about navigation. The practical methods devised by the astronomers actually work.

Do the utilitarians have a similar product to offer to the general public? Mill seems at first to be saying so. Today we are inured to the inflated claims made on behalf of dozens of high-tech systems—of cost-benefit analysis, computer-based expert systems, etc.—and from today's perspective we might suppose Mill to be engaging in an inspired bit of advertising: suggesting that utilitarianism can provide the moral agent with a foolproof Decision-making Aid. ("We have done the difficult calculations for you! All you need do is just fill in the blanks in the simple formulae provided.")

Jeremy Bentham, the founder of utilitarianism, certainly aspired to just such a "felicific calculus," complete with mnemonic jingles, just like the systems of practical celestial navigation that every sea captain memorized.

Intense, long, certain, speedy, fruitful, pure—
Such marks in *pleasures* and in *pains* endure.
Such pleasures seek if *private* be thy end:
If it be *public*, wide let them *extend*.
                    [Bentham 1789, ch. IV.]

Bentham was a cheerfully greedy reductionist—the B. F. Skinner of his day, you might say—and this myth of practicality has been part of the rhetoric of utilitarianism from the beginning. But in Mill we see already the beginning of the retreat up the ivory tower to ideality, to what is calculable "in principle" but not in practice.

Mill's idea, for instance, was that the best of the homilies and rules of thumb of everyday morality—the formulae people *actually considered* in the hectic course of their deliberations— had received (or would receive in principle) official endorsement from the full, laborious, systematic utilitarian method. The faith placed in these formulae by the average rational agent, based as it was on many lifetimes of experience accumulated in cultural memory, could be justified ("in principle") by being formally derived from the theory. But no such derivation has ever been achieved.[3]

The reason is not hard to see: it is unlikely in the extreme that there could be a *feasible algorithm* for the sort of global cost-benefit analysis that utilitarianism (or any other "consequentialist" theory) requires. Why? Because of what we might call the Three Mile Island Effect. Was the meltdown at the nuclear plant at Three Mile Island a good thing to have happened or a bad thing? If, in planning some course of action, you encountered the meltdown as a sequel of probability p, what should you assign to it as a weight? Is it a negative outcome that you should strive to avoid, or a positive outcome to be carefully fostered?[4] We can't yet say, and it is not clear that *any* particular long run would give us the answer. (Notice that this is not a problem of insufficiently *precise* measurement; we can't even determine the *sign*, positive or negative, of the value to assign to the outcome.)

Compare the problem facing us here with the problems confronting the designers of computer chess programs. One might suppose that the way to respond to the problem of real-time pressure for ethical decision-making techniques is the way one responds to time pressure in chess: heuristic search-pruning techniques. But there is no checkmate in life, no point at which we get a definitive result, positive or negative, from which we can calculate, by retrograde analysis, the actual values of the alternatives that lay along the path taken. How deep should one look before settling on a weight for a position? In chess, what looks positive from ply 5 may look disastrous from ply 7. There are ways of tuning one's heuristic search procedures to

---

3. Probably the closest anybody has come to a "result" in this field is Axelrod's (1984) derivation of Tit for Tat, but, as he himself points out, the rule's provable virtues assume conditions that are only intermittently—and controversially—realized. In particular, the "shadow of the future" must be "sufficiently great," a condition about which reasonable people might disagree indefinitely, it seems.

4. How could Three Mile Island have been a good thing? By being the *near*-catastrophe that sounded the alarm that led us away from paths that would encounter much worse misadventures—Chernobyls, for instance. Surely many people were fervently *hoping* for just such an event to happen, and might well have taken steps to ensure it, had they been in a position to act. The same moral reasoning that led Jane Fonda to create the film *The China Syndrome* (a fictional near-catastrophe at a nuclear plant) might lead someone rather differently situated to create Three Mile Island.

minimize (but not definitively) the problem of misevaluating anticipated moves. Is the anticipated capture a strongly positive future to be aimed at, or the beginning of a brilliant sacrifice for your opponent? A *principle of quiescence* will help to resolve that issue: always look a few moves beyond any flurry of exchanges to see what the board looks like when it quiets down. But in real life, there is no counterpart principle that deserves reliance. Three Mile Island has been followed by more than a decade of consolidation and quiescence (it happened in 1980), but we *still* have no idea whether it is to be counted among the good things that have happened or the bad, all things considered.

The suspicion that there is no stable and persuasive resolution to such impasses has long lain beneath the troubled surface of criticism to consequentialism, which looks to many skeptics like a thinly veiled version of the vacuous stock-market advice "Buy low and sell high"—a great idea in principle, but systematically useless as advice to follow.[5]

So not only have utilitarians never made an actual practice of determining their specific moral choices by calculating the expected utilities of (all) the alternatives (there not being time, as our original objector noted), but they have never achieved stable "off-line" *derivations* of partial results—"landmarks and direction posts," as Mill puts it—to be exploited on the fly by those who must cope with "matters of practical concernment."

What, then, of the utilitarians' chief rivals, the various sorts of Kantians? Their rhetoric has likewise paid tribute to practicality—largely via their indictments of the *im*practicality of the utilitarians.[6] What, though, do the

---

5. Judith Jarvis Thomson has objected (in a commentary on "The Moral First Aid Manual" in Ann Arbor, November 8, 1986) that neither "Buy low and sell high" nor its consequentialist counterpart, "Do more good than harm" is strictly vacuous; both presuppose something about ultimate goals, since the former would be bad advice to one who sought to lose money, and the latter would not appeal to the ultimate interests of all morally minded folk. I agree. The latter competes, for instance, with the advice the Pirate King gives to Frederick, the self-styled "slave of duty" in *Pirates of Penzance*: "Aye me lad, always do your duty—and chance the consequences!" Neither slogan is *quite* vacuous.

6. A Kantian who presses the charge of practical imponderability against utilitarianism with particular vigor and clarity is Onora O'Neill (1980). She shows how two utilitarians, Garrett Hardin and Peter Singer, armed with the same information, arrive at opposite counsels on the pressing moral dilemma of famine relief: we should take drastic steps to prevent shortsighted efforts to feed famine victims (Hardin), or we should take drastic steps to provide food for today's famine victims (Singer). For a more detailed consideration, see O'Neill 1986. An independent critic is Bernard Williams, who claims (1973, p. 137) that utilitarianism makes

enormous demands on supposed empirical information, about peoples' preferences, and that information is not only largely unavailable, but shrouded in conceptual difficulty; but that is seen in the light of a technical or practical difficulty, and utilitari-

Kantians put in the place of the unworkable consequentialist calculations? Maxim-following (often derided as rule worship) of one sort or another, such as that invoked in one of Kant's (1785) formulations of the Categorical Imperative: Act only on that maxim through which you can at the same time will that it should become a universal law. Kantian decision-making typically reveals rather different idealizations—departures from reality in other directions— doing all the work. For instance, unless some *deus ex machina* is standing by, a handy master of ceremonies to whisper suggestions in your ear, it is far from clear just how you are supposed to figure out how to limit the scope of the "maxims" of your contemplated actions before putting them to the litmus test of the Categorical Imperative. There seems to be an inexhaustible supply of candidate maxims.

Certainly the quaint Benthamite hope of a fill-in-the-blanks decision procedure for ethical problems is as foreign to the spirit of modern Kantians as it is to sophisticated utilitarians. All philosophers can agree, it seems, that real moral thinking takes insight and imagination, and is not to be achieved by any mindless application of formulae. As Mill himself puts it (1871, p. 31), still in high dudgeon, "There is no difficulty in proving any ethical standard whatever to work ill if we suppose universal idiocy conjoined with it." This bit of rhetoric is somewhat at war with his earlier analogy, however, since one of the legitimate claims of the systems of practical navigation was that just about any idiot could master them.

I do not at all intend this to be a shocking indictment, just a reminder of something quite obvious: no remotely compelling system of ethics has ever been made *computationally tractable*, even indirectly, for real-world moral problems. So, even though there has been no dearth of utilitarian (and Kantian, and contractarian, etc.) *arguments* in favor of particular policies, institutions, practices, and acts, these have all been heavily hedged with *ceteris paribus* clauses and plausibility claims about their idealizing assumptions. These hedges are designed to overcome the combinatorial explosion of calculation that threatens if one actually attempts—as theory says one must—to *consider all things*. And as arguments—not derivations—they have all been controversial (which is not to say that none of them could be sound in the last analysis).

To get a better sense of the difficulties that contribute to *actual* moral reasoning, let us give ourselves a smallish moral problem and see what we do with it. Though a few of its details are exotic, the problem I am setting exemplifies a familiar structure.

---

anism appeals to a frame of mind in which technical difficulty, even insuperable technical difficulty, is preferable to moral unclarity, no doubt because it is less alarming. (That frame of mind is in fact deeply foolish. . . .)

2. JUDGING THE COMPETITION

Your Philosophy Department has been chosen to administer a munificent bequest: a twelve-year fellowship to be awarded in open competition to the most promising graduate student in philosophy in the country. You duly announce the award and its conditions in the *Journal of Philosophy*, and then, to your dismay, you receive, by the deadline, 250,000 legal entries, complete with lengthy dossiers, samples of written work, and testimonials. A quick calculation convinces you that living up to your obligation to evaluate all the material of all the candidates by the deadline for announcing the award would not only prevent the department from performing its primary teaching mission, but— given the costs of administration and hiring additional qualified evaluators—bankrupt the award fund itself, so that all the labor of evaluation would be wasted; no one would gain.

What to do? If only you had anticipated the demand, you could have imposed tighter eligibility conditions, but it is too late for that: every one of the 250,000 candidates has, we will suppose, a right to equal consideration, and in agreeing to administer the competition you have undertaken the obligation to select the best candidate. (I don't mean to beg any questions with this formulation in terms of rights and obligations. If it makes a difference to you, recast the setting of the problem in terms of the overall disutility of violating the conditions set forth in your announcement of the competition. My point is that you would find yourself in a bind, whatever your ethical persuasion.) Before reading on, please spend a little time, as much as you think it takes, to plot your own solution to the problem (no fantasies about technological fixes, please).

When I have put this problem to colleagues and students, I find that, after a brief exploratory period, they tend to home in on one version or another of a mixed strategy, such as:

(1) choose a small number of easily checked and not entirely unsymptomatic criteria of excellence—such as Grade Point Average, number of philosophy courses completed, weight of the dossier (eliminating the too-light and the too-heavy)—and use this to make a first cut;

(2) conduct a lottery with the remaining candidates, cutting the pool down randomly to some manageably small number of finalists—say fifty or a hundred—

(3) whose dossiers will be carefully screened by a committee, which will then vote on the winner.

There is no doubt that this procedure is very unlikely to find the best candidate. Odds are, in fact, that more than a few of the losers, if given a day in court, could convince a jury that they were obviously superior to the elected winner. But, you might want to retort, that's just tough; you did the best you could. It is quite possible, of course, that you would lose the lawsuit, but you might still feel, rightly, that you could have arrived at no better decisions at the time.

My example is meant to illustrate, enlarged and in slow motion, the ubiquitous features of real-time decision-making. First, there is the simple physical impossibility of "considering all things" in the allotted time. Note that "all things" doesn't have to mean *everything* or even *everybody in the world*, but just *everything in 250,000 readily available dossiers.* You have all the information you need "at your fingertips"; there need be no talk of conducting further investigations. Second, there is the ruthless and peremptory use of some distinctly second-rate cut rules. No one thinks Grade Point Average is a remotely foolproof indicator of promise, though it is probably somewhat superior to *weight of dossier*, and clearly superior to *number of letters in surname*. There is something of a trade-off between ease of application and reliability, and if no one can *quickly* think of any easily applied criteria that one can have *some* faith in, it would be better to eliminate step (1) and proceed straight to the lottery for all candidates. Third, the lottery illustrates a partial abdication of control, giving up on a part of the task and letting something else—nature or chance—take over for a while, yet still assuming responsibility for the result. (That is the scary part.) Fourth, there is the phase where you try to salvage something presentable from the output of that wild process; having *over*simplified your task, you count on a meta-level process of self-monitoring to correct or renormalize or improve your final product to some degree. Fifth, there is the endless vulnerability to second-guessing and hindsight wisdom about what you should have done—but done is done. You let the result stand, and go on to other things. Life is short.

The decision process just described is an instance of the fundamental pattern first explicitly analyzed by Herbert Simon (1957, 1959), who named it "satisficing." Notice how the pattern repeats itself, rather like a fractal curve, as we trace down through the subdecisions, the sub-subdecisions, and so forth until the process becomes invisible. At the department meeting called to consider how to deal with this dilemma, (a) everyone is bursting with suggestions—more than can be sensibly discussed in the two hours allotted, so (b) the chairman becomes somewhat peremptory, deciding not to recognize several members who might well, of course, have some very good ideas, and then, (c) after a brief free-for-all "discussion" in which—for all anyone can tell—timing, volume, and timbre may count for more than content, (d) the chairman attempts to summarize by picking a few highlights that somehow strike him as the operative points, and the strengths and weaknesses of these are debated in a rather more orderly way, and then a vote is taken. After the meeting, (e) there are those who still think that better cut rules could have been chosen, that the department could have afforded the time to evaluate two hundred finalists (or should have restricted the number to twenty), etc., but done is done. They have learned the important lesson of how to live with the suboptimal decision-making of their colleagues, so, after a few minutes or hours of luxuriating in clever hindsight, they drop it.

"But *should I* drop it?" you ask yourself, just as you asked yourself the same question in the midst of the free-for-all when the chairman wouldn't call on you. Your head was teeming at that moment (a) with reasons why you should insist on being heard, competing with reasons why you should go along with your colleagues quietly, and all this was competing with your attempts to follow what others were saying, and so forth—more information at your fingertips than you could handle, so (b) you swiftly, arbitrarily, and unthinkingly blocked off some of it—running the risk of ignoring the most important considerations—and then (c) you gave up trying to *control* your thoughts; you relinquished meta-control and let your thoughts lead wherever they might for a while. After a bit, you somehow (d) resumed control, attempted some ordering and improving of the materials spewed up by the free-for-all, and made the decision to drop it—suffering (e) instant pangs of dubiety and toying with regret, but, because you are wise, you shrugged these off as well.

And how, precisely, did you go about dismissing that evanescent and unarticulated micro wonder ("Should I have dropped it?")? Here the processes become invisible to the naked eye of introspection, but if we look at cognitive-science models of "decision-making" and "problem-solving" *within* such swift, unconscious processes as perception and language comprehension, we see further tempting analogues of our phases in the various models of heuristic search and problem-solving.[7]

As we have seen again and again in this book, time-pressured decision-making is like that *all the way down*. Satisficing extends even back behind the fixed biological design of the decision-making agent, to the design "decisions" that Mother Nature settled for when designing us and other organisms. There may be somewhat nonarbitrary dividing lines to be drawn between biological, psychological, and cultural manifestations of this structure, but not only are the structures—and their powers and vulnerabilities—basically

---

7. The suggestion of temporal ordering in the five phases is not essential, of course. The arbitrary pruning of randomly explored search trees, the triggering of decision by a partial and nonoptimal evaluation of results, and the suppression of second-guessing need not follow the sequence in time I outline in the initial example. The process at this level is what I have described in the Multiple Drafts Model of human consciousness in Dennett 1991a.

the same; the particular contents of "deliberation" are probably not locked into any one level in the overall process but can migrate. Under suitable provocation, for instance, one can dredge up some virtually subliminal consideration and elevate it for self-conscious formulation and appreciation—it becomes an "intuition"—and then express it so that others can consider it as well. Moving in the other direction, a reason for action perennially mentioned and debated in committee can eventually "go without saying"—at least out loud—but continue to shape the thinking, both of the group and the individuals, from some more subliminal base (or bases) of operations in the process. As Donald Campbell (1975) and Richard Dawkins (1976, ch. 11) have argued, cultural institutions can sometimes be interpreted as compensations or corrections of the "decisions" made by natural selection.

The fundamentality of satisficing—the fact that it is the *basic* structure of all real decision-making, moral, prudential, economic, or even evolutionary—gives birth to a familiar and troubling slipperiness of claim that bedevils theory in several quarters. To begin with, notice that merely claiming that this structure is basic is not necessarily saying that it is best, but that conclusion is certainly invited—and inviting. We began this exploration, remember, by looking at a moral *problem* and trying to *solve* it: the problem of designing a *good* (justified, defensible, sound) candidate-evaluation process. Suppose we decide that the system we designed is about as good as it could be, given the constraints. A group of roughly rational agents—us—decide that this is the right way to design the process, and we have reasons for choosing the features we did.

Given this genealogy, we might muster the chutzpah to declare that this is optimal design—the best of all possible designs. This apparent arrogance might have been imputed to me as soon as I set the problem, for did I not propose to examine how *anyone ought* to make moral decisions by examining how *we in fact* make a particular moral decision? Who are we to set the pace? Well, who else should we trust? If we can't rely on our own good judgment, it seems we can't get started:

> Thus, what and how we do think is evidence for the principles of rationality, what and how we ought to think. This itself is a methodological principle of rationality; call it the *Factunorm Principle*. We are (implicitly) accepting the Factunorm Principle whenever we try to determine what or how we ought to think. For we must, in that very attempt, think. And unless we can think that what and how we do think there is correct— and thus is evidence for what and how we ought to think—we cannot determine what or how we ought to think. [Wertheimer 1974, pp. 110-11; see also Goodman 1965, p. 63.]

Optimality claims have a way of evaporating, however; it takes no chutzpah at all to make the modest admission that this was the best solution *we* could come up with, given our limitations. The mistake that is sometimes made is to suppose that there is or must be a single (best or highest) perspective from which to assess ideal rationality. Does the ideally rational agent have the all-too-human problem of not being able to remember certain crucial considerations when they would be most telling, most effective in resolving a quandary? If we stipulate, as a theoretical simplification, that our imagined ideal agent is immune to such disorders, then we don't get to ask the question of what the ideal way might be to cope with them.

Any such exercise presupposes that certain features—the "limitations"—are fixed, and other features are malleable; the latter are to be adjusted so as best to accommodate the former. But one can always change the perspective and ask about one of the presumably malleable features whether it is not, in fact, fixed in one position—a constraint to be accommodated. And one can ask about each of the fixed features whether it is something one would want to tamper with in any event; perhaps it is for the best as it is. Addressing that question requires one to consider still further ulterior features as fixed, in order to assess the wisdom of the feature under review. There is no Archimedean point here either; if we suppose the readers of the *Moral First Aid Manual* are *complete* idiots, our task is impossible—whereas, if we suppose they are saints, our task is too easy to shed any light.

This comes out graphically in the slippery assumptions about rationality in theoretical discussions of the Prisoner's Dilemma; there is no problem if you are entitled to assume that the players are saints; saints always cooperate, after all. Nearsighted jerks always defect, so they are hopeless. What does "the ideally rational" player do? Perhaps, as some say, he sees the rationality in adopting the meta-strategy of turning himself into a less than ideally rational player—in order to cope with the less than ideally rational players he knows he is apt to face. But, then, in what sense *is* that new player less than ideally rational? It is a mistake to suppose this instability can be made to go away if we just think carefully enough about what ideal rationality is. That is a *truly* Panglossian fallacy. (See the further reflections along these lines in Gibbard 1985 and Sturgeon 1985.)

3. THE MORAL FIRST AID MANUAL

How, then, can we hope to regulate, or at least improve, our ethical decision-making, if it is irremediably heuristic, time-pressured, and myopic? Building on the parallel between what happens in the department meeting and what happens in ourselves, we can see what the meta-problems are, and how they might be dealt with. We need to have "alert," "wise" habits of thought—or, in other words, colleagues who will regularly, if not infallibly, draw our attention in directions we will not regret in hindsight. There is no point having more than one colleague if they are clones of each other, all wanting to raise the same consideration, so we may suppose them to be specialists, each somewhat narrow-minded and preoccupied with protecting a certain set of interests (Minsky 1985).

Now, how shall we avert a cacophony of colleagues? We need some *conversation-stoppers*. In addition to our timely and appropriate generators of considerations, we need consideration-generator-squelchers. We need some ploys that will arbitrarily terminate reflections and disquisitions by our colleagues, and cut off debate independently of the specific content of current debate. Why not just a *magic word*? Magic words work fine as control-shifters in AI programs, but we're talking about controlling intelligent colleagues here, and they are not likely to be susceptible to magic words, as if they were under posthypnotic suggestion. That is, good colleagues will be reflective and rational, and open-minded within the limits imposed by their specialist narrow-mindedness. If the simplest mechanisms that compose us are *ballistic* intentional systems, as I claimed in the previous chapter, our most sophisticated subsystems, like our actual colleagues, are *indefinitely guidable* intentional systems. They need to be hit with something that will appeal to their rationality while discouraging further reflection.

It will not do at all for these people to be *endlessly* philosophizing, endlessly calling us back to first principles and demanding a justification for these apparently (and actually) quite arbitrary principles. What could possibly protect an arbitrary and somewhat second-rate conversation-stopper from such relentless scrutiny? A meta-policy that forbids discussion and reconsideration of the conversation-stoppers? But, our colleagues would want to ask, is *that* a wise policy? Can it be justified? It will not always yield the best results, surely, and . . . and so forth.

This is a matter of delicate balance, with pitfalls on both sides. On one side, we must avoid the error of thinking that the solution is *more rationality*, more rules, more justifications, for there is no end to that demand. Any policy *may* be questioned, so, unless we provide for some brute and a-rational termination of the issue, we will design a decision process that spirals fruitlessly to infinity. On the other side, no mere brute fact about the way we are built is—or should be—entirely beyond the reach of being undone by further reflection.[8]

---

8. Stephen White (1988) discusses Strawson's well-known attempt (1962) to terminate the demand for a justification of "our reactive attitudes" in a brute fact about our way of life about which "we have no choice." He shows that this conversation-stopper cannot resist a further demand for justification (which White provides in an ingeniously indirect way). See also White 1991. For a complementary (and enlightening) approach to the practical problem of ethical decision-making, see Gert 1973.

We cannot expect there to be a single stable solution to such a design problem, but, rather, a variety of uncertain and temporary equilibria, with the conversation-stoppers tending to accrete pearly layers of supporting dogma which themselves cannot withstand extended scrutiny but do actually serve on occasion, blessedly, to deflect and terminate consideration. Here are some promising examples:

"But that would do more harm than good."
"But that would be murder."
"But that would be to break a promise."
"But that would be to use someone merely as a means."
"But that would violate a person's *right*."

Bentham once rudely dismissed the doctrine of "natural and imprescriptible rights" as "nonsense upon stilts," and we might now reply that perhaps he was right. Perhaps talk of rights *is* nonsense upon stilts, but *good* nonsense—and good only because it is on stilts, only because it happens to have the "political" power to keep rising above the meta-reflections—not indefinitely, but usually "high enough"—to reassert itself as a compelling—that is, conversation-stopping—"first principle."

It might seem then that "rule worship" of a certain kind is a good thing, at least for agents designed like us. It is good not because there is a certain rule, or set of rules, which is provably the best, or which always yields the right answer, but because having rules works—somewhat—and not having rules doesn't work at all.

But this cannot be all there is to it—unless we really mean "worship"—i.e., a-rational allegiance, because just *having* rules, or *endorsing* or *accepting* rules, is no design solution at all. Having the rules, having all the information, and even having good intentions do not suffice, by themselves, to guarantee the right action; the agent must find all the right stuff and use it, even in the face of contrary rational challenges designed to penetrate his convictions.

Having, and recognizing the force of, rules is not enough, and sometimes the agent is better off with less. Douglas Hofstadter draws attention to a phenomenon he calls "reverberant doubt," which is stipulated out of existence in most idealized theoretical discussions. In what Hofstadter calls "Wolf's Dilemma," an "obvious" nondilemma is turned into a serious dilemma by nothing but the passage of time and the possibility of reverberant doubt.

Imagine that twenty people are selected from your high school graduation class, you among them. You don't know which others have been selected. . . . All you know is that they are all connected to a central computer. Each of you is in a little cubicle, seated on a chair and facing one button on an otherwise blank wall. You are given ten minutes to decide whether or not to push your button. At the end of that time, a light will go on for ten seconds, and while it is on, you may either push or refrain from pushing. All the responses will then go to the central computer, and one minute later, they will result in consequences. Fortunately, the consequences can only be good. If you pushed your button, you will get $100, no strings attached. . . . If *nobody* pushed their button, then *everybody* will get $1,000. But if there was even a single button-pusher, the refrainers will get nothing at all. [Hofstadter 1985, pp. 752–53.]

Obviously, you do not push the button, right? But what if just one person were a little bit overcautious or dubious, and began wondering whether this was obvious after all? Everyone should allow that this is an outside chance, and everyone should recognize that everyone should allow this. As Hofstadter notes (p. 753), it is a situation "in which the tiniest flicker of a doubt has become amplified into the gravest avalanche of doubt. . . . And one of the annoying things about it is that the brighter you are, the more quickly and clearly you see what there is to fear. A bunch of amiable slowpokes might well be more likely to unanimously refrain and get the big payoff than a bunch of razor-sharp logicians who all think perversely recursively reverberantly."[9]

Faced with a world in which such predicaments are not unknown, we can recognize the appeal of a little old-time religion, some unquestioning dogmatism that will render agents impervious to the subtle invasions of hyperrationality. Creating something rather like that dispositional state is indeed one of the goals of the *Moral First Aid Manual*, which, while we imagine it to be framed as *advice* to a rational, heeding audience, can also be viewed as not having achieved its end unless it has the effect of changing the "operating system"—not merely the "data" (the contents of belief or acceptance) of the agents it addresses. For it to succeed in such a special task, it will have to address its target audiences with pinpoint accuracy.

There might, then, be several different *Moral First Aid Manual*s, each effective for a different type of audience. This opens up a disagreeable prospect to philosophers, for two reasons. First, it suggests, contrary to their austere academic tastes, that there is reason to pay more attention to rhetoric and other only partly or impurely rational means of persuasion; the ideally rational *audience* to whom the ethicist may presume to address his

---

9. Robert Axelrod has pointed out to me that what Hofstadter calls "Wolf's Dilemma" is formally identical to Jean-Jacques Rousseau's Parable of the Stag Hunt, in the *Discourse on the Origin and Foundations of Inequality Among Men* (1755). For further discussion of anticipations and difficulties, see Dennett 1988b.

or her reflections is yet another dubiously fruitful idealization. And, more important, it suggests that what Bernard Williams (1985, p. 101) calls the ideal of "transparency" of a society—"the working of its ethical institutions should not depend on members of the community misunderstanding how they work"—is an ideal that may be politically inaccessible to us. Recoil as we may from elitist mythmaking, and such systematically disingenuous doctrines as the view Williams (p. 108) calls "Government House utilitarianism," we may find—this is an open empirical possibility after all—that we will be extremely lucky to find any rational and transparent route from who we are now to who we would like to be. The landscape is rugged, and it may not be possible to get to the highest peaks from where we find ourselves today.

Rethinking the *practical* design of a moral agent, via the process of writing various versions of the *Moral First Aid Manual*, might nevertheless allow us to make sense of some of the phenomena traditional ethical theories wave their hands about. For one thing, we might begin to understand our current moral position—by that I mean yours and mine, at this very moment. Here you are, devoting several hours to reading my book (and I am no doubt doing something similar). Shouldn't we both be out raising money for Oxfam or picketing the Pentagon or writing letters to our senators and representatives about various matters? Did you consciously decide, on the basis of calculations, that the time was ripe for a little sabbatical from real-world engagement, a period "off line" for a little reading? Or was your process of decision—if that is not too grand a name for it—much more a matter of your *not* tampering with some current "default" principles that virtually ensure that you will ignore all but the most galvanizing potential interruptions to your personal life, which, I am happy to say, includes periods devoted to reading rather difficult books?

If so, is that itself a lamentable feature, or something we finite beings could not conceivably do without? Consider a traditional bench-test which most systems of ethics can pass with aplomb: solving the problem of what you should do if you are walking along, minding your own business, and you hear a cry for help from a drowning man. That is the easy problem, a conveniently delimited, already well-*framed* local decision. The hard problem is: how do we get there from here? How can we *justifiably* find a route from our actual predicament to that relatively happy and straightforwardly decidable predicament? Our prior problem, it seems, is that every day, while trying desperately to mind our own business, we hear a thousand cries for help, complete with volumes of information on how we might oblige. How on Earth could anyone prioritize that cacophony? Not by any systematic process of considering all things, weighing expected utilities, and attempting to maximize. Nor by any systematic generation and testing of Kantian maxims—there are too many to consider.

Yet we do get there from here. Few of us are paralyzed by such indecision for long stretches of times. By and large, we must solve this decision problem by permitting an utterly "indefensible" set of defaults to shield our attention from all but our current projects. Disruptions of those defaults can only occur by a process that is bound to be helter-skelter heuristics, with arbitrary and unexamined conversation-stoppers bearing most of the weight.

That arena of competition encourages escalations, of course. With our strictly limited capacity for attention, the problem faced by others who want us to consider their favorite consideration is essentially a problem of advertising—of attracting the attention of the well-intentioned. This competition between memes is the same problem whether we view it in the wide-scale arena of politics or in the close-up arena of personal deliberation. The role of the traditional formulae of ethical discussion as directors of attention, or shapers of habits of moral imagination, as meta-memes *par excellence*, is thus a subject deserving further scrutiny.

CHAPTER 17: *Ethical decision-making, examined from the perspective of Darwin's dangerous idea, holds out scant hope of our ever discovering a formula or an algorithm for doing right. But that is not an occasion for despair; we have the mind-tools we need to design and redesign ourselves, ever searching for better solutions to the problems we create for ourselves and others.*

CHAPTER 18: *We come to the end of this leg of our journey through Design Space, and take stock of what we have discovered and consider where we might go from here.*

CHAPTER EIGHTEEN

The Future of an Idea

1. IN PRAISE OF BIODIVERSITY

God is in the details.

—LUDWIG MIES VAN DER ROHE, 1959

How long did it take Johann Sebastian Bach to create the *St. Matthew Passion*? An early version was performed in 1727 or 1729, but the version we listen to today dates from ten years later, and incorporates many revisions. How long did it take to create Johann Sebastian Bach? He had the benefit of forty-two years of living when the first version was heard, and more than half a century when the later version was completed. How long did it take to create the Christianity without which the *St. Matthew Passion* would have been literally inconceivable by Bach or anyone else? Roughly two millennia. How long did it take to create the social and cultural context in which Christianity could be born? Somewhere between a hundred millennia and three million years—depending on when we decide to date the birth of human culture. And how long did it take to create *Homo sapiens*? Between three and four billion years, roughly the same length of time it took to create daisies and snail darters, blue whales and spotted owls. Billions of years of *irreplaceable* design work.

We correctly intuit a kinship between the finest productions of art and science and the glories of the biosphere. William Paley was right about one thing: our need to explain how it can be that the universe contains many wonderful designed things. Darwin's dangerous idea is that they *all* exist as fruits of a single tree, the Tree of Life, and the processes that have produced each and every one of them are, at bottom, the same. The genius exhibited by Mother Nature can be disassembled into many acts of micro-genius—myopic or blind, purposeless but capable of the most minimal sort of recognition of a good (a better) thing. The genius of Bach can likewise be disassembled into many acts of micro-genius, tiny mechanical transitions between brain states, generating and testing, discarding and revising, and testing again. Then, is Bach's brain like the proverbial monkeys at the type-writers? No, because instead of generating a Vast number of alternatives, Bach's brain generated only a Vanishingly small subset of all the possibilities. His genius can be measured, if you want to measure genius, in the excellence of his particular subset of generated candidates. How did he come to be able to speed so efficiently through Design Space, never even considering the Vast neighboring regions of hopeless designs? (If you want to explore *that* territory, just sit down at a piano and try, for half an hour, to compose a good new melody.) His brain was exquisitely designed as a heuristic program for composing music, and the credit for that design must be shared; he was lucky in his genes (he did come from a famously musical family), and he was lucky to be born in a cultural milieu that filled his brain with the existing musical memes of the time. And no doubt he was lucky at many other moments in his life to be the beneficiary of one serendipitous convergence or another. Out of all this massive contingency came a unique cruise vehicle for exploring a portion of Design Space that no other vehicle could explore. No matter how many centuries or millennia of musical exploration lie ahead of us, we will never succeed in laying down tracks that make much of a mark in the Vast reaches of Design Space. Bach is precious not because he had within his brain a magic pearl of genius-stuff, a skyhook, but because he was, or contained, an utterly idiosyncratic structure of cranes, made of cranes, made of cranes, made of cranes.

Like Bach, the creation of the rest of the Tree of Life differs from the monkeys at the typewriters in having explored only a Vanishing subset of the Vast possibilities. Efficiencies of exploration have been created again and again, and they are the cranes that have sped up the lifting over the eons. Our technology now permits us to accelerate our explorations in every part of Design Space (not just gene-splicing, but computer-aided design of every imaginable thing, for instance, including this book, which I could never have written without word-processing and electronic mail), but we will never escape our finitude—or, more precisely, our tether to actuality. The Library of Babel is finite but Vast, and we will never explore all its marvels, for at every point we must build, crane-like, on the bases we have constructed to date.

Alert to the omnipresent risk of greedy reductionism, we might consider how much of what we value is explicable in terms of its designedness. A little intuition-pumping: which is worse, destroying somebody's project—even if it's a model of the Eiffel Tower made out of thousands of popsicle sticks—or destroying their supply of popsicle sticks? It all depends on the goal of the project; if the person just enjoys designing and redesigning, building and rebuilding, then destroying the supply of popsicle sticks is worse; otherwise, destroying that hard-won product of design is worse. Why is it much worse to kill a condor than to kill a cow? (I take it that, no matter how bad you think it is to kill a cow, we agree that it is much worse to kill a condor—because the loss to our actual store of design would be so much greater if the condors went extinct.) Why is it worse to kill a cow than to kill a clam? Why is it worse to kill a redwood tree than to kill an equal amount (by mass) of algae? Why do we rush to make high-fidelity copies of motion pictures, musical recordings, scores, books? Leonardo da Vinci's *Last Supper* is sadly decaying on a wall in Milan, in spite of (and sometimes because of) the efforts over the centuries to preserve it. Why would it be just as bad—maybe worse—to destroy all the old photographs of what it looked like thirty years ago as to destroy some portion of its "original" fabric today?

These questions don't have obvious and uncontroversial answers, so the Design Space perspective certainly doesn't explain everything about value, but at least it lets us see what happens when we try to unify our sense of value in a single perspective. On the one hand, it helps to explain our intuition that uniqueness or individuality is "intrinsically" valuable. On the other hand, it lets us confirm all the incommensurabilities that people talk about. Which is worth more, a human life or the *Mona Lisa*? There are many who would give their lives to save the painting from destruction, and many who would sacrifice *somebody else's life* for it, if push came to shove. (Are the guards in the Louvre armed? What steps would they take if necessary?) Is saving the spotted owl worth the abridgment of opportunities in the thousands of human lives affected? (Once again, retrospective effects loom large: if someone has invested his life chances in becoming a logger, and now we take away the opportunity to be a logger, we devalue his investment overnight, just as surely as—more surely, in fact, than—if we converted his life savings into worthless junk bonds.)

At what "point" does a human life begin or end? The Darwinian perspective lets us see with unmistakable clarity why there is no hope at all of *discovering* a telltale mark, a saltation in life's processes, that "counts." We need to draw lines; we need definitions of life and death for many important moral purposes. The layers of pearly dogma that build up in defense around these fundamentally arbitrary attempts are familiar, and in never-ending need of repair. We should abandon the fantasy that either science or religion can uncover some well-hidden fact that tells us exactly where to draw these lines. There is no "natural" way to mark the birth of a human "soul," any more than there is a "natural" way to mark the birth of a species. And, contrary to what many traditions insist, I think we all do share the intuition that there are gradations of value in the ending of human lives. Most human embryos end in spontaneous abortion—fortunately, since these are mostly *terata*, hopeless monsters whose lives are all but impossible. Is this a terri-

ble evil? Are the mothers whose bodies abort these embryos guilty of in-
voluntary manslaughter? Of course not. Which is worse, taking "heroic"
measures to keep alive a severely deformed infant, or taking the equally
"heroic" (if unsung) step of seeing to it that such an infant dies as quickly
and painlessly as possible? I do not suggest that Darwinian thinking gives us
answers to such questions; I do suggest that Darwinian thinking helps us see
why the traditional hope of solving these problems (finding a moral algorithm) is forlorn. We must cast off the myths that make these old-fashioned
solutions seem inevitable. We need to grow up, in other words.

Among the precious artifacts worth preserving are whole cultures themselves. There are still several thousand distinct languages spoken daily on
our planet, but the number is dropping fast (Diamond 1992, Hale et al.
1992). When a language goes extinct, this is the same kind of loss as the
extinction of a species, and when the culture that was carried by that
language dies, this is an even greater loss. But here, once again, we face
incommensurabilities and no easy answers.

I began this book with a song which I myself cherish, and hope will
survive "forever." I hope my grandson learns it and passes it on to his
grandson, but at the same time I do not myself believe, and do not really
want my grandson to believe, the doctrines that are so movingly expressed
in that song. They are too simple. They are, in a word, wrong—just as wrong
as the ancient Greeks' doctrines about the gods and goddesses on Mount
Olympus. Do you believe, literally, in an anthropomorphic God? If not, then
you must agree with me that the song is a beautiful, comforting falsehood.
Is that simple song nevertheless a valuable meme? I certainly think it is. It
is a modest but beautiful part of our heritage, a treasure to be preserved. But
we must face the fact that, just as there were times when tigers would not
have been viable, times are coming when they will no longer be viable,
except in zoos and other preserves, and the same is true of many of the
treasures in our cultural heritage.

The Welsh language is kept alive by artificial means, just the way condors
are. We cannot preserve *all* the features of the cultural world in which
these treasures flourished. We wouldn't want to. It took oppressive political
and social systems, rife with many evils, to create the rich soil in which
many of our greatest works of art could grow: slavery and despotism ("enlightened" though these sometimes may have been), obscene differences in
living standards between the rich and the poor—and a huge amount of
ignorance. Ignorance is a necessary condition for many excellent things.
The childish joy of seeing what Santa Claus has brought for Christmas is a
species of joy that must soon be extinguished in each child by the loss of
ignorance. When that child grows up, she can transmit that joy to her own
children, but she must also recognize a time when it has outlived its value.

The view I am expressing has clear ancestors. The philosopher George

Santayana was a Catholic atheist, if you can imagine such a thing. According to Bertrand Russell (1945, p. 811), William James once denounced Santayana's ideas as "the perfection of rottenness," and one can see why some people would be offended by his brand of aestheticism: a deep appreciation for all the formulae, ceremonies, and trappings of his religious heritage, but lacking the faith. Santayana's position was aptly caricatured: "There is no God and Mary is His Mother." But how many of us are caught in that very dilemma, loving the heritage, firmly convinced of its value, yet unable to sustain any conviction at all in its truth? We are faced with a difficult choice. Because we value it, we are eager to preserve it in a rather precarious and "denatured" state—in churches and cathedrals and synagogues, built to house huge congregations of the devout, and now on the way to being cultural museums. There is really not that much difference between the roles of the Beefeaters who stand picturesque guard at the Tower of London, and the Cardinals who march in their magnificent costumes and meet to elect the next Pope. Both are keeping alive traditions, rituals, liturgies, symbols, that otherwise would fade.

But hasn't there been a tremendous rebirth of fundamentalist faith in all these creeds? Yes, unfortunately, there has been, and I think that there are no forces on this planet more dangerous to us all than the fanaticisms of fundamentalism, of all the species: Protestantism, Catholicism, Judaism, Islam, Hinduism, and Buddhism, as well as countless smaller infections. Is there a conflict between science and religion here? There most certainly is.

Darwin's dangerous idea helps to create a condition in the memosphere that in the long run threatens to be just as toxic to these memes as civilization in general has been toxic to the large wild mammals. Save the Elephants! Yes, of course, but not *by all means*. Not by forcing the people of Africa to live nineteenth-century lives, for instance. This is not an idle comparison. The creation of the great wildlife preserves in Africa has often been accompanied by the dislocation—and ultimate destruction—of human populations. (For a chilling vision of this side effect, see Colin Turnbull 1972 on the fate of the Ik.) Those who think that we should preserve the elephants' pristine environment *at all costs* should contemplate the costs of returning the United States to the pristine conditions in which the buffaloes roam and the deer and the antelope play. We must find an accommodation.

I love the King James Version of the Bible. My own spirit recoils from a God Who is He or She in the same way my heart sinks when I see a lion pacing neurotically back and forth in a small zoo cage. I know, I know, the lion is beautiful but dangerous; if you let the lion roam free, it would kill me; safety demands that it be put in a cage. Safety demands that religions be put in cages, too—when absolutely necessary. We just can't have forced female circumcision, and the second-class status of women in Roman Catholicism and Mormonism, to say nothing of their status in Islam. The recent Supreme

Court ruling declaring unconstitutional the Florida law prohibiting the sacrificing of animals in the rituals of the Santeria sect (an Afro-Caribbean religion incorporating elements of Yoruba traditions and Roman Catholicism) is a borderline case, at least for many of us. Such rituals are offensive to many, but the protective mantle of religious tradition secures our tolerance. We are wise to respect these traditions. It is, after all, just part of respect for the biosphere.

Save the Baptists! Yes, of course, but not *by all means*. Not if it means tolerating the deliberate misinforming of children about the natural world. According to a recent poll, 48 percent of the people in the United States today believe that the book of Genesis is literally true. And 70 percent believe that "creation science" should be taught in school alongside evolution. Some recent writers recommend a policy in which parents would be able to "opt out" of materials they didn't want their children taught. Should evolution be taught in the schools? Should arithmetic be taught? Should history? Misinforming a child is a terrible offense.

A faith, like a species, must evolve or go extinct when the environment changes. It is not a gentle process in either case. We see in every Christian subspecies the battle of memes—should women be ordained? should we go back to the Latin liturgy?—and the same can also be observed in the varieties of Judaism and Islam. We must have a similar mixture of respect and self-protective caution about memes. This is already accepted practice, but we tend to avert our attention from its implications. We preach freedom of religion, but only so far. If your religion advocates slavery, or mutilation of women, or infanticide, or puts a price on Salman Rushdie's head because he has insulted it, then your religion has a feature that cannot be respected. It endangers us all.

It is nice to have grizzly bears and wolves living in the wild. They are no longer a menace; we can peacefully coexist, with a little wisdom. The same policy can be discerned in our political tolerance, in religious freedom. You are free to preserve or create any religious creed you wish, so long as it does not become a public menace. We're all on the Earth together, and we have to learn some accommodation. The Hutterite memes are "clever" not to include any memes about the virtue of destroying outsiders. If they did, we would have to combat them. We tolerate the Hutterites because they harm only themselves—though we may well insist that we have the right to impose some further openness on their schooling of their own children. Other religious memes are not so benign. The message is clear: those who will not accommodate, who will not temper, who insist on keeping only the purest and wildest strain of their heritage alive, we will be obliged, reluctantly, to cage or disarm, and we will do our best to disable the memes they fight for. Slavery is beyond the pale. Child abuse is beyond the pale. Discrimination is beyond the pale. The pronouncing of death sentences on those who blaspheme against a religion (complete with bounties or rewards for those who carry them out) is beyond the pale. It is not civilized, and it is owed no more respect in the name of religious freedom than any other incitement to cold-blooded murder.[1]

Those of us who lead fulfilling, even exciting, lives should hardly be shocked to see people in the disadvantaged world—and indeed in the drabber corners of our own world—turning to fanaticism of one brand or another. Would you settle docilely for a life of meaningless poverty, knowing what you know today about the world? The technology of the infosphere has recently made it conceivable for everybody on the globe to know roughly what you know (with a lot of distortion). Until we can provide an environment for all people in which fanaticism doesn't make sense, we can expect more and more of it. But we don't have to accept it, and we don't have to respect it. Taking a few tips from Darwinian medicine (Williams and Nesse 1991), we can take steps to conserve what is valuable in every culture without keeping alive (or virulent) all its weaknesses.

We can appreciate the bellicosity of the Spartans without wanting to re-introduce it; we can marvel at the systems of atrocities instituted by the Mayans without for one moment regretting the extinction of those practices. It must be scholarship, not human game preserves—ethnic or religious states under dictatorships—that saves superannuated cultural artifacts for posterity. Attic Greek and Latin are no longer living languages, but scholarship has preserved the art and literature of ancient Greece and Rome. Petrarch, in the fourteenth century, bragged about the volumes of Greek philosophy he had in his personal library; he couldn't read them, because the knowledge of ancient Greek had all but disappeared from the world in which he lived, but he knew their value, and strove to restore the knowledge that would unlock their secrets.

Long before there was science, or even philosophy, there were religions. They have served many purposes (it would be a mistake of greedy reductionism to look for a single purpose, a single *summum bonum* which they

---

1. Many, many Muslims agree, and we must not only listen to them, but do what we can to protect and support them, for they are bravely trying, from the inside, to reshape the tradition they cherish into something better, something ethically defensible. *That* is—or, rather, ought to be—the message of multiculturalism, not the patronizing and subtly racist hypertolerance that "respects" vicious and ignorant doctrines when they are propounded by officials of non-European states and religions. One might start by spreading the word about *For Rushdie* (Braziller, 1994), a collection of essays by Arab and Muslim writers, many critical of Rushdie, but all denouncing the unspeakably immoral "fatwa" death sentence proclaimed by the Ayatollah. Rushdie (1994) has drawn our attention to the 162 Iranian intellectuals who, with great courage, have signed a declaration in support of freedom of expression. Let us all distribute the danger by joining hands with them.

have all directly or indirectly served). They have inspired many people to lead lives that have added immeasurably to the wonders of our world, and they have inspired many more people to lead lives that were, given their circumstances, more meaningful, less painful, than they otherwise could have been. Breughel's painting *The Fall of Icarus* shows a plowman and a horse on a hillside in the foreground, a handsome sailing ship way in the background—and two almost unnoticeable white legs disappearing with a tiny splash into the sea. The painting inspired W. H. Auden to write one of my favorite poems.

MUSÉE DES BEAUX ARTS

About suffering they were never wrong,
The Old Masters: how well they understood
Its human position; how it takes place
While someone else is eating or opening a window or just walking dully
      along;
How, when the aged are reverently, passionately waiting
For the miraculous birth, there always must be
Children who did not specially want it to happen skating
On a pond at the edge of the wood:
They never forgot
That even the dreadful martyrdom must run its course
Anyhow in a corner, some untidy spot
Where the dogs go on with their doggy life and the torturer's horse
Scratches its innocent behind on a tree.

In Breughel's *Icarus*, for instance: how everything turns away
Quite leisurely from the disaster; the ploughman may
Have heard the splash, the forsaken cry,
But for him it was not an important failure; the sun shone
As it had to on the white legs disappearing into the green
Water; and the expensive delicate ship that must have seen
Something amazing, a boy falling out of the sky,
Had somewhere to get to and sailed calmly on.

That is our world, and the suffering in it matters, if anything does. Religions have brought the comfort of belonging and companionship to many who would otherwise have passed through this life all alone, without glory or adventure. At their best, religions have drawn attention to love, and made it real for people who could not otherwise see it, and ennobled the attitudes and refreshed the spirits of the world-beset. Another thing religions have accomplished, without this being thereby their *raison d'être*, is that they have kept *Homo sapiens* civilized enough, for long enough, for us to have learned how to reflect more systematically and accurately on our position in the universe. There is much more to learn. There is certainly a treasury of ill-appreciated truths embedded in the endangered cultures of the modern world, designs that have accumulated details over eons of idiosyncratic history, and we should take steps to record it, and study it, before it disappears, for, like dinosaur genomes, once it is gone, it will be virtually impossible to recover.

We should not expect this variety of respect to be satisfactory to those who wholeheartedly embody the memes we honor with our attentive—but not worshipful—scholarship. On the contrary, many of them will view anything other than enthusiastic conversion to their own views as a threat, even an intolerable threat. We must not underestimate the suffering such confrontations cause. To watch, to have to participate in, the contraction or evaporation of beloved features of one's heritage is a pain only our species can experience, and surely few pains could be more terrible. But we have no reasonable alternative, and those whose visions dictate that they cannot peacefully coexist with the rest of us we will have to quarantine as best we can, minimizing the pain and damage, trying always to leave open a path or two that may come to seem acceptable.

If you want to teach your children that they are the tools of God, you had better not teach them that they are God's rifles, or we will have to stand firmly opposed to you: your doctrine has no glory, no special rights, no intrinsic and inalienable merit. If you insist on teaching your children falsehoods—that the Earth is flat, that "Man" is not a product of evolution by natural selection—then you must expect, at the very least, that those of us who have freedom of speech will feel free to describe your teachings as the spreading of falsehoods, and will attempt to demonstrate this to your children at our earliest opportunity. Our future well-being—the well-being of all of us on the planet—depends on the education of our descendants.

What, then, of all the glories of our religious traditions? They should certainly be preserved, as should the languages, the art, the costumes, the rituals, the monuments. Zoos are now more and more being seen as second-class havens for endangered species, but at least they are havens, and what they preserve is irreplaceable. The same is true of complex memes and their phenotypic expressions. Many a fine New England church, costly to maintain, is in danger of destruction. Shall we deconsecrate these churches and turn them into museums, or retrofit them for some other use? The latter fate is at least to be preferred to their destruction. Many congregations face a cruel choice: their house of worship costs so much to maintain in all its splendor that little of their tithing is left over for the poor. The Catholic Church has faced this problem for centuries, and has maintained a position that is, I think, defensible, but not obviously so: when it spends its treasure to put gold plating on the candlesticks, instead of providing more food and better shelter for the poor of the parish, it has a different vision of what makes life worth living. Our people, it says, benefit more from having a place of splendor in which to worship than from a little more food. Any atheist or agnostic who finds this cost-benefit analysis ludicrous might pause to consider whether to support diverting all charitable and governmental support for museums, symphony orchestras, libraries, and scientific laboratories to efforts to provide more food and better living conditions for the least well off. A human life worth living is not something that can be uncontroversially measured, and that is its glory.

And there's the rub. What will happen, one may well wonder, if religion is preserved in cultural zoos, in libraries, in concerts and demonstrations? It is happening; the tourists flock to watch the Native American tribal dances, and for the onlookers it is folklore, a religious ceremony, certainly, to be treated with respect, but also an example of a meme complex on the verge of extinction, at least in its strong, ambulatory phase; it has become an invalid, barely kept alive by its custodians. Does Darwin's dangerous idea give us anything in exchange for the ideas it calls into question?

In chapter 3, I quoted the physicist Paul Davies proclaiming that the reflective power of human minds can be "no trivial detail, no minor by-product of mindless purposeless forces," and suggested that being a by-product of mindless purposeless forces was no disqualification for importance. And I have argued that Darwin has shown us how, in fact, *everything* of importance is just such a product. Spinoza called his highest being God or Nature (*Deus sive Natura*), expressing a sort of pantheism. There have been many varieties of pantheism, but they usually lack a convincing *explanation* about just how God is distributed in the whole of nature. As we saw in chapter 7, Darwin offers us one: it is in the distribution of Design throughout nature, creating, in the Tree of Life, an utterly unique and irreplaceable creation, an actual pattern in the immeasurable reaches of Design Space that could never be exactly duplicated in its many details. What is design work? It is that wonderful wedding of chance and necessity, happening in a trillion places at once, at a trillion different levels. And what miracle caused it? None. It just happened to happen, in the fullness of time. You could even say, in a way, that the Tree of Life created itself. Not in a miraculous, instantaneous whoosh, but slowly, slowly, over billions of years.

Is this Tree of Life a God one could worship? Pray to? Fear? Probably not. But it *did* make the ivy twine and the sky so blue, so perhaps the song I love tells a truth after all. The Tree of Life is neither perfect nor infinite in space or time, but it is actual, and if it is not Anselm's "Being greater than which nothing can be conceived," it is surely a being that is greater than anything any of us will ever conceive of in detail worthy of its detail. Is something sacred? Yes, say I with Nietzsche. I could not pray to it, but I can stand in affirmation of its magnificence. This world is sacred.

2. UNIVERSAL ACID: HANDLE WITH CARE

There is no denying, at this point, that Darwin's idea is a universal solvent, capable of cutting right to the heart of everything in sight. The question is: what does it leave behind? I have tried to show that once it passes through everything, we are left with stronger, sounder versions of our most important ideas. Some of the traditional details perish, and some of these are losses to be regretted, but good riddance to the rest of them. What remains is more than enough to build on.

At every stage in the tumultuous controversies that have accompanied the evolution of Darwin's dangerous idea, there has been a defiance born of fear: "You'll *never* explain *this*!" And the challenge has been taken up: "Watch me!" And in spite of—indeed, partly because of—the huge emotional investments the opponents have made in winning their sides of the argument, the picture has become clearer and clearer. We now have a *much* better sense of what a Darwinian algorithm is than Darwin ever dreamt of. Intrepid reverse engineering has brought us to the point where we can confidently assess rival claims about exactly what happened where on this planet billions of years ago. The "miracles" of life and consciousness turn out to be even better than we imagined back when we were sure they were inexplicable.

The ideas expressed in this book are just the beginning. This has been an introduction to Darwinian thinking, sacrificing details again and again to provide a better appreciation of the overall shape of Darwin's idea. But as Mies van der Rohe said, God is in the details. I urge caution alongside the enthusiasm I hope I have kindled in you. I have learned from my own embarrassing experience how easy it is to concoct remarkably persuasive Darwinian explanations that evaporate on closer inspection. The truly dangerous aspect of Darwin's idea is its seductiveness. Second-rate versions of the fundamental ideas continue to bedevil us, so we must keep a close watch, correcting each other as we go. The only way of avoiding the mistakes is to learn from the mistakes we have already made.

A meme that occurs in many guises in the world's folklore is the tale of the initially terrifying friend mistaken for an enemy. "Beauty and the Beast" is one of the best-known species of this story. Balancing it is "The Wolf in Sheep's Clothing." Now, which meme do you want to use to express your judgment of Darwinism? Is it truly a Wolf in Sheep's Clothing? Then reject it and fight on, ever more vigilant against the seductions of Darwin's idea, which is truly dangerous. Or does Darwin's idea turn out to be, in the end, just what we need in our attempt to preserve and explain the values we cherish? I have completed my case for the defense: the Beast is, in fact, a friend of Beauty, and indeed quite beautiful in its own right. You be the judge.

Appendix

Tell Me Why

Traditional

1. Tell me why the stars do shine,
2. Be - cause God made the stars to shine,

    Tell me why the i———— vy twines,
    Be - cause God made the i———— vy twine,

    Tell me why the sky's so blue.
    Be - cause God made the sky so blue.

    Then I will tell you just why I love you.
    Be - cause God made you, that's why I love you.

(The harmony line is usually sung by the higher voices an octave above the melody.)

Bibliography

ABBOTT, E. A. 1884. *Flatland: A Romance in Many Dimensions.* Reprint ed., Oxford: Blackwell, 1962.

ALEXANDER, RICHARD D. 1987. *The Biology of Moral Systems.* New York: de Gruyter.

ARAB AND MUSLIM WRITERS. 1994. *For Rushdie.* New York: Braziller.

ARBIB, MICHAEL. 1964. *Brains, Machines, and Mathematics.* New York: McGraw-Hill.

————. 1989. *The Metaphorical Brain 2: Neural Networks and Beyond.* New York: Wiley.

ARRHENIUS, S. 1908. *Worlds in the Making.* New York: Harper & Row.

ASHBY, ROSS. 1960. *Design for a Brain.* New York: Wiley.

AUSTIN, J. L. 1961. "A Plea for Excuses." In J. L. Austin, *Philosophical Papers.* Oxford: The Clarendon Press, pp. 123–52.

AXELROD, ROBERT. 1984. *The Evolution of Cooperation.* New York: Basic Books.

AXELROD, ROBERT, and HAMILTON, WILLIAM. 1981. "The Evolution of Cooperation." *Science*, vol. 211, pp. 1390–96.

AYALA, FRANCISCO J. 1982. "Beyond Darwinism? The Challenge of Macroevolution to the Synthetic Theory of Evolution." In Peter D. Asquith and Thomas Nickels, eds., *PSA 1982* (Philosophy of Science Association), vol. 2, pp. 275–91. Reprinted in Ruse 1989.

AYERS, M. 1968. *The Refutation of Determinism: An Essay in Philosophical Logic.* London: Methuen.

BABBAGE, CHARLES. 1838. *Ninth Bridgewater Treatise: A Fragment.* London: Murray.

BAK, PER; FLYVBJERG, HENRIK; and SNEPPEN, KIM. 1994. "Can We Model Darwin?" *New Scientist*, March 12, pp. 36–39.

BALDWIN, J. M. 1896. "A New Factor in Evolution." *American Naturalist*, vol. 30, pp. 441–51, 536–53.

BALL, JOHN A. 1984. "Memes as Replicators." *Ethology and Sociobiology*, vol. 5, pp. 145–61.

BARKOW, JEROME H.; COSMIDES, LEDA; and TOOBY, JOHN. 1992. *The Adapted Mind: Evolutionary Psychology and the Generation of Culture*. Oxford: Oxford University Press.

BARLOW, GEORGE W., and SILVERBERG, JAMES, eds. 1980. *Sociobiology: Beyond Nature/Nurture?* AAAS Selected Symposium. Boulder, Col.: Westview.

BARON-COHEN, SIMON. 1995. *Mindblindness and the Language of the Eyes: An Essay in Evolutionary Psychology*. Cambridge, Mass.: MIT Press.

BARRETT, P. H.; GAUTREY, P. J.; HERBERT, S.; KOHN, D.; and SMITH, S., eds. 1987. *Charles Darwin's Notebooks, 1836–44*. Cambridge: British Museum (Natural History)/Cambridge University Press.

BARROW, J. and TIPLER, F. 1988. *The Anthropic Cosmological Principle*. Oxford: Oxford University Press.

BATESON, WILLIAM. 1909. "Heredity and Variation in Modern Lights." In A. C. Seward, ed., *Darwin and Modern Science*. Cambridge: Cambridge University Press, pp. 85–101.

BEDAU, MARK. 1991. "Can Biological Teleology Be Naturalized?" *Journal of Philosophy,* vol. 88, pp. 647–57.

BENTHAM, JEREMY. 1789. *Introduction to the Principles of Morals and Legislation*. Oxford: Oxford University Press.

BETHELL, TOM. 1976. "Darwin's Mistake." *Harper's Magazine*, February, pp. 70–75.

BICKERTON, DEREK. 1993. "The Snail Wars" (review of Gould 1993d). *New York Times Book Review*, January 3, p. 5.

BONNER, JOHN TYLER. 1980. *The Evolution of Culture in Animals*. Princeton: Princeton University Press.

BORGES, JORGE LUIS. 1962. "The Library of Babel." In *Labyrinths: Selected Stories and Other Writings*. New York: New Directions. ("La Biblioteca de Babel," 1941. In *El jardin de los senderos que se bifurcan,* published as part of *Ficciones* [Buenos Aires: Emece Editores, 1956].)

———. 1993. "Poem About Quantity." Trans. Robert Mezey. *New York Review of Books*, June 24, p. 35.

BRANDON, ROBERT. 1978. "Adaptation and Evolutionary Theory." *Studies in the History and Philosophy of Science*, vol. 9, pp. 181–206.

BREUER, REINHARD. 1991. *The Anthropic Principle: Man as the Focal Point of Nature*. Boston: Birkhäuser.

BRIGGS, DEREK E. G.; FORTEY, RICHARD A.; and WILLS, MATTHEW A. 1989. "Morphological Disparity in the Cambrian." *Science*, vol. 256, pp. 1670–73.

BROOKS, RODNEY. 1991. "Intelligence Without Representation." *Artificial Intelligence Journal*, vol. 47, pp. 139–59.

BRUMBAUGH, ROBERT M., and WELLS, RULON. 1968. *The Plato Manuscripts: A New Index*. New Haven: Yale University Press.

BUSS, LEO W. 1987. *The Evolution of Individuality*. Princeton: Princeton University Press.

CAIRNS-SMITH, GRAHAM. 1982. *Genetic Takeover*. Cambridge: Cambridge University Press.

———. 1985. *Seven Clues to the Origin of Life*. Cambridge: Cambridge University Press.

CALVIN, WILLIAM. 1986. *The River That Flows Uphill: A Journey from the Big Bang to the Big Brain.* San Francisco: Sierra Club.

———. 1987. "The Brain as a Darwin Machine." *Nature*, vol. 330, pp. 33–34.

CAMPBELL, DONALD. 1975. "On the Conflicts Between Biological and Social Evolution and Between Psychology and Moral Tradition." *American Psychologist*, December, pp. 1103–26.

———. 1979. "Comments on the Sociobiology of Ethics and Moralizing." *Behavioral Science*, vol. 24, pp. 37–45.

CANN, REBECCA L.; STONEKING, MARK; and WILSON, ALLAN C. 1987. "Mitochondrial DNA and Human Evolution." *Nature*, vol. 325, pp. 31–36.

CAPOTE, TRUMAN. 1965. *In Cold Blood*. New York: Random House.

CARROLL, LEWIS. 1871. *Through the Looking Glass.* London: Macmillan.

CHANGEAUX, J.-P., and DANCHIN, A. 1976. "Selective Stabilization of Developing Synapses as a Mechanism for the Specifications of a Neuronal Networks." *Nature*, vol. 264, pp. 705–12.

CHOMSKY, NOAM. 1956. "Three Models for the Description of Language." *IRE Transactions on Information Theory IT-2(3)*, pp. 13–54.

———. 1957. *Syntactic Structures*. The Hague: Mouton.

———. 1959. Review of Skinner 1957. *Language*, vol. 35, pp. 26–58.

———. 1966. *Cartesian Linguistics*. New York: Harper & Row.

———. 1972. *Language and Mind*. Enlarged ed. New York: Harcourt Brace Jovanovich.

———. 1975. *Reflections on Language*. New York: Pantheon.

———. 1980. "Rules and Representations." *Behavioral and Brain Sciences*, vol. 3, pp. 1–15.

———. 1988. *Language and Problems of Knowledge: The Managua Lectures*. Cambridge, Mass.: MIT Press.

CHRISTENSEN, SCOTT M., and TURNER, DALE R. 1993. *Folk Psychology and the Philosophy of Mind*. Hillsdale, N.J.: Erlbaum.

CHURCHLAND, PATRICIA S., and SEJNOWSKI, TERRENCE, J. 1992. *The Computational Brain*. Cambridge, Mass.: MIT Press.

CHURCHLAND, PAUL. 1989. *A Neurocomputational Perspective: The Nature of Mind and the Structure of Science*. Cambridge, Mass.: MIT Press.

CLARK, ANDY, and KARMILOFF-SMITH, ANNETTE. 1994. "The Cognizer's Innards: A Psychological and Philosophical Perspective on the Development of Thought." *Mind & Language*, vol. 8, pp. 487–519.

CLUTTON-BROCK, T. H., and HARVEY, PAUL H. 1978. *Readings in Sociobiology*. San Francisco: Freeman.

CONWAY MORRIS, SIMON. 1989. "Burgess Shale Faunas and the Cambrian Explosion." *Science*, vol. 246, pp. 339–46.

———. 1991. "Rerunning the Tape" (review of Gould 1991b). *Times Literary Supplement*, December 13, p. 6.

————. 1992. "Burgess Shale-type Faunas in the Context of the 'Cambrian Explosion': A Review." *Journal of the Geological Society, London,* vol. 149, pp. 631–36.

COON, C. S.; GARN, S. M.; and BIRDSELL, J. B. 1950. *Races.* Springfield, Ohio: C. Thomas.

COSMIDES, LEDA, and TOOBY, JOHN. 1989, "Evolutionary Psychology and the Generation of Culture," pt. II, "Case Study: A Computational Theory of Social Exchange." *Ethology and Sociobiology,* vol. 10, pp. 51–97.

CRICHTON, MICHAEL. 1990. *Jurassic Park.* New York: Knopf.

CRICK, FRANCIS H. C. 1968. "The Origin of the Genetic Code." *Journal of Molecular Biology,* vol. 38, p. 367.

————. 1981. *Life Itself: Its Origin and Nature.* New York: Simon & Schuster.

CRICK, FRANCIS, and ORGEL, LESLIE E. 1973. "Directed Panspermia." *Icarus,* vol. 19, pp. 341–46.

CRONIN, HELENA. 1991. *The Ant and the Peacock.* Cambridge: Cambridge University Press.

CUMMINS, ROBERT. 1975. "Functional Analysis." *Journal of Philosophy,* vol. 72, pp. 741–64. Reprinted in Sober 1984b.

DALY, MARTIN. 1991. "Natural Selection Doesn't Have Goals, but It's the Reason Organisms Do" (commentary on P. J. H. Schoemaker, "The Quest for Optimality: A Positive Heuristic of Science?"). *Behaviorial and Brain Sciences,* vol. 14, pp. 219–20.

DANTO, ARTHUR. 1965. *Nietzsche as Philosopher.* New York: Macmillan.

DARWIN, CHARLES. 1859. *On the Origin of Species by Means of Natural Selection.* London: Murray.

————. 1862. *On the Various Contrivances by Which Orchids Are Fertilised by Insects.* London: Murray. 2nd ed., 1877. (2nd ed. reprint, Chicago: University of Chicago Press, 1984.)

————. 1871. *The Descent of Man, and Selection in Relation to Sex.* London: Murray. 2nd ed., 1874.

DARWIN, FRANCIS. 1911. *The Life and Letters of Charles Darwin,* 2 vols. New York: Appleton. (Originally published in 1887, in 3 vols., by Murray in London.)

DAVID, PAUL. 1985. "Clio and the Economics of QWERTY." *American Economic Review,* vol. 75, pp. 332–37.

DAVIES, PAUL. 1992. *The Mind of God.* New York: Simon & Schuster.

DAWKINS, RICHARD. 1976. *The Selfish Gene.* Oxford: Oxford University Press. (See also revised ed., Dawkins 1989a.)

————. 1981. "In Defence of Selfish Genes." *Philosophy,* vol. 54, pp. 556–73.

————. 1982. *The Extended Phenotype: The Gene as the Unit of Selection.* Oxford and San Francisco: Freeman.

————. 1983a. "Universal Darwinism." In D. S. Bendall, ed., *Evolution from Molecules to Men* (Cambridge: Cambridge University Press), pp. 403–25.

————. 1983b. "Adaptationism Was Always Predictive and Needed No Defense" (commentary on Dennett 1983). *Behavioral and Brain Sciences*, vol. 6, pp. 360–61.

———. 1986a. *The Blind Watchmaker*. London: Longmans.

———. 1986b. "Sociobiology: The New Storm in a Teacup." In Steven Rose and Lisa Appignanese, eds., *Science and Beyond* (Oxford: Blackwell), pp. 61–78.

———. 1989a. *The Selfish Gene* (2nd ed.) Oxford: Oxford University Press.

———. 1989b. "The Evolution of Evolvability." In C. Langton, ed., *Artificial Life*, vol. I (Redwood City, Calif.: Addison-Wesley), pp. 201–20.

———. 1990. Review of Gould 1989a. *Sunday Telegraph* (London), February 25.

———. 1993. "Viruses of the Mind." In Bo Dahlbom, ed., *Dennett and His Critics* (Oxford: Blackwell), pp. 13–27.

DELIUS, JUAN. 1991. "The Nature of Culture." In M. S. Dawkins, T. R. Halliday, and R. Dawkins, eds., *The Tinbergen Legacy* (London: Chapman & Hall), pp. 75–99.

DEMUS, OTTO. 1984. *The Mosaics of San Marco in Venice*, 4 vols. Chicago: University of Chicago Press.

DENNETT, DANIEL C. 1968. "Machine Traces and Protocol Statements." *Behavioral Science*, vol. 13, pp. 155–61.

———. 1969. *Content and Consciousness*. London: Routledge & Kegan Paul.

———. 1970. "The Abilities of Men and Machines." Presented at American Philosophical Association Eastern Division Meeting, December. Published in Dennett 1978, pp. 256–66.

———. 1971. "Intentional Systems." *Journal of Philosophy*, vol. 68, pp. 87–106.

———. 1972. Review of Lucas 1970. *Journal of Philosophy*, vol. 69, pp. 527–31.

———. 1975. "Why the Law of Effect Will Not Go Away." *Journal of the Theory of Social Behaviour*, vol. 5, pp. 179–87. Reprinted in Dennett 1978.

———. 1978. *Brainstorms*. Cambridge, Mass.: MIT Press/A Bradford Book.

———. 1980. "Passing the Buck to Biology." *Behavioral and Brain Sciences*, vol. 3, p. 19.

———. 1981. "Three Kinds of Intentional Psychology." In R. Healey, ed., *Reduction, Time and Reality* (Cambridge: Cambridge University Press), pp. 37–61.

———. 1983. "Intentional Systems in Cognitive Ethology: The 'Panglossian Paradigm' Defended." *Behavioral and Brain Sciences*, vol. 6, pp. 343–90.

———. 1984. *Elbow Room: The Varieties of Free Will Worth Wanting*. Cambridge, Mass.: MIT Press.

———. 1985. "Can Machines Think?" In M. Shafto, ed., *How We Know* (San Francisco: Harper & Row), pp. 121–45.

———. 1987a. "The Logical Geography of Computational Approaches: A View from the East Pole." In M. Brand and M. Harnish, eds., *Problems in the*

Representation of Knowledge (Tucson: University of Arizona Press), pp. 59–79.

———. 1987b. *The Intentional Stance*. Cambridge, Mass.: MIT Press/A Bradford Book.

———. 1988a. "When Philosophers Encounter Artificial Intelligence." *Daedalus*, vol. 117, pp. 283–95.

———. 1988b. "The Moral First Aid Manual." In Sterling M. McMurrin, ed., *Tanner Lectures on Human Values*, vol. VIII (Salt Lake City: University of Utah Press), pp. 120–47.

———. 1989a. Review of Robert J. Richards 1987. *Philosophy of Science*, vol. 56, no. 3, pp. 540–43.

———. 1989b. "Murmurs in the Cathedral" (review of Penrose 1989). *Times Literary Supplement*, September 26–October 5, pp. 1066–68.

———. 1990a. "Teaching an Old Dog New Tricks" (commentary on Schull 1990). *Behavioral and Brain Sciences*, vol. 13, pp. 76–77.

———. 1990b. "The Interpretation of Texts, People, and Other Artifacts." *Philosophy and Phenomenological Research*, vol. 50, pp. 177–94.

———. 1990c. "Memes and the Exploitation of Imagination." *Journal of Aesthetics and Art Criticism*, vol. 48, pp. 127–35.

———. 1991a. *Consciousness Explained*. Boston: Little, Brown.

———. 1991b. "Real Patterns." *Journal of Philosophy,* vol. 87, pp. 27–51.

———. 1991c. "Granny's Campaign for Safe Science." In B. Loewer and G. Rey, eds., *Meaning in Mind: Fodor and His Critics* (Oxford: Blackwell), pp. 87–94.

———. 1991d. "The Brain and Its Boundaries" (review of McGinn 1991). *Times Literary Supplement,* May 10, 1991 (corrected by erratum notice on May 24, p. 29).

———. 1991e. "Ways of Establishing Harmony." In B. McLaughlin, ed., *Dretske and His Critics* (Oxford: Blackwell); also (slightly revised) in E. Villanueva, ed., *Information, Semantics, and Epistemology*, Sociedad Filosofica Ibero-Americana (Mexico) (Oxford: Blackwell).

———. 1992. "La Compréhension artisanale." French translation of "Do-It-Yourself Understanding." In Denis Fisette, ed., *Daniel C. Dennett et les Stratégies Intentionnelles, Lekton*, vol. 11, winter, pp. 27–52.

———. 1993a. "Down with School! Up with Logoland!" (review of Papert 1993). *New Scientist*, November 6, pp. 45–46.

———. 1993b. "Confusion over Evolution: An Exchange." *New York Review of Books*, January 14, 1993, pp. 43–44.

———. 1993c. Review of John Searle 1992. *Journal of Philosophy,* vol. 90, pp. 193–205.

———. 1993d. "Living on the Edge." *Inquiry*, vol. 36, pp. 135–59.

———. 1994a. "Cognitive Science as Reverse Engineering: Several Meanings of 'Top-down' and 'Bottom-up.'" In D. Prawitz, B. Skyrms, and D. Westerståhl, eds., *Proceedings of the 9th International Congress of Logic, Methodology and Philosophy of Science* (Amsterdam: North-Holland).

———. 1994b. "Language and Intelligence." In Jean Khalfa, ed., *What Is Intelligence?* Cambridge: Cambridge University Press, pp. 161–78.

———. 1994c. "Labeling and Learning" (commentary on Clark and Karmiloff-Smith 1994). *Mind and Language*, vol. 8, pp. 540–48.

———. 1994d. "E Pluribus Unum?" *Behavioral and Brain Sciences*, vol. 17, pp. 617–18.

———. 1994e. "The Practical Requirements for Making a Conscious Robot." *Proceedings of the Royal Society.*

DENNETT, DANIEL C., and HAUGELAND, JOHN. 1987. "Intentionality." In Gregory 1987, pp. 383–86.

DENTON, MICHAEL. 1985. *Evolution: A Theory in Crisis*. London: Burnett.

DESCARTES, RENÉ. 1637. *Discourse on Method*.

DESMOND, ADRIAN, and MOORE, JAMES. 1991. *Darwin*. London: Michael Joseph.

DEUTSCH, D. 1985. "Quantum Theory, the Church-Turing Principle and the Universal Quantum Computer." *Proceedings of the Royal Society,* vol. A400, pp. 97–117.

DE VRIES, PETER. 1953. *The Vale of Laughter*. Boston: Little, Brown.

DEWDNEY, A. K. 1984. *The Planiverse*. New York: Poseidon.

DEWEY, JOHN. 1910. *The Influence of Darwin on Philosophy*. New York: Holt, 1910. Reprint ed., Bloomington: Indiana University Press, 1965.

DIAMOND, JARED. 1991. "The Saltshaker's Curse." *Natural History*, October, pp. 20–26.

———. 1992. *The Third Chimpanzee: The Evolution and Future of the Human Animal*. New York: HarperCollins.

DIDEROT, DENIS. 1749. *Letter on the Blind, for the Use of Those Who See*. Trans. and excerpted in J. Kemp, ed., *Diderot: Interpreter of Nature* (London: Lawrence and Wishart, 1937).

DIETRICH, MICHAEL. 1992. "Macromutation." In Keller and Lloyd 1992, pp. 194–201.

DIGGINS, JOHN PATRICK. 1994. Review of Marsden, *The Soul of the American University. New York Times Book Review*, April 17, p. 25.

DOBZHANSKY, THEODOSIUS. 1973. "Nothing in Biology Makes Sense Except in the Light of Evolution." *American Biology Teacher,* vol. 35, pp. 125–29.

DONALD, MERLIN. 1991. *Origins of the Modern Mind: Three Stages in the Evolution of Culture and Cognition*. Cambridge, Mass.: Harvard University Press.

DOOLITTLE, W. F., and SAPIENZA, C. 1980. "Selfish Genes, the Phenotype Paradigm and Genome Evolution." *Nature*, vol. 284, pp. 601–3.

DRETSKE, FRED. 1986. "Misrepresentation." In R. Bogdan, ed., *Belief.* Oxford: Oxford University Pres.

DREYFUS, HUBERT. 1965. "Alchemy and Artificial Intelligence." RAND Technical Report P-3244, December 1965.

———. 1972. *What Computers Can't Do: The Limits of Artificial Intelligence*. New York: Harper & Row. Revised ed., 1979.

DYSON, FREEMAN. 1979. *Disturbing the Universe*. New York: Harper & Row.

ECCLES, JOHN. 1953. *The Neurophysiological Basis of Mind*. Oxford: Clarendon.

ECKERT, SCOTT A. 1992. "Bound for Deep Water." *Natural History*, March, pp. 28–35.

EDELMAN, GERALD. 1987. *Neural Darwinism.* New York: Basic Books.

———— 1992. *Bright Air, Brilliant Fire.* New York: Basic Books.

EDWARDS, PAUL. 1965. "Professor Tillich's Confusions." *Mind*, vol. 74, pp. 192–214.

EIGEN, MANFRED. 1976. "Wie entsteht Information? Prinzipien der Selbstorganisation in der Biologie." *Berichtete der Bunsengesellschaft für Physikalische Chemie*, vol. 80, p. 1059.

————. 1983. "Self-Replication and Molecular Evolution." In D. S. Bendall, ed., *Evolution from Molecules to Men* (Cambridge: Cambridge University Press), pp. 105–30.

————. 1992. *Steps Towards Life.* Oxford: Oxford University Press.

EIGEN, M. and WINKLER-OSWATITSCH, R. 1975. *Das Spiel.* Munich. (English trans., *Laws of the Game* [New York: Knopf, 1981]).

EIGEN, M., and SCHUSTER, P. 1977. "The Hypercycle: A Principle of Natural Self-Organization. Part A: Emergence of the Hypercycle," *Naturwissenschaften*, vol. 64, pp. 541–65.

ELDREDGE, NILES. 1983. "A la recherche du Docteur Pangloss" (commentary on Dennett 1983). *Behavioral and Brain Sciences*, vol. 6, pp. 361–62.

————. 1985. *Time Frames: The Rethinking of Darwinian Evolution and the Theory of Punctuated Equilibria.* New York: Simon & Schuster.

————. 1989. *Macroevolutionary Dynamics: Species, Niches and Adaptive Peaks.* New York: McGraw-Hill.

ELDREDGE, NILES, and GOULD, S. J. 1972. "Punctuated Equilibria: An Alternative to Phyletic Gradualism." In T. J. M. Schopf, ed., *Models in Paleobiology* (San Francisco: Freeman, Cooper and Company), pp. 82–115. Reprinted in Eldredge 1985, pp. 193–223.

ELLEGÅRD, ALVAR. 1956. "The Darwinian Theory and the Argument from Design." *Lychnos*, pp. 173–92.

————. 1958. *Darwin and the General Reader.* Goteborg: Goteborg University Press.

ELLESTRAND, NORMAN. 1983. "Why Are Juveniles Smaller Than Their Parents?" *Evolution*, vol. 13, pp. 1091–94.

ELLIS, R. J., and VAN DER VIES, S. M. 1991. "Molecular Chaperones." *Annual Review of Biochemistry*, vol. 60, pp. 321–47.

ELSASSER, WALTER. 1958. *The Physical Foundations of Biology.* Oxford: Oxford University Press.

————. 1966. *Atom and Organism.* Princeton: Princeton University Press.

ENGELS, W. R. 1992. "The Origin of P Elements in *Drosoophila melanogaster.*" *BioEssays*, vol. 14, pp. 681–86.

ERESHEFSKY, MARC, ed. 1992. *The Units of Evolution: Essays on the Nature of Species.* Cambridge, Mass.: MIT Press/A Bradford Book.

ESHEL, I. 1984. "Are Intragenetic Conflicts Common in Nature? Do They Represent an Important Factor in Evolution?" *Journal of Theoretical Biology*, vol. 108, pp. 159–62.

———. 1985. "Evolutionary Genetic Stability of Mendelian Segregation and the Role of Free Recombination in the Chromosomal System." *American Naturalist*, vol. 125, pp. 412–20.

FAUSTO-STERLING, ANNE. 1985. *Myths of Gender: Biological Theories About Women and Men*. New York: Basic Books. (2nd ed., 1992.)

FEDUCCIA, ALAN. 1993. "Evidence from Claw Geometry Indicating Arboreal Habits of *Archeopteryx*." *Science*, vol. 259, pp. 790–93.

FEIGENBAUM, E. A., and FELDMAN, J. 1964. *Computers and Thought*. New York: McGraw-Hill.

FEYNMAN, RICHARD. 1988. *What do YOU Care What Other People Think?* New York: Bantam.

FISHER, DAN. 1975. "Swimming and Burrowing in *Limulus* and *Mesolimulus*." *Fossils and Strata*, vol. 4, pp. 281–90.

FISHER, R. A. 1930. *The Genetical Theory of Natural Selection*. Oxford: Clarendon.

FITCHEN, JOHN. 1961. *The Construction of Gothic Cathedrals*. Oxford: Clarendon.

———. 1986. *Building Construction Before Mechanization*. Cambridge, Mass.: MIT Press.

FODOR, JERRY. 1975. *The Language of Thought*. Hassocks, Sussex: Harvester.

———. 1980. "Methodological Solipsism Considered as a Research Strategy in Cognitive Psychology." *Behavioral and Brain Sciences*, vol. 3, pp. 63–110.

———. 1983. *The Modularity of Mind*. Cambridge, Mass.: MIT Press.

———. 1987. *Psychosemantics*. Cambridge, Mass.: MIT Press.

———. 1990. *A Theory of Content and Other Essays*. Cambridge, Mass.: MIT Press.

———. 1992. "The Big Idea: Can There Be a Science of Mind?" *Times Literary Supplement*, July 3, p. 5.

FOOTE, MIKE. 1992. "Cambrian and Recent Morphological Disparity" (response to Briggs et al. 1989). *Science*, vol. 256, p. 1670.

FORBES, GRAEME. 1983. "Thisness and Vagueness," *Synthese*, vol. 54, pp. 235–59.

———. 1984. "Two Solutions to Chisholm's Paradox." *Philosophical Studies*, vol. 46, pp. 171–87.

FOX, S. W., and DOSE, K. 1972. *Molecular Evolution and the Origin of Life*. San Francisco: Freeman.

FUTUYMA, DOUGLAS. 1982. *Science on Trial: The Case for Evolution*. New York: Pantheon.

GABBEY, ALLAN. 1993. "Descartes, Newton and Mechanics: The Disciplinary Turn." Tufts Philosophy Colloquium, November 12.

GALILEI, GALILEO. 1632. *Dialogue Concerning the Two Chief World Systems*. Florence.

GARDNER, MARTIN. 1970. "Mathematical Games." *Scientific American*, October, vol. 223, 120–23.

———. 1971. "Mathematical Games." *Scientific American*, vol. 224, February, pp. 112–17.

———. 1986. "WAP, SAP, PAP and FAP." *New York Review of Books*, May 8. (Reprinted with a postscript in Martin Gardner, *Gardner's Whys and Wherefores.* Chicago: University of Chicago Press, 1989.)

GAUTHIER, DAVID. 1986. *Morals by Agreement.* New York: Oxford University Press.

GEE, HENRY. 1992. "Something Completely Different." *Nature*, vol. 358, pp. 456–57.

GERT, BERNARD. 1973. *The Moral Rules*. New York: Harper Torchbook. Original ed., New York: Harper & Row, 1966.

GHISELIN, M. 1983. "Lloyd Morgan's Canon in Evolutionary Context." *Behavioral and Brain Sciences*, vol. 6, pp. 362–63.

GIBBARD, ALAN. 1985. "Moral Judgment and the Acceptance of Norms," and "Reply to Sturgeon." *Ethics*, vol. 96, pp. 5–41.

GILKEY, LANGDON. 1985. *Creationism on Trial: Evolution and God at Little Rock.* San Francisco: Harper & Row.

GILLE, BERTRAND. 1966. *Engineers of the Renaissance*. Cambridge, Mass.: MIT Press.

GINGERICH, PHILIP. 1983. "Rate of Evolution: Effects of Time and Temporal Scaling." *Science*, vol. 222, pp. 159–61.

———. 1984. Reply to Gould 1983c. *Science*, vol. 226, pp. 995–96.

GJERTSEN, DEREK. 1989. *Science and Philosophy: Past and Present.* London: Penguin.

GÖDEL, KURT. 1931. "Über Formal Unentscheidbare Sätze der *Principia Mathematica* und Verwandter System, I." *Monatshefte für Mathematik und Physik*, vol. 38, pp. 173–98. Tran. and published, with some discussion, as *On Formally Undecidable Propositions* (New York: Basic Books, 1962).

GODFREY-SMITH, PETER. 1993. "Spencerian Explanation and Constructivism." MIT Philosophy Colloquium, November 5.

GOLDSCHMIDT, RICHARD B. 1933. "Some Aspects of Evolution." *Science*, vol. 78, pp. 539–47.

———. 1940. *The Material Basis of Evolution*. Seattle: University of Washington Press.

GOODMAN, NELSON. 1965. *Fact, Fiction and Forecast*. 2nd ed. New York: Bobbs-Merrill.

GOODWIN, BRIAN. 1986. "Is Biology an Historical Science?" In Steven Rose and Lisa Appignanese, eds., *Science and Beyond* (Oxford: Blackwell), pp. 47–60.

GOULD, STEPHEN JAY. 1977a. *Ever Since Darwin*. New York: Norton.

———. 1977b. *Ontogeny and Phylogeny*. Cambridge: Belknap.

———. 1980a. *The Panda's Thumb*. New York: Norton.

———. 1980b. "Is a New and General Theory of Evolution Emerging?" *Paleobiology*, vol. 6, pp. 119–30.

———. 1980c. "Sociobiology and the Theory of Natural Selection." *American Association for the Advancement of Science Symposia*, vol. 35, pp. 257–69. Reprinted in Ruse 1989.

————. 1980d. "The Evolutionary Biology of Constraint." *Daedalus,* vol. 109, pp. 39–52.

————. 1981. *The Mismeasure of Man.* New York: Norton.

————. 1982a. "Darwinism and the Expansion of Evolutionary Theory." *Science,* vol. 216, pp. 380–87. Reprinted in Ruse 1989.

————. 1982b. "The Uses of Heresy: An Introduction to Richard Goldschmidt's *The Material Basis of Evolution.*" In R. Goldschmidt, *The Material Basis of Evolution* (reprinted ed., New Haven: Yale University Press).

————. 1982c. "The Meaning of Punctuated Equilibrium, and Its Role in Validating a Hierarchical Approach to Macroevolution." In R. Milkman, ed., *Perspectives on Evolution* (Sunderland, Mass.: Sinauer), pp. 83–104.

————. 1982d. "Change in Developmental Timing as a Mechanism of Macroevolution." In J. T. Bonner, ed., *Evolution and Development,* Dahlem Konferenzen (Berlin, Heidelberg, New York: Springer-Verlag).

————. 1983a. "The Hardening of the Modern Synthesis." In M. Grene, ed., *Dimensions of Darwinism.* Cambridge: Cambridge University Press, pp. 71–93.

————. 1983b. *Hen's Teeth and Horse's Toes.* New York: Norton.

————. 1983c. "Smooth Curve of Evolutionary Rate: A Psychological and Mathematical Artifact." *Science,* vol. 226, pp. 994–95.

————. 1985. *The Flamingo's Smile.* New York: Norton.

————. 1987. "Darwinism Defined: The Difference Between Fact and Theory." *Discover,* January pp. 64–70.

————. 1989a. *Wonderful Life: The Burgess Shale and the Nature of History.* New York: Norton.

————. 1989b. "Tires to Sandals." *Natural History,* April, pp. 8–15.

————. 1990. *The Individual in Darwin's World.* Edinburgh: Edinburgh University Press. (This is a direct transcription, "by Ian Wall, Ian Rolfe and Simon Gage of the City of Edinburgh District Council," of what was obviously a largely extemporaneous talk.)

————. 1991a. "The Panda's Thumb of Technology." In Gould 1991b, pp. 59–75.

————. 1991b. *Bully for Brontosaurus.* New York: Norton.

————. 1992a. "The Confusion over Evolution." *New York Review of Books,* November 19, pp. 47–54.

————. 1992b. "Life in a Punctuation." *Natural History,* vol. 101. October, pp. 10–21.

————. 1993a. "Fulfilling the Spandrels of World and Mind." In Selzer 1993, pp. 310–36.

————, ed. 1993b. *The Book of Life.* New York: Norton.

————. 1993c. "Cordelia's Dilemma." *Natural History,* vol. 103, February, pp. 10–19..

————. 1993d. *Eight Little Piggies.* New York: Norton.

————. 1993e. "Confusion over Evolution: An Exchange." *New York Review of Books,* January 14, pp. 43–44.

GOULD, S. J., and ELDREDGE, N. 1993. "Punctuated Equilibrium Comes of Age." *Nature*, vol. 366, pp. 223–27.

GOULD, S. J., and LEWONTIN, R. 1979. "The Spandrels of San Marco and the Panglossian Paradigm: A Critique of the Adaptationist Programme." *Proceedings of the Royal Society*, vol. B205, pp. 581–98.

GOULD, S. J., and VRBA, ELIZABETH. 1981. "Exaptation: A Missing Term in the Science of Form." *Paleobiology*, vol. 8, pp. 4–15.

GREENWOOD, JOHN D., ed. 1991. *The Future of Folk Psychology: Intentionality and Cognitive Science*. Cambridge: Cambridge University Press.

GREGORY, R. L. 1981. *Mind in Science: A History of Explanations in Psychology and Physics*. Cambridge: Cambridge University Press.

———, ed. 1987. *The Oxford Companion to the Mind*. Oxford: Oxford University Press.

GRICE, H. P. 1957. "Meaning." *Philosophical Review*, vol. 66, pp. 377–88.

———. 1969. "Utterer's Meaning and Intentions." *Philosophical Review*, vol. 78, pp. 147–77.

GROSSBERG, STEPHEN. 1976. "Adaptive Pattern Classification and Universal Recoding: Part I. Parallel Development and Coding of Neural Feature Detectors." *Biological Cybernetics*, vol. 23, pp. 121–34.

HADAMARD, JACQUES. 1949. *The Psychology of Inventing in the Mathematical Field*. Princeton: Princeton University Press.

HAIG, DAVID. 1992. "Genomic Imprinting and the Theory of Parent-Offspring Conflict." *Developmental Biology*, vol. 3, pp. 153–60.

———. 1993. "Genetic Conflicts in Human Pregnancy." *Quarterly Review of Biology*, vol. 68, pp. 495–532.

HAIG, DAVID, and GRAFEN, A. 1991. "Genetic Scrambling as a Defence Against Meiotic Drive." *Journal of Theoretical Biology*, vol. 153, pp. 531–58.

HAIG, DAVID, and GRAHAM, CHRIS. 1991. "Genomic Imprinting and the Strange Case of the Insulin-like Growth Factor II Receptor." *Cell*, vol. 64, pp. 1045–46.

HAIG, DAVID, and WESTOBY, M. 1989. "Parent-specific Gene Expression and the Triploid Endosperm." *American Naturalist*, vol. 134, pp. 147–55.

HALE, KEN, et al. 1992. "Endangered Languages." *Language*, vol. 68, pp. 1–42.

HAMILTON, WILLIAM. 1964. "The Genetical Evolution of Social Behavior," pts. I and II. *Journal of Theoretical Biology*, vol. 7, pp. 1–16, 17–52.

HARDIN, GARRETT. 1964. "The Art of Publishing Obscurely." In G. Hardin, ed., *Population, Evolution and Birth Control: A Collection of Controversial Readings* (San Francisco: Freeman), pp. 116–19.

———. 1968. "The Tragedy of the Commons." *Science*, vol. 162, pp. 1243–48.

HARDY, ALISTER., 1960. "Was Man More Aquatic in the Past?" *New Scientist*, pp. 642–45.

HAUGELAND, JOHN. 1985. *Artificial Intelligence: The Very Idea*. Cambridge, Mass.: MIT Press.

HAWKING, STEPHEN W. 1988. *A Brief History of Time*. New York: Bantam.

HEBB, DONALD. 1949. *The Organization of Behavior*. New York: Wiley.

HINTON, GEOFFREY E., and NOWLAND, S. J. 1987. "How Learning Can Guide Evolution." In *Complex Systems*, vol. I. Technical report CMU-CS-86-128. Carnegie-Mellon University, pp. 495–502.

HOBBES, THOMAS. 1651. *Leviathan*. London: Crooke.

HODGES, ANDREW. 1983. *Alan Turing: The Enigma*. New York: Simon & Schuster.

HOFSTADTER, DOUGLAS. 1979. *Gödel Escher Bach*. New York: Basic Books.

———. 1985. *Metamagical Themas: Questing for the Essence of Mind and Pattern*. New York: Basic Books.

HOFSTADTER, DOUGLAS R., and DENNETT, DANIEL C. 1981. *The Mind's I*. New York: Basic Books.

HOLLAND, JOHN. 1975. *Adaptation in Natural and Artificial Systems*. Ann Arbor: University of Michigan Press.

———. 1992. "Complex Adaptive Systems." *Daedalus*, Winter, p. 25.

HOLLINGDALE, R. J. 1965. *Nietzsche: The Man and His Philosophy*. London: Routledge & Kegan Paul.

HOUCK, MARILYN A.; CLARK, JONATHAN B.; PETERSON, KENNETH R.; and KIDWELL, MARGARET G. 1991. "Possible Horizontal Transfer of *Drosophila* Genes by the Mite *Proctolaelaps Regalis*." *Science*, vol. 253, pp. 1125–29.

HOUSTON, ALASDAIR. 1990. "Matching, Maximizing, and Melioration as Alternative Descriptions of Behaviour." In J. A. Meyer and S. Wilson, eds., *From Animals to Animats*. Cambridge, Mass.: MIT Press, pp. 498–509.

HOY, DAVID. 1986. "Nietzsche, Hume, and the Genealogical Method." In Y. Yovel, ed., *Nietzsche as Affirmative Thinker* (Dordrecht: Martinus Nijhoff), pp. 20–38.

HOYLE, FRED. 1964. *Of Men and Galaxies*. Seattle: University of Washington Press.

HOYLE, FRED, and WICKRAMASINGHE, CHANDRA. 1981. *Evolution from Space*. London: Dent.

HRDY, SARAH BLAFFER. 1977. *The Langurs of Abu: Female and Male Strategies of Reproduction*. Cambridge, Mass.: Harvard University Press.

HULL, DAVID. 1980. "Individuality and Selection." *Annual Review of Ecology and Systematics*, vol. 11, pp. 311–32.

———. 1982. "The Naked Meme." In H. C. Plotkin, ed., *Learning, Development and Culture* (New York: Wiley), pp. 273–327.

HUME, DAVID. 1739. *A Treatise of Human Nature*. Ed. L. A. Selby-Bigge. Oxford: Clarendon, 1964.

———. 1779. *Dialogues Concerning Natural Religion*. London.

HUMPHREY, NICHOLAS. 1976. "The Social Function of Intellect." In P. P. G. Bateson and R. A. Hinde, eds., *Growing Points in Ethology* (Cambridge: Cambridge University Press), pp. 303–17.

———. 1983. "The Adaptiveness of Mentalism?" *Behavioral and Brain Sciences*, vol. 3, p. 366.

———. 1986. *The Inner Eye*. London: Faber & Faber.

———. 1987. "Scientific Shakespeare." *Guardian* (London), August 26.

ISACK, H. A., and REYER, H.-U. 1989. "Honeyguides and Honey Gatherers: Inter-

538   BIBLIOGRAPHY

specific Communication in a Symbiotic Relationship." *Science*, vol. 243, pp. 1343–46.

ISRAEL, DAVID. 1987. *The Role of Propositional Objects of Belief in Action*. CSLI Monograph Report no. CSLI–87–72. Palo Alto: Stanford University Press.

JACKENDOFF, RAY. 1987. *Consciousness and the Computational Mind*. Cambridge, Mass.: MIT Press/A Bradford Book.

———. 1993. *Patterns in the Mind: Language and Human Nature*. London: Harvester Wheatsheaf.

JACOB, FRANÇOIS. 1982. *The Possible and the Actual*. Seattle: University of Washington Press.

———. 1983. "Molecular Tinkering in Evolution." In D. S. Bendall, ed., *Evolution from Molecules to Men* (Cambridge: Cambridge University Press), pp. 131–44.

JAMES, WILLIAM. 1880. *Lecture Notes 1880–1897*. (The quoted passage is from Sills and Merton 1991.)

JAYNES, JULIAN. 1976. *The Origins of Consciousness in the Breakdown of the Bicameral Mind*. Boston: Houghton Mifflin.

JONES, STEVE. 1993. "A Slower Kind of Bang" (review of E. O. Wilson, *The Diversity of Life*). *London Review of Books*, April, p. 20.

KANT, IMMANUEL. 1785. *Grundlegung zur Metaphysik der Sitten*. (The quoted passage is from the translation of H. J. Paton, *Groundwork of the Metaphysic of Morals* [New York: Harper & Row, 1964].)

KAUFFMAN, STUART. 1993. *The Origins of Order: Self-Organization and Selection in Evolution*. New York: Oxford University Press.

KAUFMANN, WALTER. 1950. *Nietzsche: Philosopher, Psychologist, Antichrist*. Princeton: Princeton University Press. Reprint ed., New York: Meridian paperback, 1956.

KEIL, FRANK, C. 1992. "The Origins of an Autonomous Biology." In M. Gunnar and M. Maratsos, eds., *Modularity and Constraints in Language and Cognition: The Minnesota Symposia on Child Psychology,* vol. 25. Hillsdale, N.J.: Erlbaum, pp. 103–37.

KELLER, EVELYN FOX, and LLOYD, ELISABETH A., eds. 1992. *Keywords in Evolutionary Biology*. Cambridge, Mass.: Harvard University Press.

KING'S COLLEGE SOCIOBIOLOGY GROUP. 1982. *Current Problems in Sociobiology*. Cambridge: Cambridge University Press.

KIPLING, RUDYARD. 1912. *Just So Stories*. Reprint ed., Garden City, N.Y.: Doubleday, 1952.

KIRKPATRICK, S.; GELATT, C. D.; and VECCHI, M. P. 1983. "Optimization by Simulated Annealing." *Science*, vol. 220, pp. 671–80.

KITCHER, PHILIP. 1982. *Abusing Science*. Cambridge, Mass.: MIT Press.

———. 1984. "Species." *Philosophy of Science*, vol. 51, pp. 308–33.

———. 1985a. "Darwin's Achievement." In N. Rescher, ed., *Reason and Rationality in Science* (Lanham, Md.: University Press of America), pp. 127–89.

———. 1985b. *Vaulting Ambition*. Cambridge, Mass.: MIT Press.

———. 1993. "The Evolution of Human Altruism." *Journal of Philosophy,* vol. 90, pp. 497–516.

KRAUTHEIMER, RICHARD. 1981. *Early Christian and Byzantine Architecture.* 3rd ed. London: Penguin.

KREBS, JOHN R., and DAWKINS, RICHARD. 1984. "Animal Signals: Mind-Reading and Manipulation." In J. R. Krebs and N. B. Davies, eds., *Behavioural Ecology: An Evolutionary Approach,* 2nd ed. (Oxford: Blackwell), pp. 380–402.

KÜPPERS, BERND-OLAF. 1990. *Information and the Origin of Life.* Cambridge, Mass.: MIT Press.

LANCASTER, JANE. 1975. *Primate Behavior and the Emergence of Human Culture.* New York: Holt, Rinehart and Winston.

LANDMAN, OTTO E. 1991. "The Inheritance of Acquired Characteristics." *Annual Review of Genetics,* vol. 25, pp. 1–20.

———. 1993. "Inheritance of Acquired Characteristics." *Scientific American,* March, p. 150.

LANGTON, CHRISTOPHER; TAYLOR, CHARLES; FARMER, J. DOYNE; and RASMUSSEN, STEEN. 1992. *Artificial Life II.* Redwood City, Calif.: Addison-Wesley.

LEIBNIZ, GOTTFRIED WILHELM. 1710. *Theodicy* (*Essais de Théodicée sur la bonté de Dieu, la liberté de l'homme et l'origine du mal*), Amsterdam. (George Martin Duncan translation, 1908.)

LEIGH, E. G. 1971. *Adaptation and Diversity.* San Francisco: Freeman, Cooper and Company.

LENAT, DOUGLAS B., and GUHA, R. V. 1990. *Building Large Knowledge-based Systems: Representation and Inference in the CYC Project.* Reading, Mass.: Addison-Wesley.

LESLIE, ALAN. 1992. "Pretense, Autism and the Theory-of-Mind Module." *Current Directions in Psychological Science,* vol. 1, pp. 18–21.

LESLIE, JOHN. 1989. *Universes,* London: Routledge & Kegan Paul.

LETTVIN, J. Y.; MATURANA, U.; McCULLOCH, W.; and PITTS, W. 1959. "What the Frog's Eye Tells the Frog's Brain." In *Proceedings of the IRE,* pp. 1940–51.

LEVI, PRIMO. 1984. *The Periodic Table.* New York: Schocken.

LÉVI-STRAUSS, CLAUDE. 1966. *The Savage Mind.* Chicago: University of Chicago Press.

LEWIN, ROGER. 1992. *Complexity: Life at the Edge of Chaos.* New York: Macmillan.

LEWIS, DAVID. 1986. *Philosophical Papers,* vol 2. Oxford: Oxford University Press.

LEWONTIN, RICHARD. 1980. "Adaptation." *The Encyclopedia Einaudi.* Milan: Einaudi.

———. 1983. "Elementary Errors About Evolution" (commentary on Dennett 1983). *Behavioral and Brain Sciences,* vol. 6, pp. 367–68.

———. 1987. "The Shape of Optimality." In John Dupré, ed., *The Latest on the Best: Essays on Evolution and Optimality.* Cambridge, Mass.: MIT Press.

LEWONTIN, RICHARD; ROSE, STEVEN; and KAMIN, LEON. 1984. *Not in our Genes: Biology, Ideology and Human Nature.* New York: Pantheon.

LINNAEUS, CAROLUS. 1751. *Philosophia Botanica.*

LLOYD, M., and DYBAS, H. S. 1966. "The Periodical Cicada Problem." *Evolution,* vol. 20, pp. 132–49.

LOCKE, JOHN. 1690. *Essay Concerning Human Understanding.* London.

LORD, ALBERT. 1960. *The Singer of Tales.* Cambridge, Mass.: Harvard University Press.

LORENZ, KONRAD. 1973. *Die Rückseite des Spiegels.* Munich: R. Piper. Verlag. (English trans. by Ronald Taylor, *Behind the Mirror* [New York: Harcourt Brace Jovanovich, 1977].)

LOVEJOY, ARTHUR O. 1936. *The Great Chain of Being: A Study of the History of an Idea.* New York: Harper & Row.

LUCAS, J. R. 1961. "Minds, Machines, and Gödel." *Philosophy,* vol. 36, pp. 1–12.

———. 1970. *The Freedom of the Will.* Oxford: Oxford University Press.

MACKENZIE, ROBERT BEVERLEY. 1868. *The Darwinian Theory of the Transmutation of Species Examined* (published anonymously "By a Graduate of the University of Cambridge"). Nisbet & Co. (Quoted in a review, *Athenaeum,* no. 2102, February 8, p. 217.)

MALTHUS, THOMAS. 1798. *Essay on the Principle of Population.*

MARGOLIS, HOWARD. 1987. *Patterns, Thinking and Cognition.* Chicago: University of Chicago Press.

MARGULIS, LYNN. 1981. *Symbiosis in Cell Evolution.* San Francisco: Freeman.

MARGULIS, LYNN, and SAGAN, DORION. 1986. *Microcosmos.* New York: Simon & Schuster.

———. 1987. "Bacterial Bedfellows." *Natural History,* vol. 96, March, pp. 26–33.

MARKS, JONATHAN. 1993. Review of Diamond 1992. *Journal of Human Evolution,* vol. 24, pp. 69–73.

———. 1993b. "Scientific Misconduct: Where 'Just Say No' Fails" (multiple book review). *American Scientist,* vol. 81, July–August, pp. 380–82.

MARTIN, J.; MAYHEW, M.; LANGER, T.; and HARTL, F. U. 1993. "The Reaction Cycle of GroEL and GroES in Chaperonin-assisted Protein Folding." *Nature,* vol. 366, pp. 228–33.

MASSERMAN, JULES H.; WECHKIN, STANLEY; and TERRIS, WILLIAM. 1964. " 'Altruistic' Behavior in Rhesus Monkeys." *American Journal of Psychiatry,* vol. 121, pp. 584–85.

MATTHEW, PATRICK. 1831. *Naval Timber and Arboriculture.*

———. 1860. Letter to editor, *Gardener Chronicle,* April 7.

MAYNARD SMITH, JOHN. 1958. *The Theory of Evolution.* Cambridge: Cambridge University Press. (Canto ed. 1993.)

———. 1972. *On Evolution.* Edinburgh: Edinburgh University Press.

———. 1974. "The Theory of Games and the Evolution of Animal Conflict." *Journal of Theoretical Biology,* vol. 47, pp. 209–21.

———. 1978. *The Evolution of Sex.* Cambridge: Cambridge University Press.

———. 1979. "Hypercycles and the Origin of Life." *Nature,* vol. 280, pp. 445–46. Reprinted in Maynard Smith 1982, pp. 34–38.

————. 1981. "Symbolism and Chance." In J. Agassi and R. S. Cohen, eds., *Scientific Philosophy Today* (Hingham, Mass: Kluwer). Reprinted in Maynard Smith 1988, pp. 15–21.

————. 1982. *Evolution Now: A Century After Darwin*. San Francisco: Freeman.

————. 1983. "Adaptation and Satisficing" (commentary on Dennett 1983). *Behavioral and Brain Sciences*, vol. 6, pp. 70–71.

————. 1986. "Structuralism Versus Selection—Is Darwinism Enough?" In Steven Rose and Lisa Appignanesi, eds. *Science and Beyond* (Oxford: Blackwell), pp. 39–46.

————. 1988. *Games, Sex and Evolution*. London: Harvester. (Also published under the title *Did Darwin Get it Right?*)

————. 1990. "What Can't the Computer Do?" (review of Penrose 1989). *New York Review of Books*, March 15, pp. 21-25.

————. 1991. "Dinosaur Dilemmas." *New York Review of Books*, April 25, pp. 5–7.

————. 1992. "Taking a Chance on Evolution." *New York Review of Books*, May 14, pp. 234–36.

MAYR, ERNST. 1960. "The Emergence of Evolutionary Novelties." In Sol Tax, ed., *Evolution after Darwin*, vol. 1 (Chicago: University of Chicago Press), pp. 349–80.

————. 1982. *The Growth of Biological Thought*. Cambridge, Mass.: Harvard University Press.

————. 1983. "How to Carry Out the Adaptationist Program." *American Naturalist*, vol. 121, pp. 324–34.

MAZLISH, BRUCE. 1993. *The Fourth Discontinuity: The Co-evolution of Humans and Machines*. New Haven: Yale University Press.

McCULLOCH, W. S., and PITTS, W. 1943. "A Logical Calculus of the Ideas Immanent in Nervous Activity." *Bulletin of Mathematical Biophysics*, vol. 5, pp. 115–33.

McGINN, COLIN. 1991. *The Problem of Consciousness*. Oxford: Blackwell.

————. 1993. "In and Out of the Mind" (review of Hilary Putnam, *Renewing Philosophy*). *London Review of Books*, December 2, pp. 30–31.

McLAUGHLIN, BRIAN, ed. 1991. *Dretske and His Critics*. Oxford: Blackwell.

McSHEA, DANIEL W. 1993. "Arguments, Tests, and the Burgess Shale—A Commentary on the Debate." *Paleobiology*, vol. 9, pp. 339–402.

MEDAWAR, PETER. 1977. "Unnatural Science." *The New York Review of Books*, February 3, pp. 13–18.

————. 1982. *Pluto's Republic*. Oxford: Oxford University Press.

METROPOLIS, NICHOLAS. 1992. "The Age of Computing: A Personal Memoir." *Daedalus*, Winter, pp. 119–30.

MIDGLEY, MARY. 1979. "Gene-Juggling." *Philosophy*, vol. 54, pp. 439–58.

————. 1983. "Selfish Genes and Social Darwinism." *Philosophy*, vol. 58, pp. 365–77.

MILL, JOHN STUART. 1861. *Utilitarianism*. Originally published in *Fraser's Magazine*; reprinted 1863.

MILLER, GEORGE A. 1956. "The Magical Number Seven, Plus or Minus Two." *Psychological Review*, vol. 63, pp. 81–97.

———. 1979. "A Very Personal History." MIT Cognitive Science Center Occasional Paper #1, a talk to the Cognitive Science Workshop, June 1, 1979.

MILLIKAN, RUTH. 1984. *Language, Thought and Other Biological Categories*. Cambridge, Mass.: MIT Press.

———. 1993. *White Queen Psychology and Other Essays for Alice*. Cambridge, Mass.: MIT Press.

MINSKY, MARVIN. 1985a. "Why Intelligent Aliens Will Be Intelligible." In E. Regis, ed., *Extraterrestrials* (Cambridge: Cambridge University Press), pp. 117–28.

MINSKY, MARVIN. 1985b. *The Society of Mind*. New York: Simon & Schuster.

MINSKY, MARVIN, and PAPERT, SEYMOUR. 1969. *Perceptrons*. Cambridge, Mass.: MIT Press.

MIVART, ST. GEORGE. 1871. "Darwin's *Descent of Man*." *Quarterly Review*, vol. 131, pp. 47–90.

MONOD, JACQUES. 1971. *Chance and Necessity*. New York: Knopf. Vintage paperback, 1972. (Originally published in France as *Le Hasard et la nécessité* [Paris: Editions du Seuil, 1970].)

MOORE, G. E. 1903. *Principia Ethica*. Cambridge: Cambridge University Press.

MORGAN, ELAINE. 1982. *The Aquatic Ape*. London: Souvenir.

———. 1990. *The Scars of Evolution: What Our Bodies Tell Us About Human Origins*. London: Souvenir.

MUIR, JOHN. 1972. *Original Sanscrit Texts and the Origin and History of the People of India, Their Religion and Institutions*. Delhi: Oriental. Originally published 1868–73, London: Trubner.

MURRAY, JAMES D. 1989. *Mathematical Biology*. New York: Springer-Verlag.

NEHAMAS, ALEXANDER. 1980. "The Eternal Recurrence." *Philosophical Review*, vol. 89, pp. 331–56.

NEUGEBAUER, OTTO. 1989. "A Babylonian Lunar Ephemeris from Roman Egypt." In E. Leichtz, M. de J. Ellis, and P. Gerardi, eds., *A Scientific Humanist: Studies in Honor of Abraham Sachs* (Philadelphia: The University Museum [Distributed by the Samuel North Kramer Fund]), pp. 301–4.

NEWELL, ALLEN, and SIMON, HERBERT. 1956. "The Logic Theory Machine." *IRE Transactions on Information Theory IT-2(3)*, pp. 61–79.

———. 1964. "GPS: A Program That Simulates Human Thought." In Feigenbaum and Feldman 1964.

NEWTON, ISAAC. 1726. *Philosophiae Naturalis Principia Mathematica*.

NIETZSCHE, FRIEDRICH. 1881. *Daybreak: Thoughts on the Prejudices of Morality*. (Trans. R. J. Hollingdale [Cambridge: Cambridge University Press, 1982].)

———. 1882. *Die fröliche Wissenschaft* (*The Gay Science*). (Trans. Walter Kaufmann [New York: Vintage, 1974].)

———. 1885. *Beyond Good and Evil*. (Trans. Walter Kaufmann [New York: Vintage, 1966].)

————. 1887. *On the Genealogy of Morals.* (Trans. Walter Kaufmann [New York: Vintage, 1967].)

————. 1889. *Ecce Homo.* (Trans. Walter Kaufmann [New York: Vintage, 1968].)

————. 1901. *The Will to Power.* (Trans. Walter Kaufmann and R. J. Hollingdale [New York: Vintage, 1968].)

NOWAK, MARTIN, and SIMUND, KARL. 1993. "A Strategy of Win-Stay, Lose-Shift That Outperforms Tit-for-Tat in the Prisoner's Dilemma Game." *Nature*, vol. 364, pp. 56–58.

NOZICK, ROBERT. 1981. *Philosophical Explanation.* Cambridge, Mass.: Belknap/Harvard University Press.

O'NEILL, ONORA. 1980. "The Perplexities of Famine Relief." In Tom Regan, ed., *Matters of Life and Death* (New York: Random House), pp. 26–48.

————. 1986. *Faces of Hunger.* Boston: Allen & Unwin, 1986.

ORGEL, LESLIE E., and CRICK, FRANCIS. 1980. "Selfish DNA: The Ultimate Parasite." *Nature*, vol. 284, pp. 604–607.

OTERO, CARLOS P. 1990. "The Emergence of *Homo Loquens* and the Laws of Physics." *Behavioral and Brain Sciences*, vol. 13, pp. 747–50.

PAGELS, HEINZ. 1985. "A Cozy Cosmology." *The Sciences*, March/April, pp. 34–39.

————. 1988. *The Dreams of Reason: The Computer and the Rise of the Sciences of Complexity.* New York: Simon & Schuster.

PALEY, WILLIAM. 1803. *Natural Theology: or, Evidences of the Existence and Attributes of the Deity, Collected from the Appearances of Nature.* 5th ed. London: Faulder.

PAPERT, SEYMOUR. 1980. *Mindstorms: Children, Computers and Powerful Ideas.* New York: Basic Books.

————. 1993. *The Children's Machine: Rethinking School in the Age of the Computer.* New York: Basic Books.

PAPINEAU, DAVID. 1987. *Reality and Representation.* Oxford: Blackwell.

PARFIT, DEREK. 1984. *Reasons and Persons.* Oxford: Clarendon.

PARSONS, WILLIAM BARCLAY. 1939. *Engineers and Engineering in the Renaissance.* Reprint ed., Cambridge, Mass: MIT Press, 1967.

PEACOCKE, CRISTOPHER. 1992. *A Study of Concepts.* Cambridge, Mass.: MIT Press.

PECKHAM, MORSE, ed. 1959. *The Origin of Species by Charles Darwin: A Variorum Text.* Pittsburgh: University of Pennsylvania Press.

PENROSE, ROGER. 1989. *The Emperor's New Mind Concerning Computers, Minds, and the Laws of Physics.* Oxford: Oxford University Press.

————. 1990. "The Nonalgorithmic Mind." *Behavioral and Brain Sciences*, vol. 13, pp. 692–705.

————. 1991. "Setting the Scene: The Claim and the Issues." Wolfson Lecture, January 15.

————. 1993. "Nature's Biggest Secret" (review of Weinberg 1992). *New York Review of Books*, October 21, pp. 78–82.

PIATELLI-PALMARINI, MASSIMO. 1989. "Evolution, Selection, and Cognition: From 'Learning' to Parameter Setting in Biology and the Study of Language." *Cognition*, vol. 31, pp. 1–44.

PINKER, STEVEN. 1994. *The Language Instinct*. New York: Morrow.

PINKER, STEVEN, and BLOOM, PAUL. 1990. "Natural Language and Natural Selection." *Behavioral and Brain Sciences*, vol. 13, pp. 707–84.

PITTENDRIGH, COLIN. 1958. "Adaptation, Natural Selection, and Behavior." In A. Roe and G. G. Simpson, eds., *Behavior and Evolution* (New Haven: Yale University Press), pp. 390–416.

POE, EDGAR ALLAN. 1836. "Maelzel's Chess-Player." *Southern Literary Messenger*. Reprinted in Edgar Allan Poe, *Edgar Allan Poe: Essays and Reviews* (New York: Library of America, 1984), pp. 1253–76.

————. 1836b. *Southern Literary Messenger*, suppl., July 1836. Reprinted in J. M. Walker, ed., *Edgar Allan Poe: The Critical Heritage* (New York: Routledge & Kegan Paul, 1986), pp. 89–90.

POPPER, KARL, and ECCLES, JOHN. 1977. *The Self and Its Brain*. Berlin, London: Springer-Verlag.

POUNDSTONE, WILLIAM. 1985. *The Recursive Universe: Cosmic Complexity and the Limits of Scientific Knowledge*. New York: Morrow.

————. 1992. *Prisoner's Dilemma: John von Neumann, Game Theory, and the Puzzle of the Bomb*. New York: Doubleday Anchor.

PREMACK, DAVID. 1986. *Gavagai! Or the Future History of the Animal Language Controversy*. Cambridge, Mass.: MIT Press.

PUTNAM, HILARY. 1975. *Mind, Language and Reality*. Philosophical Papers, vol. II. Cambridge: Cambridge University Press.

————. 1987. *The Faces of Realism*. LaSalle, Ill.: Open Court.

QUINE, W. V. O. 1953. "On What There Is." In *From a Logical Point of View* (Cambridge, Mass.: Harvard University Press), pp. 1–19.

————. 1960. *Word and Object*. Cambridge, Mass.: Harvard University Press.

————. 1969. "Natural Kinds." In *Ontological Relativity* (New York: Columbia University Press), pp. 114–38.

————. 1987. *Quiddities: An Intermittently Philosophical Dictionary*. Cambridge, Mass.: Harvard University Press.

RACHELS, JAMES. 1991. *Created from Animals: The Moral Implications of Darwinism*. Oxford: Oxford University Press..

RAWLS, JOHN. 1971. *A Theory of Justice*. Cambridge, Mass.: Harvard University Press.

RAY, THOMAS S. 1992. "An Approach to the Synthesis of Life." In C. G. Langton, C. Taylor, J. D. Farmer, and S. Rasmussen, eds., *Artificial Life II* (Redwood City, Calif.: Addison-Wesley), pp. 371–408.

RAYMO, CHET. 1988. "Mysterious Sleep." *Boston Globe*, September 19.

RAYMOND, ERIC S. 1993. *The New Hacker's Dictionary*. Cambridge, Mass.: MIT Press.

RÉE, PAUL. 1877. *Origin of the Moral Sensations*. (*Der Ursprung der Moralischen Empfindungen* [Chemnitz: E. Schmeitzner].)

RICHARD, MARK. 1992. *Propositional Attitudes*. Cambridge: Cambridge University Press.

RICHARDS, GRAHAM. 1991. "The Refutation That Never Was: The Reception of the Aquatic Ape Theory, 1972–1987." In Roede et al. 1991, pp. 115–26.

RICHARDS, ROBERT J. 1987. *Darwin and the Emergence of Evolutionary Theories of Mind and Behavior*. Chicago: University of Chicago Press.

RIDLEY, MARK. 1985. *The Problems of Evolution*. Oxford: Oxford University Press.
———. 1993. *Evolution*. Boston: Blackwell.

RIDLEY, MATT. 1993. *The Red Queen: Sex and the Evolution of Human Nature*. New York: Macmillan.

ROBB, CHRISTINA. 1991. "How & Why." *Boston Globe*, August 5, p. 38.

ROBBINS, TOM. 1976. *Even Cowgirls Get the Blues*. New York: Bantam.

ROEDE, MACHTELD; WIND, JAN; PATRICK, JOHN M.; and REYNOLDS, VERNON, eds. 1991. *The Aquatic Ape: Fact or Fiction*. London: Souvenir.

ROSENBLATT, FRANK. 1962. *Principles of Neurodynamics*. New York: Spartan.

ROUSSEAU, JEAN-JACQUES. 1755. *Discourse on the Origin and Foundations of Inequality Among Men*. Amsterdam: Ray.

RUMELHART, D. 1989. "The Architecture of Mind: A Connectionist Approach." In M. Posner, ed., *Foundations of Cognitive Science* (Cambridge, Mass.: MIT Press), pp. 133–59.

RUSE, MICHAEL. 1985. *Sociobiology: Sense or Nonsense?* 2nd ed. Dordrecht: Reidel.
———, ed. 1989. *Philosophy of Biology*. London: Macmillan.

RUSE, MICHAEL, and WILSON, EDWARD O. 1985. "The Evolution of Ethics." *New Scientist*, vol. 17, October, pp. 50–52. Reprinted in Ruse 1989.

RUSHDIE, SALMAN. 1994. "Born in Bombay" (letter). *London Review of Books*, April 7, p. 4. (See also Arab and Muslim Writers, 1994.)

RUSSELL, BERTRAND. 1945. *A History of Western Philosophy*. New York: Simon & Schuster.

RUTHEN, RUSSELL. 1993. "Adapting to Complexity." *Scientific American*, January, p. 138.

SAMUEL, A. L. 1964. "Some Studies in Machine Learning Using the Game of Checkers." Reprinted in Feigenbaum and Feldman 1964, pp. 71–105. Originally published in *IBM Journal of Research and Development*, July 1959, vol. 3, pp. 211–29.

SCHELLING, THOMAS. 1960. *The Strategy of Conflict*. Cambridge, Mass.: Harvard University Press.

SCHIFFER, STEPHEN. 1987. *Remnants of Meaning*. Cambridge, Mass.: MIT Press.

SCHOPF, J. WILLIAM. 1993. "Microfossils of the Early Archean Apex Chert: New Evidence of the Antiquity of Life." *Science*, vol. 260, pp. 640–46.

SCHRÖDINGER, ERNST. 1967. *What Is Life?* Cambridge: Cambridge University Press.

SCHULL, JONATHAN. 1990. "Are Species Intelligent?" *Behavioral and Brain Sciences*, vol. 13, pp. 63–108.

SEARLE, JOHN. 1980. "Minds, Brains and Programs." *Behavioral and Brain Sciences*, vol. 3, pp. 417–58.

————. 1985. *Minds, Brains and Science*. Cambridge, Mass.: Harvard University Press.

————. 1992. *The Rediscovery of the Mind*. Cambridge, Mass.: MIT Press.

SELLARS, WILFRID. 1963. *Science, Perception and Reality*. London: Routledge & Kegan Paul.

SELZER, JACK. ed. 1993. *Understanding Scientific Prose*. Madison, Wisc.: University of Wisconsin Press.

SHEA, B. T. 1977. "Eskimo Cranofacial Morphology, Cold Stress and the Maxillary Sinus." *American Journal of Physical Anthropology*, vol. 47, pp. 289–300.

SHERMAN, PAUL W.; JARVIS, JENNIFER U. M.; and ALEXANDER, RICHARD D. 1991. *The Biology of the Naked Mole-Rat*. Princeton: Princeton University Press.

SHIELDS, W. M., and SHIELDS, L. M. 1983. "Forcible Rape: An Evolutionary Perspective." *Ethology and Sociobiology*, vol. 4, pp. 115–36.

SIBLEY, C. G., and AHLQUIST, J. E. 1984. "The Phylogeny of the Hominoid Primates, as Indicated by DNA-DNA Hybridization." *Journal of Molecular Evolution*, vol. 20, pp. 2–15.

SIGMUND, KARL. 1993. *Games of Life: Explorations in Ecology, Evolution, and Behaviour*. Oxford: Oxford University Press.

SILLS, DAVID L., and ROBERT K. MERTON, eds. 1991. *The Macmillan Book of Social Science Quotations*. New York: Macmillan.

SIMON, HERBERT. 1957. *Models of Man*. New York: Wiley.

————. 1959. "Theories of Decision-Making in Economics and Behavioral Science." *American Economic Review*, vol. 49, pp. 253-83.

————. 1969. *The Sciences of the Artificial*. Cambridge, Mass.: MIT Press.

SIMON, HERBERT A., and KAPLAN, CRAIG. 1989. "Foundations of Cognitive Science." In M. Posner, ed., *Foundations of Cognitive Science* (Cambridge, Mass.: MIT Press), pp. 1–47.

SIMON, HERBERT, and NEWELL, ALLEN. 1958. "Heuristic Problem Solving: The Next Advance in Operations Research." *Operations Research*, vol. 6, pp. 1–10.

SINGER, PETER. 1981. *The Expanding Circle: Ethics and Sociobiology*. Oxford: Clarendon.

SKINNER, B. F. 1953. *Science and Human Behavior*. New York: Macmillan.

————. 1957. *Verbal Behavior*. New York: Appleton-Century-Crofts.

————. 1971. *Beyond Freedom and Dignity*. New York: Knopf.

SKYRMS, BRIAN, 1993. "Justice and Commitment" (preprint).

————. 1994a. "Sex and Justice." *Journal of Philosophy*, vol. 91, pp. 305–20.

————. 1994b. "Darwin Meets *The Logic of Decision*: Correlation in Evolutionary Game Theory." *Philosophy of Science*, December, pp. 503–28.

SMOLENSKY, PAUL. 1983. "On the Proper Treatment of Connectionism." *Behavioral and Brain Sciences*, vol. 11, pp. 1–74.

SMOLIN, LEE. 1992. "Did the Universe Evolve?" *Classical and Quantum Gravity*, vol. 9, pp. 173–91.

SNOW, C. P. 1963. *The Two Cultures, and a Second Look*. Cambridge: Cambridge University Press.

Sober, Elliot. 1981a. "Holism, Individualism, and Units of Selection. *PSA 1980* (Philosophy of Science Association), vol. 2, pp. 93–121.

———. 1981b. "The Evolution of Rationality." *Synthese*, vol. 46, pp. 95–120.

———. 1984a. *The Nature of Selection: Evolutionary Theory in Philosophical Focus*. Cambridge, Mass.: MIT Press.

———. ed., 1984b. *Conceptual Issues in Evolutionary Biology*. Cambridge, Mass.: MIT Press.

———. 1988. *Reconstructing the Past*. Cambridge, Mass.: MIT Press.

———, ed. 1994. *Conceptual Issues in Evolutionary Biology*. 2nd ed. Cambridge, Mass.: MIT Press.

Spencer, Herbert. 1870. *The Principles of Psychology*. 2nd ed. London: Williams & Norgate.

Sperber, Dan. 1985. "Anthropology and Psychology: Towards an Epidemiology of Representations." *Man*, vol. 20, pp. 73–89.

———. 1990. "The Epidemiology of Beliefs." In C. Fraser and G. Gaskell, eds., *The Social Psychological Study of Widespread Beliefs* (Oxford: Clarendon), pp. 25–44.

———. In press. "The Modularity of Thought and the Epidemiology of Representations." In Lawrence A. Hirschfeld and Susan A. Gelman, eds., *Mapping the Mind: Domain Specificity in Cognition and Culture* (Cambridge: Cambridge University Press).

Sperber, Dan, and Wilson, Deirdre. 1986. *Relevance: A Theory of Communication*. Cambridge, Mass.: Harvard University Press.

Stanley, Steven M. 1981. *The New Evolutionary Timetable: Fossils, Genes, and the Origin of Species*. New York: Basic Books.

Sterelny, Kim. 1988. Review of Dawkins 1986a. *Australasian Journal of Philosophy*, vol. 66, pp. 421–26.

———. 1992. "Punctuated Equilibrium and Macroevolution." In P. Griffiths, ed., *Trees of Life* (Norwell, Mass.: Kluwer), pp. 41–63.

———. 1994. Review of Ereshefsky, 1992, in *Philosophical Books*, vol. 35, pp. 9–29.

Sterelny, K., and Kitcher, P. 1988. "The Return of the Gene." *Journal of Philosophy*, vol. 85, pp. 339–60.

Stetter, Karl O.; Huber, R.; Blochl, E.; Kurr, M.; Eden, R. D.; Fielder, M.; Cash, H.; and Vance, I. 1993. "Hyperthermophilic Archaea Are Thriving in Deep North Sea and Alaskan Oil Reservoirs." *Nature*, vol. 365, p. 743.

Stewart, Ian, and Golubitsky, Martin. 1992. *Fearful Symmetry: Is God a Geometer?* Oxford: Blackwell.

Stove, David. 1992. "A New Religion." *Philosophy*, vol. 27, pp. 233–40.

Strawson, P. F. 1962. "Freedom and Resentment." *Proceedings of the British Academy*, vol. 48, pp. 118–39.

Sturgeon, Nicholas. 1985. "Moral Judgment and Norms." *Ethics*, vol. 96, pp. 5–41.

Symons, Donald. 1983. "FLOAT: A New Paradigm for Human Evolution." In

George M. Scherr, ed., *The Best of the Journal of Irreproducible Results* (New York: Workman).

———. 1992. "On the Use and Misuse of Darwinism in the Study of Human Behavior." In Barkow, Cosmides, and Tooby 1992, pp. 137–62.

TAIT, P. G. 1880. "Prof. Tait on the Formula of Evolution." *Nature,* vol. 23, pp. 80–82.

TEILHARD DE CHARDIN, PIERRE. 1959. *The Phenomenon of Man.* New York: Harper Brothers.

THOMPSON, D'ARCY W. 1917. *On Growth and Form.* Cambridge: Cambridge University Press.

TOOBY, JOHN, and COSMIDES, LEDA. 1992. "The Psychological Foundations of Culture." In Barkow, Cosmides, and Tooby 1992, pp. 19–136.

TRIVERS, ROBERT. 1971. "The Evolution of Reciprocal Altruism." *Quarterly Review of Biology,* vol. 4, pp. 35–57.

———. 1972. "Parental Investment and Sexual Selection." In B. Campbell, ed., *Sexual Selection and the Descent of Man* (Chicago: Aldine), pp. 136–79.

———. 1985. *Social Evolution.* Menlo Park, Calif.: Benjamin/Cummings.

TUDGE, COLIN. 1993. "Taking the Pulse of Evolution," *New Scientist,* July 24, pp. 32–36.

TURING, ALAN. 1946. *ACE Reports of 1946 and Other Papers.* Ed. B. E. Carpenter and R. W. Doran. Cambridge, Mass.: MIT Press.

———. 1950. "Computing Machinery and Intelligence." *Mind,* vol. 59, pp. 433–60.

———. 1952. "The Chemical Basis of Morphogenesis." *Philosophical Transactions of the Royal Society of London,* vol. B237, pp. 37–72.

TURNBULL, COLIN. 1972. *The Mountain People.* New York: Simon & Schuster.

ULAM, STANISLAW. 1976. *Adventures of a Mathematician.* New York: Scribner's.

UNGER, PETER. 1990. *Identity, Consciousness and Value.* New York: Oxford University Press.

UTTLEY, A. M. 1979. *Information Transmission in the Nervous System.* London: Academic.

VAN INWAGEN, PETER. 1993a. *Metaphysics.* Oxford: Oxford University Press.

———. 1993b. "Critical Study" (of Unger 1990). *Nous,* vol. 27, pp. 373–39.

VERMEIJ, GEERAT J. 1987. *Evolution and Escalation.* Princeton: Princeton University Press.

VON FRISCH, K. 1967. *A Biologist Remembers.* Oxford: Pergamon.

VON NEUMANN, JOHN. 1966. *Theory of Self-reproducing Automata.* Posthumously ed. Arthur Burks. Champaign-Urbana: University of Illinois Press.

VON NEUMANN, JOHN, and MORGENSTERN, OSKAR. 1944. *Theory of Games and Economic Behavior.* Princeton: Princeton University Press.

VRBA, ELIZABETH. 1985. "Environment and Evolution: Alternative Causes of Temporal Distribution of Evolutionary Events." *Suid-Afrikaanse Tydskif Wetens,* vol. 81, pp. 229–36.

WANG, HAO. 1974. *From Mathematics to Philosophy*. London: Routledge & Kegan Paul.

———. 1993. "On Physicalism and Algorithmism." *Philosophia Mathematica,* Series 3, vol. 1, pp. 97–138.

WASON, PETER. 1969. "Regression in Reasoning." *British Journal of Psychology*, vol. 60, pp. 471–80.

WATERS, C. KENNETH. 1990. "Why the Antireductionist Consensus Won't Survive the Case of Classical Mendelian Genetics." *PSA 1990* (Philosophy of Science Association), vol. 1, pp. 125–39. Reprinted in Sober 1994, pp. 401–17.

WECHKIN, STANLEY; MASSERMAN, JULES H.; and TERRIS, WILLIAM. 1964. "Shock to a Conspecific as an Aversive Stimulus." *Psychonomic Science*, vol. 1, pp. 47–48.

WEINBERG, STEVEN. 1992. *Dreams of a Final Theory*. New York: Pantheon.

WEISMANN, AUGUST. 1893. *The Germ Plasm: A Theory of Heredity*. English trans. London: Scott.

WERTHEIMER, ROGER. 1974. "Philosophy on Humanity." In R. L. Perkins, ed., *Abortion: Pro and Con* (Cambridge, Mass.: Schenkman).

WHEELER, JOHN ARCHIBALD. 1974. "Beyond the End of Time." In Martin Rees, Remo Ruffini, and John Archibald Wheeler, *Black Holes, Gravitational Waves and Cosmology: An Introduction to Current Research* (New York: Gordon and Breach).

WHITE, STEPHEN L. 1988. "Self-Deception and Responsibility for the Self." In B. McLaughlin and A. Rorty, eds., *Perspectives on Self-Deception* (Berkeley: University of California Press).

———. 1991. *The Unity of the Self*. Cambridge, Mass.: MIT Press.

———. "Constraints on an Evolutionary Explanation of Morality." Unpublished manuscript.

WHITFIELD, PHILIP. 1993. *From So Simple a Beginning: The Book of Evolution*. New York: Macmillan.

WILBERFORCE, SAMUEL. 1860. "Is Mr Darwin a Christian?" (review of *Origin*, published anonymously). *Quarterly Review*, vol. 108, July, pp. 225–64.

WILLIAMS, BERNARD. 1973. "A Critique of Utilitarianism." In J. J. C. Smart and Bernard Williams, *Utilitarianism: For and Against* (Cambridge: Cambridge University Press), pp. 77–150.

———. 1983. "Evolution, Ethics, and the Representation Problem." In D. S. Bendall, ed., *Evolution from Molecules to Men* (Cambridge: Cambridge University Press), pp. 555–66.

———. 1985. *Ethics and the Limits of Philosophy*. Cambridge, Mass.: Harvard University Press.

WILLIAMS, GEORGE C. 1966. *Adaptation and Natural Selection*. Princeton: Princeton University Press.

———. 1985. "A Defense of Reductionism in Evolutionary Biology." *Oxford Surveys in Evolutionary Biology*, vol. 2, pp. 1–27.

———. 1988. "Huxley's Evolution and Ethics in Sociobiological Perspective." *Zygon*, vol. 23, pp. 383–407.

————. 1992. *Natural Selection: Domains, Levels, and Challenges*. Oxford: Oxford University Press.

WILLIAMS, GEORGE C., and NESSE, RANDOLPH. 1991. "The Dawn of Darwinian Medicine." *Quarterly Review of Biology*, vol. 66, pp. 1–22.

WILSON, DAVID SLOAN, and SOBER, ELLIOT. 1994. "Re-introducing Group Selection to the Human Behavior Sciences." *Behavioral and Brain Sciences*, vol. 17, pp. 585–608.

WILSON, E. O. 1971. *The Insect Societies*. Cambridge, Mass.: Harvard University Press.

————. 1975. *Sociobiology: The New Synthesis*. Cambridge: Harvard University Press.

————. 1978. *On Human Nature*. Cambridge, Mass.: Harvard University Press.

WIMSATT, WILLIAM. 1980. "Reductionist Research Strategies and Their Biases in the Unit of Selection Controversy." In T. Nickles, ed., *Scientific Discovery: Case Studies* (Hingham, Mass.: Kluwer), pp. 213–39.

————. 1981. "Units of Selection and the Structure of the Multi-Level Genome." In P. Asquith and R. Geire, eds., *PSA-1980* (Philosophy of Science Association), vol. 2, pp. 122–83.

————. 1986. "Developmental Constraints, Generative Entrenchment, and the Innate-acquired Distinction." In W. Bechtel, ed., *Integrating Scientific Disciplines* (Dordrecht: Martinus-Nijhoff), pp. 185–208.

WIMSATT, WILLIAM, and BEARDSLEY, MONROE. "The Intentional Fallacy." In *The Verbal Icon: Studies in the Meaning of Poetry*. Lexington: University of Kentucky Press, 1954.

WITTGENSTEIN, LUDWIG. 1922. *Tractatus Logico-philosophicus*. London: Routledge & Kegan Paul.

WRIGHT, ROBERT. 1990. "The Intelligence Test" (review of Gould 1989a). *New Republic*, January 29, pp. 28–36.

WRIGHT, SEWALL. 1931. "Evolution in Mendelian Populations." *Genetics*, vol. 16, p. 97.

————. 1932. "The Roles of Mutation, Inbreeding, Crossbreeding and Selection in Evolution." *Proceedings of the XI International Congress of Genetics*, vol. 1, pp. 356–66.

————. 1967. "Comments on the Preliminary Working Papers of Eden and Waddington." In P. S. Moorehead and M. M. Kaplan, eds., *Mathematical Challenges to the Neo-Darwinian Interpretation of Evolution*, Wistar Institute Symposia, monograph no. 5. Philadelphia: Wistar Institute Press.

YOUNG, J. Z. 1965. *A Model of the Brain*. Oxford: Clarendon.

ZAHAVI, A. 1987. "The Theory of Signal Selection and Some of Its Implications." In V. P. Delfino, ed., *International Symposium on Biological Evolution, Bari, 9–14 April 1985* (Bari, Italy: Adriatici Editrici), pp. 305–27.

Index

Visit Penguin on the Internet
and browse at your leisure

◆ preview sample extracts of our forthcoming books
◆ read about your favourite authors
◆ investigate over 10,000 titles
◆ enter one of our literary quizzes
◆ win some fantastic prizes in our competitions
◆ e-mail us with your comments and book reviews
◆ instantly order any Penguin book

and masses more!

'To be recommended without reservation ... a rich and rewarding on-line experience' – Internet Magazine

www.penguin.co.uk

READ MORE IN PENGUIN

In every corner of the world, on every subject under the sun, Penguin represents quality and variety – the very best in publishing today.

For complete information about books available from Penguin – including Puffins, Penguin Classics and Arkana – and how to order them, write to us at the appropriate address below. Please note that for copyright reasons the selection of books varies from country to country.

In the United Kingdom: Please write to *Dept. EP, Penguin Books Ltd, Bath Road, Harmondsworth, West Drayton, Middlesex UB7 ODA*

In the United States: Please write to *Consumer Sales, Penguin USA, P.O. Box 999, Dept. 17109, Bergenfield, New Jersey 07621-0120*. VISA and MasterCard holders call 1-800-253-6476 to order Penguin titles

In Canada: Please write to *Penguin Books Canada Ltd, 10 Alcorn Avenue, Suite 300, Toronto, Ontario M4V 3B2*

In Australia: Please write to *Penguin Books Australia Ltd, P.O. Box 257, Ringwood, Victoria 3134*

In New Zealand: Please write to *Penguin Books (NZ) Ltd, Private Bag 102902, North Shore Mail Centre, Auckland 10*

In India: Please write to *Penguin Books India Pvt Ltd, 706 Eros Apartments, 56 Nehru Place, New Delhi 110 019*

In the Netherlands: Please write to *Penguin Books Netherlands bv, Postbus 3507, NL-1001 AH Amsterdam*

In Germany: Please write to *Penguin Books Deutschland GmbH, Metzlerstrasse 26, 60594 Frankfurt am Main*

In Spain: Please write to *Penguin Books S. A., Bravo Murillo 19, 1° B, 28015 Madrid*

In Italy: Please write to *Penguin Italia s.r.l., Via Felice Casati 20, I–20124 Milano*

In France: Please write to *Penguin France S. A., 17 rue Lejeune, F–31000 Toulouse*

In Japan: Please write to *Penguin Books Japan, Ishikiribashi Building, 2–5–4, Suido, Bunkyo-ku, Tokyo 112*

In South Africa: Please write to *Longman Penguin Southern Africa (Pty) Ltd, Private Bag X08, Bertsham 2013*

BY THE SAME AUTHOR

Consciousness Explained

'Extraordinary ... Dennett outlines an alternative view of consciousness drawn partly from the world of computers and partly from the findings of neuroscience. Our brains, he argues, are more like parallel processors than the serial processors that lie at the heart of most computers in use today. The difference is that, whereas serial processors perform one task at a time, parallel processors perform many different tasks at the same time "in parallel" ... supremely engaging and witty' – Ray Monk in the *Independent*

'Dennett's exposition is nothing short of brilliant, the best example I've seen of a science book aimed at both professionals and general readers ... clear and funny, with introspective flights of fancy worthy of Nicholson Baker' – George Johnson in *The New York Times Book Review*

'This is heady stuff, written with tremendous verve and panache ... A fabulous book, which will set the agenda for years of discussion' – Andy Clark in *The Times Higher Education Supplement*

Brainstorms
Philosophical Essays on Mind and Psychology

Invention, artificial intelligence, linguistics, dreams and free will are just some of the areas covered in Daniel C. Dennett's exhilarating investigation into the meaning of the mind and consciousness.

'An excellent book ... Throughout the collection, Dennett's enthusiasm and perceptive good sense ensure that this statement of an original and important philosophical position is a pleasure to read' – Geoffrey Hinton in *Contemporary Psychology*

also published, by Douglas R. Hofstadter and Daniel C. Dennett

The Mind's I

'What is the mind? Who am I? Can machines think?'

Anyone who confronts these questions runs headlong into perplexities. The authors conceived this book as an attempt to reveal those perplexities and make them vivid.

'A remarkable read' – *New Society*